Operator Theory
Advances and Applications
Vol. 79

Partially Specified Matrices and Operators: Classification, Completion, Applications

Israel Gohberg
Marinus A. Kaashoek
Frederik van Schagen

Birkhäuser Verlag
Basel · Boston · Berlin

Authors' addresses:

Israel Gohberg
School of Mathematical Sciences
Raymond and Beverly Sackler
Faculty of Exact Sciences
Tel Aviv University
Ramat Aviv 69978
Israel

Marinus A. Kaashoek
Faculteit Wiskunde en Informatica
Vrije Universiteit Amsterdam
De Boelelaan 1081a
1081 HV Amsterdam
The Netherlands

Frederik van Schagen
Faculteit Wiskunde en Informatica
Vrije Universiteit Amsterdam
De Boelelaan 1081a
1081 HV Amsterdam
The Netherlands

A CIP catalogue record for this book is available from the Library of Congress, Washington D.C., USA

Deutsche Bibliothek Cataloging-in-Publication Data
Gochberg, Izrail' C.:
Partially specified matrices and operators : classification,
completion, applications / Israel Gohberg ; Marinus A.
Kaashoek ; Frederik van Schagen. – Basel ; Boston ; Berlin:
Birkhäuser, 1995
(Operator theory ; Vol. 79)

NE: Kaashoek, Marinus A.:; Schagen, Frederik van:; GT

Softcover reprint of the hardcover 1st edition 1995

Printed on acid-free paper produced from chlorine-free pulp ∞
Cover design: Heinz Hiltbrunner, Basel
ISBN-13: 978-3-0348-9906-2 e-ISBN-13: 978-3-0348-9100-4
DOI:10.1007/978-3-0348-9100-4

9 8 7 6 5 4 3 2 1

Contents

Introduction

This book is devoted to a new direction in linear algebra and operator theory that deals with the invariants of partially specified matrices and operators, and with the spectral analysis of their completions. The theory developed centers around two major problems concerning matrices of which part of the entries are given and the others are unspecified. The first is a *classification problem* and aims at a simplification of the given part with the help of admissible similarities. The results here may be seen as a far reaching generalization of the Jordan canonical form. The second problem is called the *eigenvalue completion problem* and asks to describe all possible eigenvalues and their multiplicities of the matrices which one obtains by filling in the unspecified entries. Both problems are also considered in an infinite dimensional operator framework.

A large part of the book deals with applications to matrix theory and analysis, namely to stabilization problems in mathematical system theory, to problems of Wiener-Hopf factorization and interpolation for matrix polynomials and rational matrix functions, to the Kronecker structure theory of linear pencils, and to non-everywhere defined operators.

The eigenvalue completion problem has a natural associated inverse, which appears as a restriction problem. The analysis of these two problems is often simpler when a solution of the corresponding classification problem is available.

Throughout the book, the given data appear as a submatrix with no restriction on its position in the full matrix, except for the last chapter, where the given data have an other pattern and consist of all entries in the upper triangular part.

The branch of linear algebra dealing with completion problems of the type referred to above has a relatively short history. The first important paper in this direction, which concerns the eigenvalue completion problem for the case when the given submatrix is a principal submatrix, appeared in the beginning of the seventies and is due to Oliveira [1]. The full solution of this problem for principal submatrices was given by Thompson [1] and Sà [1]. Earlier (in the end of the sixties), in mathematical systems theory the Brunovski form and the solution of the pole assignment problem by Rosenbrock (see Rosenbrock [1]) appeared. The latter results can be interpreted as solutions of the above mentioned problems for the case when the given entries form a number of full columns or a number of full rows with some additional properties. Recently, the eigenvalue completion problem for such full length or full width submatrices was solved by I. Zaballa [1] without additional conditions. Also for some other classes of submatrices the eigenvalue completion problem has been solved. All these results, including the connections with mathematical system theory, appear in a systematic form in this book. The general version of the second problem remains unsolved. The solution of the first problem for arbitrary submatrices with no restriction on the size or the location and with a description of the invariants and the canonical form, which was developed by the authors in the beginning of the eighties, provides the main concepts on which a large part of this book is based. Also the authors' results which identify Wiener-Hopf factorization indices as invariants of a certain kind appearing in the solution of the first problem are presented.

The eigenvalue completion problem has also been solved for the case when the given data have another pattern and do not form a submatrix. Probably one of the first results of this kind appeared in 1958 in Mirsky [1], which considers matrices with all entries unspecified except those on the main diagonal. Some later results concern the case when the unspecified entries are all on the main diagonal (see Friedland [1], Silva [1]) or when the given data form a block diagonal (see De Oliveira [2], [3] and Silva [2]). These results are not included here. On the other hand we present the partial results from Gohberg-Rubinstein [1] and Ball-Gohberg-Rodman-Shalom [1] which deal with the eigenvalue completion problem and the problem of invariants and canonical form for the case when the given data form a triangular pattern. Mirsky's result referred to above is also included.

A part of the book concerns operator theory and deals with infinite dimensional generalizations of the results for full length and full width submatrices. Also the applications of these results to factorization problems are generalized to the operator case. This part of the book is based on the results of Gohberg-Kaashoek-Van Schagen [1], Eckstein [1] and Rowley [1].

There are many other matrix and operator theory completion problems that are not of the kind considered here. We have in mind problems that also start with a partially specified matrix but which ask for completions satisfying additional constraints, not necessary of spectral character. For instance, the contraction completion problems in which one looks for completions of a triangular matrix with norm less than one (see, e.g., Ball-Gohberg [1], Davis-Kahan-Weinberger [1], Parrott [1]), the band extension problem which consists of completing a given band to a matrix which is required to be positive definite (see, e.g., Dym-Gohberg [1], Grone-Johnson-De Sá-Wolkowicz [1]), and the finite rank extension problem, where the completion is required to have a given rank (see, e.g., Ellis-Lay [1], Kaashoek-Woerdeman [1], Woerdeman [1]). Also we mention that we do not include the infinite dimensional analogue of these and related problems for infinite Toeplitz matrices which have a long and rich history, starting with the work of Schur, Carathéodory and Toeplitz in the beginning of this century. For more information see the books Ball-Gohberg-Rodman [1], Dym [1], Foias-Frazho [1], Helton [1], and Woerdeman [2]. See also Part IX of Gohberg-Goldberg-Kaashoek [2].

The book consists of 15 chapters and an appendix. The first two chapters set up the problems in a basis free manner by viewing the submatrices as (finite dimensional) operator blocks. Some motivating examples are presented, and the notion of admissible similarity is made precise by introducing block similarities.

Also other elementary operations on operator blocks are introduced. The simplest operator blocks (which later serve as the building blocks for arbitrary ones) are described. In Chapters III and IV the two main problems (i.e., the classification problem and the eigenvalue completion problem) are solved for full length blocks. Also the associated restriction problem is solved. Applications to linear matrix pencils, to the similarity theory for non-everywhere defined linear operators, and to a certain matrix equation are also included. Chapter V derives the analogous results for full width blocks, partly by using duality arguments. Chapter VI contains the full solution of the eigenvalue completion problem for principal blocks. For such blocks the problem of invariants is trivial and the corresponding canonical form is the usual Jordan normal form. The problem of invariants and canonical form for arbitrary (finite dimensional) operator blocks is completely solved in Chapter VII. Using this result the Kronecker theorems for matrix pencils are derived and the connections between the various invariants are clarified. Also the similarity theory for non-everywhere defined linear operators modulo a subspace is presented here. The last section of this chapter gives a recent result about the eigenvalue completion problem for general blocks. Chapter VIII gives the solution of the two main problems for off-diagonal blocks by using the structure theorems for arbitrary blocks derived in the previous chapter. The ninth chapter is devoted to connections with input-output systems. No advance knowledge of this area is required from the reader. The basic notions and the first facts from mathematical system theory are introduced and identified in terms of operator blocks. The first main theorems about stabilization by state feedback are derived by using the framework developed in the previous chapters. Chapters X and XI deal with applications to matrix polynomials and rational matrix functions. Here we treat problems concerning homogeneous interpolation, Wiener-Hopf factorization, and description of factorization indices. Sections with preliminary material are included to make these chapters self-contained. Chapters XII, XIII, and XIV contain the infinite dimensional generalizations. Included are the theory of infinite dimensional operator blocks of full length and of full width, the solution of a spectral completion problem for such blocks, and applications to Wiener-Hopf factorization of operator polynomials and analytic operator functions. The last chapter treats the two main problems for triangular matrices. The admissible similarities for triangular patterns are identified, and for a generic case the corresponding invariants and canonical form are derived. A spectral radius completion problem is also solved. Related completion problems are considered too. In an appendix we give a self-contained presentation of the local spectral theory of analytic matrix functions, which is used in Chapters X and XI.

Except for Chapters XII–XIV, which deal with infinite dimensional operator blocks, the prerequisites for reading the book are basic knowledge of linear algebra. The chapters on applications to systems theory, matrix polynomials and rational matrix functions are self-contained, based only on the theory presented in the preceding chapters.

Chapter I
Main Problems and Motivation

In this chapter the eigenvalue completion problem and the associated restriction problem are introduced and illustrated on various examples. The concepts of an operator block and of block similarity, which will provide us with the main tools to reduce the eigenvalue completion problem to simpler problems, appear here. The problem of classification of blocks is introduced too.

I.1 Eigenvalue completion problems and first examples

In this book one of the main objects of study is partially specified matrices. By a *partially specified* $n \times n$ *matrix* we mean an $n \times n$ array of complex numbers in which certain entries are given and the other entries are not specified. The set of pairs (i, j), with $i \in \{1, \ldots, n\}$ and $j \in \{1, \ldots, n\}$ for which the (i, j)-th entry is given is called the *pattern* of the partially specified matrix. A partially specified matrix with pattern $\mathcal{P}$ will often be denoted by $A_{\mathcal{P}}$.

Let us give some examples. Assume that $\mathcal{P} = \{(i, j) \in \{1, \ldots, n\}^2 \mid i \leq j\}$. Then the partially specified matrix $A_{\mathcal{P}}$ with pattern $\mathcal{P}$ can be represented by

$$A_{\mathcal{P}} = \begin{pmatrix} b_{11} & b_{12} & \cdots & b_{1\ n-1} & b_{1n} \\ ? & b_{22} & \cdots & b_{2\ n-1} & b_{2n} \\ \vdots & \ddots & \ddots & \vdots & \vdots \\ ? & & \ddots & b_{n-1\ n-1} & b_{n-1\ n} \\ ? & ? & \cdots & ? & b_{nn} \end{pmatrix},$$

where the unspecified entries are denoted by question marks.

If the pattern is rectangular, that is

$$\mathcal{P} = \{(i, j) \in \{1, \ldots, n\}^2 \mid i \in \{i_1, \ldots, i_k\},\ j \in \{j_1, \ldots, j_m\}\},$$

then the corresponding partially specified matrix $A_{\mathcal{P}}$ is called a *positioned submatrix*. Let us consider some examples. First assume that $\mathcal{P} = \{1, \ldots, k\}^2 \subset \{1, \ldots, n\}^2$. Then the positioned submatrix with pattern $\mathcal{P}$ is a $k \times k$ principal submatrix,

$$A_{\mathcal{P}} = \begin{pmatrix} b_{11} & \cdots & b_{1k} & ? & \cdots & ? \\ \vdots & & \vdots & \vdots & & \vdots \\ b_{k1} & \cdots & b_{kk} & ? & \cdots & ? \\ ? & \cdots & ? & ? & \cdots & ? \\ \vdots & & \vdots & \vdots & & \vdots \\ ? & \cdots & ? & ? & \cdots & ? \end{pmatrix}.$$

Here again the question marks indicate the unspecified entries in $A_{\mathcal{P}}$. Another important example is the case when

$$\mathcal{P} = \{i_1, \dots, i_k\} \times \{1, \dots, n\} \subset \{1, \dots, n\}^2.$$

In this case the positioned submatrix is called a *full width submatrix*. In particular, if $\{i_1, \dots, i_k\} = \{1, \dots, k\}$, then

$$A_{\mathcal{P}} = \begin{pmatrix} b_{11} & \cdots & \cdots & b_{1n} \\ \vdots & & & \vdots \\ b_{k1} & \cdots & \cdots & b_{kn} \\ ? & \cdots & \cdots & ? \\ \vdots & & & \vdots \\ ? & \cdots & \cdots & ? \end{pmatrix}.$$

Let $A_{\mathcal{P}}$ be a partially specified matrix with pattern $\mathcal{P}$, and let b_{ij} denote the given entries for $(i,j) \in \mathcal{P}$. We say that the matrix $A = \left(a_{ij}\right)_{i,j=1}^{n}$ is a *completion of* $A_{\mathcal{P}}$ if $a_{ij} = b_{ij}$ for each $(i,j) \in \mathcal{P}$. One of the main problems that will be discussed is *to what extent the partially specified matrix $A_{\mathcal{P}}$ determines the eigenvalues and their multiplicities of its completions.* Such a problem will be called an *eigenvalue completion problem.*

As a first illustration consider a partially specified 2×2 matrix $A_{\mathcal{P}}$ with pattern $\mathcal{P} = \{(1,1), (1,2)\}$. So $A_{\mathcal{P}}$ is an example of a full width positioned submatrix. Write

$$A_{\mathcal{P}} = \begin{pmatrix} b_{11} & b_{12} \\ ? & ? \end{pmatrix}.$$

We consider the following eigenvalue completion problem: Describe the possible eigenvalues including algebraic and geometric multiplicities of a completion A of $A_{\mathcal{P}}$. Note that a completion A of $A_{\mathcal{P}}$ has the form

$$A = \begin{pmatrix} b_{11} & b_{12} \\ a_{21} & a_{22} \end{pmatrix},$$

and hence the characteristic polynomial of A is $(\lambda - b_{11})(\lambda - a_{22}) - a_{21}b_{12}$. If $b_{12} \neq 0$, then any monic polynomial of degree two appears by choosing appropriate values for a_{12} and a_{22}. However no eigenvalue of A can have two independent eigenvectors, since for any λ the matrix $A - \lambda I$ has a submatrix of rank 1. Thus, if $b_{12} \neq 0$, then a completion of $A_{\mathcal{P}}$ can have any two different eigenvalues or two equal eigenvalues with geometric multiplicity one. On the other hand, if $b_{12} = 0$, then the eigenvalues of A are b_{11} and a_{22}. In particular, $A_{\mathcal{P}}$ fixes the eigenvalue b_{11} for A in this case. In case $a_{22} = b_{11}$ one sees that this eigenvalue has algebraic multiplicity two with two independent eigenvectors (if $a_{21} = 0$) or only one independent eigenvector (if $a_{21} \neq 0$). So if $b_{12} = 0$, then a completion of $A_{\mathcal{P}}$ has one fixed eigenvalue equal to b_{11}.

Our next examples concern positioned submatrices of a type that will play an important role later on. In the first of these examples the pattern is $\mathcal{P} = \{2, \ldots, n\} \times \{1, \ldots, n\} \subset \{1, \ldots, n\}^2$ and the specified entries are given by

$$A_{\mathcal{P}} = \begin{pmatrix} ? & ? & \cdots & ? & ? \\ 1 & 0 & \cdots & 0 & 0 \\ 0 & 1 & & 0 & 0 \\ \vdots & & \ddots & & \vdots \\ 0 & 0 & & 1 & 0 \end{pmatrix}.$$

In this case any set of n (not necessarily different) complex numbers can occur as the set of eigenvalues of a completion of $A_{\mathcal{P}}$, but no eigenvalue can have a geometric multiplicity exceeding one. In other words, the only restriction on the Jordan canonical form of a completion of $A_{\mathcal{P}}$ is that there is one Jordan block for each eigenvalue (i.e., the completion is derogatory). To see this let $A = \left(a_{ij}\right)_{i,j=1}^{n}$ be a completion of $A_{\mathcal{P}}$. Since A is a companion type matrix, we obtain that $\det(\lambda I - A) = \lambda^n - \sum_{i=0}^{n-1} a_{1n-i}\lambda^i$. Therefore any monic polynomial of degree n can appear as the characteristic polynomial of a completion of $A_{\mathcal{P}}$. On the other hand, for an eigenvalue λ the eigenspace, which is the nullspace of $\lambda I - A$, has dimension 1, because $\lambda I - A$ has a submatrix of rank $n-1$, irrespective of λ.

The final example in this section concerns the case when the pattern $\mathcal{P} = \{2, \ldots, n\} \times \{1, \ldots, n-1\} \subset \{1, \ldots, n\}^2$ and the specified entries form an identity matrix; more precisely

$$A_{\mathcal{P}} = \begin{pmatrix} ? & ? & \cdots & ? & ? \\ 1 & 0 & \cdots & 0 & ? \\ 0 & 1 & & 0 & ? \\ \vdots & & \ddots & & \vdots \\ 0 & 0 & & 1 & ? \end{pmatrix}.$$

Like in the previous example one computes that any polynomial can occur as the characteristic polynomial of a completion of $A_{\mathcal{P}}$, and that each eigenvalue of a completion has geometric multiplicity one.

I.2 Reduction by similarity

One of our aims is to solve the eigenvalue completion problem for positioned submatrices. Since the eigenvalues and their multiplicities do not change if a similarity transformation is applied to the matrix, it is natural to try to simplify the problem by using similarity. On the other hand not each similarity can be used. To make this more precise, let $A_{\mathcal{P}}$ and $A'_{\mathcal{P}'}$ be two positioned submatrices with rectangular patterns $\mathcal{P}$ and $\mathcal{P}'$. The invertible matrix S is an *admissible similarity* of $A_{\mathcal{P}}$ and

$A'_{\mathcal{P}'}$ if for any completion A of $A_{\mathcal{P}}$ the matrix SAS^{-1} is a completion of $A'_{\mathcal{P}'}$ and for any completion A' of $A'_{\mathcal{P}'}$ the matrix $S^{-1}A'S$ is a completion of $A_{\mathcal{P}}$. This means that if S is an admissible similarity of $A_{\mathcal{P}}$ and $A'_{\mathcal{P}'}$, then for each completion A of $A_{\mathcal{P}}$ the entries a'_{ij} of SAS^{-1} for $(i,j) \in \mathcal{P}'$ depend only on S and on the entries a_{ij} with $(i,j) \in \mathcal{P}$ of $A_{\mathcal{P}}$, and not on the entries of A outside the pattern $\mathcal{P}$. Also, conversely, the entries a_{ij} of $S^{-1}A'S$ for $(i,j) \in \mathcal{P}$ depend only on S and on the entries a'_{ij} with $(i,j) \in \mathcal{P}'$ of $A'_{\mathcal{P}'}$ for any completion A' of $A'_{\mathcal{P}'}$.

As an illustration we compute the admissible similarities for some special cases. First let the patterns $\mathcal{P}$ and $\mathcal{P}'$ be given by

$$\mathcal{P} = \mathcal{P}' = \big\{(i,j) \in \{1,\dots,n\} \mid 1 \le i,j \le n-1\big\}.$$

The partially specified matrices $A_{\mathcal{P}}$ and $A'_{\mathcal{P}'}$ with pattern $\mathcal{P}$ and $\mathcal{P}'$, respectively, are positioned submatrices given by

$$A_{\mathcal{P}} = \begin{pmatrix} A_{11} & ? \\ ? & ? \end{pmatrix}, \quad A'_{\mathcal{P}'} = \begin{pmatrix} A'_{11} & ? \\ ? & ? \end{pmatrix}$$

with A_{11} and A'_{11} square matrices of order $n-1$. Let S be an invertible matrix decomposed as

$$S = \begin{pmatrix} S_{11} & S_{12} \\ S_{21} & s_{22} \end{pmatrix},$$

where S_{11} a square matrix of order $n-1$. We shall prove that in this case S is an admissible similarity if and only if $S_{12} = 0$, $S_{21} = 0$ and $A'_{11} = S_{11}A_{11}S_{11}^{-1}$.

First assume that $S_{12} = 0$, $S_{21} = 0$ and $A'_{11} = S_{11}A_{11}S_{11}^{-1}$. Let A be any completion of $A_{\mathcal{P}}$ and let A' be any matrix. Decompose A as

$$A = \begin{pmatrix} A_{11} & A_{12} \\ A_{21} & a_{22} \end{pmatrix}. \tag{2.1}$$

Then SAS^{-1} has the form

$$SAS^{-1} = \begin{pmatrix} S_{11}A_{11}S_{11}^{-1} & * \\ * & * \end{pmatrix},$$

and hence SAS^{-1} is a completion of $A'_{\mathcal{P}'}$. Now assume that

$$A' = \begin{pmatrix} A'_{11} & A'_{12} \\ A'_{21} & a'_{22} \end{pmatrix}$$

is completion of $A'_{\mathcal{P}'}$. We prove that $S^{-1}A'S$ is a completion of $A_{\mathcal{P}}$. Since $A_{11} = S_{11}^{-1}A'_{11}S_{11}$, we see that the (1,1) entry in the block decomposition of $S^{-1}A'S$ is just A_{11}. This completes the proof of the fact that in this case S is an admissible similarity.

Conversely, assume that S is an admissible similarity of $A_{\mathcal{P}}$ and $A'_{\mathcal{P}'}$. Write $T = S^{-1}$ and decompose T as

$$T = \begin{pmatrix} T_{11} & T_{12} \\ T_{21} & t_{22} \end{pmatrix},$$

where T_{11} is a square matrix of order $n-1$. Assume that A is any completion of $A_{\mathcal{P}}$ and decompose A as in (2.1) We compute $SAS^{-1} = SAT$ and get

$$SAT = \begin{pmatrix} (S_{11}A_{11} + S_{12}A_{21})T_{11} + (S_{11}A_{12} + S_{12}a_{22})T_{21} & * \\ * & * \end{pmatrix},$$

where $*$ denotes an irrelevant entry. Because SAT is a completion of $A'_{\mathcal{P}'}$, it follows that

$$S_{11}A_{11}T_{11} + S_{12}A_{21}T_{11} + S_{11}A_{12}T_{21} + S_{12}a_{22}T_{21} = A'_{11}.$$

This holds for any choice of the matrices A_{21}, A_{12}, and a_{22}. In particular, one may choose these matrices to be zero matrices. Thus $S_{11}A_{11}T_{11} = A'_{11}$, and hence

$$S_{12}A_{21}T_{11} + S_{11}A_{12}T_{21} + S_{12}a_{22}T_{21} = 0$$

for any choice of A_{21}, A_{12}, and a_{22}. Choosing $A_{21} = 0$, $A_{12} = 0$ and $a_{22} = 1$ yields that $S_{12}T_{21} = 0$. Put $A_{21} = 0$, $a_{22} = 0$ and choose for A_{12} the vectors $e_1, \dots, e_{n-1}$ from the standard basis of $\mathbb{C}^{n-1}$. Then one sees that $S_{11}\begin{pmatrix} e_1 & \cdots & e_{n-1} \end{pmatrix} T_{21} = 0$, and therefore $S_{11}T_{21} = 0$. Similarly one proves that $S_{12}T_{11} = 0$. Since T is the inverse of S, we may use $ST = I$ and $S_{12}T_{21} = 0$ to conclude that $S_{11}T_{11} = I_{n-1}$, the $(n-1) \times (n-1)$ unit matrix. Thus from $S_{12}T_{11} = 0$ one sees that $S_{12} = 0$, and hence S is a block lower triangular matrix. But then so is T and it follows from $TS = I$ that $t_{22}s_{22} = 1$. However, also $T_{21} = 0$ because $S_{11}T_{21} = 0$ and S_{11} is invertible. So T is block upper and block lower triangular, and therefore T is block diagonal. But then S is block diagonal, and thus

$$S = \begin{pmatrix} S_{11} & 0 \\ 0 & s_{22} \end{pmatrix}.$$

We conclude that $S_{12} = 0$, $S_{21} = 0$ and $A'_{11} = S_{11}A_{11}S_{11}^{-1}$.

Next, we determine the admissible similarities for certain full width positioned submatrices. We assume that the pattern is

$$\mathcal{P} = \mathcal{P}' = \big\{(i,j) \in \{1, \dots, n\} \mid 2 \le i \le n\big\}.$$

The full width positioned submatrices $A_{\mathcal{P}}$ and $A'_{\mathcal{P}'}$ thus have the form

$$A_{\mathcal{P}} = \begin{pmatrix} ? & ? \\ A_{21} & A_{22} \end{pmatrix}, \quad A'_{\mathcal{P}'} = \begin{pmatrix} ? & ? \\ A'_{21} & A'_{22} \end{pmatrix}.$$

with A_{22} and A'_{22} square matrices of order $n-1$. Let S be an invertible matrix, decomposed as

$$S = \begin{pmatrix} s_{11} & S_{12} \\ S_{21} & S_{22} \end{pmatrix},$$

where S_{22} is a square matrix of order $n-1$. Then S is an admissible similarity if and only if

$$S_{21} = 0, \quad A_{21} = S_{22}^{-1} A'_{21} s_{11}, \quad A_{22} = S_{22}^{-1}(A_{21} S_{12} S_{22}^{-1} + A'_{22}) S_{22}. \tag{2.2}$$

To prove this, first assume that (2.2) holds true. Let A' be a completion of $A'_{\mathcal{P}'}$. Then one computes that

$$S^{-1} A' S = \begin{pmatrix} * & * \\ S_{22}^{-1} A'_{21} s_{11} & S_{22}^{-1} A_{21} S_{12} + S_{22}^{-1} A'_{22} S_{22} \end{pmatrix}.$$

So it follows that $S^{-1} A' S$ is a completion of A. Next let A be completion of $A_{\mathcal{P}}$. A simple computation gives that

$$SAS^{-1} = \begin{pmatrix} * & * \\ S_{22} A_{21} s_{11}^{-1} & -S_{22} A_{21} s_{11}^{-1} S_{12} S_{22}^{-1} + S_{22} A_{22} S_{22}^{-1} \end{pmatrix} = \begin{pmatrix} * & * \\ A'_{21} & A'_{22} \end{pmatrix}.$$

We conclude that S is an admissible similarity of $A_{\mathcal{P}}$ and $A'_{\mathcal{P}'}$.

Conversely, assume that S is an admissible similarity of $A_{\mathcal{P}}$ and $A'_{\mathcal{P}'}$. Let

$$A = \begin{pmatrix} a_{11} & A_{12} \\ A_{21} & A_{22} \end{pmatrix}, \quad A_0 = \begin{pmatrix} 0 & 0 \\ A_{21} & A_{22} \end{pmatrix}$$

be completions of $A_{\mathcal{P}}$. Then SAS^{-1} and SA_0S^{-1} are completions of $A'_{\mathcal{P}'}$. Hence

$$SAS^{-1} - SA_0S^{-1} = S(A - A_0)S^{-1} = \begin{pmatrix} * & * \\ 0 & 0 \end{pmatrix},$$

and thus

$$\begin{pmatrix} s_{11} & S_{12} \\ S_{21} & S_{22} \end{pmatrix} \begin{pmatrix} a_{11} & A_{12} \\ 0 & 0 \end{pmatrix} = \begin{pmatrix} * & * \\ 0 & 0 \end{pmatrix} \begin{pmatrix} s_{11} & S_{12} \\ S_{21} & S_{22} \end{pmatrix}.$$

This equality gives that $S_{21} a_{11} = 0$, and by choosing $a_{11} \neq 0$ one obtains that $S_{21} = 0$. Since SAS^{-1} is a completion of $A'_{\mathcal{P}'}$ we found that

$$\begin{pmatrix} s_{11} & S_{12} \\ 0 & S_{22} \end{pmatrix} \begin{pmatrix} a_{11} & A_{12} \\ A_{21} & A_{22} \end{pmatrix} = \begin{pmatrix} * & * \\ A'_{21} & A'_{22} \end{pmatrix} \begin{pmatrix} s_{11} & S_{12} \\ 0 & S_{22} \end{pmatrix}.$$

From the second row in this equality we conclude the remaining two equalities in (2.2).

Now, assume that $A_{\mathcal{P}}$ is given, and let us apply an admissible similarity S that gives $A'_{\mathcal{P}}$ a simple form. We assume that the $(n-1)\times 1$ matrix $Y = A_{21}$ is a cyclic vector for the $n-1\times n-1$ matrix $X = A_{22}$. So $Y, XY, X^2Y, \dots, X^{n-2}Y$ is a basis for $\mathbb{C}^{n-1}$. Let $X^{n-1}Y + \sum_{i=0}^{n-2} \alpha_i X^i Y = 0$. We put $S_{11} = 1$, $S_{12} = (\,\alpha_{n-2} \dots \alpha_0\,)$, the $(n-1)\times 1$ submatrix S_{21} of S to be the null matrix and $S_{22} = (\,s_1 \dots s_{n-1}\,)$ where we define the vectors s_j recursively as follows. Put $s_1 = Y$ and $s_j = Xs_{j-1}+\alpha_{n-j}Y$ for $j = 2, \dots, n-1$. Then the matrix S is an admissible similarity. By a direct verification one checks that the positioned submatrix $A'_{\mathcal{P}} = S^{-1}A_{\mathcal{P}}S$ is given by

$$A'_{\mathcal{P}} = \begin{pmatrix} ? & ? & \dots & ? & ? \\ 1 & 0 & \dots & 0 & 0 \\ 0 & 1 & & 0 & 0 \\ \vdots & & \ddots & & \vdots \\ 0 & 0 & & 1 & 0 \end{pmatrix}. \tag{2.3}$$

Let us now consider the eigenvalue completion problem for $A_{\mathcal{P}}$, with $\mathcal{P}$ as in the previous paragraph. Recall that the solution of the eigenvalue completion problem for $A'_{\mathcal{P}}$ in (2.3) is given in Section 1. Since S is an admissible similarity, we know that for any completion A' of $A'_{\mathcal{P}}$ the matrix $S^{-1}A'S$ is a completion of $A_{\mathcal{P}}$ and that all completions of $A_{\mathcal{P}}$ are obtained in this way. This means that the possible eigenvalues and multiplicities of A are known now. We conclude that any monic polynomial of degree n can be the characteristic polynomial of a completion of $A_{\mathcal{P}}$, and that each eigenvalue of a completion of $A_{\mathcal{P}}$ has geometric multiplicity one.

From the above examples we see that admissible similarities can considerably simplify the eigenvalue completion problem. Therefore we shall also be interested in the following problem: *Determine invariants and canonical form of a positioned submatrix under admissible similarities.* To make this problem more precise it is useful to pass to a basis free setting. This will be done in the next two sections.

I.3 Blocks

In the previous sections we considered matrices, positioned submatrices and admissible similarities. In this and the next section we will reformulate these notions in a basis free manner. First we replace an $n \times n$ matrix by a linear operator acting on an n-dimensional linear space $\mathcal{X}$. Next we introduce on the level of linear operators the analogue of a positioned submatrix. We define an *(operator) block* to be a triple $(B; P, Q)$, where P and Q are projections of the space $\mathcal{X}$ and $B : \operatorname{Im} Q \to \operatorname{Im} P$ is a linear operator. In this case we shall refer to $\mathcal{X}$ as the *underlying space.* The underlying space is assumed to be finite dimensional without further notice. Infinite dimensional blocks will be considered in Chapters XII-XIV only. One should think of P and Q to define the position and of B to fix the entries. Sometimes, when we want to stress the dependence on the projections P and Q (and on the space $\mathcal{X}$) we speak about a (P, Q)-*block* B (on $\mathcal{X}$). Let $A : \mathcal{X} \to \mathcal{X}$ be an operator.

A (P,Q)-block B is said to be a *(P,Q)-block of A* if P and Q are projections of $\mathcal{X}$ and $B = PAQ$.

With a positioned submatrix one may associate in a canonical way an operator block, as follows. Let us assume that the $m \times p$ positioned submatrix $A_{\mathcal{P}}$ is given by the pattern

$$\mathcal{P} = \{(i,j) \in \{1,\dots,n\}^2 \mid i \in \{i_1,\dots,i_m\},\ j \in \{j_1,\dots,j_p\}\}$$

and has entries b_{ij} for $(i,j) \in \mathcal{P}$. Let $f_1,\dots,f_n$ be a basis of the space $\mathcal{X}$. Put Q the projection of $\mathcal{X}$ onto the space spanned by the vectors $\{f_{j_1},\dots,f_{j_p}\}$ and along the subspace spanned by the remaining elements of the basis $f_1,\dots,f_n$, and let P be the projection of $\mathcal{X}$ onto the space spanned by the vectors $\{f_{i_1},\dots,f_{i_m}\}$ along the subspace spanned by the remaining elements of $\{f_1,\dots,f_n\}$. Let $B : \operatorname{Im} Q \to \operatorname{Im} P$ be given by the canonical action of the matrix $\left(b_{i_k j_\ell}\right)_{k=1,\ell=1}^{m,p}$ relative to basis $f_{j_1},\dots,f_{j_p}$ of $\operatorname{Im} Q$ and the basis $f_{i_1},\dots,f_{i_m}$ of $\operatorname{Im} P$. The operator block $(B;P,Q)$ is called a *block associated with* the positioned submatrix $A_{\mathcal{P}}$. Note that in this case the matrices P and Q are diagonal with respect to the basis $f_1,\dots,f_n$ of $\mathcal{X}$.

One could pose the question whether in all cases an operator block is associated with a positioned submatrix. The answer is that a (P,Q)-block B is a block associated to a positioned submatrix if and only if its underlying space $\mathcal{X}$ has a basis such that with respect to this basis the matrices P and Q are diagonal. The necessity of this condition follows from the remark made in the previous paragraph. To prove the sufficiency, let $f_1,\dots,f_n$ be the basis of $\mathcal{X}$ such that with respect to this basis the matrices of the projections P and Q are diagonal. Let

$$\mathcal{P} = \{(i,j) \mid f_i \in \operatorname{Im} P,\ f_j \in \operatorname{Im} Q\} = \{i_1,\dots,i_m\} \times \{j_1,\dots,j_p\}.$$

Let $\left(a_{i_k j_\ell}\right)_{k,\ell=1}^{m,p}$ be the matrix of B with respect to the bases $f_{j_1},\dots,f_{j_p}$ of $\operatorname{Im} Q$ and $f_{i_1},\dots,f_{i_m}$ of $\operatorname{Im} P$. Then the positioned submatrix $A_{\mathcal{P}}$ with pattern $\mathcal{P}$ and entries a_{ij} for $(i,j) \in \mathcal{P}$ is called *the representation of the block $(B;P,Q)$ with respect to the basis $f_1,\dots,f_n$*. Clearly, the (P,Q)-block B is a block associated with the positioned submatrix $A_{\mathcal{P}}$.

There are a few trivial special cases of blocks, which nevertheless require some attention. Namely, we did not and do not want to exclude the cases when either $P = 0$ or $Q = 0$. Obviously, in these cases a block does not come from what usually is considered to be a submatrix, even if we identify the linear space $\mathcal{X}$ with $\mathbb{C}^n$ in some way.

In the next examples $\mathcal{X}$ is an n-dimensional space and $A : \mathcal{X} \to \mathcal{X}$ is a linear operator.

Assume that $P = I$ and Q are projections of $\mathcal{X}$. Consider the (P,Q)-block of A. Put $B = PAQ : \operatorname{Im} Q \to \mathcal{X}$. Now choose a basis $f_1,\dots,f_p$ for $\operatorname{Im} Q$ and extend this basis by a basis $f_{p+1},\dots,f_n$ of $\operatorname{Ker} Q$ to a basis $f_1,\dots,f_n$ of $\mathcal{X}$. The representation of B as a positioned submatrix relative to $f_1,\dots,f_n$ consists of the first p columns of the matrix of A with respect to this basis. So the representation of $(B; I, Q)$ is a positioned submatrix of full length. Therefore we call the block $(B; I, Q)$ a *block of full length.*

Next assume that P and $Q = I$ are projections of $\mathcal{X}$. Again consider the (P, Q)-block of A. So put $B = PAQ : \mathcal{X} \to \operatorname{Im} P$. Choose a basis $f_1, \dots, f_m$ for $\operatorname{Im} P$ and extend this basis by a basis $f_{m+1}, \dots, f_n$ of $\operatorname{Ker} P$ to a basis $f_1, \dots, f_n$ of $\mathcal{X}$. The representation of the (P, Q)-block B with respect to $f_1, \dots, f_n$ is a positioned submatrix that consists of the first m rows of the matrix of A with respect to $f_1, \dots, f_n$. So the block $(B; P, I)$ is represented by a positioned submatrix of full width. Therefore we call the block $(B; P, I)$ a *block of full width.*

Assume that $Q = P$ is a projection of $\mathcal{X}$. Consider the (P, Q)-block of A. Put $B = PAQ : \operatorname{Im} Q \to \operatorname{Im} P$. Choose a basis $f_1, \dots, f_m$ for $\operatorname{Im} P$ and extend this basis by a basis $f_{m+1}, \dots, f_n$ of $\operatorname{Ker} Q$ to a basis $f_1, \dots, f_n$ of $\mathcal{X}$. With respect to these bases take the matrices of A and of B. Then the matrix of B corresponds in a natural way to a submatrix of the matrix of A. This submatrix is the principal $m \times m$ submatrix of the matrix of A. So the block $(B; P, P)$ is represented by a principal submatrix. The block $(B; P, P)$ we will refer to as a *principal block.*

Assume that P and Q are projections of the finite dimensional linear space $\mathcal{X}$ such that $\operatorname{Im} Q \subset \operatorname{Ker} P$. Assume that the block $(B; P, Q)$ allows a representation $A_{\mathcal{P}}$ with respect to the basis $f_1, \dots, f_n$. Without loss of generality we now may assume that $\operatorname{Im} P = \operatorname{span}\{f_1, \dots, f_m\}$ and $\operatorname{Im} Q = \operatorname{span}\{f_{n-p+1}, \dots, f_n\}$. If $(B; P, Q)$ is the (P, Q)-block of a linear transformation A, then the positioned matrix $A_{\mathcal{P}}$ form the right upper $m \times p$ submatrix of the matrix of A with respect to the basis $f_1, \dots, f_n$. Since also $m + p \leq n$, we will call the block $(B; P, Q)$ an *off-diagonal block.*

We finish this section with an example of a block that does not allow a representation as a positioned submatrix. Put

$$P = \begin{pmatrix} 0 & 0 \\ 0 & 1 \end{pmatrix}, \quad Q = \begin{pmatrix} 1 & 1 \\ 0 & 0 \end{pmatrix}.$$

For no basis f_1, f_2 the subspaces $\operatorname{Im} P$, $\operatorname{Im} Q$, $\operatorname{Ker} P$ and $\operatorname{Ker} Q$ are spanned by vectors of the basis f_1, f_2. So for no operator A on $\mathbb{C}^2$ and no basis f_1, f_2 of $\mathbb{C}^2$ the (P, Q)-block of A is represented by a positioned 1×1 submatrix of a matrix of A with respect to f_1, f_2.

I.4 Block similarity

In this section we shall define and analyze the analogue for operator blocks of the notion of admissible similarity for submatrices. Let $(B; P, Q)$ and $(B'; P', Q')$ be blocks, where P and Q are projections on the linear space $\mathcal{X}$ and P' and Q' are projections on the linear space $\mathcal{X}'$. We will call the blocks $(B; P, Q)$ and $(B'; P', Q')$ *block similar* if there exists an invertible operator $S : \mathcal{X}' \to \mathcal{X}$ with the following properties

(i) $S[\operatorname{Ker} P'] = \operatorname{Ker} P$ and $S[\operatorname{Im} Q'] = \operatorname{Im} Q$;

(ii) $(SB' - BS)x \in \operatorname{Ker} P$ for all $x \in \operatorname{Im} Q'$.

An invertible operator S with the properties (i) and (ii) is called a *block similarity.*

Under the assumption that S is invertible (i) is equivalent to

(i') $PS(I - P') = 0$, $P'S^{-1}(I - P) = 0$ and $(I - Q)SQ' = 0$, $(I - Q')S^{-1}Q = 0$.

Note that block similarity is an equivalence relation on the set of operator blocks. In the case when the projections P and Q are both equal to the identity operator, the blocks B and B' are block similar if and only if they are similar as operators, i.e., $SB' = BS$ for an invertible operator S.

Let $S : \mathcal{X}' \to \mathcal{X}$ be a block similarity of $(B; P, Q)$ and $(B'; P', Q')$. Then $(B; P, Q)$ and $(B'; P', Q')$ are called *block similar with a block similarity of the first kind* if

(iii) $P' = SPS^{-1}$, $Q' = SQS^{-1}$ and $B' = SBS^{-1} : \operatorname{Im} Q' \to \operatorname{Im} P'$.

Note that condition (iii) implies that conditions (i) and (ii) from the definition of block similarity are fulfilled. Block similar with a block similarity of the first kind is also an equivalence relation.

There is another case of block similarity that deserves special attention, namely when the operator S in (i) and (ii) is the identity operator. This happens if and only if $\mathcal{X}' = \mathcal{X}$ and

$$\operatorname{Ker} P' = \operatorname{Ker} P, \quad \operatorname{Im} Q' = \operatorname{Im} Q, \quad B' = BP. \tag{4.1}$$

We call the operator blocks $(B; P, Q)$ and $(B'; P', Q')$ *block similar with a block similarity of the second kind* if the two blocks have the same underlying space and (4.1) holds. Also block similar with a block similarity of the second kind is an equivalence relation.

Proposition 4.1. *Every block similarity is the composition of a block similarity of the first kind and a block similarity of the second kind, and is also the composition of a block similarity of the second kind and a block similarity of the first kind.*

PROOF. Assume that $S : \mathcal{X}' \to \mathcal{X}$ is a block similarity of the blocks $(B; P, Q)$ and $(B'; P', Q')$. Put

$$P_2 = SPS^{-1}, \quad Q_2 = SQS^{-1}, \quad B_2 = SBS^{-1} : \operatorname{Im} Q_2 \to \operatorname{Im} P_2.$$

Then S is a block similarity of the first kind of $(B; P, Q)$ and (B_2, P_2, Q_2). Note that we have

$$\operatorname{Ker} P_2 = \operatorname{Ker} P', \quad \operatorname{Im} Q_2 = \operatorname{Im} Q', \quad (B' - B_2)x \in \operatorname{Ker} P'$$

for each $x \in \operatorname{Im} Q_2$. Thus $I_{\mathcal{X}'}$ is a block similarity of the blocks (B_2, P_2, Q_2) and (B', P', Q'). This proves that S is the composition of a similarity of the first kind and a similarity of the second kind and therefore the first statement of the proposition is proved. To prove the second statement we only need to apply the first statement to the block similarity S^{-1} of $(B'; P', Q')$ and $(B; P, Q)$ and to invert. □

At the end of the previous section we remarked that not every (P,Q)-block of an operator $A : \mathcal{X} \to \mathcal{X}$ has a representation as a positioned submatrix $A_{\mathcal{P}}$ with respect to a basis such that the matrix of A with respect to the same basis is a completion of $A_{\mathcal{P}}$. However we have the following proposition.

Proposition 4.2. *Let $(B; P, Q)$ be a (P,Q)-block of A with underlying space $\mathcal{X}$. Then $(B; P, Q)$ is block similar with a block similarity of the second kind to an operator block (B', P', Q') which has a representation as a positioned submatrix $A_{\mathcal{P}}$ with respect to a basis $f_1, \dots, f_n$ of $\mathcal{X}$ such that the matrix of A with respect to the same basis is a completion of $A_{\mathcal{P}}$.*

PROOF. First we construct the projections P' and Q'. Put $\operatorname{Ker} P = \operatorname{Ker} P'$ and $\operatorname{Im} Q = \operatorname{Im} Q'$. Write $\mathcal{N}_0 = \operatorname{Im} Q \cap \operatorname{Ker} P$, and define $\mathcal{N}_1$, $\mathcal{N}_2$ and $\mathcal{M}$ such that

$$\operatorname{Im} Q = \mathcal{N}_0 \oplus \mathcal{N}_1, \quad \operatorname{Ker} P = \mathcal{N}_0 \oplus \mathcal{N}_2, \quad \mathcal{X} = \mathcal{M} \oplus \mathcal{N}_1 \oplus \mathcal{N}_0 \oplus \mathcal{N}_2.$$

Now let $\operatorname{Ker} Q' = \mathcal{N}_2 \oplus \mathcal{M}$ and $\operatorname{Im} P' = \mathcal{N}_1 \oplus \mathcal{M}$. Then the projections P' and Q' are well defined. Note that the blocks $(B; P, Q)$ and $(P'B; P', Q')$ allow a block similarity of the second kind. Take bases for $\mathcal{M}$, $\mathcal{N}_0$, $\mathcal{N}_1$ and $\mathcal{N}_2$, and let $f_1, \dots, f_n$ be the union of these bases. Then $f_1, \dots, f_n$ is a basis of $\mathcal{X}$, and $(P'B; P', Q')$ has representation $A_{\mathcal{P}}$ as a positioned submatrix with respect to this basis. Since $P'B = P'A|_{\operatorname{Im} Q'}$ it follows that the matrix of A is a completion of $A_{\mathcal{P}}$. □

We illustrate this proposition on the next example. Let

$$P = \begin{pmatrix} 1 & 0 \\ 0 & 0 \end{pmatrix}, \quad Q = \begin{pmatrix} 1 & 1 \\ 0 & 0 \end{pmatrix}, \quad A = \begin{pmatrix} 1 & 3 \\ 2 & 4 \end{pmatrix},$$

define operators on $\mathbb{C}^2$. We write e_1, e_2 for the standard basis of $\mathbb{C}^2$. Let $B = PA|_{\operatorname{Im} Q}$. Then $(B; P, Q)$ is the (P,Q)-block of the operator A. This block can not be represented by a submatrix of the matrix of A with respect to a basis of $\mathbb{C}^2$. We construct a block that is block similar to this block. Like in the proof of Proposition 4.1 we put $\mathcal{N}_0 = \operatorname{Im} Q \cap \operatorname{Ker} P = (0)$, $\mathcal{N}_1 = \operatorname{Im} Q$, $\mathcal{N}_2 = \operatorname{Ker} P$, and then $\mathcal{X} = \mathcal{N}_1 \oplus \mathcal{N}_2$. Put $f_1 = e_1$ and $f_2 = e_2$. Define P' by $\operatorname{Im} P' = \mathcal{N}_1$, $\operatorname{Ker} P' = \operatorname{Ker} P$ and Q' by $\operatorname{Im} Q' = \operatorname{Im} Q$ and $\operatorname{Ker} Q' = \mathcal{N}_2$. So with respect to the basis f_1, f_2 we find the matrix representations

$$P' = \begin{pmatrix} 1 & 0 \\ 0 & 0 \end{pmatrix}, \quad Q' = \begin{pmatrix} 1 & 0 \\ 0 & 0 \end{pmatrix}.$$

The operator $B = P'A|_{\operatorname{Im} Q'}$ is given by the 1×1 left upper submatrix of the matrix of A with respect to the basis f_1, f_2 and the blocks $(B; P, Q)$ and $(B'; P', Q')$ are block similar with a block similarity of the second kind.

We give some examples of similarities of the second kind that will be of use later on. Let P and Q be projections of the linear space $\mathcal{X}$ such that $\mathcal{X} =$

$\operatorname{Im} Q \oplus \operatorname{Ker} P$. Then the (P, Q)-block of an operator A is block similar to a principal block of the operator A. Indeed, let P' be the projection onto $\operatorname{Im} Q$ and along $\operatorname{Ker} P$. So $\operatorname{Im} Q = \operatorname{Im} P'$ and $\operatorname{Ker} P = \operatorname{Ker} P'$. Then the blocks (PAQ, P, Q) and $(P'AP'; P', P')$ are block similar with a block similarity of the second kind.

Let P and Q be projections of the linear space $\mathcal{X}$. Assume that $\operatorname{Im} Q \subset \operatorname{Ker} P$. Then the (P, Q)-block $(B; P, Q)$ is similar to an off-diagonal block. To see this use Proposition 4.2 to find a similar block (B', P', Q') and a basis such that (B', P', Q') has a representation as a positioned submatrix with respect to this basis. Since (B', P', Q') is block similar to $(B; P, Q)$, we have that $\operatorname{Im} Q' \subset \operatorname{Ker} P'$, and thus (B', P', Q') is an off-diagonal block.

Our next goal is to describe the connection between admissible similarity for positioned submatrices and block similarity for blocks of operators. The key result for this is the following theorem.

Theorem 4.3. *Let $(B; P, Q)$ and (B', P', Q') be operator blocks with underlying space $\mathcal{X}$ such that the corresponding projections are non-zero. The invertible operator $S : \mathcal{X}' \to \mathcal{X}$ is a block similarity of the blocks $(B; P, Q)$ and (B', P', Q') if and only if the following two conditions hold:*

(1) *$B' = P'S^{-1}ASQ'$ for each $A : \mathcal{X} \to \mathcal{X}$ such that $B = PAQ : \operatorname{Im} Q \to \operatorname{Im} P$;*

(2) *$B = PSA'S^{-1}Q$ for each $A' : \mathcal{X}' \to \mathcal{X}'$ such that $B' = P'A'Q' : \operatorname{Im} Q' \to \operatorname{Im} P'$.*

Before we prove this result we remark that (1) means that the operator B' depends on B and S only, and is independent of the extension A that enters in (1). Also, conversely, according to (2) B depends on B' and S only.

Proof. First, assume that conditions (1) and (2) are fulfilled. From condition (1) we obtain that $PAQ = PA_1Q$ implies that $P'S^{-1}ASQ' = P'S^{-1}A_1SQ'$. So for any A_0 such that $PA_0Q = 0$ it follows that $P'S^{-1}A_0SQ' = 0$. Now let A be an arbitrary operator on $\mathcal{X}$, and put $A_0 = (I - P)A$. From $PA_0Q = 0$ we conclude that $P'S^{-1}A_0SQ' = 0$, and therefore $P'S^{-1}(I - P)ASQ' = 0$ for any operator A. Since $SQ' \neq 0$, it follows that $P'S^{-1}(I - P) = 0$ and thus $S^{-1}[\operatorname{Ker} P] \subset \operatorname{Ker} P'$. From condition (2) we obtain the converse inclusion in the same way. Now take $A_0 = A(I-Q)$ for some operator A. We get from $PA_0Q = 0$ that $P'S^{-1}A_0SQ' = 0$, and thus $P'S^{-1}A(I - Q)SQ' = 0$ for any operator A. Since $P'S \neq 0$, it follows that $(I - Q)SQ' = 0$, which is equivalent to $S[\operatorname{Im} Q'] \subset \operatorname{Im} Q$. Again condition (2) provides the converse inclusion. We proved the property (i) from the definition of block similarity. In particular, we have $PSP' = PS$. Now let A be an operator such that $B = PAQ : \operatorname{Im} Q \to \operatorname{Im} P$. Then $B' = P'S^{-1}ASQ'$ by condition (2) and we compute for $x \in \operatorname{Im} Q'$ that

$$\begin{aligned} P(BS - S'B')x &= P(PAQS - SP'S^{-1}ASQ')Q'x \\ &= (PAQS - PSS^{-1}ASQ')Q'x = 0. \end{aligned}$$

This proves that S is a block similarity.

Next, assume that S is a block similarity. First note that $P'S^{-1}P = P'S^{-1}$ and $QSQ' = SQ'$. To prove (1), take A such that $B = PAQ$. For $x \in \operatorname{Im} Q'$ we compute

$$B'x = P'S^{-1}BSx = P'S^{-1}PAQSQ'x = P'S^{-1}ASQ'x.$$

So $B' = P'S^{-1}ASQ' : \operatorname{Im} Q' \to \operatorname{Im} P'$. By applying the previous results with S^{-1} one obtains (2). □

Let

$$\mathcal{P} = \{i_1, \ldots, i_m\} \times \{j_1, \ldots, j_p\} \subset \{1, \ldots, n\}^2$$
$$\mathcal{P}' = \{i'_1, \ldots, i'_m\} \times \{j'_1, \ldots, j'_p\} \subset \{1, \ldots, n\}^2$$

be the patterns of the positioned submatrices $A_{\mathcal{P}}$ and $A'_{\mathcal{P}'}$. Let $(B; P, Q)$ be the block associated with $A_{\mathcal{P}}$ and $(B'; P', Q')$ be the block associated with $A'_{\mathcal{P}'}$. We will show that in this situation the notion of admissible similarity of $A_{\mathcal{P}}$ and $A'_{\mathcal{P}'}$ and the notion of block similarity of $(B; P, Q)$ and $(B'; P', Q')$ coincide.

Proposition 4.4. *Let $(B; P, Q)$ and (B', P', Q') be operator blocks with underlying spaces $\mathcal{X}$ and $\mathcal{X}'$, respectively. Assume that $A_{\mathcal{P}}$ ($A'_{\mathcal{P}'}$) is the representation as a positioned submatrix of $(B; P, Q)$ ((B', P', Q')) with respect to the basis $f_1, \ldots, f_n$ of $\mathcal{X}$ ($f'_1, \ldots, f'_n$ of $\mathcal{X}'$). Let $S : \mathcal{X}' \to \mathcal{X}$ be an invertible operator, and let M_S be its matrix representation relative to the bases $f'_1, \ldots, f'_n$ and $f_1, \ldots, f_n$. Then S is a block similarity of $(B; P, Q)$ and (B', P', Q') if and only if M_S is an admissible similarity of $A_{\mathcal{P}}$ and $A'_{\mathcal{P}'}$.*

PROOF. Assume that M_S is an admissible similarity of $A_{\mathcal{P}}$ and $A'_{\mathcal{P}'}$. Let $A : \mathcal{X} \to \mathcal{X}$ be such that $B = PAQ : \operatorname{Im} Q \to \operatorname{Im} P$. Since $A_{\mathcal{P}}$ is a matrix representation of $(B; P, Q)$, the matrix M_A of A with respect to the basis $f_1, \ldots, f_n$ is a completion of the positioned submatrix $A_{\mathcal{P}}$. The matrix $M_S^{-1} M_A M_S$ is a completion of $A'_{\mathcal{P}'}$ because M_S is an admissible similarity of $A_{\mathcal{P}}$ and $A'_{\mathcal{P}'}$. The matrix $M_S^{-1} M_A M_S$ is the matrix of $S^{-1}AS$ with respect to the basis $f'_1, \ldots, f'_n$. Because $A'_{\mathcal{P}'}$ is a representation of (B', P', Q'), it follows that $B' = P'S^{-1}ASQ' : \operatorname{Im} Q' \to \operatorname{Im} P'$. This proves that condition (1) of Theorem 4.3 is fulfilled. Next, let $A' : \mathcal{X}' \to \mathcal{X}'$ be an operator such that $B' = P'A'Q' : \operatorname{Im} Q' \to \operatorname{Im} P'$. A similar reasoning as the one above leads to the conclusion that $B = PSA'S^{-1}Q : \operatorname{Im} Q \to \operatorname{Im} P$. So also condition (2) is satisfied. We conclude that S is a block similarity.

Conversely, assume that S is a block similarity. Let M_A be a completion of $A_{\mathcal{P}}$. With respect to the basis $f_1, \ldots, f_n$ the matrix M_A defines an operator A such that $B = PAQ : \operatorname{Im} Q \to \operatorname{Im} P$. The operator S has property (1) from Theorem 4.3 and hence $B' = P'S^{-1}ASQ' : \operatorname{Im} Q' \to \operatorname{Im} P'$. Therefore the matrix of $S^{-1}AS$ is a completion of $A'_{\mathcal{P}'}$. This shows that $M_S^{-1} M_A M_S$ is a completion of $A'_{\mathcal{P}'}$. Similarly one proves, using (2) from Theorem 4.3, that if $M'_{A'}$ is a completion of $A'_{\mathcal{P}'}$, then $M_S M_{A'} M_S^{-1}$ is a completion of $A_{\mathcal{P}}$. By definition it follows that M_S is an admissible similarity of $A_{\mathcal{P}}$ and $A'_{\mathcal{P}'}$. □

Let us return to the special classes of operator blocks we introduced in Section 3. Recall that we identified the class of the blocks $(B; P, Q)$ on $\mathcal{X}$ with $\operatorname{Im} Q \oplus \operatorname{Ker} P = \mathcal{X}$ as the class of blocks that are block similar to principal blocks. Also we showed that if $\operatorname{Im} Q \subset \operatorname{Ker} P$, then $(B; P, Q)$ is block similar to an off-diagonal block. The next result states that the four important subclasses of blocks considered in Section 3 are closed under applying block similarity.

Theorem 4.5. *The class of full length blocks, the class of full width blocks, the class of the blocks $(B; P, Q)$ on $\mathcal{X}$ with $\operatorname{Im} Q \oplus \operatorname{Ker} P = \mathcal{X}$, and the class of blocks with $\operatorname{Im} Q \subset \operatorname{Ker} P$ are closed under block similarity transformations.*

PROOF. Assume that S is a block similarity of the block $(B; P, Q)$ with a full length block $(B'; I', Q')$. Then $\operatorname{Ker} P = S[\operatorname{Ker} I'] = \{0\}$ and hence $P = I$. So $(B; P, Q)$ is a full length block. Next assume that S is a block similarity of $(B; P, Q)$ with a full width block $(B'; P', I')$. Then $\operatorname{Im} Q = S[\operatorname{Im} I']$ and hence $Q = I$, which shows that $(B; P, Q)$ is a full width block. Now assume that S is a block similarity of $(B; P, Q)$ with a block $(B'; P', Q')$ such that $\operatorname{Im} Q' \oplus \operatorname{Ker} P' = \mathcal{X}'$. Note that $(B'; P', Q')$ is block similar with a block similarity of the second kind to a principal block. Then

$$\mathcal{X} = S[\mathcal{X}'] = S[\operatorname{Im} Q' \oplus \operatorname{Ker} P'] = \operatorname{Im} Q \oplus \operatorname{Ker} P.$$

So again $(B; P, Q)$ is in the same class as $(B'; P', Q')$. Finally, assume that S is a block similarity of $(B; P, Q)$ with a block $(B'; P', Q')$ such that $\operatorname{Im} Q' \subset \operatorname{Ker} P'$. Then

$$\operatorname{Im} Q = S[\operatorname{Im} Q'] \subset S[\operatorname{Ker} P'] = \operatorname{Ker} P.$$

This proves that $(B; P, Q)$ is in the same class as $(B'; P', Q')$. □

I.5 Special cases of block similarity

In this section we show what block similarity means for each of the subclasses considered in Theorem 4.5.

First we discuss block similarity for blocks of full length. Consider the full length blocks $(B; I, Q)$ and $(B'; I', Q')$ with underlying spaces $\mathcal{X}$ and $\mathcal{X}'$, respectively. Here I (I') denotes the indentity operator on $\mathcal{X}$ $(\mathcal{X}')$. Assume that $S : \mathcal{X}' \to \mathcal{X}$ is a block similarity of $(B; I, Q)$ and $(B'; I', Q')$. Thus

(i) $S[\operatorname{Im} Q'] = \operatorname{Im} Q$,

(ii) $(SB' - BS)x = 0$ for all $x \in \operatorname{Im} Q'$.

Write $\mathcal{X} = \operatorname{Im} Q \oplus \operatorname{Ker} Q$ and $\mathcal{X}' = \operatorname{Im} Q' \oplus \operatorname{Ker} Q'$. Rewrite the operators as operator matrices with respect to these decompositions. So

$$S = \begin{pmatrix} S_{11} & S_{12} \\ S_{21} & S_{22} \end{pmatrix}, \quad B = \begin{pmatrix} B_1 \\ B_2 \end{pmatrix}, \quad B' = \begin{pmatrix} B_1' \\ B_2' \end{pmatrix}.$$

From (i) above it follows that $S_{21} = 0$. So S_{11} and S_{22} are invertible. Since (ii) gives that

$$S_{11}B_1' + S_{12}B_2' = B_1S_{11}, \quad S_{22}B_2' = B_2S_{11},$$

the invertibility of S_{11} yields

$$B_1 = S_{11}(B_1' + S_{11}^{-1}S_{12}B_2')S_{11}^{-1}, \quad B_2 = S_{22}B_2'S_{11}^{-1}. \tag{5.1}$$

The relations (5.1) are known in systems theory as output injection equivalence for the pairs (B_1, B_2) and (B_1', B_2'). The notion of output injection will be discussed in Chapter IX.

The second special case concerns blocks of full width. Let the blocks $(B; P, I)$ and $(B'; P', I')$ be of full width with underlying spaces $\mathcal{X}$ and $\mathcal{X}'$, respectively. Assume that $S : \mathcal{X}' \to \mathcal{X}$ is a block similarity of $(B; P, I)$ and $(B'; P', I')$. Thus

(i) $S[\operatorname{Ker} P'] = \operatorname{Ker} P$,

(ii) $(SB' - BS)x \in \operatorname{Ker} P$ for all x.

Write $\mathcal{X} = \operatorname{Im} P \oplus \operatorname{Ker} P$ and $\mathcal{X}' = \operatorname{Im} P' \oplus \operatorname{Ker} P'$. With respect to these decomposition rewrite the operators as

$$S = \begin{pmatrix} S_{11} & S_{12} \\ S_{21} & S_{22} \end{pmatrix}, \quad B = \begin{pmatrix} B_1 & B_2 \end{pmatrix}, \quad B' = \begin{pmatrix} B_1' & B_2' \end{pmatrix}.$$

From (i) above it follows that $S_{12} = 0$. So S_{11} and S_{22} are invertible. Rewrite (ii) as

$$S_{11}B_1' = B_1S_{11} + B_2S_{21}, \quad S_{11}B_2' = B_2S_{22},$$

and use that S_{11} is invertible to obtain

$$B_1' = S_{11}^{-1}(B_1 + B_2S_{21}S_{11}^{-1})S_{11}, \quad B_2' = S_{11}^{-1}B_2S_{22}. \tag{5.2}$$

The relations (5.2) are known in systems theory as state feedback equivalence for the pairs (B_1, B_2) and (B_1', B_2'). In Chapter IX we will explain in detail the relation between block similarity of full width blocks and state feedback equivalence.

For the third special case let P and Q be projections of the linear space $\mathcal{X}$ and let P' and Q' be projections of the linear space $\mathcal{X}'$. Assume that $\mathcal{X} = \operatorname{Im} Q \oplus \operatorname{Ker} P$ and that $\mathcal{X}' = \operatorname{Im} Q' \oplus \operatorname{Ker} P'$. Then the blocks $(B; P, Q)$ and $(B'; P', Q')$ are similar to principal blocks. Assume that $S : \mathcal{X}' \to \mathcal{X}$ is a block similarity of $(B; P, Q)$ and $(B'; P', Q')$. Then

(i) $S[\operatorname{Ker} P'] = \operatorname{Ker} P, \quad S[\operatorname{Im} Q'] = \operatorname{Im} Q$,

(ii) $P(SB' - BS)x = 0$ for all $x \in \operatorname{Im} Q'$.

With respect to the decompositions $\mathcal{X} = \operatorname{Im} P \oplus \operatorname{Ker} Q$ and $\mathcal{X}' = \operatorname{Im} P' \oplus \operatorname{Ker} Q'$. we rewrite the operator S as

$$S = \begin{pmatrix} S_{11} & S_{12} \\ S_{21} & S_{22} \end{pmatrix}.$$

From (i) above we conclude that $S_{21} = 0$ and $S_{12} = 0$. Since S is invertible, so are S_{11} and S_{22}. Moreover from (ii) it follows that $S_{11}B'x = BS_{11}x$ for all $x \in \operatorname{Im} Q'$. We conclude that in the case when the spaces $\mathcal{X} = \operatorname{Im} Q \oplus \operatorname{Ker} P$ and $\mathcal{X}' = \operatorname{Im} Q' \oplus \operatorname{Ker} P'$ are of equal finite dimension, the blocks $(B; P, Q)$ and $(B'; P', Q')$ are block similar if and only if there exists an invertible $S_{11} : \operatorname{Im} Q' \to \operatorname{Im} Q$ such that $B' = S_{11}BS_{11}^{-1}$.

Our last example concerns off-diagonal blocks. Assume that $S : \mathcal{X}' \to \mathcal{X}$ is a block similarity of $(B; I - Q, Q)$ on $\mathcal{X}$ and $(B'; I - Q', Q')$ on $\mathcal{X}'$. Thus

(i) $S[\operatorname{Im} Q'] = \operatorname{Im} Q$,

(ii) $(I - Q)SB'x = (I - Q)BSx$ for all $x \in \operatorname{Im} Q'$.

Write $\mathcal{X} = \operatorname{Im} Q \oplus \operatorname{Im}(I-Q)$ and $\mathcal{X}' = \operatorname{Im} Q' \oplus \operatorname{Im}(I-Q')$, and rewrite the operator S with respect to these decompositions as

$$S = \begin{pmatrix} S_{11} & S_{12} \\ S_{21} & S_{22} \end{pmatrix},$$

From (i) above we conclude that $S_{21} = 0$. The invertibility of S yields that S_{11} and S_{22} are invertible. Compute the right hand side of (ii). Because $Sx \in \operatorname{Im} Q$ for $x \in \operatorname{Im} Q'$, one has $Sx = S_{11}x$ and $(I - Q)BSx = (I - Q)BS_{11}x = BS_{11}x$. Next compute the left hand side of (ii). Since $B'x \in \operatorname{Im}(I - Q')$ and $(I - Q)S : \mathcal{X} \to \operatorname{Im}(I - Q)$, one finds that $(I - Q)SB'x = S_{22}B'x$. So (ii) is equivalent to $S_{22}B' = BS_{11}$. It follows that if the blocks $(B; I - Q, Q)$ and $(B'; I - Q', Q')$ are block similar, then $\operatorname{rank} B' = \operatorname{rank} B$.

Conversely, if $\operatorname{rank} B' = \operatorname{rank} B$ then there exist invertible $S_{11} : \operatorname{Im} Q' \to \operatorname{Im} Q$ and $S_{22} : \operatorname{Im}(I - Q') \to \operatorname{Im}(I - Q)$ such that $S_{22}B' = BS_{11}$. Define the operator S as the operator matrix

$$S = \begin{pmatrix} S_{11} & 0 \\ 0 & S_{22} \end{pmatrix},$$

with respect to the decompositions

$$\mathcal{X} = \operatorname{Im} Q \oplus \operatorname{Im}(I - Q), \quad \mathcal{X}' = \operatorname{Im} Q' \oplus \operatorname{Im}(I - Q').$$

Then obviously (i) is fulfilled. But in this case $S_{22}B' = BS_{11}$ is equivalent to (ii), which proves that S is a block similarity for the blocks $(B; I - Q, Q)$ and $(B'; I - Q', Q')$.

We conclude that the blocks $(B; I - Q, Q)$ and $(B'; I - Q', Q')$ are block similar if and only if $\operatorname{rank} B' = \operatorname{rank} B$.

I.6 Eigenvalue completion and restriction problems

Let $(B; P, Q)$ be an operator block with underlying space $\mathcal{X}$. An operator A on $\mathcal{X}$ is called *a completion* of the block $(B; P, Q)$ if $B = PAQ : \operatorname{Im} Q \to \operatorname{Im} P$. We now can state one of the main problems discussed in this book.

PROBLEM A. *Describe the eigenvalues with their multiplicities of the completions of the block* $(B; P, Q)$.

The next theorem gives a first result about this problem.

Theorem 6.1. *If* $(B; P, Q)$ *and* $(B'; P', Q')$ *are block similar, then for each completion* A *of* $(B; P, Q)$ *there exists a completion* A' *of* $(B'; P', Q')$, *such that* $A' = S^{-1}AS$ *for some invertible operator* S.

PROOF. Let $S : \mathcal{X}' \to \mathcal{X}$ be a block similarity of $(B; P, Q)$ and $(B'; P', Q')$. From Theorem 4.3 we conclude that $A' = S^{-1}AS$ is a completion of $(B'; P', Q')$. □

Theorem 6.1 implies that the solution of Problem A does not depend on the particular block $(B; P, Q)$, but only on the similarity class of $(B; P, Q)$. Therefore Problem A may be restated in the following equivalent form.

PROBLEM A'. *Describe the eigenvalues with their multiplicities of the completions of all the blocks in the block similarity class of the block* $(B; P, Q)$.

We shall refer to Problem A (or A') as the *eigenvalue completion problem.* To find the solution of this problem it is important to know how one may simplify an operator block by block similarity. This will be one of the main topics of the next chapters.

Let us consider a converse problem. Assume that projections P and Q are given and that we are looking for all (P, Q)-blocks which have a completion with prescribed eigenvalues (including multiplicities). If $(B; P, Q)$ and $(B'; P', Q')$ have a completion with the same eigenvalues and multiplicities, then they do not have to be block similar. However, we have the following result.

Theorem 6.2. *If* A *is a completion of the block* $(B; P, Q)$ *and* A' *is similar to* A, *then* A' *is a completion of a block that is block similar to* $(B; P, Q)$.

PROOF. Assume that $SAS^{-1} = A'$. Let $(B'; P', Q')$ be the block that is block similar to $(B; P, Q)$ with first kind block similarity equal to S. Thus $P' = S^{-1}PS$, $Q' = S^{-1}QS$ and $B' = SBS^{-1} : \operatorname{Im} Q' \to \operatorname{Im} P'$. Then A' is a completion of $(B'; P', Q')$. □

In view of this theorem we can restate the above mentioned converse problem as follows.

PROBLEM B. *Given a set of eigenvalues and multiplicities, describe all the block similarity classes of blocks that appear as a block of an operator with these eigenvalues and multiplicities.*

This problem we will call the *eigenvalue restriction problem.*

We finish this section with an example in which we determine (the block similarity classes of) full length blocks of operators on $\mathbb{C}^2$ with eigenvalues 1 and 0. In view of Theorem 6.2 we may fix the operator to be

$$A = \begin{pmatrix} 1 & 0 \\ 0 & 0 \end{pmatrix},$$

and ask for all (P, I) blocks of A with P a rank 1 projection. So P factorizes as

$$P = \begin{pmatrix} a & 0 \\ b & 0 \end{pmatrix} \begin{pmatrix} c & d \\ 0 & 0 \end{pmatrix}$$

with $ac + bd = 1$. First we compute that

$$PA = \begin{pmatrix} ac & 0 \\ bc & 0 \end{pmatrix}.$$

Since we consider $PA : \mathbb{C}^2 \to \operatorname{Im} P$, we take as a basis of $\mathbb{C}^2$ the vectors $\begin{pmatrix} a & b \end{pmatrix}^T \in \operatorname{Im} P$ and $\begin{pmatrix} -d & c \end{pmatrix}^T \in \operatorname{Ker} P$ and compute the image of of these base vectors. So

$$PA \begin{pmatrix} a \\ b \end{pmatrix} = ac \begin{pmatrix} a \\ b \end{pmatrix}, \quad PA \begin{pmatrix} -d \\ c \end{pmatrix} = -cd \begin{pmatrix} a \\ b \end{pmatrix}.$$

Thus with respect to this basis and the basis $\begin{pmatrix} a & b \end{pmatrix}^T$ of $\operatorname{Im} P$, the matrix of PA is $\begin{pmatrix} ac & -cd \end{pmatrix}$. Clearly any matrix of a full width block can be written in this form. Therefore any full width block on $\mathbb{C}^2$ is a full width block of A. To determine all these similarity classes we note that each class contains a block with a representation as a full width positioned submatrix with first row $(\beta_1 \quad \beta_2)$. Now apply formula (5.2) which determines the block similarity of such positioned submatrices. So the class represented by $(\beta_1 \quad \beta_2)$ is equal to the class represented by $(\beta_1' \quad \beta_2')$ if and only if there exists an invertible matrix

$$S = \begin{pmatrix} s_{11} & s_{12} \\ s_{21} & s_{22} \end{pmatrix}$$

such that

$$\beta_1' = \beta_1 + \beta_2 s_{21} s_{11}^{-1}, \quad \beta_2' = \beta_2 s_{22} s_{11}^{-1}. \tag{6.1}$$

If $\beta_2 = 0$, then $\beta_1' = \beta_1$ and $\beta_2' = 0$. So in this case the class is uniquely determined by the value of β_1. If $\beta_2 \neq 0$, then $\beta_2' \neq 0$ and irrespective of the values of β_1, β_1', $\beta_2 \neq 0$ and $\beta_2' \neq 0$ there exists a matrix S that fulfills (6.1). So for $\beta_2 \neq 0$ all positioned submatrices considered represent the same block similarity class.

Notes

The concept of an operator block and of block similarity is taken from the paper Gohberg-Kaashoek-Van Schagen [1]. The eigenvalue completion problem in the form as it is stated here can be found in Gohberg-Kaashoek-Van Schagen [4]. Earlier versions of this problem, with restrictions on the location of the given submatrix, have been treated by De Oliveira [1], Thompson [1], De Sá [1] and Zaballa [1]. Also the pole shifting theorem in mathematical systems theory, which appears in Rosenbrock [1], may be viewed as a solution of a special case of the eigenvalue completion problem. All these different versions will be considered in the coming chapters. For some other versions of the eigenvalue completion problem involving other patterns see Mirsky [1], De Oliveira [2], [3], Friedland [1] and Silva [1], and also Chapter XV.

Chapter II
Elementary Operations on Blocks

In this chapter we introduce the notions of block-invariant subspace and of direct sum of blocks. Together with block similarity these notions provide the main tools for analyzing the structure of block decompositions. A list (which will later be shown to be a full list) of indecomposable blocks is given.

II.1 Block-invariant subspaces

Let the block $(B; P, Q)$ be given on the space $\mathcal{X}$. A subspace $\mathcal{M}$ of $\mathcal{X}$ is called *block-invariant for* $(B; P, Q)$ or simply $(B; P, Q)$*-invariant* if

$$B[\mathcal{M} \cap \operatorname{Im} Q] \subset \mathcal{M} + \operatorname{Ker} P. \tag{1.1}$$

In the case when both P and Q are equal to the identity operator on $\mathcal{X}$, the operator B is a linear operator B acting on the full space $\mathcal{X}$ and (1.1) reduces to $B[\mathcal{M}] \subset \mathcal{M}$. We see that $\mathcal{M}$ is a $(B; I, I)$-invariant subspace of $\mathcal{X}$ if and only if $\mathcal{M}$ is an invariant subspace for the operator B. The relation between the notions of block similarity and block invariance is given by the next lemma.

Lemma 1.1. *Let $S : \mathcal{X}' \to \mathcal{X}$ be a block similarity of the blocks $(B; P, Q)$ on $\mathcal{X}$ and $(B'; P', Q')$ on $\mathcal{X}'$. If $\mathcal{M}' \subset \mathcal{X}'$ is a $(B'; P', Q')$-invariant subspace of $\mathcal{X}'$, then $S[\mathcal{M}']$ is a $(B; P, Q)$-invariant subspace of $\mathcal{X}$.*

PROOF. Take $x \in S[\mathcal{M}'] \cap \operatorname{Im} Q$. Then $x = Sy$. Since S is invertible and $S^{-1}[\operatorname{Im} Q] = \operatorname{Im} Q'$, we have that $y \in \mathcal{M}' \cap \operatorname{Im} Q'$. Because $\mathcal{M}'$ is $(B'; P', Q')$-invariant, it follows that $B'y \in \mathcal{M}' + \operatorname{Ker} P'$. Using that S is a block similarity, we get that $SB'y \in S[\mathcal{M}'] + \operatorname{Ker} P$ and that $BSy - SB'y \in \operatorname{Ker} P$. So $Bx = BSy \in S[\mathcal{M}'] + \operatorname{Ker} P$, which proves that $S[\mathcal{M}']$ is a $(B; P, Q)$-invariant subspace of $\mathcal{X}$. □

The $(B; P, Q)$-invariant subspace $\mathcal{M}$ of $\mathcal{X}$ is called *regularly block-invariant for* $(B; P, Q)$ or *regularly* $(B; P, Q)$*-invariant* if

$$P[\mathcal{M}] \subset \mathcal{M}, \quad Q[\mathcal{M}] \subset \mathcal{M}. \tag{1.2}$$

If $\mathcal{M}$ is regularly $(B; P, Q)$-invariant and $x \in \mathcal{M} \cap \operatorname{Im} Q$, then $Bx = PBx \in P[\mathcal{M} + \operatorname{Ker} P] \subset \mathcal{M}$. So the condition (1.1) implies that

$$B[\mathcal{M} \cap \operatorname{Im} Q] \subset \mathcal{M}. \tag{1.3}$$

Conversely (1.3) always implies (1.1). We conclude that $\mathcal{M}$ is regularly $(B; P, Q)$-invariant if and only if the conditions (1.2) and (1.3) hold.

Lemma 1.2. *If $\mathcal{M}$ is invariant for the block $(B;P,Q)$ on $\mathcal{X}$, then there exists a block $(B';P',Q')$ on $\mathcal{X}$ such that $\mathcal{M}$ is a regularly $(B';P',Q')$-invariant subspace of $\mathcal{X}$, and $(B;P,Q)$ and $(B';P',Q')$ are block-similar with a block similarity of the second kind.*

PROOF. Let $\mathcal{N}_0$ be such that $\operatorname{Ker} P = (\mathcal{M} \cap \operatorname{Ker} P) \oplus \mathcal{N}_0$ and $\mathcal{N}_1$ be such that $\mathcal{M} = (\mathcal{M} \cap \operatorname{Ker} P) \oplus \mathcal{N}_1$. Choose $\mathcal{N}_2$ such that $\mathcal{X} = (\mathcal{M} + \operatorname{Ker} P) \oplus \mathcal{N}_2$. So

$$\mathcal{X} = (\mathcal{M} \cap \operatorname{Ker} P) \oplus \mathcal{N}_0 \oplus \mathcal{N}_1 \oplus \mathcal{N}_2.$$

We define the projection P' by setting $\operatorname{Ker} P' = \operatorname{Ker} P$ and $\operatorname{Im} P' = \mathcal{N}_1 \oplus \mathcal{N}_2$. Let $\mathcal{M}_0$ be such that $\operatorname{Im} Q = (\mathcal{M} \cap \operatorname{Im} Q) \oplus \mathcal{M}_0$ and $\mathcal{M}_1$ be such that $\mathcal{M} = (\mathcal{M} \cap \operatorname{Im} Q) \oplus \mathcal{M}_1$. Choose $\mathcal{M}_2$ such that $\mathcal{X} = (\mathcal{M} + \operatorname{Im} Q) \oplus \mathcal{M}_2$. So

$$\mathcal{X} = (\mathcal{M} \cap \operatorname{Im} Q) \oplus \mathcal{M}_0 \oplus \mathcal{M}_1 \oplus \mathcal{M}_2.$$

We define the projection Q' by setting $\operatorname{Im} Q' = \operatorname{Im} Q$ and $\operatorname{Ker} Q' = \mathcal{M}_1 \oplus \mathcal{M}_2$. Then $\mathcal{M}$ is an invariant subspace for P' and Q'. Now put $B' = P'B : \operatorname{Im} Q' \to \operatorname{Im} P'$. Note that $(B';P',Q')$ is block similar to $(B;P,Q)$ with a block similarity of the second kind. If $x \in \mathcal{M} \cap \operatorname{Im} Q'$, then $x \in \mathcal{M} \cap \operatorname{Im} Q$ and hence

$$B'x = P'Bx \in P'[\mathcal{M} + \operatorname{Ker} P] \subset \mathcal{M}.$$

Thus by the remark made at the end of the paragraph preceding the present lemma, the space $\mathcal{M}$ is regularly $(B';P',Q')$-invariant. □

For the regularly block-invariant subspace $\mathcal{M}$ for $(B;P,Q)$ we define the *regular restriction of* $(B;P,Q)$ *to* $\mathcal{M}$ to be the block $(B_1;P_1,Q_1)$, where P_1 the restriction of P to $\mathcal{M}$, Q_1 the restriction of Q to $\mathcal{M}$ and

$$B_1 = B|_{\mathcal{M} \cap \operatorname{Im} Q} : \mathcal{M} \cap \operatorname{Im} Q \to \mathcal{M} \cap \operatorname{Im} P.$$

If the subspace $\mathcal{M}$ of $\mathcal{X}$ is just invariant for $(B;P,Q)$, then there exists a block $(B';P',Q')$ such that $\mathcal{M}$ is a regularly $(B';P',Q')$-invariant subspace, and the blocks $(B;P,Q)$ and $(B';P',Q')$ are block-similar with a block similarity of the second kind. The regular restriction of $(B';P',Q')$ to $\mathcal{M}$ is called a *restriction of* $(B;P,Q)$ *to* $\mathcal{M}$. Note that a restriction of a block $(B;P,Q)$ to $\mathcal{M}$ is not unique. However we have the following theorem.

Theorem 1.3. *For $i = 1,2$ let $(B_i;P_i,Q_i)$ be an operator block with underlying space $\mathcal{X}_i$. Let $S : \mathcal{X}_1 \to \mathcal{X}_2$ be a block similarity of $(B_1;P_1,Q_1)$ and $(B_2;P_2,Q_2)$. Let $\mathcal{M}_1$ be a $(B_1;P_1,Q_1)$-invariant subspace of $\mathcal{X}_1$. Then $\mathcal{M}_2 = S[\mathcal{M}_1]$ is a $(B_2;P_2,Q_2)$-invariant subspace of $\mathcal{X}_2$ and any restriction of $(B_1;P_1,Q_1)$ to $\mathcal{M}_1$ is block-similar to any restriction of $(B_2;P_2,Q_2)$ to $\mathcal{M}_2$.*

PROOF. Lemma 1.1 gives that $\mathcal{M}_2$ is $(B_2;P_2,Q_2)$-invariant. According to Lemma 1.2 there exist for $i = 1,2$ blocks $(B_i';P_i',Q_i')$ block similar of the second kind to $(B_i;P_i,Q_i)$ and such that $\mathcal{M}_i$ is a regularly invariant subspace of

$(B_i'; P_i', Q_i')$. From the definition of similarity of the second kind it follows that $S: \mathcal{X}_1 \to \mathcal{X}_2$ is a block similarity of $(B_1'; P_1', Q_1')$ and $(B_2'; P_2', Q_2')$. It remains to prove that the blocks $(B_1'|_{\mathcal{M}_1 \cap \operatorname{Im} Q_1}, P_1'|_{\mathcal{M}_1}, Q_1'|_{\mathcal{M}_1})$ and $(B_2'|_{\mathcal{M}_2 \cap \operatorname{Im} Q_2}, P_2'|_{\mathcal{M}_2}, Q_2'|_{\mathcal{M}_2})$ are block-similar. Since $\mathcal{M}_i$ is regularly $(B_i'; P_i', Q_i')$-invariant, the operator

$$B_i'|_{\mathcal{M}_i \cap \operatorname{Im} Q_i} : \mathcal{M}_i \cap \operatorname{Im} Q_i' \to \mathcal{M}_i \cap \operatorname{Im} P_i'$$

is well defined for $i = 1, 2$. We will prove that

$$S|_{\mathcal{M}_1}[\operatorname{Ker}(P_1'|_{\mathcal{M}_1})] = \operatorname{Ker}(P_2'|_{\mathcal{M}_2}), \quad S|_{\mathcal{M}_1}[\operatorname{Im}(Q_1'|_{\mathcal{M}_1})] = \operatorname{Im}(Q_2'|_{\mathcal{M}_2}), \tag{1.4}$$

$$(S|_{\mathcal{M}_1} B_1'|_{\mathcal{M}_1} - B_2'|_{\mathcal{M}_1} S|_{\mathcal{M}_1}) m_1 \in \operatorname{Ker}(P_2'|_{\mathcal{M}_2}) \tag{1.5}$$

for each $m_1 \in \operatorname{Im}(Q_1'|_{\mathcal{M}_1})$. The equalities in (1.4) follow from the similarity of the blocks $(B_1'; P_1', Q_1')$ and $(B_2'; P_2', Q_2')$ and the fact that $\mathcal{M}_i$ is an invariant subspace for P_i' and Q_i'. For (1.5) we compute

$$(S|_{\mathcal{M}_1} B_1'|_{\mathcal{M}_1} - B_2'|_{\mathcal{M}_1} S|_{\mathcal{M}_1}) m_1 = (SB_1' - B_2'S) m_1.$$

The right hand side is in $\operatorname{Ker} P_2'$ since S is a block-similarity of $(B_1'; P_1', Q_1')$ and $(B_2'; P_2', Q_2')$. It is also in $\mathcal{M}_2$ since $SB_1'm_1 \in S[\mathcal{M}_1] = \mathcal{M}_2$ and $B_2'Sm_1 \in B_2'[\mathcal{M}_2] \subset \mathcal{M}_2$. Finally $\mathcal{M}_2 \cap \operatorname{Ker} P_2' = \operatorname{Ker}(P_2'|_{\mathcal{M}_2})$ because $\mathcal{M}_2$ is regularly $(B_2'; P_2', Q_2')$-invariant. □

Corollary 1.4. *The restriction of a block to a block-invariant subspace is unique up to a block similarity of the second kind.*

PROOF. We specialize Theorem 1.3 to the case when

$$(B_1; P_1, Q_1) = (B_2; P_2, Q_2)$$

and $S = I$, and conclude that any two restrictions of a block to a block-invariant subspace are block similar with a block similarity of the second kind.

Conversely, assume that $\mathcal{M} \subset \mathcal{X}$ is $(B; P, Q)$-block invariant and $(B_0; P_0, Q_0)$ is a restriction of $(B; P, Q)$ to $\mathcal{M}$. If the $(B_0'; P_0', Q_0')$ on $\mathcal{M}$ is second kind block similar to $(B_0; P_0, Q_0)$, then $(B_0'; P_0', Q_0')$ is a restriction of $(B; P, Q)$ to $\mathcal{M}$. To see this first note that there are projections P' and Q' of $\mathcal{X}$ such that $\operatorname{Ker} P' = \operatorname{Ker} P$, $\operatorname{Im} P' \cap \mathcal{M} = \operatorname{Im} P_0'$, $\operatorname{Im} Q' = \operatorname{Im} Q$ and $\operatorname{Ker} Q' \cap \mathcal{M} = \operatorname{Ker} Q_0'$. Then $(P'B; P', Q')$ is second kind similar to $(B; P, Q)$ and $(B_0'; P_0', Q_0')$ is a regular restriction of $(P'B; P', Q')$ to $\mathcal{M}$ and hence $(B_0'; P_0', Q_0')$ is a restriction of $(B; P, Q)$ to $\mathcal{M}$. □

II.2 Direct sums of blocks and decomposable blocks

We will call a set of subspaces $\mathcal{M}_1, \dots, \mathcal{M}_k$ a set of *(P, Q)-complementary subspaces* if $\mathcal{X}$ is the direct sum of $\mathcal{M}_1, \dots, \mathcal{M}_k$, $\mathcal{X} = \mathcal{M}_1 \oplus \dots \oplus \mathcal{M}_k$, and

$$\operatorname{Ker} P = (\mathcal{M}_1 \cap \operatorname{Ker} P) \oplus \dots \oplus (\mathcal{M}_k \cap \operatorname{Ker} P), \tag{2.1a}$$

$$\operatorname{Im} Q = (\mathcal{M}_1 \cap \operatorname{Im} Q) \oplus \dots \oplus (\mathcal{M}_k \cap \operatorname{Im} Q). \tag{2.1b}$$

In the case when the projections P and Q are both equal to the identity operator, the subspaces $\mathcal{M}_1, \dots, \mathcal{M}_k$ are (P, Q)-complementary if and only if $\mathcal{X} = \mathcal{M}_1 \oplus \dots \oplus \mathcal{M}_k$. The next lemma characterizes a set of (P, Q)-complementary subspaces.

Lemma 2.1. *Let $T_1, \dots, T_k$ be projections of $\mathcal{X}$ such that $\sum_{i=1}^k T_i = I$ and $T_i T_j = 0$ if $i \neq j$. Then $\operatorname{Im} T_1, \dots, \operatorname{Im} T_k$ is a set of (P, Q)-complementary subspaces of $\mathcal{X}$ if and only if*

$$PT_i(I - P) = 0, \quad i = 1, \dots, k, \tag{2.2a}$$

$$(I - Q)T_i Q = 0, \quad i = 1, \dots, k. \tag{2.2b}$$

PROOF. Remark that if $x \in \mathcal{X}$, then $x = (\sum_{i=1}^k T_i)x = \sum_{i=1}^k T_i x$. So $\mathcal{X} = \operatorname{Im} T_1 + \dots + \operatorname{Im} T_k$. On the other hand, if $\sum_{i=1}^k T_i x = 0$, then $T_j x = \sum_{i=1}^k T_j T_i x = T_j \sum_{i=1}^k T_i x = 0$ for $j = 1, \dots, k$. So

$$\mathcal{X} = \operatorname{Im} T_1 \oplus \dots \oplus \operatorname{Im} T_k.$$

It remains to prove that (2.1a) and (2.2a) are equivalent, and that (2.1b) is equivalent to (2.2b). Assume that (2.1a) holds true with $\mathcal{M}_i = \operatorname{Im} T_i$ for $i = 1, \dots, k$. Let $(I - P)x = T_1 m_1 + \dots + T_k m_k$, with $T_i m_i \in \operatorname{Im} T_i \cap \operatorname{Ker} P$. Then $T_i(I - P)x = T_i m_i$ and thus $PT_i(I - P)x = 0$. Since this holds for any $x \in \mathcal{X}$, we proved (2.2a). Conversely, assume that (2.2a) holds true. Take $x \in \operatorname{Ker} P$. Since $x = T_1 x + \dots + T_k x$, it is sufficient to prove that $T_i x \in \operatorname{Ker} P$. Now $T_i x = T_i(I - P)x$ because $x \in \operatorname{Ker} P$. So $PT_i x = PT_i(I - P)x$, and using the assumption that $PT_i(I - P) = 0$, we get that $T_i x \in \operatorname{Ker} P$.

The equivalence of (2.1b) and (2.2b) one obtains by applying the equivalence of (2.1a) and (2.2a) with P replaced by $(I - Q)$ □

A set $\mathcal{M}_1, \dots, \mathcal{M}_k$ of (P, Q)-complementary subspaces is called a set of *regularly (P, Q)-complementary subspaces* if $\mathcal{M}_1, \dots, \mathcal{M}_k$ are invariant subspaces for both P and Q. In the spirit of Lemma 2.1 we can formulate the following result.

Lemma 2.2. *Let $T_1, \dots, T_k$ be projections of $\mathcal{X}$ such that $\sum_{i=1}^k T_i = I$ and $T_i T_j = 0$ if $i \neq j$. Then $\operatorname{Im} T_1, \dots, \operatorname{Im} T_k$ is a set of regularly (P, Q)-complementary subspaces of $\mathcal{X}$ if and only if $PT_i = T_i P$ and $T_i Q = QT_i$, for $i = 1, \dots, k$.*

PROOF. Assume that $\operatorname{Im} T_1, \dots, \operatorname{Im} T_k$ are regularly (P, Q)-complementary subspaces. So $P[\operatorname{Im} T_i] \subset \operatorname{Im} T_i$, for $i = 1, \dots, k$. For any $x \in \mathcal{X}$ one has $Px =$

$PT_1x + \cdots + PT_kx$ and since $PT_ix \in \operatorname{Im} T_i$ it follows that $T_jPx = T_jPT_1x + \cdots + T_jPT_kx = PT_jx$. We see that $PT_j = T_jP$. Similarly one proves that $T_jQ = QT_j$, for $j = 1, \ldots, k$.

Conversely, assume that $PT_i = T_iP$ and $T_iQ = QT_i$ for $i = 1, \ldots, k$. Remark that this implies that $PT_i(I - P) = 0$ and $(I - Q)T_iQ = 0$, for $i = 1, \ldots, k$. From Lemma 2.1 we obtain that $\operatorname{Im} T_1, \ldots, \operatorname{Im} T_k$ is a set of (P, Q)-complementary subspaces of $\mathcal{X}$. Since $PT_ix = T_iPx \in \operatorname{Im} T_i$ and $QT_ix = T_iQx \in \operatorname{Im} T_i$ for any $x \in \mathcal{X}$, it follows that $P[\operatorname{Im} T_i] \subset \operatorname{Im} T_i$ and $Q[\operatorname{Im} T_i] \subset \operatorname{Im} T_i$ for $i = 1, \ldots, k$. This proves that $\operatorname{Im} T_1, \ldots, \operatorname{Im} T_k$ are regularly (P, Q)-complementary subspaces. □

The block $(B; P, Q)$ will be called *decomposable* if there exists a set

$$\{\mathcal{M}_1, \ldots, \mathcal{M}_k\}, \quad k \geq 2,$$

of nonzero (P, Q)-complementary $(B; P, Q)$-invariant subspaces. This set will be called a set of *decomposing subspaces* of the block $(B; P, Q)$. Note that $\mathcal{M}_1$ and $\mathcal{M}_2 \oplus \cdots \oplus \mathcal{M}_k$ are $(B; P, Q)$-invariant. Moreover, $\mathcal{M}_1$ and $\mathcal{M}_2 \oplus \cdots \oplus \mathcal{M}_k$ are (P, Q)-complementary subspaces. Hence the block $(B; P, Q)$ is decomposable if and only if there exist $(B; P, Q)$-invariant subspaces $\mathcal{M}_1$ and $\mathcal{M}_2$, which are (P, Q)-complementary.

Lemma 2.3. *Let $T_1, \ldots, T_k$, $k \geq 2$, be projections of $\mathcal{X}$ such that $\sum_{i=1}^k T_i = I$ and $T_iT_j = 0$ if $i \neq j$. Then $\operatorname{Im} T_1, \ldots, \operatorname{Im} T_k$ is a set of decomposing subspaces of the block $(B; P, Q)$ if and only if $T_i \neq 0$, for $i = 1, \ldots, k$,*

$$PT_i(I - P) = 0, \quad (I - Q)T_iQ = 0, \quad i = 1, \ldots, k, \tag{2.3}$$

and

$$PT_iBT_jQ = 0, \quad i \neq j, \quad i, j = 1, \ldots, k. \tag{2.4}$$

PROOF. Assume that $\operatorname{Im} T_1, \ldots, \operatorname{Im} T_k$ is a set of decomposing subspaces of the block $(B; P, Q)$. Then clearly $T_i \neq 0$ for $i = 1, \ldots, k$, and from Lemma 2.1 one sees that (2.3) holds true. It remains to prove (2.4). For any $x \in \mathcal{X}$ one has $T_jQx \in \operatorname{Im} T_j$, and from (2.3) one gets $T_jQx = QT_jQx$. So also $T_jQx \in \operatorname{Im} Q$. Thus $T_jQx \in \operatorname{Im} T_j \cap \operatorname{Im} Q$. Since $\operatorname{Im} T_j$ is $(B; P, Q)$-invariant, it follows that $BT_jQx \in \operatorname{Im} T_j + \operatorname{Ker} P$. Let $i \neq j$. Then $T_iBT_jQx \in T_i[\operatorname{Ker} P]$. Since $PT_i(I - P) = 0$, one gets that $PT_iBT_jQx = 0$ for $i \neq j$ and each $x \in \mathcal{X}$. This proves (2.4)

Conversely, assume that $T_i \neq 0$, for $i = 1, \ldots, k$, and that (2.3) and (2.4) hold true. From Lemma 2.1 we obtain that $\{\operatorname{Im} T_1, \ldots, \operatorname{Im} T_k\}$ is a set of (P, Q)-complementary subspaces. It remains to show that $\operatorname{Im} T_j$ is a $(B; P, Q)$-invariant subspace. Take $x \in \operatorname{Im} T_j \cap \operatorname{Im} Q$. Then $Qx = x$ and $T_jQx = T_jx = x$. So $Bx = BT_jQx$. From (2.4) we get that $PT_iBT_jQ = 0$, for $i \neq j$. This gives that $PT_iBx = 0$ if $i \neq j$. It follows that $P(I - T_j)Bx = 0$. Using this we compute that

$$Bx = PBx = T_jBx - (I - P)T_jBx \in \operatorname{Im} T_j + \operatorname{Ker} P.$$

This finishes the proof. □

The block $(B;P,Q)$ will be called *regularly decomposable* if there exists a set of at least two of nonzero regularly (P,Q)-complementary $(B;P,Q)$-invariant subspaces. Note that this implies that these subspaces are regularly $(B;P,Q)$-invariant. Assume that $\{\mathcal{M}_1,\dots,\mathcal{M}_k\}$ is a set of regularly (P,Q)-complementary $(B;P,Q)$-invariant subspaces. Then $\{\mathcal{M}_1,\mathcal{M}_2\oplus\cdots\oplus\mathcal{M}_k\}$ is again a set regularly (P,Q)-complementary $(B;P,Q)$-invariant subspaces. So we can say that $(B;P,Q)$ is regularly decomposable if and only there exist two nonzero regularly (P,Q)-complementary $(B;P,Q)$-invariant subspaces.

From the definitions we know that a $(B;P,Q)$-invariant subspace that is both P- and Q-invariant is a regularly $(B;P,Q)$-invariant subspace. So for regularly decomposable block $(B;P,Q)$ with decomposing subspaces $\mathcal{M}_1,\dots,\mathcal{M}_k$, there exist regular restrictions $(B_1;P_1,Q_1),\dots,(B_k;P_k,Q_k)$ to $\mathcal{M}_1,\dots,\mathcal{M}_k$, respectively. For $x\in\mathcal{X}$ there are unique $m_i\in\mathcal{M}_i$ for $i=1,\dots,k$, such that $x=m_1+\cdots+m_k$. It follows that $Bx=B_1m_1+\cdots+B_km_k$, $Px=P_1m_1+\cdots+P_km_k$ and $Qx=Q_1m_1+\cdots+Q_km_k$. We write $(B;P,Q)=(B_1;P_1,Q_1)\oplus\cdots\oplus(B_k;P_k,Q_k)$, and we refer to $(B;P,Q)$ as a *direct sum* of the blocks $(B_1;P_1,Q_1),\dots,(B_k;P_k,Q_k)$.

With an example we show that not each decomposable block is regularly decomposable. Let

$$P=\begin{pmatrix}1&0\\1&0\end{pmatrix},\quad Q=\begin{pmatrix}1&1\\0&0\end{pmatrix}$$

and let $B=0:\operatorname{Im}Q\to\operatorname{Im}P$. Then there is no nontrivial subspace of $\mathbb{C}^2$ invariant for both Q and P. So this block is not regularly decomposable. Write $\{e_1,e_2\}$ for the standard basis of $\mathbb{C}^2$. Let $\mathcal{M}_1=\operatorname{span}\{e_1\}$ and $\mathcal{M}_2=\operatorname{span}\{e_2\}$. Then these spaces are (P,Q)-complementary $(B;P,Q)$-invariant subspaces of $\mathbb{C}^2$. Thus the block $(B;P,Q)$ is decomposable.

The next result combines Lemma 2.2 and Lemma 2.3.

Proposition 2.4. *The block $(B;P,Q)$ is regularly decomposable if and only if there exists a set of nonzero projections $\{T_1,\dots,T_k\}$, $k\geq 2$, such that $\sum_{i=1}^k T_i=I$, $T_iT_j=0$ if $i\neq j$,*

$$PT_i=T_iP,\quad T_iQ=QT_i,\quad i=1,\dots,k, \tag{2.5}$$

and

$$PT_iBT_jQ=0,\quad i\neq j,\quad i,j=1,\dots,k. \tag{2.6}$$

Proof. Assume that $(B;P,Q)$ is regularly decomposable. Then there exists a set $\{\mathcal{M}_1,\dots,\mathcal{M}_k\}$ of regularly (P,Q)-complementary $(B;P,Q)$-invariant subspaces. Put T_i the projection onto $\mathcal{M}_i$ along $\dot{\bigoplus}_{j\neq i}\mathcal{M}_j$. Then $\sum_{i=1}^k T_i=I$ and $T_iT_j=0$ if $i\neq j$. Apply Lemma 2.2 and Lemma 2.3 to obtain (2.5) and (2.6), respectively.

Conversely, assume that the projections $T_1,\dots,T_k$ are such that $\sum_{i=1}^k T_i=I$, $T_iT_j=0$ if $i\neq j$ and (2.5) and (2.6) are fulfilled. Then $\operatorname{Im}T_1,\dots,\operatorname{Im}T_k$ are regularly (P,Q)-complementary subspaces according to Lemma 2.2 and $(B;P,Q)$-invariant subspaces according to Lemma 2.3. □

For operators $A : \mathcal{X} \to \mathcal{Y}$ and $A' : \mathcal{X}' \to \mathcal{Y}'$ as usual the direct sum $A \oplus A' : \mathcal{X} \times \mathcal{X}' \to \mathcal{Y} \times \mathcal{Y}'$ is defined by

$$(A \oplus A') \begin{pmatrix} x \\ x' \end{pmatrix} = \begin{pmatrix} Ax \\ A'x' \end{pmatrix}.$$

Note that $\operatorname{Im}(A \oplus A') = \operatorname{Im} A \times \operatorname{Im} A'$ and $\operatorname{Ker}(A \oplus A') = \operatorname{Ker} A \times \operatorname{Ker} A'$. If $A : \mathcal{X} \to \mathcal{X}$ and $A' : \mathcal{X}' \to \mathcal{X}'$ then $A \oplus A'$ is a projection of $\mathcal{X} \times \mathcal{X}'$ if and only if both A and A' are projections.

Let $(B; P, Q)$ and $(B'; P', Q')$ be blocks with underlying spaces $\mathcal{X}$ and $\mathcal{X}'$, respectively. We define the *formal direct sum* of $(B; P, Q)$ and $(B'; P', Q')$ by

$$(B \oplus B'; P \oplus P', Q \oplus Q').$$

Then $(B \oplus B'; P \oplus P', Q \oplus Q')$ is a block with underlying space $\mathcal{X} \times \mathcal{X}'$. Similarly we can define the formal direct sum of m blocks. From the definition of a formal direct sum it is clear that a formal direct sum is regularly decomposable. In this way we can construct large blocks from small blocks.

We give an example. In the one dimensional space $\mathbb{C}^1$ we take $P = Q = B = I$ and $P' = 0$, $Q' = I$ and $B' = 0$. Then the formal direct sum of the blocks $(B; P, Q)$ and $(B'; P', Q')$ is $(B \oplus B'; P \oplus P', Q \oplus Q')$. With respect to the natural basis in the space $\mathbb{C}^1 \times \mathbb{C}^1$ we find

$$P \oplus P' = \begin{pmatrix} 1 & 0 \\ 0 & 0 \end{pmatrix}, Q \oplus Q' = \begin{pmatrix} 1 & 0 \\ 0 & 1 \end{pmatrix}, B \oplus B' \begin{pmatrix} x \\ x' \end{pmatrix} = \begin{pmatrix} x \\ 0 \end{pmatrix}.$$

So the block $(B \oplus B'; P \oplus P', Q \oplus Q')$ is represented by the positioned submatrix, which consists of the first row of the matrix

$$\begin{pmatrix} 1 & 0 \\ ? & ? \end{pmatrix}.$$

In the above example we took the formal direct sum of a block $(B; P, Q)$ and a block $(B'; 0, Q')$. On the level of positioned submatrices this comes down to extending a $p \times q$ positioned submatrix in an $n \times n$ matrix with q' zero columns to a $p \times (q+q')$ positioned submatrix in an $(n+n') \times (n+n')$ matrix. In the same way we can interpret the formal direct sum of a block $(B; P, Q)$ and a block $(B'; P', 0)$ in terms of positioned submatrices. It means extending a $p \times q$ positioned submatrix in an $n \times n$ matrix with p' zero rows to a $(p + p') \times q$ positioned submatrix in an $(n + n') \times (n + n')$ matrix. Remember however that not every (P, Q)-block can be represented by a positioned submatrix.

Theorem 2.5. *The operator block $(B; P, Q)$ is decomposable if and only if it is block similar with a block similarity of the second kind to a regularly decomposable block.*

PROOF. Let $(B;P,Q)$ be a block on the space $\mathcal{X}$. Assume that $(B;P,Q)$ is decomposable, and that $\mathcal{M}_1$ and $\mathcal{M}_2$ are (P,Q)-complementary $(B;P,Q)$-invariant subspaces of $\mathcal{X}$. We choose P' such that $\operatorname{Ker} P' = \operatorname{Ker} P$ and $P'[\mathcal{M}_i] \subset \mathcal{M}_i$ for $i = 1,2$. This can be done by fixing complements of $\mathcal{M}_1 \cap \operatorname{Ker} P$ in $\mathcal{M}_1$ and of $\mathcal{M}_2 \cap \operatorname{Ker} P$ in $\mathcal{M}_2$ and putting $\operatorname{Im} P'$ their direct sum. Next we choose Q' such that $\operatorname{Im} Q' = \operatorname{Im} Q$ and $Q'[\mathcal{M}_i] \subset \mathcal{M}_i$ for $i = 1,2$, by fixing complements of $\mathcal{M}_1 \cap \operatorname{Im} Q$ in $\mathcal{M}_1$ and of $\mathcal{M}_2 \cap \operatorname{Im} Q$ in $\mathcal{M}_2$ and putting $\operatorname{Ker} Q'$ their direct sum. Take $(B';P',Q') = (P'B;P',Q')$. Then $(B;P,Q)$ and $(B';P',Q')$ are similar with a similarity of the second kind. Also $\mathcal{M}_1$ and $\mathcal{M}_2$ are regularly (P',Q')-complementary $(B';P',Q')$-invariant subspaces of $\mathcal{X}$, and hence $(B';P',Q')$ is a regularly decomposable block.

Conversely, assume that there exists a similarity of the second kind of $(B;P,Q)$ with $(B';P',Q')$ which is a regularly decomposable block. So there exist (P',Q')-complementary $(B';P',Q')$-invariant subspaces $\mathcal{M}_1$ and $\mathcal{M}_2$ of $\mathcal{X}$. Since $\operatorname{Ker} P = \operatorname{Ker} P'$ and $\operatorname{Im} Q = \operatorname{Im} Q'$, it is clear that $\mathcal{M}_1$ and $\mathcal{M}_2$ are (P,Q)-complementary. Also using that $B = PB'$ one sees that $\mathcal{M}_1$ and $\mathcal{M}_2$ are $(B;P,Q)$-invariant subspaces. This shows that $(B;P,Q)$ is decomposable. □

Note that the class of decomposable blocks is closed under taking block similarities. If in two direct sums of two blocks the first terms are block similar and the second terms are block similar, then the two direct sums are block similar. In fact the similarity is the direct sum of the two given similarities. The converse is also true, provided that an obvious extra condition is satisfied. This is the content of the next result.

Theorem 2.6. *Let $S : \mathcal{X}_{11} \oplus \mathcal{X}_{12} \to \mathcal{X}_{21} \oplus \mathcal{X}_{22}$ be a block similarity of the blocks $(B_{11};P_{11},Q_{11}) \oplus (B_{12};P_{12},Q_{12})$ on $\mathcal{X}_{11} \oplus \mathcal{X}_{12}$ and $(B_{21};P_{21},Q_{21}) \oplus (B_{22};P_{22},Q_{22})$ on $\mathcal{X}_{21} \oplus \mathcal{X}_{22}$. If $S\mathcal{X}_{11} = \mathcal{X}_{21}$, then $(B_{11};P_{11},Q_{11})$ is block similar to $(B_{21};P_{21},Q_{21})$ and $(B_{12};P_{12},Q_{12})$ is block similar to $(B_{22};P_{22},Q_{22})$.*

PROOF. Remark that $\mathcal{X}_{11}$ is $(B_{11};P_{11},Q_{11}) \oplus (B_{12};P_{12},Q_{12})$-invariant, S is a block similarity and $\mathcal{X}_{21} = S[\mathcal{X}_{11}]$. Hence it follows form Theorem 1.3 that the restriction of $(B_{11};P_{11},Q_{11}) \oplus (B_{12};P_{12},Q_{12})$ to $\mathcal{X}_{11}$ is block similar to the restriction of $(B_{21};P_{21},Q_{21}) \oplus (B_{22};P_{22},Q_{22})$ to $\mathcal{X}_{21}$. So the block $(B_{11};P_{11},Q_{11})$ is block similar to $(B_{21};P_{21},Q_{21})$. Write

$$S = \begin{pmatrix} S_1 & S_3 \\ 0 & S_2 \end{pmatrix} : \mathcal{X}_{11} \oplus \mathcal{X}_{12} \to \mathcal{X}_{21} \oplus \mathcal{X}_{22} .$$

The zero in the left lower corner of the matrix representation of S reflects the fact that $S\mathcal{X}_{11} = \mathcal{X}_{21}$. Let $x_{12} \in \operatorname{Ker} P_{12}$. Since $S[\operatorname{Ker} P_{11} \oplus \operatorname{Ker} P_{12}] = \operatorname{Ker} P_{21} \oplus \operatorname{Ker} P_{22}$, we have that

$$S \begin{pmatrix} 0 \\ x_{12} \end{pmatrix} \in \operatorname{Ker} P_{21} \oplus \operatorname{Ker} P_{22} .$$

We see that $S_2x_{12} \in \operatorname{Ker} P_{22}$. It follows that $S_2[\operatorname{Ker} P_{12}] = \operatorname{Ker} P_{22}$. Similarly we obtain from $S[\operatorname{Im} Q_{11} \oplus \operatorname{Im} Q_{12}] = \operatorname{Im} Q_{21} \oplus \operatorname{Im} Q_{22}$ that $S_2[\operatorname{Im} Q_{12}] = \operatorname{Im} Q_{22}$. Because S is a block similarity,

$$\begin{pmatrix} P_{21} & 0 \\ 0 & P_{22} \end{pmatrix} \begin{pmatrix} S_1 & S_3 \\ 0 & S_2 \end{pmatrix} \begin{pmatrix} B_{11} & 0 \\ 0 & B_{12} \end{pmatrix} \begin{pmatrix} 0 \\ x_{12} \end{pmatrix} =$$

$$\begin{pmatrix} P_{21} & 0 \\ 0 & P_{22} \end{pmatrix} \begin{pmatrix} B_{21} & 0 \\ 0 & B_{22} \end{pmatrix} \begin{pmatrix} S_1 & S_3 \\ 0 & S_2 \end{pmatrix} \begin{pmatrix} 0 \\ x_{12} \end{pmatrix},$$

for each $x_{12} \in \operatorname{Im} Q_{12}$. The second component in this equality yields that

$$P_{22}S_2B_{12}x_{12} = P_{22}B_{22}S_2x_{12}.$$

Since we already proved that $S_2[\operatorname{Ker} P_{12}] = \operatorname{Ker} P_{22}$ and that $S_2[\operatorname{Im} Q_{12}] = \operatorname{Im} Q_{22}$, this shows that S_2 is a block similarity of $(B_{12}; P_{12}, Q_{12})$ and $(B_{22}; P_{22}, Q_{22})$. □

II.3 Indecomposable blocks

Let $(B; P, Q)$ be an operator block on the space $\mathcal{X}$. A block is called *indecomposable* if it is not decomposable. We will give some important examples of indecomposable blocks.

For the first of these examples we consider the full width positioned $(n-1)\times n$ submatrix

$$A = \begin{pmatrix} ? & ? & \cdots & ? & ? \\ 1 & 0 & \cdots & 0 & 0 \\ 0 & 1 & & 0 & 0 \\ \vdots & & \ddots & & \vdots \\ 0 & 0 & & 1 & 0 \end{pmatrix}.$$

Let $(B; P, Q)$ be the associated block. Then the projection Q is the identity operator on $\mathbb{C}^n$, the projection P acts on $\mathbb{C}^n$ along the span of e_1 onto the span of $\{e_2, \ldots, e_n\}$, and B is given by $Be_i = e_{i+1}$ for $i = 1, \ldots, n-1$, $Be_n = 0$. In order to prove that $(B; P, Q)$ is indecomposable, we assume that $\mathcal{M}_1$ and $\mathcal{M}_2$ are (P, Q)-complementary and $(B; P, Q)$-invariant subspaces of $\mathbb{C}^n$. So

$$\operatorname{Ker} P = (\mathcal{M}_1 \cap \operatorname{Ker} P) \oplus (\mathcal{M}_2 \cap \operatorname{Ker} P), \quad \mathbb{C}^n = \mathcal{M}_1 \oplus \mathcal{M}_2.$$

Since the dimension of $\operatorname{Ker} P$ is one, we know that either $\mathcal{M}_1 \cap \operatorname{Ker} P = \operatorname{Ker} P$ or $\mathcal{M}_2 \cap \operatorname{Ker} P = \operatorname{Ker} P$. Without loss of generality we may assume that $\mathcal{M}_1 \cap \operatorname{Ker} P = \operatorname{Ker} P$. This means that $\operatorname{Ker} P \subset \mathcal{M}_1$ and thus $\mathcal{M}_1 + \operatorname{Ker} P = \mathcal{M}_1$. Since $\mathcal{M}_1$ is $(B; P, Q)$-invariant and $\operatorname{Im} Q = \mathcal{X}$, we now have that $B[\mathcal{M}_1] \subset \mathcal{M}_1$. It follows that $e_j = B^{j-1}e_1 \in \mathcal{M}_1$ for $j = 2, \ldots, n$. Together with $e_1 \in \operatorname{Ker} P \subset \mathcal{M}_1$ this proves that $e_j \in \mathcal{M}_1$ for $j = 1, 2, \ldots, n$, and hence $\mathcal{M}_1 = \mathbb{C}^n$. Therefore the block $(B; P, Q)$ is indecomposable. This block $(B; P, Q)$ is called a *shift of the first kind* with *index* $n-1$.

In the second example we consider the full length positioned $n \times (n-1)$ submatrix

$$A = \begin{pmatrix} 0 & 0 & \cdots & 0 & ? \\ 1 & 0 & \cdots & 0 & ? \\ 0 & 1 & & 0 & ? \\ \vdots & & \ddots & & \vdots \\ 0 & 0 & & 1 & ? \end{pmatrix}.$$

Let $(B;P,Q)$ be the associated block. Then the projection P is the identity operator on $\mathbb{C}^n$ and Q is the projection of $\mathbb{C}^n$ along e_n onto the span of $e_1,\ldots,e_{n-1}$. Furthermore, $Be_i = e_{i+1}$ for $i=1,\ldots,n-1$. Again we assume that $\mathcal{M}_1$ and $\mathcal{M}_2$ are (P,Q)-complementary and $(B;P,Q)$-invariant subspaces of $\mathbb{C}^n$. Then we have that

$$\operatorname{Im} Q = (\mathcal{M}_1 \cap \operatorname{Im} Q) \oplus (\mathcal{M}_2 \cap \operatorname{Im} Q)\ ,\ \ \mathbb{C}^n = \mathcal{M}_1 \oplus \mathcal{M}_2\ .$$

Since the dimension of $\operatorname{Im} Q$ is $n-1$, we conclude that either $\mathcal{M}_1 \cap \operatorname{Im} Q = \mathcal{M}_1$ or $\mathcal{M}_2 \cap \operatorname{Im} Q = \mathcal{M}_2$. Without loss of generality we may assume that $\mathcal{M}_1 \cap \operatorname{Im} Q = \mathcal{M}_1$. This means that $\mathcal{M}_1 \subset \operatorname{Im} Q$. Since $\mathcal{M}_1$ is $(B;P,Q)$-invariant and $\operatorname{Ker} P = \{0\}$, we have that $Bx \in \mathcal{M}_1$ for each $x \in \mathcal{M}_1$. Let $x \in \mathcal{M}_1$, and let $x = \sum_{i=1}^{j} \alpha_i e_i$ for some $j \le n-1$. Then $B^{n-j}x$ is well defined and $B^{n-j}x = \sum_{i=n-j+1}^{n} \alpha_{i-n+j} e_i$. The right hand side of the previous expression is in $\mathcal{M}_1$ and therefore in $\operatorname{Im} Q$. So we find that the coefficient of e_n is equal to 0, i.e. $\alpha_j = 0$. We conclude that $x = 0$ and thus $\mathcal{M}_1 = \{0\}$. Therefore the block $(B;P,Q)$ is indecomposable. We call this block $(B;P,Q)$ *a shift of the third kind* with *index* $n-1$.

The third example concerns the positioned $(n-1) \times (n-1)$ submatrix

$$A = \begin{pmatrix} ? & ? & \cdots & ? & ? \\ 1 & 0 & \cdots & 0 & ? \\ 0 & 1 & & 0 & ? \\ \vdots & & \ddots & & \vdots \\ 0 & 0 & & 1 & ? \end{pmatrix}.$$

Let $(B;P,Q)$ be the associated block. Then P is the projection of $\mathbb{C}^n$ along the span of e_1 onto the span of $\{e_2,\ldots,e_n\}$, the projection Q acts on $\mathbb{C}^n$ along e_n onto the span of $e_1,\ldots,e_{n-1}$, and $Be_i = e_{i+1}$ for $i = 1,\ldots,n-1$. We shall prove that $(B;P,Q)$ is indecomposable. Therefore, let $\mathcal{M}_1$ and $\mathcal{M}_2$ be (P,Q)-complementary and $(B;P,Q)$-invariant subspaces of $\mathbb{C}^n$. This means that

$$\operatorname{Im} Q = (\mathcal{M}_1 \cap \operatorname{Im} Q) \oplus (\mathcal{M}_2 \cap \operatorname{Im} Q),$$

$$\operatorname{Ker} P = (\mathcal{M}_1 \cap \operatorname{Ker} P) \oplus (\mathcal{M}_2 \cap \operatorname{Ker} P),$$

$$\mathcal{X} = \mathcal{M}_1 \oplus \mathcal{M}_2.$$

Since the dimension of $\operatorname{Ker} P$ is one, we know that either $\mathcal{M}_1 \cap \operatorname{Ker} P = \operatorname{Ker} P$ or $\mathcal{M}_2 \cap \operatorname{Ker} P = \operatorname{Ker} P$. Without loss of generality we may assume that $\mathcal{M}_1 \cap \operatorname{Ker} P =$

$\operatorname{Ker} P$. So $\operatorname{Ker} P \subset \mathcal{M}_1$ and thus $\mathcal{M}_1 + \operatorname{Ker} P = \mathcal{M}_1$. Because $\mathcal{M}_1$ is $(B; P, Q)$-invariant, we now have that $B[\mathcal{M}_1 \cap \operatorname{Im} Q] \subset \mathcal{M}_1$. Moreover, $e_1 \in \operatorname{Ker} P \subset \mathcal{M}_1$. We will prove by induction that $e_j \in \mathcal{M}_1$ for $j = 1, \ldots, n$. So we assume that for $j \leq n-1$ the vector $e_j \in \mathcal{M}_1$. Then $e_j \in \mathcal{M}_1 \cap \operatorname{Im} Q$ and $e_{j+1} = Be_j \in \mathcal{M}_1$. We conclude that $e_j \in \mathcal{M}_1$ for $j = 1, \ldots, n$, and hence $\mathcal{M}_1 = \mathbb{C}^n$. This proves that the block $(B; P, Q)$ is indecomposable. This block $(B; P, Q)$ will be called a *shift of the second kind* with *index* $n-1$.

Our final example of indecomposable blocks concerns the case when the positioned submatrix is a full $n \times n$ matrix A. The associated block is then $(B; I, I)$, where $B : \mathbb{C}^n \to \mathbb{C}^n$ is defined in the canonical way by the matrix A. In this case the notion of block-invariant subspace coincides with the notion of invariant subspace for the operator B. Also block-complementary subspaces are just ordinary complementary subspaces. So in this case the indecomposable blocks are just the operators that have, with respect to a suitable basis in $\mathbb{C}^n$, a matrix that consists of a single Jordan block.

Theorem 3.1. *Shifts of the first, second and third kind and single Jordan blocks are up to block similarity the only indecomposable blocks.*

The proof of this theorem will be one of the main topics of the Chapters III, V-VII. Chapters III, V, VI concern the special classes of blocks mentioned earlier. The proof of Theorem 3.1 for general blocks appears in Chapter VII.

II.4 Duality of blocks

Let $\mathcal{X}$ be a vector space. We denote by $\mathcal{X}^*$ the dual space, that is the space of all linear functions $f : \mathcal{X} \to \mathbb{C}$. If $e_1, \ldots, e_n$ is a basis of $\mathcal{X}$, then the linear functions $e_1^*, \ldots, e_n^*$ in $\mathcal{X}^*$, fixed by $e_j^*(e_i) = 0$ if $i \neq j$ and $e_i^*(e_i) = 1$ for $i, j = 1, \ldots, n$, form a basis of $\mathcal{X}^*$. Indeed, if $f \in \mathcal{X}^*$, then it is easy to check that $f(x) = \sum_{i=1}^n f(e_i) e_i^*(x)$. We consider the dual $\mathcal{X}^{**}$ of $\mathcal{X}^*$. So $x^{**} \in \mathcal{X}^{**}$ is a linear function $\mathcal{X}^* \to \mathbb{C}$. We construct a linear transformation $N : \mathcal{X} \to \mathcal{X}^{**}$ by defining the element $N(x) \in \mathcal{X}^{**}$ by $N(x)(f) = f(x)$ for each $f \in \mathcal{X}^*$. Then N is a linear transformation. Recall that $f(x) = 0$ for all $f \in \mathcal{X}^*$ implies that $x = 0$. So, if $N(x) = 0$, then $x = 0$, and therefore the transformation N is injective. Since $\dim \mathcal{X}^{**} = \dim \mathcal{X}^* = \dim \mathcal{X}$, the map N is a linear isomorphism of $\mathcal{X}$ and $\mathcal{X}^{**}$. This transformation will be called the *natural isomorphism* from $\mathcal{X}$ to $\mathcal{X}^{**}$.

With each linear operator $L : \mathcal{X}_1 \to \mathcal{X}_2$ we have a *dual operator* $L^* : \mathcal{X}_2^* \to \mathcal{X}_1^*$, which is defined on each element $f \in \mathcal{X}_2^*$ by $(L^* f)(x) = f(Lx)$ for each $x \in \mathcal{X}_1$. The dual operator of $L^* : \mathcal{X}_2^* \to \mathcal{X}_1^*$ is denoted as $L^{**} : \mathcal{X}_1^{**} \to \mathcal{X}_2^{**}$. One checks that $L^{**} N_1 = N_2 L$, where $N_1 : \mathcal{X}_1 \to \mathcal{X}_1^{**}$ and $N_2 : \mathcal{X}_2 \to \mathcal{X}_2^{**}$ are the natural isomorphisms. Remark that if $L : \mathcal{X}_1 \to \mathcal{X}_2$ and $M : \mathcal{X}_2 \to \mathcal{X}_3$ are linear operators, then $(ML)^* = L^* M^*$. So if S is an invertible operator, then $(S^*)^{-1} = (S^{-1})^*$. Let $P : \mathcal{X} \to \mathcal{X}$ be a projection. Then $P^* = (P^2)^* = (P^*)^2$. So also $P^* : \mathcal{X}^* \to \mathcal{X}^*$ is a projection.

For a subspace $\mathcal{M}$ of the linear space $\mathcal{X}$ we define the subspace $\mathcal{M}^\perp$ of $\mathcal{X}^*$, called the *annihilator of* $\mathcal{M}$, by $\mathcal{M}^\perp = \{f \in \mathcal{X}^* \mid f|_\mathcal{M} = 0\}$. For a linear operator $L : \mathcal{X}_1 \to \mathcal{X}_2$ we have

$$\operatorname{Ker} L^* = \{f \in \mathcal{X}_2^* \mid f|_{\operatorname{Im} L} = 0\} = (\operatorname{Im} L)^\perp, \tag{4.1}$$

$$\operatorname{Im} L^* = \{g \in \mathcal{X}_1^* \mid g|_{\operatorname{Ker} L} = 0\} = (\operatorname{Ker} L)^\perp. \tag{4.2}$$

The identity (4.1) is trivial. To prove (4.2) first assume that $f \in \operatorname{Im} L^*$. Then there exists a $g \in \mathcal{X}_1$ such that $f = L^*g$. So for $x \in \operatorname{Ker} L$ we have that $f(x) = L^*g(x) = g(Lx) = 0$. Hence $f \in (\operatorname{Ker} L)^\perp$. Conversely, for $f \in (\operatorname{Ker} L)^\perp$ one has that $f(x) = 0$ for each $x \in \operatorname{Ker} L$. Therefore there exists a $g \in \mathcal{X}^*$ such that $f = gL$. So $f(x) = g(Lx)$ for each $x \in \mathcal{X}_1$. This proves that $f = L^*g$, and thus $f \in \operatorname{Im} L^*$ and (4.2) is proved. From (4.1) and (4.2) it follows that L^* is invertible if and only L is invertible.

We will introduce now duality for blocks. Let $(B; P, Q)$ be a block on the n-dimensional space $\mathcal{X}$. Assume that $A : \mathcal{X} \to \mathcal{X}$ is a completion of the block. Thus for each $x \in \operatorname{Im} Q$ one has that $PAx = Bx$. Let A^*, P^* and Q^* be the dual operators of A, P and Q, respectively. Write $B^* = Q^*A^*|_{\operatorname{Im} P^*} : \operatorname{Im} P^* \to \operatorname{Im} Q^*$. Then $(B^*; Q^*, P^*)$ is called *the dual block of* $(B; P, Q)$. We will prove that the dual block is independent of the choice of the completion A. Assume that A_0 is also a completion of $(B; P, Q)$. Then $PAQ = PA_0Q$. Hence $Q^*A^*P^* = Q^*A_0^*P^*$, which shows that indeed the notion of dual block is well defined. Remark that A^* is a completion of $(B^*; Q^*, P^*)$. Put $B^{**} = P^{**}A^{**}|_{\operatorname{Im} Q^{**}} : \operatorname{Im} Q^{**} \to \operatorname{Im} P^{**}$. Then $(B^{**}; P^{**}, Q^{**})$ is the dual of $(B^*; Q^*, P^*)$. One easily checks that the natural isomorphism from $\mathcal{X}$ to $\mathcal{X}^{**}$ establishes a similarity of the first kind of $(B; P, Q)$ and $(B^{**}; P^{**}, Q^{**})$.

Let $(B; I, Q)$ be a full length block. Then the dual block $(B^*; Q^*, I^*)$ is a full width block. Likewise, the dual block of the full width block $(B; P, I)$ is a full length block.

The next results establishes the relation between duality and block similarity.

Proposition 4.1. *Assume that the blocks $(B_1; P_1, Q_1)$ on $\mathcal{X}_1$ and $(B_2; P_2, Q_2)$ on $\mathcal{X}_2$ are given. Then $S : \mathcal{X}_1 \to \mathcal{X}_2$ is a block similarity of $(B_1; P_1, Q_1)$ and $(B_2; P_2, Q_2)$ if and only if the operator $S^* : \mathcal{X}_2^* \to \mathcal{X}_1^*$ is a block similarity of the dual blocks $(B_2^*; Q_2^*, P_2^*)$ and $(B_1^*; Q_1^*, P_1^*)$.*

PROOF. Let A_1 be a completion of $(B_1; P_1, Q_1)$ and A_2 a completion of $(B_2; P_2, Q_2)$. Recall that S is a block similarity of $(B_1; P_1, Q_1)$ and $(B_2; P_2, Q_2)$ if and only if

(i) $P_1(S(I - P_2) = 0$, $P_2S^{-1}(I - P_1) = 0$, $(I - Q_1)SQ_2 = 0$, $(I - Q_2)S^{-1}Q_1 = 0$,

(ii) $P_1(B_1S - SB_2)Q_2 = 0$.

Since $B_1 = P_1A_1Q_1|_{\operatorname{Im} Q_1}$ and $B_2 = P_2A_2Q_2|_{\operatorname{Im} Q_2}$, we can use condition (i) to rewrite condition (ii) in the following form:

(iii) $P_1(P_1A_1Q_1S - SP_2A_2Q_2)Q_2 = 0$.

We apply duality and get that (i) and (iii) are equivalent to

(i') $(I^* - P_2^*)S^*P_1^* = 0$, $(I^* - P_1^*)(S^*)^{-1}P_2^* = 0$,
$Q_2^*S^*(I^* - Q_1^*) = 0$, $Q_1^*(S^*)^{-1}(I^* - Q_2^*) = 0$,

(iii') $Q_2^*(Q_2^*A_2^*P_2^*S^* - S^*Q_1^*A_1^*P_1^*)P_1^* = 0$.

Use

$$B_1^* = Q_1^*A_1^*P_1^*|_{\operatorname{Im} P_1^*} : \operatorname{Im} P_1^* \to \operatorname{Im} Q_1^*, \quad B_2^* = Q_2^*A_2^*P_2^*|_{\operatorname{Im} P_2^*} : \operatorname{Im} P_2^* \to \operatorname{Im} Q_2^*,$$

to rewrite the equality (iii') as

(ii') $Q_2^*(B_2^*S^* - S^*B_1^*)P_1^* = 0$.

The identities in (i') and (ii') mean that S^* is a block similarity of $(B_2^*; Q_2^*, P_2^*)$ and $(B_1^*; Q_1^*, P_1^*)$. We conclude that $S : \mathcal{X}_1 \to \mathcal{X}_2$ is a block similarity of $(B_1; P_1, Q_1)$ and $(B_2; P_2, Q_2)$ if and only if $S^* : \mathcal{X}_2^* \to \mathcal{X}_1^*$ is a block similarity of $(B_2^*; Q_2^*, P_2^*)$ and $(B_1^*; Q_1^*, P_1^*)$. □

Let $(B; P, Q)$ be a block on the space $\mathcal{X}$. If the subspaces $\mathcal{M}_1$ and $\mathcal{M}_2$ of $\mathcal{X}$ are (P, Q)-complementary, then $\mathcal{M}_1^\perp$ and $\mathcal{M}_2^\perp$ are (Q^*, P^*)-complementary subspaces of $\mathcal{X}^*$. To prove this we have to prove that

$$\operatorname{Im} P^* = (\operatorname{Im} P^* \cap \mathcal{M}_1^\perp) \oplus (\operatorname{Im} P^* \cap \mathcal{M}_2^\perp). \tag{4.3}$$

Let $f \in \operatorname{Im} P^*$, and write $f = f_1 + f_2$ with $f_i \in \mathcal{M}_i^\perp$ for $i = 1, 2$. If $x \in \operatorname{Ker} P \cap \mathcal{M}_1$, then $f(x) = 0$ and $f_1(x) = 0$, and therefore also $f_2(x) = 0$. Similarly we see that $f(x) = 0$, $f_2(x) = 0$ and $f_1(x) = 0$ for $x \in \operatorname{Ker} P \cap \mathcal{M}_2$. If $x \in \operatorname{Ker} P$, then, according to our hypotheses, $x = x_1 + x_2$ with $x_i \in \operatorname{Ker} P \cap \mathcal{M}_i$ for $i = 1, 2$. Thus if $x \in \operatorname{Ker} P$, then $f(x) = 0$, $f_1(x) = 0$ and $f_2(x) = 0$. So we have proved that $f_i \in \operatorname{Im} P^* \cap \mathcal{M}_i^\perp$ for $i = 1, 2$. This proves (4.3). Similarly we prove that

$$\operatorname{Ker} Q^* = (\operatorname{Ker} Q^* \cap \mathcal{M}_1^\perp) \oplus (\operatorname{Ker} Q^* \cap \mathcal{M}_2^\perp).$$

Since $\mathcal{M}$ is P-invariant if and only if $\mathcal{M}^\perp$ is P^*-invariant, we also conclude that subspaces $\mathcal{M}_1$ and $\mathcal{M}_2$ of $\mathcal{X}$ are regularly (P, Q)-complementary if and only if $\mathcal{M}_1^\perp$ and $\mathcal{M}_2^\perp$ are regularly (Q^*, P^*)-complementary.

Proposition 4.2. *The block $(B; P, Q)$ on the linear space $\mathcal{X}$ is decomposable if and only if the dual block $(B^*; P^*, Q^*)^*$ on $\mathcal{X}^*$ is decomposable. The block $(B; P, Q)$ is a direct sum if and only if the dual block $(B^*; P^*, Q^*)$ is a direct sum.*

PROOF. From Lemma 2.3 it follows that the block $(B; P, Q)$ is decomposable if and only if there exists a projection T of $\mathcal{X}$ such that

$$PT(I - P) = 0, \ (I - Q)TQ = 0, \ P(I - T)BTQ = 0, \ PTB(I - T)Q = 0.$$

This is equivalent to

$$(I^* - P^*)T^*P^* = 0, \ Q^*T^*(I - Q^*) = 0,$$

$$Q^*T^*(PBQ)^*(I^* - T^*)P^* = 0, \ (I^* - Q^*)(PBQ)^*T^*P^* = 0.$$

We conclude that $(B; P, Q)$ is decomposable with decomposing subspaces $\operatorname{Im} T$ and $\operatorname{Ker} T$ if and only if $(B^*; P^*, Q^*)$ is decomposable with decomposing subspaces $\operatorname{Ker} T^*$ and $\operatorname{Im} T^*$.

According to Proposition 2.4 the block $(B; P, Q)$ is a direct sum if and only if there exists a projection T of $\mathcal{X}$ such that $PT = TP$, $QT = TQ$, $P(I-T)BTQ = 0$ and $PTB(I - Q) = 0$. This is equivalent to $P^*T^* = T^*P^*$, $Q^*T^* = T^*Q^*$, $Q^*T^*(PBQ)^*(I^* - T^*)P^* = 0$ and $(I^* - Q^*)(PBQ)^*T^*P^* = 0$. So it follows that $(B; P, Q)$ is a direct sum if and only if the dual block $(B^*; P^*, Q^*)$ is a direct sum. □

Notes

The text of this chapter originates from Gohberg-Kaashoek-Van Schagen [1]. The material has been improved in presentation and various details.

Chapter III
Full Length Blocks

In this chapter we give the decomposition in indecomposable blocks and describe the invariants under block similarity for full length blocks. Applications to matrix pencils and to non-everywhere defined linear operators on finite dimensional spaces are included. In the last section we bring together some results on non-increasing sequences of nonnegative numbers, that are used in several places in this chapter and elsewhere in the book.

III.1 Structure theorems for full length blocks

In this section the block similarity classes of full length blocks will be described. Each class will be characterized by a set of invariants. To describe the results, we need to recall that an operator block $(B; I, Q)$ with underlying space $\mathbb{C}^{\mu+1}$ is a shift of the third kind with index μ if with respect to the standard basis $e_1, \ldots, e_{\mu+1}$ of $\mathbb{C}^{\mu+1}$ the representation of the block is such that

(i) $\operatorname{Im} Q = \operatorname{span}\{e_i\}_{i=1}^{\mu}$, $\operatorname{Ker} Q = \operatorname{span}\{e_{\mu+1}\}$;

(ii) $Be_i = e_{i+1}$, for $i = 1, \ldots, \mu$.

If $\mu = 0$, then $\operatorname{Im} Q = \{0\}$ and the second condition is void. We showed in Section II.3 that a shift of the third kind is an irreducible operator block. If we omit the condition that $\operatorname{Ker} Q = \operatorname{span}\{e_{\mu+1}\}$, then the block is similar to a shift of the third kind. The similarity can be chosen to be of the second kind. Conversely, if a block is similar to a shift of the third kind with index μ, then there exists a basis $\{f_i\}_{i=1}^{\mu+1}$ of $\mathcal{X}$ such that $\operatorname{Im} Q = \operatorname{span}\{f_i\}_{i=1}^{\mu}$ and $Bf_i = f_{i+1}$, for $i = 1, \ldots, \mu$. We shall show that shifts of the third kind are important building blocks of full length blocks.

From the definition of a full length block it follows directly that the class of full length blocks is closed under taking direct sums, under the application of block similarity (see Theorem I.4.5), and under restriction to a block invariant subspace. So, if a block is block similar to a direct sum of shifts of the third kind and a block $(J; I, I)$, then it is a full length block. The first result in this section states that the converse is also true.

Theorem 1.1. *An operator block $(B; I, Q)$ is block similar to a direct sum of an operator J and shifts of the third kind with indices $\mu_1 \geq \cdots \geq \mu_q \geq 0$. The number q of shifts of the third kind in this direct sum is equal to $\dim \operatorname{Ker} Q$. Furthermore, the operator J and the indices $\mu_1 \geq \cdots \geq \mu_q$ are uniquely determined by the block $(B; I, Q)$.*

The numbers $\mu_1 \geq \cdots \geq \mu_q \geq 0$ will be called *indices of the third kind of the block* $(B; I, Q)$. The invariant polynomials of the operator J will be called *the invariant polynomials of the block* $(B; I, Q)$. We order the invariant polynomials with decreasing degrees. If useful, we extend the set of invariant polynomials with polynomials constant and equal to 1. In particular, in the case of a block with an n-dimensional underlying space $\mathcal{X}$ we often consider the set of invariant polynomials of the block to contain n elements.

We will give formulas for the indices of the third kind of a block $(B; I, Q)$. In order to do so we first introduce a sequence of subspaces for an arbitrary block (B, P, Q). Let

$$\mathcal{D}_0 = \mathcal{X}, \quad \mathcal{D}_1 = \operatorname{Im} Q, \quad \mathcal{D}_i = \{x \in \mathcal{D}_{i-1} \mid Bx \in \mathcal{D}_{i-1} + \operatorname{Ker} P\}, \; i = 2, 3, \ldots .$$

The space $\mathcal{D}_k$ will be called the *k-th definition space* of the block. Obviously $\mathcal{D}_{i+1} \subset \mathcal{D}_i$, and hence the sequence $(\mathcal{D}_i)_{i=1}^{\infty}$ is descending. Therefore there exists a number μ such that $\mathcal{D}_{\mu-1} \neq \mathcal{D}_\mu = \mathcal{D}_{\mu+1}$. It follows that $\mathcal{D}_j = \mathcal{D}_\mu$ for $j \geq \mu$. We call $\mathcal{D}_\infty = \mathcal{D}_\mu$ the *residual subspace* of the block $(B; P, Q)$. The subspace $\mathcal{D}_\infty$ of $\mathcal{X}$ is $(B; P, Q)$-block-invariant. Indeed $B[\mathcal{D}_{\mu+1}] \subset \mathcal{D}_\mu + \operatorname{Ker} P$, and hence $B[\mathcal{D}_\infty \cap \operatorname{Im} Q] = B[\mathcal{D}_\infty] \subset \mathcal{D}_\infty + \operatorname{Ker} P$. On the other hand, if $\mathcal{Y} \subset \operatorname{Im} Q$ and $\mathcal{Y}$ is $(B; P, Q)$-block-invariant, then $\mathcal{Y} \subset \mathcal{D}_\infty$ To see this note that $\mathcal{Y} \subset \mathcal{D}_1$. Moreover, if $\mathcal{Y} \subset \mathcal{D}_i$, then

$$B[\mathcal{Y}] \subset \mathcal{Y} + \operatorname{Ker} P \subset \mathcal{D}_i + \operatorname{Ker} P,$$

and thus $\mathcal{Y} \subset \mathcal{D}_{i+1}$. We conclude that the residual subspace $\mathcal{D}_\infty$ is the largest $(B; P, Q)$-block-invariant subspace in $\operatorname{Im} Q$.

Let $(B; I, Q)$ be a full length block. Then the j-th definition space of $(B; I, Q)$ is simpler to describe. In fact, we have

$$\mathcal{D}_0 = \mathcal{X}, \quad \mathcal{D}_1 = \operatorname{Im} Q, \quad \mathcal{D}_i = \{x \in \mathcal{D}_{i-1} \mid Bx \in \mathcal{D}_{i-1}\}, \quad i = 2, 3, \ldots . \tag{1.1}$$

Thus for a full length block $(B; I, Q)$ the space $\mathcal{D}_j$ is simply the subspace of all $x \in \mathcal{X}$ for which $B^j x$ is defined. The subspace $\mathcal{D}_\infty$ is the largest B-invariant subspace contained in $\operatorname{Im} Q$. The restriction $B|_{\mathcal{D}_\infty}$ of B to $\mathcal{D}_\infty$ is called the *invariant operator of the block* $(B; I, Q)$. The next result describes how to compute the indices of the third kind and the invariant polynomials of a block in terms of the spaces $\mathcal{D}_j$.

Theorem 1.2. *Let $(B; I, Q)$ be a full length block with underlying space $\mathcal{X}$. Put*

$$\mathcal{D}_0 = \mathcal{X}, \quad \mathcal{D}_1 = \operatorname{Im} Q, \quad \mathcal{D}_i = \{x \in \mathcal{D}_{i-1} \mid Bx \in \mathcal{D}_{i-1}\}, \quad i = 2, 3, \ldots,$$

and $\mathcal{D}_\infty = \bigcap_{k=1}^{\infty} \mathcal{D}_k$. Set $\mu = \min\{j \mid \mathcal{D}_j = \mathcal{D}_{j+1}\}$, and

$$q_k = \dim(\mathcal{D}_k / \mathcal{D}_{k+1}) \tag{1.2}$$

for $k = 0, \dots, \mu - 1$. Then the indices of the third kind of the block $(B; I, Q)$ are the first q_0 elements of the dual sequence of $q_1 \geq q_2 \geq \cdots$, i.e., they are given by

$$\mu_j = \#\{k \geq 1 \mid q_k \geq j\}, \quad j = 1, \dots, q_0. \tag{1.3}$$

The invariant polynomials of the block are the invariant polynomials of the restriction of B to the invariant subspace $\mathcal{D}_\infty$.

For the definition and properties of dual sequences we refer to Section 4 of this chapter, where we bring together some of the main properties of this notion.

The proof of both Theorems 1.1 and 1.2 will make use of a special basis of the space $\mathcal{X}$. The existence and properties of this basis are the subject of the following result.

Proposition 1.3. *Let $(B; I, Q)$ be an operator block on the space $\mathcal{X}$, and let $\mathcal{D}_\infty$ be the residual subspace of the block. Then with the numbers $q = \dim \operatorname{Ker} Q$ and $\mu_1 \geq \cdots \geq \mu_q \geq 0$ defined by (1.3) there exists a linearly independent set of vectors $\{f_{1j}, \dots, f_{\mu_j+1\ j} \mid j = 1, \dots, q\}$ in $\mathcal{X}$ such that:*

(i) $\mathcal{X} = \operatorname{span}\{f_{ij}\}_{i=1,j=1}^{\mu_j+1,q} \oplus \mathcal{D}_\infty$;

(ii) $\operatorname{Im} Q = \operatorname{span}\{f_{ij}\}_{i=1,j=1}^{\mu_j,q} \oplus \mathcal{D}_\infty$;

(iii) $Bf_{ij} = f_{i+1\ j}$, *for $i = 1, \dots, \mu_j$ and $j = 1, \dots, q$.*

PROOF. If $\operatorname{Im} Q = \mathcal{X}$, then $\mathcal{D}_\infty = \mathcal{X}$, and the proposition is clear in this case. So, assume that $\operatorname{Im} Q \neq \mathcal{X}$. Let for $j = 0, 1, \dots$ the subspaces $\mathcal{D}_j$ be defined by (1.1). Recall that $\mathcal{D}_\infty = \bigcap_{j=1}^\infty \mathcal{D}_j$, and let the number μ be such that $\mathcal{D}_\infty = \mathcal{D}_\mu \neq \mathcal{D}_{\mu-1}$. Let $1 \leq k \leq \mu$. If $x \in \mathcal{D}_k$, then it follows from (1.1) that $Bx \in \mathcal{D}_{k-1}$. So B defines a linear operator $B|_{\mathcal{D}_k} : \mathcal{D}_k \to \mathcal{D}_{k-1}$. Since $\mathcal{D}_k \subset \mathcal{D}_{k-1}$ and $B[\mathcal{D}_{k+1}] \subset \mathcal{D}_k$, there exists an induced linear operator

$$[B]_k : \mathcal{D}_k/\mathcal{D}_{k+1} \to \mathcal{D}_{k-1}/\mathcal{D}_k, \quad [B]_k([x]) = [Bx].$$

Note that if $y \in \mathcal{D}_k$ and $By \in \mathcal{D}_k$, then $y \in \mathcal{D}_{k+1}$. This shows that $[B]_k$ is injective. Define q_k by formula (1.2). From the injectivity of $[B]_k$ we get that $q = q_0 \geq q_1 \geq \cdots \geq q_{\mu-1}$. The numbers $\mu_1 \geq \cdots \geq \mu_q \geq \theta$ are the first elements of the dual sequence associated with $q_1 \geq q_2 \geq \cdots$. In other words (see Lemma 4.1)

$$\mu_j = k - 1, \quad j = q_k + 1, \dots, q_{k-1}.$$

We are now ready to introduce the desired linearly independent set of vectors. First, choose $f_{11}, \dots, f_{1q_{\mu-1}}$ to be a basis of $\mathcal{D}_{\mu-1}$ modulo $\mathcal{D}_\mu$, and define

$$f_{ij} = B^i f_{1j}, \quad i = 1, \dots, \mu = \mu_j + 1, \quad j = 1, \dots, q_{\mu-1}.$$

Note that the vectors $f_{21}, \dots, f_{2q_{\mu-1}}$ are in $\mathcal{D}_{\mu-2}$, and they are linearly independent modulo $\mathcal{D}_{\mu-1}$, because the linear map $[B]_{\mu-1}$ is injective. Hence we may choose vectors $f_{1\ q_{\mu-1}+1}, \dots, f_{1q_{\mu-2}}$ in $\mathcal{D}_{\mu-2}$ so that

$$f_{21}, \dots, f_{2q_{\mu-1}}, f_{1\ q_{\mu-1}+1}, \dots, f_{1q_{\mu-2}}$$

is a basis of $\mathcal{D}_{\mu-2}$ modulo $\mathcal{D}_{\mu-1}$. Put

$$f_{ij} = B^i f_{1j}, \quad i = 1, \dots, \mu - 1 = \mu_j + 1, \quad j = q_{\mu-1} + 1, \dots, q_{\mu-2}.$$

Since $[B]_{\mu-2}$ is injective, the vectors

$$f_{31}, \dots, f_{3q_{\mu-1}}, f_{2\ q_{\mu-1}+1}, \dots, f_{2q_{\mu-2}}$$

form a set of vectors in $\mathcal{D}_{\mu-3}$ which is linearly independent modulo $\mathcal{D}_{\mu-2}$, and hence this set can be extended to a basis of $\mathcal{D}_{\mu-3}$ modulo $\mathcal{D}_{\mu-2}$ by choosing vectors $f_{1\ q_{\mu-2}+1}, \dots, f_{1q_{\mu-3}}$ in $\mathcal{D}_{\mu-3}$. Now put

$$f_{ij} = B^i f_{1j}, \quad i = 1, \dots, \mu - 2 = \mu_j + 1, \quad j = q_{\mu-2} + 1, \dots, q_{\mu-3}$$

and repeat the above construction. After a finite number of steps we have constructed a linearly independent set of vectors $\{f_{1j}, \dots, f_{\mu_j+1,j} \mid j = 1, \dots, q\}$ satisfying the properties (i), (ii) and (iii). □

We illustrate the preceding proof on an example. Let $\mathcal{X} = \mathbb{C}^7$ and Q the projection onto the space spanned by e_1, e_2, e_3, e_4, e_6 along the space spanned by e_5, e_7. Here $e_1, \dots, e_7$ denotes the standard basis of $\mathbb{C}^7$. The matrix of $B : \operatorname{Im} Q \to \mathcal{X}$ with respect to the basis e_1, e_2, e_3, e_4, e_6 of $\operatorname{Im} Q$ and the basis $e_1, \dots, e_7$ of $\mathcal{X}$ is given by

$$B = \begin{pmatrix} 0 & 1 & 0 & 0 & 0 \\ -1 & 2 & 0 & 0 & 0 \\ -1 & 2 & 0 & 0 & 0 \\ -1 & 1 & 1 & 0 & 0 \\ 0 & 0 & -\frac{1}{2} & 1 & -\frac{1}{2} \\ -1 & 1 & 1 & 0 & 0 \\ 0 & 0 & \frac{1}{2} & 0 & -\frac{1}{2} \end{pmatrix}.$$

We will compute the subspaces $\mathcal{D}_j$. The subspaces $\mathcal{D}_0 = \mathcal{X}$ and $\mathcal{D}_1 = \operatorname{Im} Q$ are given. To compute $\mathcal{D}_2 = \{x \in \mathcal{D}_1 \mid Bx \in \mathcal{D}_1\}$ note that $x \in \mathcal{D}_2$ means that the fifth and the seventh coordinate of Bx are equal to 0. This leads to $\mathcal{D}_2 = \operatorname{span}\{e_1, e_2, e_3 + e_4 + e_6\}$. Next we compute $\mathcal{D}_3 = \{x \in \mathcal{D}_2 \mid Bx \in \mathcal{D}_2\}$. So $\mathcal{D}_3$ consists of all combinations of e_1, e_2 and $e_3 + e_4 + e_6$ that are such that their B-images have equal third, fourth and sixth coordinates. One obtains that $\mathcal{D}_3 = \operatorname{span}\{e_1 + e_2 + e_3 + e_4 + e_6, e_2 + e_3 + e_4 + e_6\}$. Finally one checks that $\mathcal{D}_4 = \mathcal{D}_3$. Thus $\mathcal{D}_\infty = \mathcal{D}_3$. Now compute the numbers $q_k = \dim \mathcal{D}_k - \dim \mathcal{D}_{k+1}$. One checks that $q_0 = 2, q_1 = 2, q_2 = 1, q_3 = 0$. The dual sequence associated with

$q_1 = 2, q_2 = 1$ is given by $\mu_1 = 2$ and $\mu_2 = 1$. Next choose $f_{11} = e_3 + e_4 + e_6$ in the complement of $\mathcal{D}_3$ in $\mathcal{D}_2$. Then $f_{21} = Bf_{11} = e_4 + e_6$ is in the complement of $\mathcal{D}_2$ in $\mathcal{D}_1$. Next choose $f_{12} = e_6$ in the complement of $\mathcal{D}_2 + \operatorname{span}\{e_4 + e_6\}$ in $\mathcal{D}_1$. Finally we get $f_{31} = Bf_{21} = (e_5 + e_7)/2$ and $f_{22} = Bf_{12} = (-e_5 + e_7)/2$.

PROOF OF THEOREM 1.1. Let $\mathcal{D}_\infty$ be the residual subspace and choose vectors $\{f_{ij}\}_{i=1,j=1}^{\mu_j+1\;,q}$ with the properties (i),(ii), and (iii). Put $\mathcal{X}_j = \operatorname{span}\{f_{ij}\}_{i=1}^{\mu_j+1}$, for $j = 1, \ldots, q$. Then $\mathcal{X} = \mathcal{X}_1 \oplus \cdots \oplus \mathcal{X}_q \oplus \mathcal{D}_\infty$ and $\mathcal{X}_j$ is a regular $(B; I, Q)$-invariant subspace of $\mathcal{X}$, for each i. Define the operator block $(B_j; I_j, Q_j)$ to be the regular restriction of $(B; I, Q)$ to $\mathcal{X}_j$. The block $(B_j; I_j, Q_j)$ has the property that $\operatorname{Im} Q_j = \operatorname{span}\{f_{ij}\}_{j=1}^{\mu_j}$ and $Bf_{ij} = f_{i+1\;j}$ for $i = 1, \ldots, \mu_j$. Thus $(B_j; I_j, Q_j)$ is block similar to a shift of the third kind with index μ_j. Put J the restriction of B to the invariant subspace $\mathcal{D}_\infty$. So by definition J is the invariant operator of the block $(B; I, Q)$. Then it is clear that

$$(B; I, Q) = (B_1; I_1, Q_1) \oplus \cdots \oplus (B_q; I_q, Q_q) \oplus (J, I, I).$$

Since each of the $(B_j; I_j, Q_j)$ is block similar to a shift of the third kind, it follows that $(B; I, Q)$ is block similar to a direct sum of shifts of the third kind and an operator J. This proves the first part of the theorem.

Next we prove the uniqueness of the numbers $\mu_1, \ldots, \mu_q$ and the operator J. Assume that there are subspaces $\mathcal{Y}_1, \ldots, \mathcal{Y}_p$ of $\mathcal{X}$, a subspace $\mathcal{Y}_\infty$ of $\mathcal{X}$, blocks $(S_j; I_j, Q_j)$ on $\mathcal{Y}_j$, which are similar to a shift of the third kind with index ν_j $(j = 1, \ldots, p)$, and a block $(M_\infty; I_\infty, I_\infty)$ on $\mathcal{Y}_\infty$ such that

$$(B; I, Q) = (S_1; I_1, Q_1) \oplus \cdots \oplus (S_p; I_p, Q_p) \oplus (M_\infty; I_\infty; I_\infty).$$

We also assume that $\nu_1 \geq \cdots \geq \nu_p \geq 0$. Remark that

$$\dim \operatorname{Ker} Q = \sum_{j=1}^{p} \dim \operatorname{Ker} Q_j.$$

This proves that $p = q$. For $j = 1, \ldots, q$ define

$$\mathcal{D}_{0j} = \mathcal{Y}_j, \quad \mathcal{D}_{1j} = \operatorname{Im} Q_j, \quad \mathcal{D}_{ij} = \{x \in \mathcal{D}_{i-1\;j} \mid Bx \in \mathcal{D}_{i-1\;j}\}, \quad i = 2, 3, \ldots.$$

It follows that $\dim \mathcal{D}_{kj} = \max\{\nu_j + 1 - k, 0\}$. Let $\mathcal{D}_j$ be given by (1.1). Thus we see that $\mathcal{D}_k = \mathcal{D}_{k1} \oplus \cdots \oplus \mathcal{D}_{kq} \oplus \mathcal{Y}_\infty$ and

$$\begin{aligned} q_k &= \dim \mathcal{D}_k - \dim \mathcal{D}_{k+1} = \\ &\sum_{j=1}^{q} \max\{\nu_j + 1 - k, 0\} - \max\{\nu_j - k, 0\} = \#\{j \mid \nu_j \geq k\}. \end{aligned} \tag{1.4}$$

This implies that the sequence (ν_j) is dual to the sequence (q_k) and hence $\mu_j = \nu_j$ for $j = 1, \ldots, q$. So the indices are uniquely determined by the block $(B; I, Q)$. Finally we see that $\mathcal{Y}_\infty$ coincides with the subspace $\mathcal{D}_\infty = \bigcap_{k=0}^{\infty} \mathcal{D}_k$. This proves that $M_\infty = J$, and hence J is uniquely determined by the block. □

We continue the example following the proof of Proposition 1.3. Here we shall construct the corresponding shifts and the invariant polynomials of the block. First we construct a basis for $\mathcal{D}_\infty$. Remark that $B(e_1+e_2+e_3+e_4+e_6) = e_1+e_2+e_3+e_4+e_6$ and that $B(e_2+e_3+e_4+e_6) = (e_1+e_2+e_3+e_4+e_6)+(e_2+e_3+e_4+e_6)$. Put $g_1 = e_1+e_2+e_3+e_4+e_6$ and $g_2 = e_2+e_3+e_4+e_6$. Then $\{g_1, g_2\}$ is a basis for $\mathcal{D}_\infty$ and with respect to this basis the restriction of B to $\mathcal{D}_\infty$ is in Jordan canonical form. So we see that the invariant polynomials of the block are

$$p_1(\lambda) = (\lambda-1)^2, p_2(\lambda) = 1, \dots, p_7(\lambda) = 1.$$

The final step is to give the basis of $\mathcal{X}$ and a projection Q' with $\operatorname{Im} Q' = \operatorname{Im} Q$ such that the block $(B; I, Q')$ has is a direct sum of the Jordan matrix, that we already found, a shift of the third kind of index 2 and a shift of the third kind of index 1. We choose the basis $g_1, g_2, f_{11}, f_{21}, f_{31}, f_{12}, f_{22}$ of $\mathcal{X}$, with $\operatorname{Im} Q' = \operatorname{span}\{g_1, g_2, f_{11}, f_{21}, f_{12}\}$ and $\operatorname{Ker} Q' = \operatorname{span}\{f_{31}, f_{22}\}$ Then $(B; I, Q')$ has with respect to this basis the following representation as a positioned submatrix

$$\begin{pmatrix} 1 & 1 & 0 & 0 & ? & 0 & ? \\ 0 & 1 & 0 & 0 & ? & 0 & ? \\ 0 & 0 & 0 & 0 & ? & 0 & ? \\ 0 & 0 & 1 & 0 & ? & 0 & ? \\ 0 & 0 & 0 & 1 & ? & 0 & ? \\ 0 & 0 & 0 & 0 & ? & 0 & ? \\ 0 & 0 & 0 & 0 & ? & 1 & ? \end{pmatrix}.$$

Hence it is clear that $(B; I, Q)$ is block similar to a direct sum of the Jordan matrix, a shift of the third kind of index 2, and a shift of the third kind of index 1.

PROOF OF THEOREM 1.2. In the proofs of Proposition 1.3 and of Theorem 1.1 we obtained that the indices of the third kind of the block $(B; I, Q)$ are the dual sequence of the sequence $q_1 \geq q_2 \geq \cdots$. So we already proved the first statement. The given characterization of the invariant polynomials coincides with the definition of the invariant polynomials of the block. □

Theorem 1.4. *The blocks $(B_1; I_1, Q_1)$ on $\mathcal{X}_1$ and $(B_2; I_2, Q_2)$ on $\mathcal{X}_2$ are block similar if and only if they have the same indices of the third kind and the same invariant polynomials.*

PROOF. Assume that the blocks $(B_1; I_1, Q_1)$ and $(B_2; I_2, Q_2)$ are block similar. Let $S : \mathcal{X}_2 \to \mathcal{X}_1$ be the block similarity. For $j = 1, 2$ put

$$\mathcal{D}_{0j} = \mathcal{X}_j, \quad \mathcal{D}_{1j} = \operatorname{Im} Q_j, \quad \mathcal{D}_{ij} = \{x \in \mathcal{D}_{i-1\ j} \mid Bx \in \mathcal{D}_{i-1\ j}\}, \quad i = 2, 3, \dots .$$

In view of Theorem 1.2 it is sufficient to prove that $S[\mathcal{D}_{k2}] = \mathcal{D}_{k1}$ for all k. This is immediate from the definition of $\mathcal{D}_{kj}$ and the fact that $S[\operatorname{Im} Q_2] = \operatorname{Im} Q_1$ and $SB_2 = B_1 S$. The converse statement follows from the facts that shifts of the third kind with the same index are block similar and that direct sums of block similar blocks are block similar. □

Let $(B;I,Q)$ be a full length block on the space $\mathcal{X}$. Put $\mathcal{X}' = \operatorname{Im} Q$ and let Q' be a projection of $\mathcal{X}'$ onto the subspace $\mathcal{D}_2 = \{x \in \mathcal{X}' \mid Bx \in \mathcal{X}'\}$. Write $B' = B|_{\operatorname{Im} Q'} : \operatorname{Im} Q' \to \mathcal{X}'$. The block $(B'; I_{\mathcal{X}'}, Q')$ is called *a compression to* $\operatorname{Im} Q$ of $(B;I,Q)$. Note that the freedom in the choice of the projection Q' only lies in the choice of the kernel of Q'. Hence different choices of q' lead to compressions $(B'; I_{\mathcal{X}'}, Q')$ which are block similar with a block similarity of the second kind. We illustrate the notion of compression in the following example. Assume that $(B;I,Q)$ is the shift of the third kind on $\mathbb{C}^n$ with $n \geq 2$. Thus $\operatorname{Im} Q = \operatorname{span}\{e_1, \ldots, e_{n-1}\}$, $\operatorname{Ker} Q = \operatorname{span}\{e_n\}$, and $Be_i = e_{i+1}$ for $i = 1, \ldots, n-1$. Then $\mathcal{X}' = \operatorname{Im} Q$ and $\mathcal{D}_2 = \operatorname{span}\{e_1, \ldots, e_{n-2}\}$. We define Q' to be the projection of $\mathcal{X}'$ onto $\mathcal{D}_2$ along e_{n-1}. Furthermore $B' = B|_{\mathcal{D}_2}$ and hence $B'e_i = e_{i+1}$ for $i = 1, \ldots, n-2$. We see that $(B'; I_{\mathcal{X}'}, Q')$ is block similar to shift of the third kind with index $n-2$. We conclude that the compression to $\operatorname{Im} Q$ of the shift of the third kind $(B;I,Q)$ with index $n-1$ is block similar to a shift of the third kind with index $n-2$. The next theorem shows that such phenomena appear in general.

Theorem 1.5. *Let $(B;I,Q)$ be a block on the space $\mathcal{X}$ with indices of the third kind $\mu_1 \geq \cdots \geq \mu_p > 0 = \mu_{p+1} = \cdots = \mu_q$ and invariant polynomials $p_1(\lambda), \ldots, p_n(\lambda)$. Then a compression of $(B;I,Q)$ to $\operatorname{Im} Q$ has invariant polynomials $p_1(\lambda), \ldots, p_n(\lambda)$ and indices of the third kind $\mu_1 - 1 \geq \cdots \geq \mu_p - 1 \geq 0$.*

Proof. Let $(B'; I_{\mathcal{X}'}, Q')$ be a compression of $(B;I,Q)$ to $\operatorname{Im} Q$. We define the subspaces $\mathcal{D}_i$ of $\mathcal{X}$ by formula (1.1) and the subspaces $\mathcal{D}_i'$ of $\mathcal{X}'$ by

$$\mathcal{D}_0' = \mathcal{X}', \quad \mathcal{D}_1' = \operatorname{Im} Q', \quad \mathcal{D}_i' = \{x \in \mathcal{D}_{i-1}' \mid B'x \in \mathcal{D}_{i-1}'\}, \quad i = 2, 3, \ldots.$$

Since $\mathcal{X}' = \mathcal{D}_1'$ and $B'x = Bx$ if $x \in \mathcal{D}_2$, we see that $\mathcal{D}_i' = \mathcal{D}_{i+1}$ for $i = 0, 1, \ldots$. Now it is clear that $\bigcap_{i=1}^{\infty} \mathcal{D}_i = \bigcap_{i=1}^{\infty} \mathcal{D}_i'$. We conclude that $\mathcal{D}_\infty = \mathcal{D}_\infty'$ and thus $B'|_{\mathcal{D}_\infty'} = B|_{\mathcal{D}_\infty}$. This proves that $(B;I,Q)$ and $(B'; I_{\mathcal{X}'}, Q')$ have the same invariant polynomials. Put $q_k = \dim(\mathcal{D}_k/\mathcal{D}_{k+1})$ and $q_k' = \dim(\mathcal{D}_k'/\mathcal{D}_{k+1}')$ and note that $q_k' = q_{k+1}$. According to Lemma 4.2 the dual sequence of $(q_k')_{k=1}^{\infty}$ is $\mu_1 - 1 \geq \cdots \geq \mu_p - 1 \geq 0$. □

Corollary 1.6. *The blocks $(B_1; I_1, Q_1)$ on $\mathcal{X}_1$ and $(B_2; I_2, Q_2)$ on $\mathcal{X}_2$ are block similar if and only if $\dim \mathcal{X}_1 = \dim \mathcal{X}_2$ and a compression to $\operatorname{Im} Q_1$ of $(B_1; I_1, Q_1)$ is block similar to a compression to $\operatorname{Im} Q_2$ of $(B_2; I_2, Q_2)$.*

Proof. Let $(B_1'; I_1', Q_1')$ be a compression of $(B_1; I_1, Q_1)$ to $\operatorname{Im} Q_1$, and let $(B_2'; I_2', Q_2')$ be a compression of $(B_2; I_2, Q_2)$ to $\operatorname{Im} Q_2$. Then, for $i = 1, 2$ the projection I_i' is the identity on $\mathcal{X}_i' = \operatorname{Im} Q_i$, the map Q_i' is a projection of $\mathcal{X}_i'$ with $\operatorname{Im} Q_i' = \{x \in \mathcal{X}_i' \mid B_i x \in \mathcal{X}_i'\}$ and $B_i' = B_i|_{\operatorname{Im} Q_i'} : \operatorname{Im} Q_i' \to \mathcal{X}_i'$.

First assume that $(B_1; I_1, Q_1)$ and $(B_2; I_2, Q_2)$ are block similar. Thus the indices of the third kind and the invariant polynomials of these blocks are equal. Form Theorem 1.5 we obtain that the indices of the third kind and the invariant polynomials of $(B_1'; I_1', Q_1')$ and $(B_2'; I_2', Q_2')$ are equal. So $(B_1'; I_1', Q_1')$ and $(B_2'; I_2', Q_2')$ are block similar.

Now conversely, assume that $\dim \mathcal{X}_1 = \dim \mathcal{X}_2$ and that the indices of the third kind and the invariant polynomials of $(B_1'; I_1', Q_1')$ and $(B_2'; I_2', Q_2')$ are equal. Then the invariant polynomials and the nonzero indices of the third kind of $(B_1; I_1, Q_1)$ and $(B_2; I_2, Q_2)$ are equal, according to Theorem 1.5. Since, for $i = 1, 2$, the total number of indices of the third kind of $(B_i; I_i, Q_i)$ is equal to $\dim \mathcal{X}_i - \dim \operatorname{Im} Q_i = \dim \mathcal{X}_i - \dim \mathcal{X}_i'$, the blocks $(B_1; I_1, Q_1)$ and $(B_2; I_2, Q_2)$ also have the same number of indices of the third kind that are equal to 0. So $(B_1; I_1, Q_1)$ and $(B_2; I_2, Q_2)$ have the same indices of the third kind and the same invariant polynomials. Apply Theorem 1.4 to conclude that $(B_1; I_1, Q_1)$ and $(B_2; I_2, Q_2)$ are block similar. □

III.2 Finite dimensional operator pencils

An operator pencil is a linear operator function $\lambda G - H : \mathcal{X} \to \mathcal{Y}$, where $G : \mathcal{X} \to \mathcal{Y}$ and $H : \mathcal{X} \to \mathcal{Y}$ are linear operators. In this section we consider a special class of operator pencils, namely operator pencils of the form

$$L(\lambda) = \begin{pmatrix} \lambda I_{\mathcal{X}} - F \\ -H \end{pmatrix} = \lambda \begin{pmatrix} I_{\mathcal{X}} \\ 0 \end{pmatrix} - \begin{pmatrix} F \\ H \end{pmatrix} , \tag{2.1}$$

where $F : \mathcal{X} \to \mathcal{X}$ is an operator on the space $\mathcal{X}$, the operator $I_{\mathcal{X}}$ denotes the identity on $\mathcal{X}$, and $H : \mathcal{X} \to \mathcal{Y}$. Moreover, the spaces $\mathcal{X}$ and $\mathcal{Y}$ are assumed to be finite dimensional. A simple example of such a pencil is

$$\lambda T_\mu - S_\mu = \begin{pmatrix} \lambda & 0 & 0 & \dots & 0 \\ -1 & \lambda & 0 & \dots & 0 \\ 0 & -1 & \lambda & \dots & 0 \\ \vdots & \vdots & \vdots & \ddots & \vdots \\ 0 & 0 & 0 & \dots & \lambda \\ 0 & 0 & 0 & \dots & -1 \end{pmatrix} : \mathbb{C}^\mu \to \mathbb{C}^{\mu+1} = \mathbb{C}^\mu \oplus \mathbb{C}^1. \tag{2.2}$$

Then

$$T_\mu = \begin{pmatrix} I_\mu \\ 0 \end{pmatrix} : \mathbb{C}^\mu \to \mathbb{C}^\mu \oplus \mathbb{C}^1, \quad S_\mu = \begin{pmatrix} F \\ H \end{pmatrix} : \mathbb{C}^\mu \to \mathbb{C}^\mu \oplus \mathbb{C}^1, \tag{2.3}$$

where $Fe_i = e_{i+1}$ for $i = 1, \dots, \mu - 1$, $Fe_\mu = 0$ and $He_i = 0$ for $i = 1, \dots, \mu - 1$, $He_\mu = 1$. As before $e_1, \dots, e_\mu$ denotes the standard basis in $\mathbb{C}^\mu$. The pencil (2.2) is called an *elementary pencil with minimal row index* μ and the number μ is called its *minimal row index*. (We allow that $\mu = 0$.) A second example occurs in the case when $\mathcal{Y} = \{0\}$. In this case we identify $\mathcal{X} \oplus \{0\}$ with $\mathcal{X}$ and the pencil takes the form $\lambda I - F$. One of the main results in this section is that any pencil of the form (2.1) is the direct sum of a pencil $\lambda I - F$, and elementary pencils of the form (2.2) with $\mu = 0$ included. Here the *direct sum* of the pencils $\lambda G_1 - H_1 : \mathcal{X}_1 \to \mathcal{Y}_1$ and $\lambda G_2 - H_2 : \mathcal{X}_2 \to \mathcal{Y}_2$ is by definition the pencil $\lambda(G_1 \oplus G_2) - (H_1 \oplus H_2) : \mathcal{X}_1 \oplus \mathcal{X}_2 \to \mathcal{Y}_1 \oplus \mathcal{Y}_2$.

Theorem 2.1. *Let the linear space $\mathcal{X}$ be n-dimensional, the linear space $\mathcal{Y}$ be m-dimensional, and let $L(\lambda)$ be an operator pencil given by (2.1). Then $L(\lambda)$ is strictly equivalent to a pencil $L'(\lambda)$ which is the direct sum of elementary pencils with minimal row indices $\mu_1 \geq \cdots \geq \mu_m$, and a pencil $\lambda I - J$.*

If in the pencil $L'(\lambda)$ the matrix J is in Jordan canonical form, then the pencil $L'(\lambda)$ is called the *Kronecker canonical form* of the pencil $L(\lambda)$. The numbers $\mu_1 \geq \cdots \geq \mu_p > \mu_{p+1} = \cdots = \mu_m = 0$ are called *the minimal row indices* of the pencil $L(\lambda)$. The invariant polynomials of $\lambda I_s - J$ are called *the invariant polynomials* of the pencil $L(\lambda)$.

Proof of Theorem 2.1. Let Q be a projection of $\mathcal{X} \oplus \mathcal{Y}$ along $\mathcal{Y}$ onto $\mathcal{X}$. Put

$$B = \begin{pmatrix} F \\ H \end{pmatrix} : \mathcal{X} \to \mathcal{X} \oplus \mathcal{Y}. \tag{2.4}$$

We apply Proposition 1.3 to the block $(B; I_{\mathcal{X}\oplus\mathcal{Y}}, Q)$ in order to construct a basis $\{f_{1j}, \ldots, f_{\mu_j+1\ j} \mid j = 1, \ldots, q\}$ of $\mathcal{X} \oplus \mathcal{Y}$ with the following properties.

(i) $\mathcal{X} \oplus \mathcal{Y} = \operatorname{span}\{f_{ij}\}_{i=1,j=1}^{\mu_j+1,\ q} \oplus \mathcal{D}_\infty$;

(ii) $\mathcal{X} = \operatorname{span}\{f_{ij}\}_{i=1,j=1}^{\mu_j,\ q} \oplus \mathcal{D}_\infty$;

(iii) $B[\mathcal{D}_\infty] \subset \mathcal{D}_\infty$;

(iv) $Bf_{ij} = f_{i+1\ j}$, for $i = 1, \ldots, \mu_j$, and $j = 1, \ldots, q$.

Choose a basis $\{e_i\}_{i=1}^s$ for $\mathcal{D}_\infty$. Assume that $\mu_1 \geq \cdots \geq \mu_p > 0 = \mu_{p+1} = \cdots = \mu_m$. Then

$$e_1, \ldots, e_s, f_{11}, \ldots, f_{\mu_1+1\ 1}, \ldots, f_{1p}, \ldots, f_{\mu_p+1\ p}, f_{1\ p+1}, \ldots, f_{1\ m} \tag{2.5}$$

is a basis of $\mathcal{X} \oplus \mathcal{Y}$, and

$$e_1, \ldots, e_s, f_{11}, \ldots, f_{\mu_1 1}, \ldots, f_{1p}, \ldots, f_{\mu_p p} \tag{2.6}$$

is a basis of $\mathcal{X}$. Put $\mathcal{X}_j = \operatorname{span}\{f_{1j}, \ldots, f_{\mu_j j}\}$, $\mathcal{Z}_j = \operatorname{span}\{f_{1j} \ldots, f_{\mu_j+1\ j}\}$ for $j = 1, \ldots, p$, and $\mathcal{Z}_0 = \operatorname{span}\{f_{1\ p+1}, \ldots, f_{1m}\}$. From (ii) above it follows that $B[\mathcal{X}_j] \subset \mathcal{Z}_j$. We denote by R_j the map that embeds $\mathcal{X}_j$ into $\mathcal{Z}_j$, and set $B_j = B|_{\mathcal{X}_j} : \mathcal{X}_j \to \mathcal{Z}_j$. Then $\mathcal{X} \oplus \mathcal{Y} = \mathcal{D}_\infty \oplus (\bigoplus_{j=1}^p \mathcal{Z}_j) \oplus \mathcal{Z}_0$ and $\mathcal{X} = \mathcal{D}_\infty \oplus (\bigoplus_{j=1}^p \mathcal{X}_j)$. With respect to these direct sum decomposition $L(\lambda)$ has a block diagonal representation

$$(\lambda I_{\mathcal{D}_\infty} - B|_{\mathcal{D}_\infty}) \oplus (\lambda R_1 - B_1) \oplus \cdots \oplus (\lambda R_p - B_p) \oplus O_{m-p}. \tag{2.7}$$

Remark that the matrix of $\lambda R_j - B_j$ with respect to the given bases of $\mathcal{X}_j$ and $\mathcal{Z}_j$ has the form (2.2) and hence is an elementary pencil with minimal row index μ_j. Finally, the pencil O_{m-p} is a direct sum of $m - p$ pencils of the form (2.2) with minimal row index 0. □

We give an example which illustrates how to find the special basis appearing in the theorem. Let

$$L(\lambda)=\begin{pmatrix} \lambda & -1 & 0 & 0 & 0\\ 1 & \lambda-2 & 0 & 0 & 0\\ 1 & -2 & \lambda & 0 & 0\\ 1 & -1 & -1 & \lambda & 0\\ 1 & -1 & -1 & 0 & \lambda\\ 0 & 0 & 1/2 & -1 & 1/2\\ 0 & 0 & -1/2 & 0 & 1/2 \end{pmatrix}.$$

Put $B=-L(0)$ and Q the projection of $\mathbb{C}^7$ along the span of $\{e_6,e_7\}$ and onto the space spanned by $\{e_1,\dots,e_5\}$. The block $(B;I,Q)$ is very much alike the block appearing in the example following the proof of Proposition 1.3. In fact only the roles of e_5 and e_6 are interchanged. With this simple change in mind we repeat the computation and end up with a basis of $\operatorname{Im} Q$ and an extension to $\mathbb{C}^7$ such that the matrix with respect to these bases is

$$\begin{pmatrix} \lambda-1 & -1 & 0 & 0 & 0\\ 0 & \lambda-1 & 0 & 0 & 0\\ 0 & 0 & \lambda & 0 & 0\\ 0 & 0 & -1 & \lambda & 0\\ 0 & 0 & 0 & -1 & 0\\ 0 & 0 & 0 & 0 & \lambda\\ 0 & 0 & 0 & 0 & -1 \end{pmatrix}.$$

To be precise, these bases are

$$e_1+e_2+e_3+e_4+e_5, e_2+e_3+e_4+e_5,\ e_3+e_4+e_5,\ e_4+e_5,\ e_5$$

in $\mathbb{C}^5$ and

$$e_1+e_2+e_3+e_4+e_5,\ e_2+e_3+e_4+e_5,\ e_3+e_4+e_5,\\ e_4+e_5,\ (e_6+e_7)/2,\ e_5,\ (-e_6+e_7)/2$$

in $\mathbb{C}^7$.

Two operator pencils $\lambda G_1-H_1:\mathcal{X}_1\to\mathcal{Y}_1$ and $\lambda G_2-H_2:\mathcal{X}_2\to\mathcal{Y}_2$ are called *strictly equivalent* if there exist invertible operators $S:\mathcal{X}_1\to\mathcal{X}_2$ and $T:\mathcal{Y}_1\to\mathcal{Y}_2$ such that $\lambda G_2-H_2=T(\lambda G_1-H_1)S^{-1}$.

In the proof of Theorem 2.1 we established a direct connection between the pencil $L(\lambda)$ and a block. The next result gives an important property of this relation.

Theorem 2.2. *Let*

$$L_1(\lambda) = \begin{pmatrix} \lambda I_{\mathcal{X}_1} - F_1 \\ -H_1 \end{pmatrix} : \mathcal{X}_1 \to \mathcal{X}_1 \oplus \mathcal{Y}_1, \quad L_2(\lambda) = \begin{pmatrix} \lambda I_{\mathcal{X}_2} - F_2 \\ -H_2 \end{pmatrix} : \mathcal{X}_2 \to \mathcal{X}_2 \oplus \mathcal{Y}_2 \tag{2.8}$$

be operator pencils, and let the blocks $(B_1; I_{\mathcal{X}_1 \oplus \mathcal{Y}_1}, Q_1)$ *and* $(B_2; I_{\mathcal{X}_2 \oplus \mathcal{Y}_2}, Q_2)$ *with*

$$B_1 = \begin{pmatrix} F_1 \\ H_1 \end{pmatrix} : \mathcal{X}_1 \to \mathcal{X}_1 \oplus \mathcal{Y}_1, \quad B_2 = \begin{pmatrix} F_2 \\ H_2 \end{pmatrix} : \mathcal{X}_2 \to \mathcal{X}_2 \oplus \mathcal{Y}_2,$$

and with Q_1 (Q_2) *the projection of* $\mathcal{X}_1 \oplus \mathcal{Y}_1$ $(\mathcal{X}_2 \oplus \mathcal{Y}_2)$ *along* $\mathcal{Y}_1$ $(\mathcal{Y}_2)$ *onto* $\mathcal{X}_1$ $(\mathcal{X}_2)$. *Then* $L_1(\lambda)$ *and* $L_2(\lambda)$ *are strictly equivalent if and only if* $(B_1; I_{\mathcal{X}_1 \oplus \mathcal{Y}_1}, Q_1)$ *and* $(B_2; I_{\mathcal{X}_2 \oplus \mathcal{Y}_2}, Q_2)$ *are block similar.*

PROOF. First assume that $L_1(\lambda)$ and $L_2(\lambda)$ are strictly equivalent. Thus there are invertible operators $S : \mathcal{X}_1 \to \mathcal{X}_2$ and

$$T = \begin{pmatrix} T_{11} & T_{12} \\ T_{21} & T_{22} \end{pmatrix} : \mathcal{X}_1 \oplus \mathcal{Y}_1 \to \mathcal{X}_2 \oplus \mathcal{Y}_2$$

such that $L_2(\lambda) = T L_1(\lambda) S^{-1}$. From comparing the coefficients of λ one obtains that $T_{21} = 0$ and that $T_{11} = S$. So we have that $T[\operatorname{Im} Q_1] = \operatorname{Im} Q_2$ and that

$$\begin{pmatrix} T_{11} & T_{12} \\ 0 & T_{22} \end{pmatrix} \begin{pmatrix} F_1 \\ H_1 \end{pmatrix} = \begin{pmatrix} F_2 \\ H_2 \end{pmatrix} S.$$

So we see that the blocks $(B_1; I_{\mathcal{X}_1 \oplus \mathcal{Y}_1}, Q_1)$ and $(B_2; I_{\mathcal{X}_2 \oplus \mathcal{Y}_2}, Q_2)$ are block similar.

Secondly assume that $(B_1; I_{\mathcal{X}_1 \oplus \mathcal{Y}_1}, Q_1)$ and $(B_2; I_{\mathcal{X}_2 \oplus \mathcal{Y}_2}, Q_2)$ are block similar. Thus there exists an invertible operator

$$T = \begin{pmatrix} T_{11} & T_{12} \\ T_{21} & T_{22} \end{pmatrix} : \mathcal{X}_1 \oplus \mathcal{Y}_1 \to \mathcal{X}_2 \oplus \mathcal{Y}_2$$

such that $T[\operatorname{Im} Q_1] = \operatorname{Im} Q_2$ and $(TB_1 - B_2 T)x = 0$ for all $x \in \operatorname{Im} Q_1$. It follows that $T_{21} = 0$ and that $L_2(\lambda) T_{11} = T L_1(\lambda)$. From this the strict equivalence of the pencils $L_1(\lambda)$ and $L_2(\lambda)$ is obvious. □

The next result follows directly from the proof of Theorem 2.1.

Corollary 2.3. *The minimal row indices and the invariant polynomials of*

$$L(\lambda) = \begin{pmatrix} \lambda I_{\mathcal{X}} - F \\ -H \end{pmatrix} : \mathcal{X} \to \mathcal{X} \oplus \mathcal{Y}$$

are the indices of the third kind and the invariant polynomials of the block

$$\left(\begin{pmatrix} F \\ H \end{pmatrix}; \begin{pmatrix} I & 0 \\ 0 & I \end{pmatrix}, \begin{pmatrix} I & 0 \\ 0 & 0 \end{pmatrix} \right)$$

on $\mathcal{X} \oplus \mathcal{Y}$.

From Corollary 2.3 and Theorem 1.1 it follows that the minimal row indices and invariant polynomial are uniquely determined by the pencil. From Corollary 2.3 and Theorem 1.4 one deduces the next result.

Corollary 2.4. *The pencils $L_1(\lambda)$ and $L_2(\lambda)$ given by formula (2.8) are strictly equivalent if and only if they have the same minimal row indices and the same invariant polynomials.*

We will say that the pencil $L(\lambda)$ has no invariant polynomials if $\dim \mathcal{D}_\infty = 0$. The last theorem of this section characterizes the pencils without invariant polynomials.

Theorem 2.5. *Let the linear space $\mathcal{X}$ be n-dimensional and the linear space $\mathcal{Y}$ be m-dimensional. Let*

$$L(\lambda) = \begin{pmatrix} \lambda I_{\mathcal{X}} - F \\ -H \end{pmatrix} : \mathcal{X} \to \mathcal{X} \oplus \mathcal{Y}$$

be an operator pencil. Then the following three statements are equivalent:

(1) *$L(\lambda)$ has no invariant polynomials;*

(2) $\cap_{j=0}^{\infty} \operatorname{Ker} HF^j = \{0\}$;

(3) $\operatorname{rank} L(\lambda) = n$ *for all* λ.

PROOF. Put

$$B = \begin{pmatrix} F \\ H \end{pmatrix} : \mathcal{X} \to \mathcal{X} \oplus \mathcal{Y}\,, \quad Q = \begin{pmatrix} I & 0 \\ 0 & 0 \end{pmatrix} : \mathcal{X} \oplus \mathcal{Y} \to \mathcal{X} \oplus \mathcal{Y}\,,$$

and let $\mathcal{D}_k$ be the k-th definition space of the block. We write $\mathcal{D}_\infty$ for the residual space of the block.

We show that $\mathcal{D}_k = \cap_{j=0}^{k-2} \operatorname{Ker} HF^j$, for $k = 2, 3, \ldots$. Firstly remark that $\mathcal{D}_2 = \{x \in \mathcal{X} \mid Bx \in \mathcal{X}\}$. The restriction $Bx \in \mathcal{X}$ means $x \in \operatorname{Ker} H$. So we see that $\mathcal{D}_2 = \operatorname{Ker} H$. Next assume that $\mathcal{D}_{k-1} = \cap_{j=0}^{k-3} \operatorname{Ker} HF^j$. Then $x \in \mathcal{D}_{k-1}$ and $Bx \in \mathcal{D}_{k-1}$ if and only if $x \in \operatorname{Ker} H$ and $HF^j(Fx) = 0$ for $j = 0, \ldots, k-3$. Thus we see that $x \in \mathcal{D}_k$ if and only if $HF^j x = 0$, for $j = 0, \ldots, k-2$.

We conclude that $\mathcal{D}_\infty = \{0\}$ if and only if $\cap_{j=0}^{\infty} \operatorname{Ker} HF^j = \{0\}$. By definition the polynomial $L(\lambda)$ has no invariant polynomials if $\mathcal{D}_\infty = \{0\}$. This shows the equivalence of (1) and (2).

To see the equivalence of (1) and (3) note that the representation (2.7) of $L(\lambda)$ has rank n for all λ if and only if $\lambda I_{\mathcal{D}_\infty} - B|_{\mathcal{D}_\infty}$ has full rank for all λ. This can only happen if $\dim \mathcal{D}_\infty = 0$. So $L(\lambda)$ has rank n for all λ if and only if $L(\lambda)$ has no invariant polynomials. □

III.3 Similarity of non-everywhere defined linear operators

In this section we treat operators of the type $B : \mathcal{D}(B)$ of $\mathcal{X}$, where $\mathcal{D}(B) \subset \mathcal{X}$ is a linear subspace of $\mathcal{X}$. Such an operator will be called a *non-everywhere defined linear operator in* $\mathcal{X}$, and will be denoted as $B(\mathcal{X} \to \mathcal{X})$. Two non-everywhere defined operators $B_1(\mathcal{X}_1 \to \mathcal{X}_1)$ and $B_2(\mathcal{X}_2 \to \mathcal{X}_2)$ will be called *similar* if there exists an invertible linear operator $S : \mathcal{X}_1 \to \mathcal{X}_2$ such that $S[\mathcal{D}(B_1)] = \mathcal{D}(B_2)$ and $SB_1y = B_2Sy$ for each $y \in \mathcal{D}(B_1)$. In that case the linear operator S is called a *similarity*. The direct sum of $B_1(\mathcal{X}_1 \to \mathcal{X}_1)$ and $B_2(\mathcal{X}_2 \to \mathcal{X}_2)$ is by definition the operator $B_1 \oplus B_2(\mathcal{X}_1 \oplus \mathcal{X}_2 \to \mathcal{X}_1 \oplus \mathcal{X}_2)$ with $\mathcal{D}(B_1 \oplus B_2) = \mathcal{D}(B_1) \oplus \mathcal{D}(B_2)$.

A simple example of a non-everywhere defined operator is the following. Let $e_1, \ldots, e_n$ be the standard basis of $\mathbb{C}^n$. Define $\mathcal{D}(S) = \operatorname{span}\{e_1, \ldots, e_{n-1}\}$ and $S : \mathcal{D}(S) \to \mathbb{C}^n$ by $Se_i = e_{i+1}$ for $i = 1, \ldots, n-1$. The non-everywhere defined operator $S(\mathbb{C}^n \to \mathbb{C}^n)$ is called a *non-everywhere defined shift.*

Let $B(\mathcal{X} \to \mathcal{X})$ be a non-everywhere defined operator. We construct a sequence of subspaces by putting

$$\begin{aligned} &\mathcal{D}_0(B) = \mathcal{X}, \quad \mathcal{D}_1(B) = \mathcal{D}(B), \\ &\mathcal{D}_k(B) = \{x \in \mathcal{D}_{k-1}(B) \mid Bx \in \mathcal{D}_{k-1}(B)\}, \quad k = 2, 3, \ldots \end{aligned} \tag{3.1}$$

Write $\mathcal{D}_\infty(B) = \bigcap_{k=0}^{\infty} \mathcal{D}_k(B)$. The subspace $\mathcal{D}_\infty(B)$ is the largest invariant subspace for B. The restriction $B|_{\mathcal{D}_\infty(B)} : \mathcal{D}_\infty(B) \to \mathcal{D}_\infty(B)$ of B to $\mathcal{D}_\infty(B)$ is called the *everywhere defined part* of $B(\mathcal{X} \to \mathcal{X})$. Now we are able to state the main result of this section.

Theorem 3.1. *A non-everywhere defined operator $B(\mathcal{X} \to \mathcal{X})$ is similar to a direct sum of its everywhere defined part and $\dim \mathcal{X} - \dim \mathcal{D}(B)$ non-everywhere defined shifts.*

Before we prove the theorem we make a connection between non-everywhere defined operators and operator blocks. Let $B(\mathcal{X} \to \mathcal{X})$ be given, and let Q be a projection onto $\mathcal{D}(B)$. Then $B : \operatorname{Im} Q \to \mathcal{X}$, and thus $(B; I, Q)$ is a well defined block. The block $(B; I, Q)$ is not uniquely determined by $B(\mathcal{X} \to \mathcal{X})$, but since $\operatorname{Im} Q$ is uniquely determined, the blocks defined by different choices of Q are block similar with a block similarity of the second kind. We refer to $(B; I, Q)$ as a *block associated to* $B(\mathcal{X} \to \mathcal{X})$. Conversely, assume that a block $(B; I, Q)$ is given. Then $B(\mathcal{X} \to \mathcal{X})$ with $\mathcal{D}(B) = \operatorname{Im} Q$ is a uniquely determined non-everywhere defined operator. Therefore $B(\mathcal{X} \to \mathcal{X})$ is called *the non-everywhere defined operator given by the block* $(B; I, Q)$.

Proof of Theorem 3.1. The idea of the proof is to show that this result can be seen as a translation of Theorem 1.1 into the language of non-everywhere defined operators.

Assume that $(B_1; I_1, Q_1)$ and $(B_2; I_2, Q_2)$ are blocks associated to $B_1(\mathcal{X}_1 \to \mathcal{X}_1)$ and $B_2(\mathcal{X}_2 \to \mathcal{X}_2)$, respectively. Then $B_1(\mathcal{X}_1 \to \mathcal{X}_1)$ and $B_2(\mathcal{X}_2 \to \mathcal{X}_2)$ are similar if and only if $(B_1; I_1, Q_1)$ and $(B_2; I_2, Q_2)$ are block similar. This follows immediately from the definitions of block similarity and similarity of non-everywhere defined operators. Also directly from the definitions one obtains that $(B_1; I_1, Q_1) \oplus (B_2; I_2, Q_2)$ is a block associated to the non-everywhere defined operator $B_1 \oplus B_2(\mathcal{X}_1 \oplus \mathcal{X}_1 \to \mathcal{X}_2 \oplus \mathcal{X}_2)$. Furthermore for a non-everywhere defined operator $B(\mathcal{X} \to \mathcal{X})$ the subspace $\mathcal{D}_\infty(B)$ is equal to the residual subspace of an associated block $(B; I, Q)$. This follows from the formulas (3.1) and (1.1). Hence the everywhere defined part of $B(\mathcal{X} \to \mathcal{X})$ is the invariant operator of the block. Next remark that a shift of the third kind of index μ is a block associated to a non-everywhere defined shift of rank μ. Finally remark that the operator J appearing in Theorem 1.1 is the invariant operator of the block $(B; I, Q)$. □

For $B(\mathcal{X} \to \mathcal{X})$ we define numbers s_k, $k = 0, 1, 2, \ldots$, by setting $s_k = \dim \mathcal{D}_k(B) - \dim \mathcal{D}_{k+1}(B)$. Here $\mathcal{D}_k(B)$ is given by (3.1). The numbers $s_0, s_1, \ldots$ are called the *defect dimensions* of $B(\mathcal{X} \to \mathcal{X})$.

Theorem 3.2. *Two non-everywhere defined operators are similar if and only if they have the same sequence of defect dimensions and their everywhere defined parts are similar operators.*

It is not difficult to give a proof of this theorem based on the Theorems 1.2 and 1.4 and on the translation to blocks presented in the proof of Theorem 3.1. We choose however to present a proof using the idea of compression appearing in Corollary 1.6. For this purpose we need the following definition. For $B(\mathcal{X} \to \mathcal{X})$ put $\mathcal{Y} = \mathcal{D}(B)$ and $\mathcal{D}(B_1) = \{x \in \mathcal{Y} \mid Bx \in \mathcal{Y}\}$. Define $B_1 : \mathcal{D}(B_1) \to \mathcal{Y}$ by $B_1 x = Bx$. Then $B_1(\mathcal{Y} \to \mathcal{Y})$ is a non-everywhere defined operator, which is called the *part of B in $\mathcal{D}(B)$*.

Lemma 3.3. *The non-everywhere defined operators $B(\mathcal{X} \to \mathcal{X})$ and $B'(\mathcal{X}' \to \mathcal{X}')$ are similar if and only if the part of $B(\mathcal{X} \to \mathcal{X})$ in $\mathcal{D}(B)$ is similar to the part of $B'(\mathcal{X}' \to \mathcal{X}')$ in $\mathcal{D}(B')$ and $\dim \mathcal{X} = \dim \mathcal{X}'$.*

We will give a direct proof of this lemma, but note that the result is just a translation of Corollary 1.6 to the language of non-everywhere defined operators.

PROOF OF LEMMA 3.3. Write $\mathcal{Y} = \mathcal{D}(B)$, $\mathcal{Y}' = \mathcal{D}(B')$, $\mathcal{D}(B_1) = \{x \in \mathcal{Y} \mid Bx \in \mathcal{Y}\}$, and $\mathcal{D}(B_1') = \{x \in \mathcal{Y}' \mid B'x \in \mathcal{Y}'\}$. The part of $B(\mathcal{X} \to \mathcal{X})$ in $\mathcal{D}(B)$ is then $B_1(\mathcal{Y} \to \mathcal{Y})$ and the part of $B'(\mathcal{X}' \to \mathcal{X}')$ in $\mathcal{D}(B')$ is $B_1'(\mathcal{Y}' \to \mathcal{Y}')$.

First assume that $B(\mathcal{X} \to \mathcal{X})$ and $B'(\mathcal{X}' \to \mathcal{X}')$ are similar. Thus there exists an invertible linear operator $S : \mathcal{X} \to \mathcal{X}'$ such that $S[\mathcal{Y}] = \mathcal{Y}'$ and $B'Sy = SBy$ for all $y \in \mathcal{Y}$. If $y \in \mathcal{D}(B_1)$ then $y \in \mathcal{Y}$ and $By \in \mathcal{Y}$. So it follows that $Sy \in \mathcal{Y}'$ and $B'Sy = SBy \in \mathcal{Y}'$. This proves that $Sy \in \mathcal{D}(B_1')$ and thus $S[\mathcal{D}(B_1)] \subset \mathcal{D}(B_1')$. From the symmetry of the similarity relation the converse inclusion follows also, and this proves that $S[\mathcal{D}(B_1)] = \mathcal{D}(B_1')$. Since obviously $B_1'Sy = SB_1y$ for each $y \in \mathcal{Y}_1$ and $\dim \mathcal{X} = \dim \mathcal{X}'$, we finished the first part of the proof.

Conversely, assume that an invertible linear operator $S_1 : \mathcal{Y} \to \mathcal{Y}'$ is given such that $S_1[\mathcal{D}(B_1)] = \mathcal{D}(B_1')$ and $B_1'S_1y = S_1B_1y$ for all $y \in \mathcal{D}(B_1)$. For $y \in \mathcal{Y}$ we put $Sy = S_1y$. Now assume that $x = By$. Then we define $Sx = B'(S_1y)$. To see that S is well defined assume that $By = Bz$. Then $B(y-z) = 0$, and therefore $y - z \in \mathcal{D}(B_1)$. Thus $B'S_1(y-z) = S_1B(y-z) = 0$ and $B'S_1y = B'S_1z + B'S_1(y-z) = B'S_1z$. Next remark that if $x = By \in \mathcal{Y}$, then the two definitions that we gave for Sx coincide. Indeed, in this case $y \in \mathcal{D}(B_1)$, and $S_1x = S_1By = B'S_1y$. To prove that S is injective we have to show that if $x = By$ and $B'S_1y = 0$, then $x = 0$. Indeed, if $B'S_1y = 0$, then $S_1y \in \mathcal{D}(B_1')$, and thus $y \in \mathcal{D}(B_1)$. Thus $0 = B'S_1y = S_1By = S_1x$. Since S_1 is invertible, it follows that $x = 0$. We proved $S : \mathcal{Y} + B[\mathcal{Y}] \to \mathcal{Y}' + B'[\mathcal{Y}']$ is injective. Clearly this operator is also surjective. Finally we are able to extend this operator to an invertible linear operator $S : \mathcal{X} \to \mathcal{X}'$ since $\dim \mathcal{X} = \dim \mathcal{X}'$. □

PROOF OF THEOREM 3.2. Let $B(\mathcal{X} \to \mathcal{X})$ and $B'(\mathcal{X}' \to \mathcal{X}')$ be the non-everywhere defined operators. Let $\mathcal{D}_k(B)$ be given by (3.1) and $\mathcal{D}_k(B')$ be defined analogously for $B'(\mathcal{X}' \to \mathcal{X}')$. Since the sequences of spaces $\mathcal{D}_k(B)$ and $\mathcal{D}_k(B')$ are descending, there is a number q such that $\mathcal{D}_q(B) = \cap_{k=0}^{\infty}\mathcal{D}_k(B) = \mathcal{D}_\infty(B)$ and $\mathcal{D}_q(B') = \cap_{k=0}^{\infty}\mathcal{D}_k(B') = \mathcal{D}_\infty(B')$. Put $B_1 = B$. For each k the non-everywhere defined operator $B_k\big(\mathcal{D}_{k-1}(B) \to \mathcal{D}_{k-1}(B)\big)$ is defined as the part of B_{k-1} in $\mathcal{D}_{k-1}(B)$. Analogously we define B_k'. We apply Lemma 3.3 to obtain that B_k is similar to B_k' if and only if $\dim \mathcal{D}_k(B) = \dim \mathcal{D}_k(B')$ and B_{k+1} is similar to B_{k+1}'. So we obtained that B is similar to B' if and only if $B|_{\mathcal{D}_q(B)}$ is similar to $B'|_{\mathcal{D}_q(B')}$ and

$$\dim \mathcal{D}_0(B) = \dim \mathcal{D}_0(B'), \quad \dots \quad , \quad \dim \mathcal{D}_q(B) = \dim \mathcal{D}_q(B').$$

Thus B and B' are similar if and only if their everywhere defined parts are similar and their defect numbers are equal. □

Corollary 3.4. *Let $B(\mathcal{X} \to \mathcal{X})$ be a non-everywhere defined linear operator of which the strictly positive defect dimensions are $s_0, s_1, \dots, s_q$. Let Q be a projection of $\mathcal{X}$ such that $\operatorname{Im} Q = \mathcal{D}(B)$. Then $\mu_1, \mu_2, \dots$, the indices of the third kind of the associated block $(B; I, Q)$, is the dual sequence of the sequence $s_1, s_2, \dots, s_q$, and the invariant polynomials of the block are the invariant polynomials of the everywhere defined part of $B(\mathcal{X} \to \mathcal{X})$.*

PROOF. The first statement follows from the formulas (3.1), (1.1), (1.2), and (1.3), and the definition of defect dimensions. The statement on the invariant polynomials follows from the observation made in the proof of Theorem 3.1 that the everywhere defined part of $B(\mathcal{X} \to \mathcal{X})$ is the invariant operator of the block. □

III.4 Dual sequences

In this section we present some results that are useful on several places in this and the next chapters. Let $\mu_1 \geq \mu_2 \geq \cdots \geq \mu_q \geq \cdots$ be a nonincreasing sequence of nonnegative integers of which only a finite number are nonzero. We call such a sequence for short an *index sequence*. We define the *dual sequence* $(\mu_j^\#)_{j=1}^\infty$ of $(\mu_k)_{k=1}^\infty$ by putting

$$\mu_j^\# = \#\{k \geq 1 \mid \mu_k \geq j\}, \quad j = 1, 2, \ldots . \tag{4.1}$$

In words, $\mu_j^\#$ is the number of elements of the sequence that is larger than or equal to j. Since $(\mu_k)_{k=1}^\infty$ is a nonincreasing sequence, one also has

$$\mu_j^\# = \max\{k \geq 1 \mid \mu_k \geq j\}, \quad j = 1, 2, \ldots .$$

The sequence $(\mu_j^\#)_{j=1}^\infty$ is again a nonincreasing sequence of integers with only finitely many nonzero elements. The duality relation can be understood as follows. Represent the index sequence as the graph of the function f defined by $f(x) = \mu_k$ for $k - 1 < x \leq k$ as is done in figure 1a for the sequence $7, 6, 4, 4, 3, 1, 1, 0, \ldots$. The dual sequence is then $7, 5, 5, 4, 2, 2, 1, 0, \ldots$. The graph of the function $f^\#$, where $f^\#(x) = \mu_j^\#$ for $j - 1 < x \leq j$, is given in figure 1b. Note that in this case the graph of $f^\#$ can be obtained from the graph of f by reflection in the diagonal of the first quadrant. This reflection principle for dual sequences holds true in general and provides the idea behind many of the proofs in this section. A formal justification of the reflection principle will be based on Proposition 4.1.

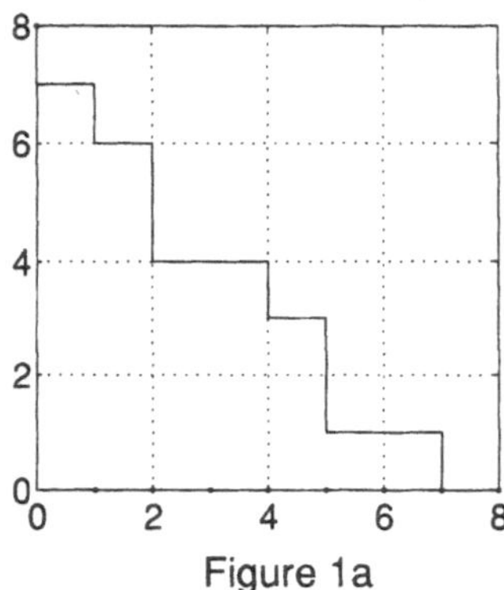

Figure 1a

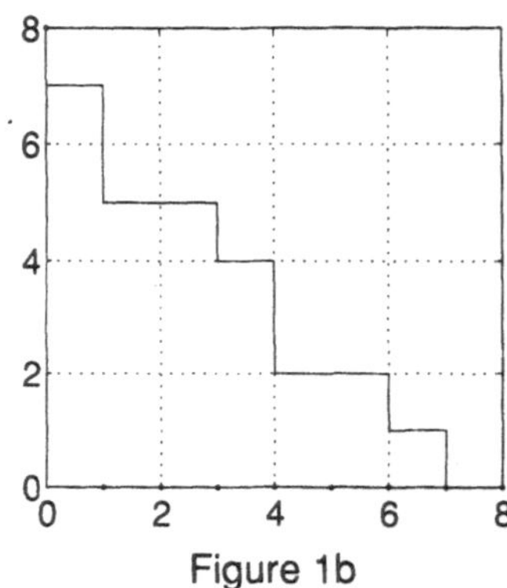

Figure 1b

The first proposition gives some properties of sequences that could also be used as definition of the dual sequence. In particular the proposition shows that the dual sequence of a dual sequence is the original sequence we started with.

Proposition 4.1. *Let $(\mu_k)_{k=1}^\infty$ be an index sequence. Then the following statements are equivalent.*

(1) *$(\mu_j^\#)_{j=1}^\infty$ is the dual sequence of $(\mu_k)_{k=1}^\infty$;*

(2) *$(\mu_k)_{k=1}^\infty$ is the dual sequence of $(\mu_j^\#)_{j=1}^\infty$;*

(3) *for a pair of positive integers* (k, j) *the equality* $\mu_k = j$ *holds if and only if* $\mu^{\#}_{j+1} < k \leq \mu^{\#}_j$;

(4) *for a pair of positive integers* (k, j) *the equality* $\mu^{\#}_j = k$ *holds if and only if* $\mu_{k+1} < j \leq \mu_k$;

(5) *for a pair of positive integers* (k, j) *the inequality* $\mu_k \geq j$ *holds if and only if the inequality* $\mu^{\#}_j \geq k$ *holds.*

PROOF. First we prove that (1) implies (3). From (4.1) for $j+1$ we see that $\mu_k < j+1$ if $k > \mu^{\#}_{j+1}$. Also, if $k \leq \mu^{\#}_j$ then $\mu_k \geq j$. So we get $j \leq \mu_k \leq j+1$ if $\mu^{\#}_{j+1} < k \leq \mu^{\#}_j$. On the other hand, if $k \leq \mu^{\#}_{j+1}$, then $\mu_k \geq j+1$, and if $k > \mu^{\#}_j$, then $\mu_k < j$. We conclude that $\mu_k = j$ if and only if $\mu^{\#}_{j+1} < k \leq \mu^{\#}_j$. This proves (3).

Next we show that (3) implies (5). From (3) and the fact that the sequences are nonincreasing, it is clear that $\mu_k \geq j$ if $k \leq \mu^{\#}_j$, and that $\mu_k < j$ if $k > \mu^{\#}_j$. Hence, $\mu_k \geq j$ if and only if $\mu^{\#}_j \geq k$.

To prove that (5) implies (1) note that from (5) it follows that $\mu^{\#}_j \geq k+1$ if and only if $\mu_{k+1} \geq j$. Hence $\mu^{\#}_j < k+1$ if and only if $\mu_{k+1} < j$. Combine this with the fact that $\mu^{\#}_j \geq k$ if and only if $\mu_k \geq j$, to obtain that $\mu^{\#}_j = k$ if and only if $\mu_1 \geq \cdots \geq \mu_k \geq j > \mu_{k+1}$. This shows that $\mu^{\#}_j = \#\{k \geq 1 \mid \mu_k \geq j\}$.

We proved that (1), (3) and (5) are equivalent. Since the property (5) is symmetric in the sequences $(\mu^{\#}_j)_{j=1}^{\infty}$ and $(\mu_k)_{k=1}^{\infty}$, we conclude that also (2), (4) and (5) are equivalent statements. □

Now we are able to explain the reflection principle for dual index sequences. Let $(\mu^{\#}_j)_{j=1}^{\infty}$ be the dual sequence of $(\mu_k)_{k=1}^{\infty}$. Let f be defined by $f(x) = \mu_k$ for $k-1 < x \leq k$ and $f^{\#}$ be defined by $f^{\#}(x) = \mu^{\#}_j$ for $j-1 < x \leq j$. From (3) in Proposition 4.1 we get that $f(x) = j$ if $\mu^{\#}_{j+1} < x \leq \mu^{\#}_j$. This means that $f(x) = j$ if and only if $f^{\#}(j-) < x \leq f^{\#}(j)$. In other words $f(x) = j$ if and only if x is in the gap of the jump discontinuity of $f^{\#}$ at j. Conversely, $f^{\#}(y) = k$ if and only if y is in the gap of the jump discontinuity of f at k. These two statements provide the formal basis for the reflection principle for index sequences.

The next result is easily understood from the reflection principle.

Proposition 4.2. *Let* $(\mu_k)_{k=1}^{\infty}$ *be an index sequence, and let* $(\mu^{\#}_j)_{j=1}^{\infty}$ *be its dual sequence. Then*

$$\sum_{j=1}^{\infty} \mu^{\#}_j = \sum_{k=1}^{\infty} \mu_k. \tag{4.2}$$

Furthermore, if $\mu_k(\ell) = \mu_{k+\ell}$ *for* $k = 1, 2, \dots$ *, then*

$$\mu_j(\ell)^\# = \max\{0, \mu_j^\# - \ell\}, \tag{4.3}$$

where $(\mu_j(\ell)^\#)_{j=1}^\infty$ *is the dual sequence of* $(\mu_k(\ell))_{k=1}^\infty$.

PROOF. First note that $\sum_{k=1}^\infty \mu_k = \sum_{j=1}^\infty \sum_{\mu_k = j} \mu_k$. Now $\mu_k = j$ if and only if $\mu_{j+1}^\# < k \le \mu_j^\#$, and hence

$$\sum_{k=1}^\infty \mu_k = \sum_{j=1}^\infty \sum_{k=\mu_{j+1}^\# + 1}^{\mu_j^\#} j = \sum_{j=1}^\infty j(\mu_j^\# - \mu_{j+1}^\#) = \sum_{j=1}^\infty \mu_j^\#.$$

If $\mu_{\ell+1} \ge j$, then

$$\begin{aligned} \mu_j(\ell)^\# =& \#\{k \ge 1 \mid \mu_k(\ell) \ge j\} = \#\{k \ge 1 \mid \mu_{k+\ell} \ge j\} = \\ & \#\{k \ge \ell + 1 \mid \mu_k \ge j\} = \#\{k \ge 1 \mid \mu_k \ge j\} - \ell = \mu_j^\# - \ell > 0. \end{aligned}$$

So in this case we proved (4.3). On the other hand, if $\mu_{\ell+1} < j$, then $\mu_1(\ell) < j$ and hence $\mu_j(\ell)^\# = 1$. Since in this case also $\mu_j^\# \le \ell$, again (4.3) holds true. □

The following result is the basis for the proof of Proposition 4.4.

Proposition 4.3. *Let* $(\mu_k)_{k=1}^\infty$ *and* $(\nu_k)_{k=1}^\infty$ *be index sequences with dual sequences* $(\mu_j^\#)_{j=1}^\infty$ *and* $(\nu_j^\#)_{j=1}^\infty$. *Then*

$$\max\{\mu_j, \nu_j\}^\# = \max\{\mu_j^\#, \nu_j^\#\} \quad j = 1, 2, \dots \ , \tag{4.4}$$

where the sequence $(\max\{\mu_j, \nu_j\}^\#)_{j=1}^\infty$ *is the dual sequence of the index sequence* $(\max\{\mu_k, \nu_k\})_{k=1}^\infty$.

PROOF. In order to prove (4.4) assume that $\max\{\mu_j^\#, \nu_j^\#\} = \ell$. Without loss of generality we may take $\nu_j^\# = \ell$ and $\mu_j^\# \le \ell$. Then $\nu_\ell \ge j$ by Proposition 4.1 (5), and $\nu_{\ell+1} < j$ and $\mu_{\ell+1} < j$ follow from Proposition 4.1 (4). Thus we see

$$\max\{\nu_{\ell+1}, \mu_{\ell+1}\} < j \le \max\{\nu_\ell, \mu_\ell\}.$$

Apply again Proposition 4.1 (3) to conclude that $(\max\{\mu_j, \nu_j\})^\# = \ell$. □

The next result will be used in the proof of Proposition IV.4.3.

Proposition 4.4. *Let $(\mu_k)_{k=1}^{\infty}$ and $(\nu_k)_{k=1}^{\infty}$ be index sequences. For $\ell \geq 1$ put*

$$\gamma_\ell = \sum_{k=1}^{\infty} \max\{\mu_k, \nu_{k+\ell-1}\}.$$

Then

(i) $\gamma_\ell \geq \gamma_{\ell+1}$;

(ii) $\gamma_\ell - \gamma_{\ell+1} \leq \nu_\ell$;

(iii) $\gamma_\ell - \gamma_{\ell+1} \geq \gamma_{\ell+1} - \gamma_{\ell+2}$.

PROOF. Put $\nu_k(\ell-1) = \nu_{k+\ell-1}$ for $k = 1, 2, \dots$ and $\ell \geq 1$. We have that $\gamma_\ell = \sum_{k=1}^{\infty} \max\{\mu_k, \nu_k(\ell-1)\}$. Use Proposition 4.3 and Proposition 4.2 to get

$$\begin{aligned}\gamma_\ell &= \sum_{k=1}^{\infty} \max\{\mu_k, \nu_k(\ell-1)\}^{\#} = \sum_{k=1}^{\infty} \max\{\mu_k^{\#}, \nu_k(\ell-1)^{\#}\} = \\ &= \sum_{k=1}^{\infty} \max\{\mu_k^{\#}, \nu_k^{\#} - (\ell-1)\}.\end{aligned}$$

Then

$$\gamma_\ell - \gamma_{\ell+1} = \sum_{k=1}^{\infty} \max\{\mu_k^{\#}, \nu_k^{\#} - (\ell-1)\} - \max\{\mu_k^{\#}, \nu_k^{\#} - \ell\}. \tag{4.5}$$

Clearly the terms on the right hand side are nonnegative, and hence (i) holds. Next

$$\begin{aligned}\gamma_\ell - \gamma_{\ell+1} &= \#\{k \mid \nu_k^{\#} \geq \ell\} - \{k \mid 0 \leq \nu_k^{\#} - \ell < \mu_k^{\#}\} \\ &= \nu_\ell - \#\{k \mid 0 \leq \nu_k^{\#} - \ell < \mu_k^{\#}\}.\end{aligned}$$

This shows that $\gamma_\ell - \gamma_{\ell+1} \leq \nu_\ell$. Finally, note that the terms on the right hand side of (4.5) differ by 1 if $\nu_k^{\#} - (\ell-1) > \mu_k^{\#}$, and hence

$$\gamma_\ell - \gamma_{\ell+1} = \#\{k \mid \nu_k^{\#} - (\ell-1) > \mu_k^{\#}\}.$$

This proves (iii). □

Notes

The theorems about full length blocks in Section 1 and their applications in Section 2 are improved versions of the analogous results in Gohberg-Kaashoek-Van Schagen [1]. In that paper the results on full length blocks appear as special cases of the general theory for arbitrary blocks. The results on operator pencils provide a special case of the Kronecker canonical form. The general theory of arbitrary blocks and pencils will be treated in Chapter VII. For another presentation of some of the results in Section 3 see Ferrer-Puerta [1]. Dual sequences play often, implicitly or explicitly, a role in eigenvalue multiplicity theorems. More results on dual sequences can be found in the book Marshall-Olkin [1].

Chapter IV
The Eigenvalue Completion Problem for Full Length Blocks

In this chapter we give the solution of the eigenvalue completion problem for full length blocks. The proof is based on a reduction to a matrix polynomial completion problem. The related full length restriction problem is also solved. Applications to matrix pencils and non-everywhere defined operators on finite dimensional spaces are included. Another treatment of the problem, in terms of a certain matrix equation, is also presented.

IV.1 Main theorems

Let $(B; I, Q)$ be a full length block with underlying space $\mathcal{X}$, i.e., $B : \operatorname{Im} Q \to \mathcal{X}$. Recall that an operator $A : \mathcal{X} \to \mathcal{X}$ is called a completion of $(B; I, Q)$ if $B = AQ|_{\operatorname{Im} Q}$. In this chapter we are interested in the eigenvalues of all possible completions A. From Theorem I.6.1 we know that, in order to solve this problem, we may replace $(B; I, Q)$ by a block similar block. From Theorem III.1.1 we see that without loss of generality we may assume that $(B; I, Q)$ is a direct sum of shifts of the third kind with indices $\mu_1, \ldots, \mu_m$ and a Jordan matrix. For example, $(B; I, Q)$ might be represented by the positioned submatrix

$$\begin{pmatrix} 0 & 0 & 0 & ? & & ? & & & \\ 1 & 0 & 0 & ? & & ? & & & \\ 0 & 1 & 0 & ? & & ? & & & \\ 0 & 0 & 1 & ? & & ? & & & \\ & & & ? & 0 & ? & & & \\ & & & ? & 1 & ? & & & \\ & & & ? & & ? & 2 & 1 & 0 \\ & & & ? & & ? & 0 & 2 & 0 \\ & & & ? & & ? & 0 & 0 & 1 \end{pmatrix}.$$

If the block $(B; I, Q)$ is in canonical form (as for example in the above picture), any completion of the block $(B; I, Q)$ is block upper triangular with the Jordan part in the right lower corner. It follows that *each eigenvalue of the Jordan part is an eigenvalue of the completion. Furthermore, we see that there are* $\sum_{i=1}^{m}(\mu_i + 1)$ *eigenvalues free to choose.* To describe the possible eigenvalues together with their multiplicities is a more difficult problem.

The next theorem gives for full length blocks the solution of the eigenvalue completion problem (Problem A in Section I.6).

Theorem 1.1. *Let $(B; I, Q)$ be a full length block on the n dimensional space $\mathcal{X}$, with indices of the third kind $\mu_1 \geq \cdots \geq \mu_m \geq 0$ and invariant polynomials $p_n | \cdots | p_1$. Consider polynomials $q_n | \cdots | q_1$ with $\sum_{i=1}^{n} \deg q_i = n$. Then there exists a completion A of the block $(B; I, Q)$ which has $q_1, \dots, q_n$ as invariant polynomials if and only if*

(1) $p_i | q_i$ *for* $i = 1, \dots, n$, *and* $q_{i+m} | p_i$ *for* $i = 1, \dots, n-m$;

(2) $\sum_{i=1}^{j} \deg s_i \geq \sum_{i=1}^{j} (\mu_i + 1)$ *for* $j = 1, \dots, m$, *with equality holding for* $j = m$.

The polynomials s_j in (2) are defined as $s_j = t_j / t_{j+1}$, for $j = 1, \dots, m$, where

$$t_j = \prod_{i=j}^{n} \text{l.c.m.}(p_{i-j+1}, q_i) \ .$$

The proof is complicated and will be given in the next sections. We first illustrate the result with an example. Consider the block $(B; I, Q)$ on $\mathbb{C}^9$ given by the positioned submatrix

$$\begin{pmatrix}
0 & 0 & 0 & ? & & ? & & & \\
1 & 0 & 0 & ? & & ? & & & \\
0 & 1 & 0 & ? & & ? & & & \\
0 & 0 & 1 & ? & & ? & & & \\
 & & & ? & 0 & ? & & & \\
 & & & ? & 1 & ? & & & \\
 & & & ? & & ? & 0 & 1 & 0 \\
 & & & ? & & ? & 0 & 0 & 0 \\
 & & & ? & & ? & 0 & 0 & 0
\end{pmatrix}.$$

Here the open places stand for zero entries and the question marks denote the unspecified entries. First we ask if it is possible to complete the block $(B; I, Q)$ to a matrix similar to the following matrix

$$A = \begin{pmatrix}
1 & 1 & & & & & & & \\
 & 1 & & & & & & & \\
 & & 1 & & & & & & \\
 & & & 1 & & & & & \\
 & & & & 0 & 1 & & & \\
 & & & & & 0 & & & \\
 & & & & & & 0 & 1 & \\
 & & & & & & & 0 & \\
 & & & & & & & & 0
\end{pmatrix}.$$

The blank spots in the matrix A denote zero entries. To answer the question we compute the invariant polynomials of the block (B, I, Q) and of the matrix A. The invariant polynomials of the block are $p_1(\lambda) = \lambda^2$, $p_2(\lambda) = \lambda$ and the other invariant polynomials of (B, I, Q) are of degree zero. The invariant polynomials of A are $q_1(\lambda) = \lambda^2(\lambda-1)^2$, $q_2(\lambda) = \lambda^2(\lambda-1)$, $q_3(\lambda) = \lambda(\lambda-1)$, and the other invariant polynomials of A are constants. First check condition (1) in Theorem 1.1. Notice that indeed $p_i | q_i$ for $i = 1, 2, 3, \ldots$. Next we compute that $m = 2$ and hence we have to verify that $q_3 | p_1$. This condition is violated and thus the answer to the question is negative: there does not exist a completion of (B, I, Q) similar to A.

Next, we ask if it is possible to complete (B, I, Q) to a matrix similar to

$$A' = \begin{pmatrix} 1 &&&&&&&& \\ & 1 &&&&&&& \\ && 0 & 1 &&&&& \\ &&& 0 &&&&& \\ &&&& 0 & 1 &&& \\ &&&&& 0 &&& \\ &&&&&& 0 & 1 & \\ &&&&&&& 0 & \\ &&&&&&&& 0 \end{pmatrix}.$$

Again blank spots in the matrix A' denote zero entries. The invariant polynomials of A' are $q_1(\lambda) = \lambda^2(\lambda-1)^2$, $q_2(\lambda) = \lambda^2(\lambda-1)$, $q_3(\lambda) = \lambda^2$, $q_4(\lambda) = \lambda$ and $q_i(\lambda) = 1$ for $i = 5, \ldots, 9$. Condition (1) of Theorem 1.1 is fulfilled. Let us compute the polynomials $t_i(\lambda)$. Because $p_i | q_i$ we get $t_1 = q_1 q_2 q_3 q_4$, and because $q_{i+2} | p_i$ we get $t_3 = p_1 p_2$. Furthermore, $t_i = t_3$ if $i \geq 3$. Finally, one checks that $t_2 = q_2 q_3 q_4$ and thus $s_1 = q_1$ and $s_2 = q_2$. Next we remark that the block (B, I, Q) is in canonical form, and hence we read of that the indices are $\mu_1 = 3$ and $\mu_2 = 1$. We find that $\deg s_1 = 3 < \mu_1 + 1$. So the condition (2) is not fulfilled. The conclusion is that there does not exist a completion of (B, I, Q) which is similar to A'.

An important special case occurs when there are no invariant polynomials for the block $(B; I, Q)$. Then Theorem 1.1 yields the following result.

Corollary 1.2. *Let $(B; I, Q)$ be a full length block on the n dimensional space $\mathcal{X}$, with indices of the third kind $\mu_1 \geq \cdots \geq \mu_m \geq 0$ and no invariant polynomials. Consider polynomials $q_n | \cdots | q_1$ with $\sum_{i=1}^n \deg q_i = n$. Then there exists a completion A of the block $(B; I, Q)$ which has $q_1, \ldots, q_n$ as invariant polynomials if and only if*

$$\sum_{i=1}^{j} \deg q_i \geq \sum_{i=1}^{j} (\mu_i + 1), \quad j = 1, \ldots, m, \tag{1.1}$$

with equality holding for $j = m$.

PROOF. First assume that (1.1) is fulfilled. It follows that $\deg q_i = 0$ if $i > m$. The block has no invariant polynomials. This means that all invariant polynomials of the block are constants. So $p_i | q_i$, and since $\deg q_i = 0$ for $i > m$, we also have $q_{i+m} | p_i$. Thus condition (1) of Theorem 1.1 is fulfilled. To check condition

(2) one observes that $t_j = \prod_{i=j}^{m} q_i$, and hence $s_j = q_j$. Therefore condition (2) reduces to the inequalities (1.1). Hence the polynomials $q_1, \dots, q_n$ are the invariant polynomials of a completion of $(B; I, Q)$.

Conversely, assume that the polynomials $q_n | \cdots | q_1$ are the invariant polynomials of a completion of $(B; I, Q)$. Since $s_j = q_j$, it follows that condition (2) reduces to (1.1). □

We will now give the solution of the eigenvalue restriction problem for full length blocks. Recall that a full length block $(B; I, Q)$ is called a *block restriction of* the operator $A : \mathcal{X} \to \mathcal{X}$ if $B = A|_{\operatorname{Im} Q}$.

Theorem 1.3. *Let $q_n | \cdots | q_1$ be the invariant polynomials of the operator $A : \mathcal{X} \to \mathcal{X}$. Let $\mu_1 \geq \cdots \geq \mu_m \geq 0$ be integers and $p_n | \cdots | p_1$ be polynomials. Then A has a block of full length with indices of the third kind $\mu_1 \geq \cdots \geq \mu_m$ and invariant polynomials $p_1, \dots, p_n$ if and only if*

(0) $\sum_{i=1}^{m} (\mu_i + 1) + \sum_{i=1}^{n} \deg p_i = n$;

(1) $p_i | q_i$, *for* $i = 1, \dots, n$, *and* $q_{i+m} | p_i$, *for* $i = 1, \dots, n - m$;

(2) $\sum_{i=1}^{j} \deg s_i \geq \sum_{i=1}^{j} (\mu_i + 1)$ *for* $j = 1, \dots, m$, *with equality holding for* $j = m$.

Here the polynomials s_j are defined by $s_j = t_j / t_{j+1}$ for $j = 1, \dots, m$, where

$$t_j = \prod_{i=j}^{n} \text{l.c.m.}(p_{i-j+1}, q_i) .$$

Proof. Assume that the conditions (0), (1), (2) are satisfied. Condition (0) guarantees that full length blocks $(B; I, Q)$ with indices of the third kind $\mu_1 \geq \cdots \geq \mu_m$ and invariant polynomials $p_1, \dots, p_n$ do exist. Let $(B; I, Q)$ be such a block. By Theorem 1.1 and conditions (1) and (2) the block $(B; I, Q)$ has a completion A' with invariant polynomials $q_1, \dots, q_n$. Then A' is similar to A and therefore there exists an invertible operator S such that $A'S = SA$. Put $Q'' = S^{-1}QS$. It follows that $(B; I, Q)$ is block similar to $(A|_{\operatorname{Im} Q''}; I, Q'')$, since $\operatorname{Im} Q = S[\operatorname{Im} Q'']$ and $BSx = A'Sx = SAx = SA|_{\operatorname{Im} Q''}x$ for each $x \in \operatorname{Im} Q''$. Thus we obtain that the operator A has a block restriction $(A|_{\operatorname{Im} Q''}; I, Q'')$ with indices of the third kind $\mu_1 \geq \cdots \geq \mu_m$ and invariant polynomials $p_1, \dots, p_n$.

Conversely, assume that A has a restriction $(A|_{\operatorname{Im} Q''}; I, Q'')$ with indices of the third kind $\mu_1 \geq \cdots \geq \mu_m$ and invariant polynomials $p_1, \dots, p_n$. By Theorem III.1.4 there exists a block similarity S of $(A|_{\operatorname{Im} Q''}; I, Q'')$ and $(B; I, Q)$. So $S[\operatorname{Im} Q''] = \operatorname{Im} Q$ and $(SA|_{\operatorname{Im} Q''} - BS)x = 0$ for each $x \in \operatorname{Im} Q''$. Put $A' = SAS^{-1}$. Then A' has invariant polynomials $q_1, \dots, q_n$ and for $x \in \operatorname{Im} Q$ we obtain that $A'x = A'SS^{-1}x = SAS^{-1}x = BSS^{-1}x = Bx$. This proves that A' is a completion of $(B; I, Q)$ with invariant polynomials $q_1, \dots, q_n$. □

IV.2 Reduction to a problem on matrix polynomials

An $m \times m$ matrix polynomial $D(\lambda) = (\delta_{ij}(\lambda))_{i,j=1}^{m}$ is called *degree diagonally dominated* if

$$\deg \delta_{ii} > \deg \delta_{ij}, \quad j \neq i, \quad j = 1, \ldots, m, \quad i = 1, \ldots, m. \tag{2.1}$$

The next theorem is the main result of this section. It tells us that the eigenvalue completion problem for a block $(B; I, Q)$ is equivalent to the problem of finding matrix polynomials $C(\lambda)$ and $D(\lambda)$ with certain properties.

Theorem 2.1. *Let $(B; I, Q)$ be a block on the n-dimensional space $\mathcal{X}$ with invariant polynomials $p_n|\cdots|p_1$ and indices of the third kind $\mu_1 \geq \cdots \geq \mu_m \geq 0$. Let $k = n - m - (\mu_1 + \cdots + \mu_m)$, and let M be a $k \times k$ matrix with invariant polynomials $p_n|\cdots|p_1$. Then there exists a completion A of $(B; I, Q)$ with invariant polynomials $q_n|\cdots|q_1$ if and only if there exists a $k \times m$ matrix polynomial $C(\lambda)$ and an $m \times m$ degree diagonally dominated matrix polynomial $D(\lambda) = (\delta_{ij}(\lambda))_{i,j=1}^{m}$, with $\deg \delta_{ii} = \mu_i + 1$ for $i = 1, \ldots, m$, such that*

$$\begin{pmatrix} M - \lambda I_k & C(\lambda) \\ 0 & D(\lambda) \end{pmatrix}$$

has invariant polynomials $q_{m+k}(\lambda)|\cdots|q_1(\lambda)$.

Before we proceed with the proof of Theorem 2.1, we recall that two matrix polynomials $X(\lambda)$ and $Y(\lambda)$ are called *polynomially equivalent* if there exist unimodular matrix polynomials $E(\lambda)$ and $F(\lambda)$ (i.e., matrix polynomials with constant non-zero determinants), such that $X(\lambda) = E(\lambda)Y(\lambda)F(\lambda)$. If $X(\lambda)$ and $Y(\lambda)$ are polynomially equivalent, then $X(\lambda)$ and $Y(\lambda)$ have the same invariant polynomials.

In the proof of Theorem 2.1 the following lemma will be used.

Lemma 2.2. *Let A be the $n \times n$ matrix*

$$A = \begin{pmatrix} N & 0 & \cdots & 0 & C_0 \\ 0 & J_1 & & 0 & C_1 \\ \vdots & & \ddots & & \vdots \\ 0 & 0 & & J_q & C_q \\ 0 & B_1 & \cdots & B_q & D \end{pmatrix} \tag{2.2}$$

where for $i = 1, \ldots, q$ the matrix J_i is a $\mu_i \times \mu_i$ lower triangular Jordan cell with zero eigenvalue, the $m \times \mu_i$ matrix B_i differs from the zero matrix only in the entry on place (i, μ_i), where B_i has an entry 1,

$$D = (d_{\ell j})_{\ell,j=1}^{m}, \quad C_i = \left(c_{\ell j}^{(i)}\right)_{\ell,j=1}^{\mu_i, m}, i = 1, \ldots, q. \tag{2.3}$$

Then the matrix polynomial $A - \lambda I_n$ is polynomially equivalent to the matrix polynomial

$$P(\lambda) = \begin{pmatrix} I & 0 & 0 \\ 0 & N - \lambda I & C_0 \\ 0 & 0 & D(\lambda) \end{pmatrix}, \tag{2.4}$$

where $D(\lambda) = (\delta_{ij}(\lambda))_{i,j=1}^{m}$ is given by

$$\delta_{ij}(\lambda) = \begin{cases} d_{ij}\lambda^{\mu_i} + \sum_{\ell=1}^{\mu_i} c_{\ell j}^{(i)} \lambda^{\ell-1}, & i \neq j \\ (d_{ii} - \lambda)\lambda^{\mu_i} + \sum_{\ell=1}^{\mu_i} c_{\ell i}^{(i)} \lambda^{\ell-1}, & i = j \end{cases}. \tag{2.5}$$

PROOF. The proof will consist of five steps.

STEP 1. In this step we show that $A - \lambda I$ is equivalent to a pencil

$$P_1(\lambda) = \begin{pmatrix} N - \lambda I & 0 & \cdots & 0 & C_0 \\ 0 & J_1(\lambda) & & 0 & T_1(\lambda) \\ \vdots & & \ddots & & \vdots \\ 0 & 0 & & J_q(\lambda) & T_q(\lambda) \\ 0 & B_1 & \cdots & B_q & D - \lambda I \end{pmatrix},$$

where the $\mu_i \times \mu_i$ matrix $J_i(\lambda)$ is given by

$$J_i(\lambda) = \begin{pmatrix} 0 & & & & -\lambda^{\mu_i} \\ 1 & 0 & & & \cdot \\ & \cdot & \ddots & & \vdots \\ & & \ddots & \ddots & \vdots \\ & & & \cdot\ 0 & \cdot \\ & & & 1 & -\lambda \end{pmatrix},$$

and $T_i(\lambda) = \left(t_{\ell j}^{(i)}(\lambda)\right)_{\ell,j=1}^{\mu_i,m}$, with $t_{1j}^{(i)}(\lambda) = \sum_{\ell=1}^{\mu_i} c_{\ell j}^{(i)} \lambda^{\ell-1}$ for $j = 1, \ldots, m$, and for $\ell > 1$ and $j = 1, \ldots, m$ the expression $t_{\ell j}^{(i)}(\lambda)$ is a polynomial of degree strictly less than μ_i.

To see this we put

$$S_i(\lambda) = \begin{pmatrix} 1 & \lambda & & \cdots & \lambda^{\mu_i - 1} \\ & 1 & \cdot & & \\ & & \ddots & \ddots & \vdots \\ & & & \ddots & \cdot \\ & & & 1 & \lambda \\ & & & & 1 \end{pmatrix}.$$

Remark that $J_i(\lambda) = S_i(\lambda)(J_i - \lambda I)$, and put $T_i(\lambda) = S_i(\lambda)C_i$. Then $T_i(\lambda)$ has the described properties. So

$$P_1(\lambda) = \begin{pmatrix} I & & & & \\ & S_1(\lambda) & & & \\ & & \ddots & & \\ & & & S_q(\lambda) & \\ & & & & I \end{pmatrix} (A - \lambda I) \tag{2.6}$$

establishes the desired equivalence.

STEP 2. We show that $A - \lambda I$ is polynomially equivalent with the matrix polynomial

$$P_2(\lambda) = \begin{pmatrix} N - \lambda I & 0 & \cdots & 0 & C_0 \\ 0 & J_1(\lambda) & & 0 & T_{10}(\lambda) \\ \vdots & & \ddots & & \vdots \\ 0 & 0 & & J_q(\lambda) & T_{q0}(\lambda) \\ 0 & B_1 & \cdots & B_q & D - \lambda I \end{pmatrix},$$

with

$$T_{i0}(\lambda) = \begin{pmatrix} t_{11}^{(i)}(\lambda) & \cdots & t_{1m}^{(i)}(\lambda) \\ 0 & \cdots & 0 \\ \vdots & & \vdots \\ 0 & \cdots & 0 \end{pmatrix}$$

and $t_{1j}^{(i)}(\lambda) = \sum_{\ell=1}^{\mu_i} c_{\ell j}^{(i)} \lambda^{\ell-1}$.

To see this put $X_i(\lambda) = \left(x_{\ell j}^{(i)}(\lambda)\right)_{\ell,j=1}^{\mu_i,m}$, with $x_{\ell j}^{(i)}(\lambda) = -t_{\ell+1\ j}^{(i)}(\lambda)$ for $\ell = 1, \ldots, \mu_i - 1$ and $x_{\mu_i\ j}^{(i)} = 0$, for $j = 1, \ldots, m$. Here $t_{\ell+1\ j}^{(i)}(\lambda)$ are the entries of the matrices $T_i(\lambda)$. Then

$$P_2(\lambda) = P_1(\lambda) \begin{pmatrix} I & 0 & \cdots & 0 & 0 \\ 0 & I & & 0 & X_1(\lambda) \\ \vdots & & \ddots & & \vdots \\ 0 & 0 & & I & X_q(\lambda) \\ 0 & 0 & \cdots & 0 & I \end{pmatrix}.$$

In combination with (2.6) this yields the desired equivalence.

STEP 3. We show that $A - \lambda I$ is polynomially equivalent with the matrix polynomial

$$P_3(\lambda) = \begin{pmatrix} N - \lambda I & 0 & \cdots & 0 & C_0 \\ 0 & J_1 & & 0 & D_1(\lambda) \\ \vdots & & \ddots & & \vdots \\ 0 & 0 & & J_q & D_q(\lambda) \\ 0 & B_1 & \cdots & B_q & D - \lambda I \end{pmatrix},$$

where

$$D_i(\lambda) = \begin{pmatrix} d_{i1}(\lambda) & \cdots & d_{im}(\lambda) \\ 0 & \cdots & 0 \\ \vdots & & \vdots \\ 0 & \cdots & 0 \end{pmatrix}$$

Here $d_{ij}(\lambda) = d_{ij}\lambda^{\mu_i} + t_{1j}^{(i)}(\lambda)$ for $i \neq j$ and $d_{ii}(\lambda) = (d_{ii} - \lambda)\lambda^{\mu_i} + t_{1i}^{(i)}(\lambda)$ with $t_{1j}^{(i)}(\lambda)$ as defined in step 2 and d_{ij} the entries of the matrix D.

To obtain this result we put $Y_i(\lambda)$ to be the $\mu_i \times m$ matrix with all entries equal to 0 except for the entry on place $(1, i)$ where there is an entry λ^{μ_i}. Put

$$Z_i(\lambda) = \begin{pmatrix} 1 & & & \lambda^{\mu_i - 1} \\ & \ddots & & \vdots \\ & & 1 & \lambda \\ & & 0 & 1 \end{pmatrix} .$$

Then

$$P_3(\lambda) = \begin{pmatrix} I & 0 & \cdots & 0 & 0 \\ 0 & I & & 0 & Y_1(\lambda) \\ \vdots & & \ddots & & \vdots \\ 0 & 0 & & I & Y_q(\lambda) \\ 0 & 0 & \cdots & 0 & I \end{pmatrix} P_2(\lambda) \begin{pmatrix} I & 0 & \cdots & 0 & 0 \\ 0 & Z_1(\lambda) & & 0 & 0 \\ \vdots & & \ddots & & \vdots \\ 0 & 0 & & Z_q(\lambda) & 0 \\ 0 & 0 & \cdots & 0 & I \end{pmatrix} .$$

STEP 4. We show that $A - \lambda I$ is polynomially equivalent with the matrix polynomial

$$P_4(\lambda) = \begin{pmatrix} N - \lambda I & 0 & \cdots & 0 & C_0 \\ 0 & J_1 & & 0 & D_1(\lambda) \\ \vdots & & \ddots & & \vdots \\ 0 & 0 & & J_q & D_q(\lambda) \\ 0 & B_1 & \cdots & B_q & D_0(\lambda) \end{pmatrix} ,$$

where $D_0(\lambda)$ partitions as

$$D_0(\lambda) = \begin{pmatrix} 0 & 0 \\ D_{21} & D_{22} - \lambda I_{m-q} \end{pmatrix} .$$

We put $U_i(\lambda)$ the $\mu_i \times m$ matrix with all entries equal to 0 except for the entries in the last row, where we insert the i-th row of $D - \lambda I$. Then

$$P_4(\lambda) = P_3(\lambda) \begin{pmatrix} I & 0 & \cdots & 0 & 0 \\ 0 & I & & 0 & -U_1(\lambda) \\ \vdots & & \ddots & & \vdots \\ 0 & 0 & & I & -U_q(\lambda) \\ 0 & 0 & \cdots & 0 & I \end{pmatrix} .$$

STEP 5. We show that $A - \lambda I$ is polynomially equivalent with the $n \times n$ matrix polynomial $P(\lambda)$ of the form (2.4).

To achieve this take $P_4(\lambda)$ and first reorder the rows to get

$$\begin{pmatrix} N-\lambda I & 0 & \cdots & 0 & C_0 \\ 0 & I & & 0 & 0 \\ \vdots & & \ddots & & \vdots \\ 0 & 0 & & I & 0 \\ 0 & 0 & \cdots & 0 & D(\lambda) \end{pmatrix}.$$

Then apply a similarity by a permutation matrix to obtain the result. □

PROOF OF THEOREM 2.1. First we fix a basis for the space $\mathcal{X}$. We apply Proposition III.1.3 to the block $(B; I, Q)$. We define the number q by $\mu_1 \geq \cdots \geq \mu_q > 0 = \mu_{q+1} = \cdots = \mu_m$ and choose a basis $g_1, \ldots, g_k$ for the residual subspace $\mathcal{D}_\infty$. Thus we obtain a basis of the space $\mathcal{X}$

$$g_1, \ldots, g_k, f_{11}, \ldots, f_{\mu_1+1\ 1}, \ldots, f_{1q}, \ldots, f_{\mu_q+1\ q}, f_{1\ q+1}, \ldots, f_{1m} \ ,$$

such that

$$g_1, \ldots, g_k, f_{11}, \ldots, f_{\mu_1\ 1}, \ldots, f_{1q}, \ldots, f_{\mu_q\ q}, f_{1\ q+1}, \ldots, f_{1m}$$

is a basis for $\operatorname{Im} Q$, and with respect to these bases, the matrix of $B : \operatorname{Im} Q \to \mathcal{X}$ has the following $(q+2) \times (q+1)$ block matrix representation:

$$B = \begin{pmatrix} N & 0 & \cdots & 0 \\ 0 & S_1 & & 0 \\ \vdots & & \ddots & \\ 0 & 0 & \cdots & S_q \\ O_0 & O_1 & \cdots & O_q \end{pmatrix},$$

with N a square $k \times k$ matrix, the $(\mu_i + 1) \times \mu_i$ matrix S_i given by

$$S_i = \begin{pmatrix} 0 & & & & \\ 1 & 0 & & & \\ & \cdot & \ddots & & \\ & & \ddots & \ddots & \\ & & & \cdot & 0 \\ & & & & 1 \end{pmatrix},$$

with O_0 the $(m-q) \times k$ zero matrix and O_j the $(m-q) \times \mu_j$ zero matrix for $j = 1, \ldots, q$. We reorder this basis to

$$g_1, \ldots, g_k, f_{11}, \ldots, f_{\mu_1\ 1}, \ldots, f_{1q}, \ldots, f_{\mu_q\ q}, f_{\mu_1+1\ 1}, \ldots, f_{\mu_q+1\ q}, f_{1\ q+1}, \ldots, f_{1m} \ . \tag{2.7}$$

Assume that A is a completion of $(B; I, Q)$. Then the matrix of A with respect to the basis (2.7), has the form (2.2). We apply Lemma 2.2 to obtain that A has invariant polynomials $q_1(\lambda), \ldots, q_n(\lambda)$, if and only if the matrix polynomial given by (2.4) has invariant polynomials $q_1(\lambda), \ldots, q_n(\lambda)$, and hence

$$P(\lambda) = \begin{pmatrix} N - \lambda I & C_0 \\ 0 & D(\lambda) \end{pmatrix},$$

has invariant polynomials $q_1(\lambda), \ldots, q_{m+k}(\lambda)$ and $q_{m+k+1}(\lambda) = \cdots = q_n(\lambda) = 1$.

Since N and M have the same invariant polynomials and the same order, there exists an invertible matrix S such that $M = SNS^{-1}$. Put $C(\lambda) = SC_0$. Then

$$\begin{pmatrix} M - \lambda I_k & C(\lambda) \\ 0 & D(\lambda) \end{pmatrix} \tag{2.8}$$

has the desired properties.

Conversely, assume that the $k \times m$ matrix polynomial $C(\lambda)$ and the $m \times m$ matrix polynomial $D(\lambda) = (\delta_{ij}(\lambda))_{i,j=1}^m$ with $\deg \delta_{ii} = \mu_i + 1$ for $i = 1, \ldots, m$ and $\deg \delta_{ij} \leq \mu_i$ for $j \neq i$, $j = 1, \ldots, m$ and $i = 1, \ldots, m$ are given such that the matrix polynomial (2.8) has invariant polynomials $q_1(\lambda), \ldots, q_{m+k}(\lambda)$. Take S such that $N = S^{-1}MS$. Then

$$\begin{pmatrix} N - \lambda I_k & S^{-1}C(\lambda) \\ 0 & D(\lambda) \end{pmatrix}$$

has invariant polynomials $q_1(\lambda), \ldots, q_{m+k}(\lambda)$. Put $S^{-1}C(\lambda) = C_0 + (N - \lambda I_k)R(\lambda)$. Then

$$\begin{pmatrix} N - \lambda I_k & C_0 \\ 0 & D(\lambda) \end{pmatrix} = \begin{pmatrix} N - \lambda I_k & S^{-1}C(\lambda) \\ 0 & D(\lambda) \end{pmatrix} \begin{pmatrix} I & -R(\lambda) \\ 0 & I \end{pmatrix}$$

and thus we see that

$$\begin{pmatrix} N - \lambda I_k & C_0 \\ 0 & D(\lambda) \end{pmatrix}$$

has invariant polynomials $q_1(\lambda), \ldots, q_{m+k}(\lambda)$. The form of $D(\lambda)$ is such that the entries of $D(\lambda)$ can be represented in the form (2.5). Now (2.3) can be used to construct matrices C_i, $i = 1, \ldots, q$, and D and hence a matrix A in the form (2.2) is constructed. According to Lemma 2.2 the matrix polynomial $A - \lambda I_n$ is then polynomially equivalent to the matrix polynomial $P(\lambda)$ given by (2.4). Thus A is the matrix of a completion of $(B; I, Q)$ with respect to the basis (2.7), and A has invariant polynomials $q_1(\lambda), \ldots, q_n(\lambda)$ with $q_{m+k+1}(\lambda) = \cdots = q_n(\lambda) = 1$. □

IV.3 A one column completion problem for matrix polynomials

The result of the previous section shows that solving the eigenvalue completion problem is equivalent to the construction of matrix polynomials with certain properties. The existence and construction of such matrix polynomials will depend on the following result.

Theorem 3.1. *Let $M(\lambda)$ be a $m \times k$ matrix polynomial with $k < m$ and invariant polynomials $p_k | \cdots | p_1$. Then there exists an $m \times 1$ matrix polynomial $M_1(\lambda)$ such that $q_{k+1} | \cdots | q_1$ are the invariant polynomials of $\begin{pmatrix} M(\lambda) & M_1(\lambda) \end{pmatrix}$ if and only if $p_i | q_i$ and $q_{i+1} | p_i$, for $i = 1, \dots, k$.*

PROOF. PART 1: NECESSITY. Write $A(\lambda) = \begin{pmatrix} M(\lambda) & M_1(\lambda) \end{pmatrix}$. Assume that $M_1(\lambda) = (c_i(\lambda))_{i=1}^m$. Since we are only interested in the invariant polynomials of $A(\lambda)$ and $M(\lambda)$, we may multiply $A(\lambda)$ with matrix polynomials with constant nonzero determinants from the left and the right freely. Therefore we may assume that $M(\lambda)$ is in Smith canonical form. Furthermore, by applying elementary row operations and the Euclidean algorithm, we can achieve that the column $M_1(\lambda) = (a_i(\lambda))_{i=1}^m$ is such that $a_i(\lambda) = c_i(\lambda)$ for $i = 1, \dots, k$, $a_{k+1}(\lambda) = \text{g.c.d.}\{c_i(\lambda) \mid i = k+1, \dots, m\}$, and $a_i(\lambda) = 0$ for $i = k+2, \dots, m$. Put $a_0(\lambda) = a_{k+1}(\lambda)$. Then

$$A(\lambda) = \begin{pmatrix} M(\lambda) & M_1(\lambda) \end{pmatrix} = \begin{pmatrix} p_1(\lambda) & & 0 & a_1(\lambda) \\ & \ddots & & \vdots \\ 0 & & p_k(\lambda) & a_k(\lambda) \\ 0 & \dots & 0 & a_0(\lambda) \\ 0 & \dots & 0 & 0 \\ \vdots & & \vdots & \vdots \\ 0 & \dots & 0 & 0 \end{pmatrix}.$$

We may assume that $a_i(\lambda) \not\equiv 0$ for $i = 0, \dots, k$. (If necessary add a multiple of p_i to a_i.) Let $\lambda_0 \in \mathbb{C}$, and write

$$q_i(\lambda) = (\lambda - \lambda_0)^{\beta_i} t_i(\lambda), \quad p_i(\lambda) = (\lambda - \lambda_0)^{\rho_i} s_i(\lambda), \quad a_i(\lambda) = (\lambda - \lambda_0)^{\alpha_i} b_i(\lambda),$$

with $t_i(\lambda)$, $s_i(\lambda)$ and $b_i(\lambda)$ polynomials which are nonzero at λ_0. The product $q_{j+1} \cdots q_{k+1}$, $j \leq k$, is the greatest common divisor of all determinants of submatrices of order $k+1-j$ of $A(\lambda)$. Each of these determinants is just a product of entries of $A(\lambda)$ (and not a sum of products). So $q_{j+1} \cdots q_{k+1}$ is the g.c.d. of a certain set of polynomials of the type $(\lambda - \lambda_0)^\gamma c(\lambda)$ with $c(\lambda_0) \neq 0$. To compute the multiplicity of the factor $(\lambda - \lambda_0)$ in this g.c.d. it is sufficient to look at the multiplicity of the factor $\lambda - \lambda_0$ for each of the products in the set. This set of multiple factors $\lambda - \lambda_0$ is precisely the set of determinants of the submatrices of order $k+1-j$ of the matrix

$$A_0(\lambda) = \begin{pmatrix} (\lambda-\lambda_0)^{\rho_1} & & 0 & (\lambda-\lambda_0)^{\alpha_1} \\ & \ddots & & \vdots \\ 0 & & (\lambda-\lambda_0)^{\rho_k} & (\lambda-\lambda_0)^{\alpha_k} \\ 0 & \dots & 0 & (\lambda-\lambda_0)^{\alpha_0} \\ 0 & \dots & 0 & 0 \\ \vdots & & \vdots & \vdots \\ 0 & \dots & 0 & 0 \end{pmatrix}.$$

Therefore we compute the invariant polynomials of $A_0(\lambda)$ to obtain the exponents β_i. To see that $p_i|q_i$ and $q_{i+1}|p_i$ it is sufficient to see for an arbitrary λ_0 that $\rho_i \le \beta_i$ and $\beta_{i+1} \le \rho_i$.

Before we compute the invariant polynomials of $A_0(\lambda)$ we apply some elementary row and column operations on $A_0(\lambda)$ in order to get that

$$0 \le \alpha_i \le \rho_i, \quad 0 \le \alpha_i \le \alpha_0, \quad i = 1, \dots, k. \tag{3.1}$$

Note that taking the greatest common divisor of a set of monomials amounts to taking the minimal exponent among the occurring exponents. We compute the exponent of the greatest common divisor of the determinants of the $(k-j+1) \times (k-j+1)$ submatrices of $A_0(\lambda)$. We neglect those determinants which are, in view of (3.1), obvious multiples of others and obtain

$$\beta_{j+1} + \cdots + \beta_{k+1} = \min\Big\{\rho_{j+1} + \cdots + \rho_k + \min\{\alpha_1, \dots, \alpha_j\}, \gamma_{j+1}\Big\}. \tag{3.2}$$

Here γ_j is defined by

$$\gamma_j = \min\{\rho_{j-1} + \cdots + \rho_k + (\alpha_l - \rho_l) \mid l = j, \dots, k\}$$

for $j = 2, \dots, k$. Thus we also have

$$\beta_j + \cdots + \beta_{k+1} = \min\{\rho_j + \cdots + \rho_k + \min\{\alpha_1, \dots, \alpha_{j-1}\}, \gamma_j\} \tag{3.3}$$

for $j = 2, \dots, k+1$. We note that

$$\rho_j + \cdots + \rho_k + \min\{\alpha_1, \dots, \alpha_{j-1}\} = \rho_j + (\rho_{j+1} + \cdots + \rho_k + \min\{\alpha_1, \dots, \alpha_{j-1}\}),$$

and

$$\gamma_j = \rho_{j-1} + \min\{\rho_{j+1} + \cdots + \rho_k + \alpha_j, \gamma_{j+1}\}.$$

Then we substitute these equalities in (3.3) and obtain

$$\begin{aligned} \beta_j + \cdots + \beta_{k+1} = \\ \min\Big\{&\rho_j + (\rho_{j+1} + \cdots + \rho_k + \min\{\alpha_1, \dots, \alpha_{j-1}\}), \\ &\rho_{j-1} + \min\{\rho_{j+1} + \cdots + \rho_k + \alpha_j, \gamma_{j+1}\}\Big\} \end{aligned} \tag{3.4}$$

Use $\rho_{j-1} \geq \rho_j$ to see that

$$\beta_j + \cdots + \beta_{k+1} \leq \rho_{j-1} + \min\left\{\rho_{j+1} + \cdots + \rho_k + \min\{\alpha_1, \ldots, \alpha_j\}, \gamma_{j+1}\right\}$$

Compare this with (3.2) and note that it follows that $\beta_j \leq \rho_{j-1}$. This holds for $j = 2, \ldots, k+1$. Again from (3.4) and $\rho_{j-1} \geq \rho_j$ we see that

$$\beta_j + \cdots + \beta_{k+1} \geq \rho_j + \min\left\{\rho_{j+1} + \cdots + \rho_k + \min\{\alpha_1, \ldots, \alpha_j\}, \gamma_{j+1}\right\}$$

Combined with (3.2) this proves that $\beta_j \geq \rho_j$ for $j = 2, \ldots, k$. It remains to prove that $\beta_1 \geq \rho_1$. Note that

$$\beta_1 + \cdots + \beta_{k+1} = \alpha_0 + \rho_1 + \cdots + \rho_k$$

and

$$\beta_2 + \cdots + \beta_{k+1} = \min\{\rho_2 + \cdots + \rho_k + \alpha_1, \gamma_2\}$$

and therefore

$$\beta_2 + \cdots + \beta_{k+1} = \rho_1 + \cdots + \rho_k + \min\{(\alpha_l - \rho_l) \mid l = 1, \ldots, k\}.$$

We observe that

$$\begin{aligned} \beta_1 = \alpha_0 - \min\{(\alpha_l - \rho_l) \mid l = 1, \ldots, k\} = \\ \alpha_0 + \max\{(\rho_l - \alpha_l) \mid l = 1, \ldots, k\} = \\ \max\{\rho_l + (\alpha_0 - \alpha_l) \mid l = 1, \ldots, k\} \geq \rho_1 \end{aligned}$$

since $\alpha_i \leq \alpha_0$ for all l.

PART 2: SUFFICIENCY. Let

$$A(\lambda) = \begin{pmatrix} p_1(\lambda & & 0 & a_1(\lambda) \\ \vdots & \ddots & & \vdots \\ 0 & & p_k(\lambda) & a_k(\lambda) \\ 0 & \ldots & 0 & a_0(\lambda) \end{pmatrix},$$

with

$$a_i(\lambda) = \frac{q_{i+1}(\lambda) \cdots q_{k+1}(\lambda)}{p_{i+1}(\lambda) \cdots p_k(\lambda)}, \quad i = 0, \ldots, k.$$

Then $a_i(\lambda)$ is a polynomial. The matrix polynomial $A(\lambda)$ has invariant polynomials $q_{k+1} | \cdots | q_1$. We will show this. Note that $a_i | a_{i-1}$ for $i = 1, \ldots, k$. We compute the greatest common divisor x_j the determinants of all $(k+1-j) \times (k+1-j)$

submatrices of $A(\lambda)$. In the computation we leave out all those determinants that are obviously multiples of other determinants. We obtain, for $j = 1, \ldots, k$, that

$$\begin{aligned} x_j &= \text{g.c.d.}\{p_j \cdots p_k a_i / p_i \mid i = j, \ldots, k\} \\ &= \text{g.c.d.}\{p_j \cdots p_{i-1} q_{i+1} \ldots q_{k+1} \mid i = j, \ldots, k\}. \end{aligned}$$

If we put $p_0 = 1$, then the formula also holds for $j = 0$. Now use that $q_{\ell+1} | p_\ell$, for $\ell = 1, \ldots, k$ to see that $p_j \cdots p_{i-1} q_{i+1} \cdots q_{k+1}$ is divisible by $q_{j+1} \cdots q_{k+1}$. Remark that if $i = j$, then $q_{j+1} \cdots q_{k+1} = p_j \cdots p_{i-1} q_{i+1} \cdots q_{k+1}$. This proves that $x_j = q_{j+1} \cdots q_{k+1}$, and therefore $A(\lambda)$ has invariant polynomials $q_{k+1} | \cdots | q_1$. Next note that

$$M(\lambda) = E(\lambda) \begin{pmatrix} p_1(\lambda & & 0 \\ \vdots & \ddots & \vdots \\ 0 & & p_k(\lambda) \\ 0 & \ldots & 0 \\ \vdots & & \vdots \\ 0 & \ldots & 0 \end{pmatrix} F(\lambda),$$

with $E(\lambda)$ and $F(\lambda)$ matrix polynomials with constant nonzero determinant. Put

$$(\, M(\lambda) \quad M_1(\lambda)\,) = E(\lambda) \begin{pmatrix} p_1(\lambda & & 0 & a_1(\lambda) \\ \vdots & \ddots & & \vdots \\ 0 & & p_k(\lambda) & a_k(\lambda) \\ 0 & \ldots & 0 & a_0(\lambda) \\ 0 & \ldots & 0 & 0 \\ \vdots & & & \vdots \\ 0 & \ldots & 0 & 0 \end{pmatrix} \begin{pmatrix} F(\lambda) & 0 \\ 0 & 1 \end{pmatrix}.$$

The invariant polynomials of this polynomial are the invariant polynomials of the middle factor on the right hand side. So the polynomials $q_{k+1} | \cdots | q_1$ are the invariant polynomials of $(\, M(\lambda) \quad M_1(\lambda)\,)$. □

IV.4 Proof of the first main theorem

In this section we prove Theorem 1.1. The proof will be based on the following result.

Theorem 4.1. *Let $p_k | \cdots | p_1$ be the invariant polynomials of the matrix pencil $N - \lambda I_k$, and put $p_j = 1$ if $j > k$. Let $0 \le \nu_1 \le \cdots \le \nu_m$. There exists an $(m+k) \times (m+k)$ matrix polynomial*

$$P(\lambda) = \begin{pmatrix} N - \lambda I_k & C(\lambda) \\ 0 & D(\lambda) \end{pmatrix},$$

with

(a) $D(\lambda) = (d_{ij}(\lambda))_{i,j=1}^{m}$, *a degree diagonally dominated matrix polynomial, such that for* $i = 1, \dots, m$ *the diagonal element* $d_{ii}(\lambda)$ *has degree* $\nu_i + 1$,

(b) *the invariant polynomials of* $P(\lambda)$ *equal to* $q_{m+k}| \cdots |q_1$,

if and only if the following two conditions hold true:

(1) $p_i|q_i$ *for* $i = 1, \dots, m+k$, *and* $q_{i+m}|p_i$ *for* $i = 1, \dots, k$;

(2) $\sum_{i=1}^{j} \deg v^{(i)} \leq \sum_{i=1}^{j} (\nu_i + 1)$ *for* $j = 1, \dots, m$, *with equality holding for* $j = m$.

The polynomials $v^{(i)}$ *in* (2) *are defined as* $v^{(j)} = w_j / w_{j-1}$, *for* $j = 1, \dots, m$, *where*

$$w_j = \prod_{i=1}^{k+j} \text{l.c.m.}\,(p_i, q_{i+m-j}), \quad j = 0, \dots, m.$$

The next lemma is a key result about degree diagonally dominated matrix polynomials. It will play an important role in the proof of the sufficiency of the conditions (1) and (2) in Theorem 4.1.

Lemma 4.2. *Let* $D(\lambda) = (d_{ij}(\lambda))_{i,j=1}^{m}$ *be a degree diagonally dominated matrix polynomial such that* $\deg d_{ii}(\lambda) = \kappa_i$ *for* $i = 1, \dots, m$. *If* $\kappa_p + 1 \leq \kappa_q - 1$, *then there exists a degree diagonally dominated matrix polynomial* $E(\lambda) = (e_{ij}(\lambda))_{i,j=1}^{m}$, *polynomially equivalent with* $D(\lambda)$, *and such that* $\deg e_{ii} = \sigma_i$ *for* $i = 1, \dots, m$. *Here* $\sigma_i = \kappa_i$ *for* $i \neq p$ *and* $i \neq q$ *and* $\sigma_p = \kappa_p + 1$, $\sigma_q = \kappa_q - 1$.

PROOF. We shall show how to construct $E(\lambda)$ from $D(\lambda)$ by applying elementary row and column operations and multiplications by constant matrices. First add λ times column p to column q. We obtain $D_1(\lambda) = \left(d_{ij}^{(1)}(\lambda)\right)_{i,j=1}^{m}$. Then $\deg d_{pq}^{(1)} = \kappa_p + 1$. Let α_{pq} be the leading coefficient of $d_{pq}^{(1)}(\lambda)$, and let α_{qq} be the coefficient of λ^{κ_q} in $d_{qq}^{(1)}(\lambda)$. Now subtract $(\alpha_{qq}/\alpha_{pq})\lambda^{\kappa_q - \kappa_p - 1}$ times row p from row q to obtain $D_2(\lambda) = \left(d_{ij}^{(2)}(\lambda)\right)_{i,j=1}^{m}$. Then

$$\deg d_{pq}^{(2)} = \kappa_p + 1, \quad \deg d_{pp}^{(2)} = \kappa_p, \quad \deg d_{qq}^{(2)} \leq \kappa_q - 1, \quad \deg d_{qp}^{(2)} \leq \kappa_q - 1,$$

$$\deg d_{pj}^{(2)} = \kappa_p - 1, \quad \deg d_{qj}^{(2)} \leq \kappa_q - 2, \quad \deg d_{iq}^{(2)} \leq \kappa_i \quad \text{for all } i, j \neq p, q.$$

Also $\deg d_{ij}^{(2)} < \deg d_{ii}^{(2)} = \kappa_i$ for $i \neq p, q$ and $j \neq q, i$. So the degree of the i-th row in $D_2(\lambda)$ is at most σ_i. We know that $\deg \det D_2(\lambda) = \deg \det D(\lambda)$, and thus $\deg \det D_2(\lambda) = \sum_{i=1}^{m} \sigma_i$. Consider the matrix

$$K = \lim_{\lambda \to \infty} \operatorname{diag}(\lambda^{-\sigma_i}) D_2(\lambda).$$

The matrix K is invertible since $\det K$ is the leading coefficient of $\det D_2(\lambda)$. Put $E(\lambda) = D_2(\lambda) K^{-1}$. From $\lim_{\lambda \to \infty} \operatorname{diag}(\lambda^{-\sigma_i}) E(\lambda) = I_m$ it follows that $\deg e_{ii} = \sigma_i$ and $\deg e_{ij} < \sigma_i$ for $i \neq j$. □

PROOF OF THEOREM 4.1. We divide the proof into several parts.

PART 1. We first prove the necessity of (1). Put $n = m + k$. Put $P_j(\lambda)$ be the submatrix of $P(\lambda)$ that consists of the first $k + j$ columns of $P(\lambda)$. Let $u_n^{(j)}|\cdots|u_1^{(j)}$ be the invariant polynomials of $P_j(\lambda)$. From Theorem 3.1 we obtain for $j = 0, \dots, m-1$ that $u_i^{(j)}|u_i^{(j+1)}$ for $i = 1, \dots, n$ and $u_{i+1}^{(j+1)}(\lambda)|u_i^{(j)}(\lambda)$ for $i = 1, \dots, n-1$. So $p_i = u_i^{(0)}|u_i^{(m)} = q_i$ for $i = 1, \dots, n$ and $q_{i+m} = u_{i+m}^{(m)}|u_i^{(0)} = p_i$ for $i = 1, \dots, k$. This proves (1).

PART 2. Next we prove the necessity of (2). For $0 \leq j \leq m$ we obtain that $q_{i+m-j} = u_{i+m-j}^{(j+m-j)}|u_i^{(j)}$ and $p_i|u_i^{(j)}$. Put

$$v_i^{(j)} = \text{l.c.m.}(p_i, q_{i+m-j}), \quad i = 1, \dots, k+j. \tag{4.1}$$

Then $v_i^{(j)}|u_i^{(j)}$. It follows from condition (1) that $v_i^{(0)} = \text{l.c.m.}(p_i, q_{i+m}) = p_i$ and

$$v_i^{(m)} = \text{l.c.m.}(p_i, q_i) = q_i. \tag{4.2}$$

Put

$$u^{(j)} = \frac{\prod_{i=1}^{k+j} u_i^{(j)}}{\prod_{i=1}^{k+j-1} u_i^{(j-1)}}, \quad v^{(j)} = \frac{\prod_{i=1}^{k+j} v_i^{(j)}}{\prod_{i=1}^{k+j-1} v_i^{(j-1)}}.$$

Then

$$\prod_{i=1}^{j} v^{(i)} = \frac{\prod_{i=1}^{k+j} v_i^{(j)}}{\prod_{i=1}^{k} p_i}, \quad \prod_{i=1}^{j} u^{(i)} = \frac{\prod_{i=1}^{k+j} u_i^{(j)}}{\prod_{i=1}^{k} p_i}.$$

So

$$\left(\prod_{i=1}^{j} v^{(i)}\right) \Big| \left(\prod_{i=1}^{j} u^{(i)}\right).$$

We are interested in the degree of the left hand side of the latter expression. Let us compute the degree of the right hand side. We have

$$\deg \prod_{i=1}^{j} u^{(i)} = \deg \prod_{i=1}^{k+j} u_i^{(j)} - k. \tag{4.3}$$

The first term in the right hand side of (4.3) is the product of the invariant polynomials of $P_j(\lambda)$. This product is the greatest common divisor of all determinants of $(k+j) \times (k+j)$ submatrices of $P_j(\lambda)$. In particular, it is a divisor of the determinant of the submatrix of $P_j(\lambda)$ that consists of the first $k+j$ rows. Now this determinant has a degree equal to $k + \sum_{i=1}^{j}(\nu_i + 1)$, since it is the product of $\det(N - \lambda I_k)$ and the determinant of the left upper $j \times j$ submatrix of $D(\lambda)$. From (4.3) we obtain

$$\deg \prod_{i=1}^{j} u^{(i)} \leq \sum_{i=1}^{j} (\nu_i + 1).$$

So certainly

$$\deg \prod_{i=1}^{j} v^{(i)} \leq \sum_{i=1}^{j} (\nu_i + 1).$$

PART 3. In this part we construct a matrix polynomial with invariant polynomials $q_n|\cdots|q_1$. Let $v_i^{(j)}$ be given by formula (4.1). Condition (1) implies that $v_{i+1}^{(j)}|v_i^{(j)}$ for $i = 1, \ldots, k+j-1$, and $v_i^{(j)}|v_i^{(j+1)}$ and $v_{i+1}^{(j+1)}|v_i^{(j)}$ for $i = 1, \ldots, k+j$. Assume that we have constructed a square matrix polynomial

$$Q_{j-1}(\lambda) = \begin{pmatrix} N - \lambda I_k & C_{j-1}(\lambda) \\ 0 & D_{j-1}(\lambda) \end{pmatrix}$$

with invariant polynomials $v_{k+j-1}^{(j-1)}|\cdots|v_1^{(j-1)}$ and $D_{j-1}(\lambda)$ upper triangular. Note that the invariant polynomials of

$$\begin{pmatrix} Q_{j-1}(\lambda) \\ 0 \end{pmatrix}$$

are $v_{k+j-1}^{(j-1)}|\cdots|v_1^{(j-1)}$. Next use the division relations of the polynomials $v_i^{(j)}$ and apply Theorem 3.1. We obtain the existence of a vector polynomial $c_j(\lambda)$ and a polynomial $d_j(\lambda)$ such that

$$Q_j(\lambda) = \begin{pmatrix} Q_{j-1}(\lambda) & c_j(\lambda) \\ 0 & d_j(\lambda) \end{pmatrix}$$

has invariant polynomials $v_{k+j}^{(j)}|\cdots|v_1^{(j)}$. Then $Q_0(\lambda) = N - \lambda I_k$ and $Q_m(\lambda)$ has invariant polynomials $v_{k+m}^{(m)}|\cdots|v_1^{(m)}$. Now remember that equation (4.2) gives that $v_i^{(m)} = q_i$. So $Q_m(\lambda)$ has the desired invariant polynomials. Notice that

$$d_j(\lambda) = \frac{\det Q_j(\lambda)}{\det Q_{j-1}(\lambda)} = \frac{\prod_{i=1}^{k+j} v_i^{(j)}(\lambda)}{\prod_{i=1}^{k+j-1} v_i^{(j-1)}(\lambda)} = \frac{w_j(\lambda)}{w_{j-1}(\lambda)} = v^{(j)}(\lambda).$$

The next step is to reduce the degrees of the off-diagonal elements in row j of $D_m(\lambda)$ to a value with a degree less than $\deg v^{(j)}$ by elementary column operations. To do this, form the last row up to the first and apply division with remainder by the diagonal elements.

PART 4. The only property that the matrix polynomial $Q_m(\lambda)$ still misses is that the degrees of the rows are not $\nu_1 + 1, \ldots, \nu_m + 1$ but $\deg v^{(1)}, \ldots, \deg v^{(m)}$. We will correct this in a finite number of steps by applying in each step Lemma 4.2.

Let $E_0(\lambda) = D_m(\lambda)$. Assume that we have constructed

$$E_s(\lambda) = \left(e_{ij}^{(s)}(\lambda)\right)_{i,j=1}^{m}$$

such that for $i \neq j$

$$\deg e_{ij}^{(s)} < \deg e_{ii}^{(s)} \leq \nu_i + 1.$$

and $E_s(\lambda) = F_s(\lambda)D_m(\lambda)G_s(\lambda)$ with $F_s(\lambda)$ and $G_s(\lambda)$ unimodular matrix polynomials. Let $a_s = \sum_{i=1}^m |\nu_i + 1 - \deg e_{ii}^{(s)}|$. If $a_s = 0$, then $\deg e_{ii}^{(s)} = \nu_i + 1$ for $i = 1, \dots, m$ and we are done. Assume that $a_s > 0$. We will construct $E_{s+1}(\lambda)$ with the same properties as $E_s(\lambda)$ but with $a_{s+1} = a_s - 2$. If for $j = 1, \dots, m$

$$\sum_{i=1}^{j} \deg e_{ii}^{(s)} = \sum_{i=1}^{j} (\nu_i + 1),$$

then $\deg e_{ii}^{(s)} = \nu_i + 1$ for $i = 1, \dots, m$, which contradicts $a_s > 0$. So there exists a number r such that $\sum_{i=1}^r \deg e_{ii}^{(s)} < \sum_{i=1}^r (\nu_i + 1)$, and thus $\sum_{i=r+1}^m \deg e_{ii}^{(s)} > \sum_{i=r=1}^m (\nu_i + 1)$. There exist numbers $p \leq r$ and $q > r$ such that

$$\deg e_{pp}^{(s)} < \nu_p + 1 \leq \nu_q + 1 < \deg e_{qq}^{(s)}.$$

It follows that $\deg e_{pp}^{(s)} + 1 \leq \deg e_{qq}^{(s)} - 1$. Now apply the Lemma 4.2 and construct $E_{s+1}(\lambda)$, with $\deg e_{pj}^{(s+1)} < \deg e_{pp}^{(s+1)} = \deg e_{pp}^{(s)}(\lambda) + 1$ for $j \neq p$ and $\deg e_{qj}^{(s+1)} < \deg e_{qq}^{(s+1)} = \deg e_{qq}^{(s)}(\lambda) - 1$ for $j \neq q$ and the degrees of the other rows equal to the degrees of the corresponding rows of $E_s(\lambda)$, keeping the property that the diagonal element has a strictly larger degree than the other elements in the same row. Clearly $a_{s+1} = a_s - 2$. □

Note that the ordering of the numbers $\nu_1, \dots, \nu_m$ is used for the sufficiency of the conditions (1) and (2) only.

PROOF OF THEOREM 1.1. First note that the conditions (1) of Theorem 1.1 and Theorem 4.1 coincide. Secondly, we show that condition (2) of Theorem 4.1 is equivalent to condition (2) of Theorem 1.1. Put $\nu_i = \mu_{m+1-i}$. By a simple renumbering one sees that condition (2) in Theorem 4.1 may be rewritten as

$$\sum_{i=1}^{j} \deg v^{(m+1-i)} \geq \sum_{i=1}^{j} (\mu_i + 1), \quad j = 1, \dots, m.$$

By a straightforward renumbering one obtains from the definitions of $v^{(m+1-j)}$ and s_j that $v^{(m+1-j)} = s_j$. This shows that the condition (2) of Theorem 4.1 coincides with the condition (2) from Theorem 1.1. Thirdly remark that we may reorder the rows and columns of $D(\lambda)$ in Theorem 4.1 in the opposite ordering. Finally one uses Theorem 2.1 to see that now Theorem 1.1 follows from Theorem 4.1. □

The next proposition presents some of the properties of the polynomials s_j appearing in Theorem 1.1.

Proposition 4.3. *Let $q_1, \dots, q_n$ and $p_1, \dots, p_n$ be sequences of polynomials such that $q_{j+1}|q_j$ and $p_{j+1}|p_j$ for $j = 1, \dots, n-1$. Assume that for a number $m < n$ we have $q_{j+m}|p_j|q_j$ for $j = 1, \dots, n-m$ and that $p_{n-m+1} = 1$. Put*

$$t_j = \prod_{i=j}^{n} \text{l.c.m.}(p_{i-j+1}, q_i), \quad \text{for} \quad j = 1, \dots, m+1 \tag{4.4a}$$

and

$$s_j = t_j / t_{j+1} \quad \text{for} \quad j = 1, \dots, m. \tag{4.4b}$$

Then

(i) s_j *is a polynomial for* $j = 1, 2, \dots, m$;

(ii) $s_j|q_j$, *for* $j = 1, 2, \dots, m$;

(iii) $s_{j+1}|s_j$, *for* $j = 1, 2, \dots, m-1$.

PROOF. First we note that the statement is about division relations only. So we may check it for each zero of the polynomials involved separately. Therefore we may, without loss of generality, assume that $q_i(\lambda) = \lambda^{\alpha_i}$ and $p_i(\lambda) = \lambda^{\beta_i}$ for $i = 1, 2, \dots, n$. Then $t_j(\lambda) = \lambda^{\gamma_j}$, with $\gamma_j = \sum_{i=j}^{n} \max\{\beta_{i+1-j}, \alpha_i\} = \sum_{k=1}^{n-j+1} \max\{\beta_k, \alpha_{k+j-1}\}$. Remark that if $k > n - j + 1$, then $\beta_k = 0$ and $\alpha_{k+j-1} = 0$. Hence $\gamma_j = \sum_{k=1}^{\infty} \max\{\beta_k, \alpha_{k+j-1}\}$, where we put $\beta_k = 0$ if $k > n$, and $\alpha_k = 0$ if $k > n - m + 1$. We apply Proposition III.4.4 and obtain that

(i') $\gamma_j \geq \gamma_{j+1}$ for $j = 1, 2, \dots, m$;

(ii') $\gamma_j - \gamma_{j+1} \leq \alpha_j$ for $j = 1, 2, \dots, m$;

(iii') $\gamma_j - \gamma_{j+1} \geq \gamma_{j+1} - \gamma_{j+2}$ for $j = 1, 2, \dots, m-1$.

These three statements are equivalent to (i), (ii) and (iii) above in the case when $q_i(\lambda) = \lambda^{\alpha_i}$ and $p_i(\lambda) = \lambda^{\beta_i}$. □

IV.5 Some applications of the restriction problem

In this section we present two theorems which are derived from Theorem 1.3 and a result on the existence of similarities of the first kind for full length blocks. The first result is an application of Theorem 1.3 to matrix pencils.

Theorem 5.1. *Let F be an $n \times n$ matrix with invariant polynomials $q_n|\cdots|q_1$. There exists a $k \times n$ matrix H of rank m such that the pencil*

$$L(\lambda) = \lambda \begin{pmatrix} I \\ 0 \end{pmatrix} - \begin{pmatrix} F \\ H \end{pmatrix}$$

has invariant polynomials $p_n|\cdots|p_1$ and Kronecker row indices $\mu_1 \geq \cdots \geq \mu_m > 0 = \mu_{m+1} = \cdots = \mu_k$ if and only if

(1) $p_i|q_i$, *for* $i = 1, \dots, n$, *and* $q_{i+m}|p_i$, *for* $i = 1, \dots, n-m$;

(2) $\sum_{i=1}^{j} \deg s_i \geq \sum_{i=1}^{j} \mu_i$ *for* $j = 1, \dots, m$, *with equality holding for* $j = m$.

The polynomials s_j *are defined as* $s_j = t_j/t_{j+1}$, *for* $j = 1, \dots, m$, *where*

$$t_j = \prod_{i=j}^{n} \text{l.c.m.}(p_{i-j+1}, q_i), \quad j = 1, \dots, m+1.$$

PROOF. From Corollary III.2.3 we know that the invariant polynomials and the Kronecker row indices of $L(\lambda)$ are the invariant polynomials and indices of the third kind of the block

$$\Theta = (\begin{pmatrix} F \\ H \end{pmatrix}; \begin{pmatrix} I & 0 \\ 0 & I \end{pmatrix}, \begin{pmatrix} I \\ 0 \end{pmatrix})$$

on $\mathbb{C}^n \oplus \mathbb{C}^m$. Assume that this block has invariant polynomials $p_n|\cdots|p_1$ and indices of the third kind $\mu_1 \geq \cdots \geq \mu_m > 0 = \mu_{k+1} = \cdots = \mu_k$. Let Q be any projection of $\mathbb{C}^n$ such that $\operatorname{Im} Q = \operatorname{Ker} H$. Then it follows from Theorem III.1.5 that $(F|_{\operatorname{Im} Q}; I, Q)$ has invariant polynomials $p_1, \dots, p_n$ and indices of the third kind $\mu_1 - 1, \dots, \mu_m - 1$. Also, conversely, if we have a block $(F|_{\operatorname{Im} Q}; I, Q)$ with invariant polynomials $p_1, \dots, p_n$ and indices of the third kind $\mu_1 - 1, \dots, \mu_m - 1$ and $\operatorname{Ker} H = \operatorname{Im} Q$, then the invariant polynomials of Θ are $p_1, \dots, p_n$ and the indices of the third kind of Θ are $\mu_1 \geq \cdots \geq \mu_m > 0 = \mu_{k+1} = \cdots = \mu_k$. So to find the desired H it is necessary and sufficient to find a projection Q of $\mathbb{C}^n$ such that the block $(F|_{\operatorname{Im} Q}; I, Q)$ has invariant polynomials $p_1, \dots, p_n$ and indices of the third kind $\mu_1 - 1, \dots, \mu_m - 1$. Apply Theorem 1.3 to see that this is possible if and only if the conditions (1) and (2) hold true. □

The next theorem is a direct translation of Theorem 1.3 to the language of non-everywhere defined operators.

Theorem 5.2. *Let* $q_n|\cdots|q_1$ *be the invariant polynomials of the operator* $A : \mathcal{X} \to \mathcal{X}$. *Let* $\mu_1 \geq \cdots \geq \mu_m \geq 0$ *be integers and* $p_n|\cdots|p_1$ *be polynomials. Then there exists a subspace* $\mathcal{Y}$ *of* $\mathcal{X}$ *such that the restriction* $A|_{\mathcal{Y}}(\mathcal{X} \to \mathcal{X})$, *viewed as a non-everywhere defined operator with domain* $\mathcal{Y}$, *is a direct sum of a square part with invariant polynomials* $p_1, \dots, p_n$ *and non-everywhere defined shifts of rank* $\mu_1 \geq \cdots \geq \mu_m$ *if and only if*

(0) $\sum_{i=1}^{m} (\mu_i + 1) + \sum_{i=1}^{n} \deg p_i = n$;

(1) $p_i|q_i$, *for* $i = 1, \dots, n$, *and* $q_{i+m}|p_i$, *for* $i = 1, \dots, n-m$;

(2) $\sum_{i=1}^{j} \deg s_i \geq \sum_{i=1}^{j} (\mu_i + 1)$ *for* $j = 1, \dots, m$, *with equality holding for* $j = m$,

where the polynomials s_j are defined as $s_j = t_j/t_{j+1}$, for $j = 1, \dots, m$, where

$$t_j = \prod_{i=j}^{n} \text{l.c.m.}(p_{i-j+1}, q_i), \quad j = 1, \dots, m+1.$$

IV.6 A matrix equation

Let $A : \mathbb{C}^n \to \mathbb{C}^n$ and $C : \mathbb{C}^n \to \mathbb{C}^r$ be linear operators. We consider the operator equation

$$TA + GC = I_n. \tag{6.1}$$

The aim is to find all solutions T, G of this equation. In order to formulate the solution we introduce a notation. For a polynomial p of degree n with $p(0) \neq 0$ we write $p^{\#}(\lambda) = \lambda^n p(\lambda^{-1})/p(0)$. We shall prove the following result.

Theorem 6.1. *Let $A : \mathbb{C}^n \to \mathbb{C}^n$ and $C : \mathbb{C}^n \to \mathbb{C}^r$ be linear operators such that*

$$\begin{pmatrix} \lambda I - A \\ -C \end{pmatrix}$$

has invariant polynomials $p_1, \ldots, p_n$ and Kronecker row indices $\mu_1 \geq \cdots \geq \mu_m > 0 = \mu_{m+1} = \cdots = \mu_r$. Let $q_1, \ldots, q_n$ be monic polynomials such that $q_n | \cdots | q_1$ and $\sum_{j=1}^n \deg q_j = n$. Then there exists a transformation $T : \mathbb{C}^n \to \mathbb{C}^n$ with invariant polynomials $q_1, \ldots, q_n$ and an operator G such that $TA + GC = I$ if and only if

(0) $p_1(0) \neq 0$;

(1) $p_i^{\#} | q_i$, *for* $i = 1, \ldots, n$, *and* $q_{i+m} | p_i^{\#}$, *for* $i = 1, \ldots, n - m$;

(2) $\sum_{i=1}^j \deg s_i \geq \sum_{i=1}^j \mu_i$ *for* $j = 1, \ldots, m$, *with equality holding for* $j = m$,

where the polynomials s_j are defined as $s_j = t_j / t_{j+1}$, for $j = 1, \ldots, m$, with

$$t_j = \prod_{i=j}^{n} \text{l.c.m.}(p_{i-j+1}^{\#}, q_i) \quad j = 1, \ldots, m + 1.$$

To prove this theorem we need some auxiliary results. First we reduce the problem to the setting of operator blocks.

Lemma 6.2. *Let $A : \mathbb{C}^n \to \mathbb{C}^n, T : \mathbb{C}^n \to \mathbb{C}^n$ and $C : \mathbb{C}^n \to \mathbb{C}^r$ be linear operators, and let $Q : \mathbb{C}^n \to \mathbb{C}^n$ be a projection such that $\operatorname{Im} Q = \operatorname{Ker} C$ and $B = A|_{\operatorname{Im} Q}$. Then $TA + GC = I$ for some operator G if and only if $TBx = x$ for each $x \in \operatorname{Im} Q$.*

PROOF. Assume that $TA + GC = I$. Then we see that $TAQ = Q$. So if $x \in \operatorname{Im} Q$, then $TBx = x$. Conversely, assume that $TBx = x$, for each $x \in \operatorname{Im} Q$. Then $TAQ = Q$. So $I = TA + (I - Q) - TA(I - Q) = TA + (I - TA)(I - Q)$. Now for the second term in the right hand side of this equality one has $(I - TA)(I - Q)x = 0$ for each $x \in \operatorname{Ker} C$. This means that there exist an operator G such that $(I - TA)(I - Q) = GC$. So $I = TA + GC$. □

The next result describes the eigenvalues and their multiplicities for the solution of the equation $TB = I_{\operatorname{Im} Q}$ in the case when the block $(B; I, Q)$ fulfills an extra condition.

Proposition 6.3. *Let $(B; I, Q)$ be a block on $\mathbb{C}^n$ with indices of the third kind $\mu_1 \geq \cdots \geq \mu_m$ and no (non-constant) invariant polynomials. Furthermore, let $q_1, \ldots, q_n$ be monic polynomials such that $q_n | \cdots | q_1$ and $\sum_{j=1}^{n} \deg q_j = n$. Then there exists a transformation $T : \mathbb{C}^n \to \mathbb{C}^n$ such that $TBx = x$, for each $x \in \operatorname{Im} Q$, and the invariant polynomials of T are $q_1, \ldots, q_n$ if and only if $\sum_{j=1}^{k}(\mu_j + 1) \leq \sum_{j=1}^{k} \deg q_j$ for $k = 1, \ldots, m$.*

PROOF. We apply Proposition III.1.3 to obtain a basis $\{f_{ij}\}_{i=1,j=1}^{\mu_i+1,m}$ of $\mathbb{C}^n$ such that $\operatorname{Im} Q = \operatorname{span}\{f_{ij}\}_{i=1,j=1}^{\mu_i,m}$ and $Bf_{ij} = f_{i+1\,j}$ for $i = 1, \ldots, \mu_j$ and $j = 1, \ldots, m$. Now, clearly, to satisfy $TBx = x$ for each $x \in \operatorname{Im} Q$, it is necessary and sufficient that $Tf_{i+1\,j} = f_{ij}$, for $i = 1, \ldots, \mu_j$ and $j = 1, \ldots, m$. Define Q_1 to be a projection onto $\operatorname{span}\{f_{ij}\}_{i=2,j=1}^{\mu_i+1,m}$, and define $T_1 : \operatorname{Im} Q_1 \to \mathbb{C}^n$ by $T_1 f_{i+1\,j} = f_{ij}$, for $i = 1, \ldots, \mu_j$ and $j = 1, \ldots, m$. Then T is a linear transformation $T : \mathbb{C}^n \to \mathbb{C}^n$ such that $TBx = x$ for each $x \in \operatorname{Im} Q$, and the invariant polynomials of T are $q_1, \ldots, q_n$ if and only if T is a completion of the block $(T_1; I, Q_1)$. Since the indices of the third kind of this block are $\mu_1 \geq \cdots \geq \mu_m$ and this block does not have invariant polynomials, this is possible if and only if $\sum_{j=1}^{k}(\mu_j + 1) \leq \sum_{j=1}^{k} \deg q_j$ for $k = 1, \ldots, m$. □

We show the existence of solutions of equation (6.1) for a special case.

Theorem 6.4. *Let $A : \mathbb{C}^n \to \mathbb{C}^n$ and $C : \mathbb{C}^n \to \mathbb{C}^r$ be linear operators such that*

$$\begin{pmatrix} \lambda I - A \\ -C \end{pmatrix}$$

has no invariant polynomials and Kronecker row indices $\mu_1 \geq \cdots \geq \mu_r$. Let $q_1, \ldots, q_n$ be monic polynomials such that $q_n | \cdots | q_1$ and $\sum_{j=1}^{n} \deg q_j = n$. Then there exists an operator $T : \mathbb{C}^n \to \mathbb{C}^n$ with invariant polynomials $q_1, \ldots, q_n$ and an operator G such that $TA + GC = I$ if and only if $\sum_{j=1}^{k} \mu_j \leq \sum_{j=1}^{k} \deg q_j$ for $k = 1, \ldots, r$.

PROOF. Define a projection $Q : \mathbb{C}^n \to \mathbb{C}^n$ by requiring that $\operatorname{Im} Q = \operatorname{Ker} C$. Let $(B; I, Q)$ be the block given by $B = A|_{\operatorname{Im} Q}$. Lemma 6.2 gives that $TA + GC = I$ for some G if and only if $TBx = x$ for each $x \in \operatorname{Im} Q$. We apply Proposition 6.3. It remains to express the indices of the third kind of the block $(B; I, Q)$ in the Kronecker row indices $\mu_1 \geq \cdots \geq \mu_r$. Apply Corollary III.2.3 and Theorem III.1.5 to see that the indices of the block are $\mu_1 - 1, \ldots, \mu_m - 1$, where m is the largest number such $\mu_m > 0$. This translates the necessary and sufficient condition of Proposition 6.3 into $\sum_{j=1}^{k} \mu_j \leq \sum_{j=1}^{k} \deg q_j$ for $k = 1, \ldots, r$. □

Also in the general case we give a result on operator blocks first.

Proposition 6.5. *Let $(B; I, Q)$ be a block on $\mathbb{C}^n$ with indices of the third kind $\mu_1 \geq \cdots \geq \mu_m$ and invariant polynomials $p_n|\cdots|p_1$. Furthermore, let $q_1, \ldots, q_n$ be monic polynomials such that $q_n|\cdots|q_1$ and $\sum_{j=1}^n \deg q_j = n$. Then there exists an operator $T : \mathbb{C}^n \to \mathbb{C}^n$ such that $TBx = x$ for each $x \in \operatorname{Im} Q$ and T has invariant polynomials $q_1, \ldots, q_n$ if and only if*

(0) $p_1(0) \neq 0$;

(1) $p_i^\#|q_i$, *for* $i = 1, \ldots, n$, *and* $q_{i+m}|p_i^\#$, *for* $i = 1, \ldots, n-m$;

(2) $\sum_{i=1}^j \deg s_i \geq \sum_{i=1}^j (\mu_i + 1)$ *for* $j = 1, \ldots, m$, *with equality holding for* $j = m$,

where the polynomials s_j are defined as $s_j = t_j / t_{j+1}$, for $j = 1, \ldots, m$, with

$$t_j = \prod_{i=j}^n \text{l.c.m.}(p_{i-j+1}^\#, q_i), \quad j = 1, \ldots, m+1.$$

Proof. We apply Proposition III.1.3 to obtain a linearly independent set of vectors $\{f_{ij}\}_{i=1,j=1}^{\mu_j+1\;,m}$ and a subspace $\mathcal{D}_\infty$ of $\mathcal{X}$ such that:

(i) $\mathcal{X} = \operatorname{span}\{f_{ij}\}_{i=1,j=1}^{\mu_j+1,m} \oplus \mathcal{D}_\infty$;

(ii) $\operatorname{Im} Q = \operatorname{span}\{f_{ij}\}_{i=1,j=1}^{\mu_j,m} \oplus \mathcal{D}_\infty$;

(iii) $B[\mathcal{D}_\infty] \subset \mathcal{D}_\infty$;

(iv) $Bf_{ij} = f_{i+1\;j}$ for $i = 1, \ldots, \mu_j$ and $j = 1, \ldots, m$.

For the existence of an operator T such $TBx = x$ for each $x \in \operatorname{Im} Q$ it is necessary and sufficient that $B|_{\mathcal{D}_\infty}$ is invertible. This condition can be restated as $p_1(0) \neq 0$, which means that zero is not an eigenvalue of $B|_{\mathcal{D}_\infty}$. We take a projection Q_1 of $\mathbb{C}^n$ such that $\operatorname{Im} Q_1 = \operatorname{span}\{f_{ij}\}_{i=2,j=1}^{\mu_j+1,m} \oplus \mathcal{D}_\infty$, and define an operator $T_1 : \operatorname{Im} Q_1 \to \mathbb{C}^n$ by $T_1|_{\mathcal{D}_\infty} = (B|_{\mathcal{D}_\infty})^{-1}$ and $T_1 f_{i+1\;j} = f_{ij}$, for $i = 1, \ldots, \mu_j$ and $j = 1, \ldots, m$. So in order that $TBx = x$, for each $x \in \operatorname{Im} Q$ it is necessary and sufficient that $T|_{\operatorname{Im} Q_1} = T_1$. This means that T is a completion of the block $(T_1; I, Q_1)$. Remark that the indices of the third kind of the block $(T_1; I, Q_1)$ are $\mu_1 \geq \cdots \geq \mu_m$. The invariant polynomials are the invariant polynomials of $(B|_{\mathcal{D}_\infty})^{-1}$ supplemented by a number of constant polynomials equal to 1, that is, they are $p_1^\#, \ldots, p_n^\#$. Apply Theorem 1.1 to see that a completion of $(T_1; I, Q_1)$ with invariant polynomials $q_1, \ldots, q_n$ exist if and only if the conditions (1) and (2) are fulfilled. □

Proof of Theorem 6.1. The proof is essentially the same as the proof of Theorem 6.4. The only difference is that Proposition 6.5 replaces Proposition 6.3 and that one has to notice that the invariant polynomials of the pencil and the block are the same. □

Notes

The solution of the eigenvalue completion problem for full length submatrices is due to Zaballa [1]. The proof presented here is based on the ideas of Zaballa's original paper and uses some elements of the theory presented in the previous chapter. An alternative proof, which does not use the reduction to a matrix polynomial problem and which is completely within the framework of operator blocks, is not known. Theorem 3.1 goes back to Thompson [1]. Theorems 1.1 and 4.1 appear in Zaballa [1]; the first of these two theorems may be viewed as a generalization of Rosenbrock's theorem on pole assignment (see Zaballa [2] and Section IX.6 below). Lemma 4.2 is Lemma 4.1 in Chapter V of Rosenbrock [1]. Proposition 4.3 is due to Gohberg-Kaashoek-Van Schagen [4]. The full length restriction problem and its solution (Theorem 1.3) are taken from Gohberg-Kaashoek-Van Schagen [5]. Theorem 5.1 with a different proof and without the connections with the restriction problem appears in Zaballa [3]. The results in Section 6 are taken from Gohberg-Kaashoek-Ran [1]. Connections between the results of this chapter and mathematical systems theory will be discussed in Chapter X.

Chapter V
Full Width Blocks

This chapter presents the analogues of the results of the two previous chapters for full width blocks. The first three sections extend the results of Chapter III. The fourth section explains the duality between full length and full width blocks, and provides the tools to transfer the results of Chapter IV to full width blocks, which is done in the last two sections.

V.1 Structure theorems for full width blocks

The block similarity classes of full width blocks will be described in this section. In the description shifts of the first kind play an important role. Recall (see Section II.3) that a block $(B; P, I)$ on the space $\mathbb{C}^{\kappa+1}$ is called a shift of the first kind with index κ if with respect to the standard basis $\{e_1, \dots, e_{\kappa+1}\}$ of $\mathbb{C}^{\kappa+1}$ the block we have

(i) $\operatorname{Ker} P = \operatorname{span}\{e_1\}$, $\operatorname{Im} P = \operatorname{span}\{e_2, \dots, e_{\kappa+1}\}$,

(ii) $Be_i = e_{i+1}$, for $i = 1, \dots, \kappa$, and $Be_{\kappa+1} = 0$.

If $\kappa = 0$, then $\operatorname{Im} P = \{0\}$, and therefore the second condition is automatically fulfilled.

Such a shift of the first kind is an irreducible block, as was shown in Section II.3. Let $\operatorname{span}\{f_1, \dots, f_{\kappa+1}\}$ be a basis of $\mathcal{X}$, and assume that the block $(B_1; P_1, I_1)$ is such that

(i) $\operatorname{Ker} P = \operatorname{span}\{f_1\}$,

(ii) $Bf_i = f_{i+1}$ for $i = 1, \dots, \kappa$, and $Be_{\kappa+1} = 0$.

Then $(B_1; P_1, I_1)$ is block similar to a shift of the first kind with index κ. Indeed define $Sf_i = e_i$. Then $S[\operatorname{Ker} P_1] = \operatorname{span}\{e_1\}$, and $(SB_1 - BS)f_i = 0$ for $i = 1 \dots, \kappa + 1$.

The class of full width blocks is closed under taking direct sums, under application of block similarity and under restriction to a block invariant subspace. This follows immediately from the definitions of these notions. Therefore, a block that is block similar to a direct sum of shifts of the first kind and a block $(J; I, I)$ is a full width block. The main result of this section is that the converse is also true.

Theorem 1.1. *An operator block $(B; P, I)$ is block similar to a direct sum of an operator J and shifts of the first kind with indices $\kappa_1 \geq \cdots \geq \kappa_q \geq 0$. In this case*

$q = \dim \operatorname{Ker} P$. *Furthermore, the invariant polynomials of the operator* J *and the indices* $\kappa_1 \geq \cdots \geq \kappa_q$, *are uniquely determined by the block* $(B; P, I)$.

The proof of this theorem will be given later on in this section. First we introduce some terminology and make some preparations. The numbers $\kappa_1 \geq \cdots \geq \kappa_q$ in Theorem 1.1 are called *the indices of the first kind of the block* $(B; P, I)$. The invariant polynomials of the operator J are called *the invariant polynomials of the block* $(B; P, I)$. As before we order the invariant polynomials with decreasing degrees. If useful, we will extend the set of invariant polynomials by constant polynomials equal to 1. In this way we may for instance assume the number of invariant polynomials to be equal to $\dim \mathcal{X}$.

To describe the indices of the first kind we introduce a second sequence of subspaces. Let (B, P, Q) be an arbitrary block. For (B, P, Q) we introduce the following subspaces:

$$\mathcal{F}_0 = \{0\}, \quad \mathcal{F}_1 = \operatorname{Ker} P, \quad \mathcal{F}_{i+1} = B[\mathcal{F}_i \cap \operatorname{Im} Q] \oplus \operatorname{Ker} P, \ i = 1, 2, \ldots .$$

We call $\mathcal{F}_j$ the *j-th iterated image* of $(B; P, Q)$. Obviously $\mathcal{F}_j \subset \mathcal{F}_{j+1}$, and if $\mathcal{F}_j = \mathcal{F}_{j+1}$, then $\mathcal{F}_j = \mathcal{F}_i$ for all $i \geq j + 1$. Since the sequence is ascending in a finite dimensional space, we can find a number κ such that $\mathcal{F}_{\kappa-1} \neq \mathcal{F}_\kappa = \mathcal{F}_{\kappa+1}$. Put $\mathcal{F}_\infty = \mathcal{F}_\kappa$. We call $\mathcal{F}_\infty$ the *(final) iterated image* of $(B; P, Q)$. The subspace $\mathcal{F}_\infty$ is $(B; P, Q)$-block-invariant. To see this note that

$$\begin{aligned} B[\mathcal{F}_\infty \cap \operatorname{Im} Q] &= B[\mathcal{F}_\kappa \cap \operatorname{Im} Q] \\ &\subset B[\mathcal{F}_\kappa \cap \operatorname{Im} Q] \oplus \operatorname{Ker} P = \mathcal{F}_{\kappa+1} = \mathcal{F}_\kappa = \mathcal{F}_\infty \\ &\subset \mathcal{F}_\infty \oplus \operatorname{Ker} P. \end{aligned}$$

If $\mathcal{Y}$ is a $(B; P, Q)$-block-invariant subspace that contains $\operatorname{Ker} P$, then $\mathcal{Y} \supset \mathcal{F}_\infty$. To prove this note that $\mathcal{Y} \supset \mathcal{F}_1$. Next remark that if $\mathcal{Y} \supset \mathcal{F}_i$, then

$$\mathcal{F}_{i+1} = B[\mathcal{F}_i \cap \operatorname{Im} Q] \oplus \operatorname{Ker} P \subset B[\mathcal{Y} \cap \operatorname{Im} Q] \oplus \operatorname{Ker} P \subset \mathcal{Y}.$$

Here the last inclusion follows from the assumption that $\mathcal{Y}$ is $(B; P, Q)$-block-invariant. So $\mathcal{F}_\infty$ is the smallest $(B; P, Q)$-block-invariant subspace that contains $\operatorname{Ker} P$.

If the block is a full width block $(B; P, I)$, then the j-th iterated image is given by

$$\mathcal{F}_0 = \{0\}, \ \mathcal{F}_1 = \operatorname{Ker} P, \ \mathcal{F}_j = \operatorname{Ker} P \oplus B[\mathcal{F}_{j-1}], \tag{1.1}$$

for $j = 1, 2, \ldots$. The next result expresses the indices of the first kind of the block $(B; P, I)$ in terms of the spaces $\mathcal{F}_j$.

Theorem 1.2. *Let* $(B; P, I)$ *be an operator block on the space* $\mathcal{X}$. *Put*

$$\mathcal{F}_0 = \{0\}, \ \mathcal{F}_1 = \operatorname{Ker} P, \ \mathcal{F}_j = \operatorname{Ker} P \oplus B[\mathcal{F}_{j-1}], \quad j \geq 1.$$

Set $\kappa = \min\{j \mid \mathcal{F}_j = \mathcal{F}_{j+1}\}$, *and*

$$p_k = \dim(\mathcal{F}_{k+1}/\mathcal{F}_k) \tag{1.2}$$

for $k = 0, \dots, \kappa - 1$. *Then the indices of the first kind of the block are the first* p_0 *elements of the dual sequence of* $p_1 \geq p_2 \geq \cdots$, *i.e., they are given by*

$$\kappa_j = \#\{k \geq 1 \mid p_k \geq j\}\,, \quad j = 1, \dots, p_0. \tag{1.3}$$

Moreover, there exists a $(B; P, I)$*-invariant complement* $\mathcal{G}_\infty$ *of* $\mathcal{F}_\kappa$, *and the invariant polynomials of* $(B; P, I)$ *are the invariant polynomials of the operator* $PB|_{G\infty} : \mathcal{G}_\infty \to \mathcal{G}_\infty$.

The proofs of Theorem 1.1 and Theorem 1.2 will be based on the following result, which describes a special basis for $\mathcal{X}$.

Proposition 1.3. *Let* $(B; P, I)$ *be an operator block on the space* $\mathcal{X}$. *Then there exist numbers* $\kappa_1 \geq \cdots \geq \kappa_q \geq 0$, *with* $q = \dim \operatorname{Ker} P$, *an independent set of vectors* $\{e_{1j}, \dots, e_{\kappa_j+1\ j} \mid j = 1, \dots, q\}$, *and a subspace* $\mathcal{G}_\infty$ *of* $\mathcal{X}$ *such that:*

(i) $\mathcal{X} = \operatorname{span}\{e_{ij}\}_{i=1,j=1}^{\kappa_j+1,q} \oplus \mathcal{G}_\infty$;

(ii) $\operatorname{Ker} P = \operatorname{span}\{e_{11}, \dots, e_{1q}\}$;

(iii) $B[\mathcal{G}_\infty] \dot{\subset} \mathcal{G}_\infty \oplus \operatorname{Ker} P$;

(iv) $Be_{ij} = Pe_{i+1\ j}$, *for* $i = 1, \dots, \kappa_j$, *and* $Be_{\kappa_j+1\ j} = 0$, *for* $j = 1, \dots, q$.

Proof. Let the spaces $\mathcal{F}_j$ be defined by (1.1), and let κ be the smallest integer such that $\mathcal{F}_\kappa = \mathcal{F}_{\kappa+1}$. We will construct a $(B; P, I)$-invariant complement $\mathcal{G}_\infty$ of $\mathcal{F}_\kappa$. Choose a complement $\mathcal{G}_\kappa$ of $\mathcal{F}_\kappa$ in $\mathcal{X}$. Then one has for $j = \kappa$ that

(1) $\mathcal{X} = \mathcal{F}_\kappa \oplus \mathcal{G}_j$;

(2) $B[\mathcal{G}_j] \subset \mathcal{F}_j \oplus \mathcal{G}_j$.

We want to find the subspace $\mathcal{G}_\infty = \mathcal{G}_1$ that fulfills the conditions (1) and (2) for $j = 1$.

So assume that $\mathcal{G}_j$ satisfies (1) and (2) for some j with $2 \leq j \leq \kappa$. We construct $\mathcal{G}_{j-1}$. For $y \in \mathcal{G}_j$ one has $By = f_1 + Bf_{j-1} + y'$ with $f_1 \in \mathcal{F}_1$, $f_{j-1} \in \mathcal{F}_{j-1}$ and $y' \in \mathcal{G}_j$. It follows that there exists a linear transformations $T_j : \mathcal{G}_j \to \mathcal{F}_1$, $R_j : \mathcal{G}_j \to \mathcal{G}_j$ and $S_j : \mathcal{G}_j \to \mathcal{F}_{j-1}$ such that $By = (T_j + BS_j + R_j)y$ for each $y \in \mathcal{G}_j$. Then $B(I - S_j)y = (T_j + S_j R_j)y + (I - S_j)R_j y$ for each $y \in \mathcal{G}_j$. We put $\mathcal{G}_{j-1} = (I - S_j)\mathcal{G}_j$. It follows that $B[\mathcal{G}_{j-1}] \subset \mathcal{F}_{j-1} + \mathcal{G}_{j-1}$ and that $\mathcal{X} = \mathcal{F}_\kappa \oplus \mathcal{G}_{j-1}$. From property (2) above for $j = 1$ we obtain that condition (iii) is fulfilled.

It remains to construct the special basis $\{e_{ij}\}_{i=1,j=1}^{\kappa_j+1\ ,q}$ of $\mathcal{F}_\kappa$. If this is done, then condition (i) is also met. So this basis has to fulfill the conditions (ii) and (iv) to complete the proof of the proposition. We make some preliminary remarks for the construction. First note that $B[\mathcal{F}_{i-1}] \subset \mathcal{F}_i$ and hence the operator $B : \mathcal{F}_{i-1} \to \mathcal{F}_i$ induces a quotient operator

$$[B]_i : \mathcal{F}_{i-1}/\mathcal{F}_{i-2} \to \mathcal{F}_i/\mathcal{F}_{i-1}, \quad i = 2, 3, \dots, \kappa.$$

Let us prove that $[B]_i$ is surjective. Denote the class of f in $\mathcal{F}_i/\mathcal{F}_{i-1}$ by $[f]_i$. Assume that $f \in \mathcal{F}_i$ and $f \notin \mathcal{F}_{i-1}$. Then $f = f_0 + Bg$ with $f_0 \in \operatorname{Ker} P$ and $g \in \mathcal{F}_{i-1}$. Thus $Pf = Pf_0 + PBg = Bg$. Since $(I-P)f \in \mathcal{F}_{i-1}$, one has that $[f]_i = [Pf]_i = [Bg]_i = [B]_i[g]_{i-1}$. It is useful here to make another remark. Assume that $[f]_{i-1} \in \operatorname{Ker}[B]_i$. Then $f \in \mathcal{F}_{i-1}$ and $Bf \in \mathcal{F}_{i-1}$. Thus $Bf = f_0 + Bg$ with $f_0 \in \operatorname{Ker} P$ and $g \in \mathcal{F}_{i-2}$. Hence $f-g$ is such that $B(f-g) = PB(f-g) = Pf_0 = 0$. So with $h = f - g$ we have that $[h]_{i-1} = [f]_{i-1} \in \mathcal{F}_{i-1}/\mathcal{F}_{i-2}$ and $Bh = 0$. We conclude that each class in $\operatorname{Ker}[B]_i$ has a representative in $\operatorname{Ker} B$.

Let p_k be given by (1.2). The first step in the induction procedure to construct the required set $\{e_{ij}\}_{i=1,j=1}^{\kappa_j+1\;,q}$ is to choose a basis $[e_{\kappa 1}], \dots, [e_{\kappa p_{\kappa-1}}]$ of $\mathcal{F}_\kappa/\mathcal{F}_{\kappa-1}$. Since $\mathcal{F}_{\kappa+1}/\mathcal{F}_\kappa = \{0\}$, the vectors $[e_{\kappa 1}], \dots, [e_{\kappa p_{\kappa-1}}]$ are in $\operatorname{Ker}[B]_\kappa$. So we may assume that $Be_{\kappa j} = 0$ for $j = 1, \dots, p_{\kappa-1}$.

The second step is the induction step. Assume that $[e_{k1}], \dots, [e_{kp_{k-1}}]$ is a basis of $\mathcal{F}_k/\mathcal{F}_{k-1}$. Choose a set of vectors $e_{k-1\;1}, \dots, e_{k-1 p_{k-1}}$ in $\mathcal{F}_{k-1}$ such that $Be_{k-1\;j} = Pe_{kj}$ for $j = 1, \dots, p_{k-1}$. Hence $[e_{k-1\;1}], \dots, [e_{k-1\;p_{k-1}}]$ is independent in $\mathcal{F}_{k-1}/\mathcal{F}_{k-2}$ and $[B]_k[e_{k-1\;j}] = [e_{kj}]$. Next choose vectors $e_{k-1\;p_{k-1}+1}$, $\dots, e_{k-1\;p_{k-2}}$ in $\mathcal{F}_{k-1}$ such that $Be_{k-1\;j} = 0$ for $j = p_{k-1}+1, \dots, p_{k-2}$ and $[e_{k-1\;p_{k-1}+1}], \dots, [e_{k-1\;p_{k-2}}]$ is a basis of $\operatorname{Ker}[B]_k$. So $[e_{k-1\;1}], \dots, [e_{k-1\;p_{k-2}}]$ is a basis of $\mathcal{F}_{k-1}/\mathcal{F}_{k-2}$.

At $k = 2$ we end up with a basis $e_{11}, \dots, e_{1p_0}$ of $\operatorname{Ker} P$. Note that $q = p_0$ and that (ii) is fulfilled by the set $\{e_{ij}\}_{i=1\;j=1}^{\kappa\;p_{i-1}}$. Moreover this set is linearly independent in $\mathcal{F}_\kappa$. Since $\mathcal{F}_0 = \{0\}$, we have that

$$\dim \mathcal{F}_\kappa = \sum_{j=1}^{\kappa} \dim(\mathcal{F}_j/\mathcal{F}_{j-1}) = \sum_{j=1}^{\kappa} p_{j-1},$$

which shows that the number of vectors in the set $\{e_{ij}\}_{i=1\;j=1}^{\kappa\;p_{i-1}}$ is equal to $\dim \mathcal{F}_\kappa$. Hence $\{e_{ij}\}_{i=1\;j=1}^{\kappa\;p_{i-1}}$ is a basis of $\mathcal{F}_\kappa$. To complete the proof it suffices to remark that with κ_j given by (1.3) we have

$$\{e_{ij}\}_{i=1\;j=1}^{\kappa\;p_{i-1}} = \{e_{ij}\}_{i=1\;j=1}^{\kappa_j+1\;q}.$$

To see this note that $1 \le j \le p_{i-1}$ is equivalent to $1 \le i \le \kappa_j + 1$. □

One could wonder if it is possible to choose the subspace $\mathcal{G}_\infty$ and the independent set of vectors $\{e_{ij}\}_{i=1,j=1}^{\kappa_j+1\;,q}$ in Proposition 1.3 such that either $Be_{ij} = e_{i+1\;j}$ for $i = 1, \dots, \kappa_j$ or $B[\mathcal{G}_\infty] \subset \mathcal{G}_\infty$. The next example shows that this not possible in general.

Let e_1, e_2, e_3 be the standard basis of $\mathbb{C}^3$. Choose the projection

$$P = \begin{pmatrix} 0 & 0 & 0 \\ 0 & 1 & 0 \\ 0 & 0 & 1 \end{pmatrix} ,$$

and let $B : \mathbb{C}^3 \to \operatorname{Im} P$ be defined by $Be_1 = e_2$, $Be_2 = e_2$ and $Be_3 = e_2 + e_3$. Consider the block $(B; P, I)$ on $\mathbb{C}^3$. We are going to construct the independent set of vectors $\{e_{ij}\}_{i=1,j=1}^{\kappa_j+1\ ,q}$ and the subspace $\mathcal{G}_\infty$ as given in Proposition 1.3. First note that $\operatorname{Ker} P = \operatorname{span}\{e_1\}$ and therefore we have that $q = 1$ and $e_{11} = e_1$. Now $Be_1 = e_2$ and $Be_2 = e_2$. This forces us to choose $e_{21} = -e_1 + e_2$ in order to obtain $Be_{11} = Pe_{21}$ and $Be_{21} = 0$, as required by condition (iv). One sees that $Be_{11} \neq e_{21}$. Next we try to find $\mathcal{G}_\infty$. Condition (i) implies that $\mathcal{G}_\infty = \operatorname{span}\{e_3 + \alpha_1 e_1 + \alpha_2 e_2\}$ for some constants α_1 and α_2. We have

$$B(e_3 + \alpha_1 e_1 + \alpha_2 e_2) = e_3 + (1 + \alpha_1 + \alpha_2)e_2 \ .$$

The right hand side cannot be a multiple of $e_3 + \alpha_1 e_1 + \alpha_2 e_2$, since this would give that the multiplication factor would be 1, and hence $\alpha_1 = 0$ and $\alpha_2 = 1 + \alpha_2$. We conclude that it is impossible to choose α_1 and α_2 in such a way that condition (i) is fulfilled and $B[\mathcal{G}_\infty] \subset \mathcal{G}_\infty$. Next let us give a choice of $\mathcal{G}_\infty$ that fulfills the conditions (i) and (iii). For example, take $\mathcal{G}_\infty = \operatorname{span}\{-e_1 + e_2 + e_3\}$. Then $B[\mathcal{G}_\infty] = \operatorname{span}\{e_2 + e_3\}$ and $B[\mathcal{G}_\infty]$ is a subset of $\mathcal{G}_\infty \oplus \operatorname{Ker} P$.

In the next example we carry out the construction described in the proof of Proposition 1.3 for a concrete case. Let $\mathcal{X} = \mathbb{C}^7$, and let P be the projection along the space spanned by e_5, e_7 onto the space spanned by e_1, e_2, e_3, e_4, e_6. Here $e_1, \ldots, e_7$ denotes the standard basis of $\mathbb{C}^7$. The matrix of $B : \mathcal{X} \to \operatorname{Im} P$ with respect to the basis $e_1, \ldots, e_7$ of $\mathcal{X}$ and the basis e_1, e_2, e_3, e_4, e_6 of $\operatorname{Im} P$ is given by

$$B = \begin{pmatrix} 0 & -1 & -1 & -1 & 0 & -1 & 0 \\ 1 & 2 & 2 & 1 & 0 & 1 & 0 \\ 0 & 0 & 0 & 1 & -\frac{1}{2} & 1 & \frac{1}{2} \\ 0 & 0 & 0 & 0 & 1 & 0 & 0 \\ 0 & 0 & 0 & 0 & -\frac{1}{2} & 0 & -\frac{1}{2} \end{pmatrix} .$$

So $(B; P, I)$ is the block associated with the positioned submatrix

$$\begin{pmatrix} 0 & -1 & -1 & -1 & 0 & -1 & 0 \\ 1 & 2 & 2 & 1 & 0 & 1 & 0 \\ 0 & 0 & 0 & 1 & -\frac{1}{2} & 1 & \frac{1}{2} \\ 0 & 0 & 0 & 0 & 1 & 0 & 0 \\ ? & ? & ? & ? & ? & ? & ? \\ 0 & 0 & 0 & 0 & -\frac{1}{2} & 0 & -\frac{1}{2} \\ ? & ? & ? & ? & ? & ? & ? \end{pmatrix} . \tag{1.4}$$

The first step is to compute the subspaces $\mathcal{F}_j$. We get $\mathcal{F}_1 = \operatorname{Ker} P$, which can be rewritten as $\mathcal{F} = \operatorname{span}\{e_5, e_7\}$. Next one computes that $\mathcal{F}_2 = \mathcal{F}_1 + B[\mathcal{F}_1]$, which can be written as

$$\mathcal{F}_2 = \operatorname{Im} \begin{pmatrix} 0 & 0 & 0 & 0 \\ 0 & 0 & 0 & 0 \\ 0 & 0 & -\frac{1}{2} & \frac{1}{2} \\ 0 & 0 & 1 & 0 \\ 1 & 0 & 0 & 0 \\ 0 & 0 & -\frac{1}{2} & -\frac{1}{2} \\ 0 & 1 & 0 & 0 \end{pmatrix} = \begin{pmatrix} e_5 & e_7 & f_3 & f_4 \end{pmatrix} .$$

Note that $Be_5 = f_3$ and $Be_7 = f_4$. The space $\mathcal{F}_3 = \mathcal{F}_1 + B[\mathcal{F}_2]$ is now given by

$$\mathcal{F}_3 = \operatorname{Im} \begin{pmatrix} 0 & 0 & 0 & 0 & 0 \\ 0 & 0 & 0 & 0 & \frac{1}{2} \\ 0 & 0 & -\frac{1}{2} & \frac{1}{2} & -\frac{1}{2} \\ 0 & 0 & 1 & 0 & 0 \\ 1 & 0 & 0 & 0 & 0 \\ 0 & 0 & -\frac{1}{2} & -\frac{1}{2} & 0 \\ 0 & 1 & 0 & 0 & 0 \end{pmatrix} = \begin{pmatrix} e_5 & e_7 & f_3 & f_4 & f_5 \end{pmatrix} .$$

Remark that $-Bf_3 = Bf_4 = f_5$. It is easy to check that $\mathcal{F}_4 = \mathcal{F}_3$, and therefore the smallest number j such that $\mathcal{F}_{j+1} = \mathcal{F}_j$ is 3. So $\kappa = 3$.

The next step is to compute the subspace $\mathcal{G}_\infty$. We start with a complement $\mathcal{G}_3$ to $\mathcal{F}_3$. Choose $\mathcal{G}_3 = \operatorname{span}\{e_1, e_4\}$. Compute that

$$Be_1 = e_2 = 2f_5 + f_4 - f_3 + e_4 = B(2f_4 + e_7 - e_5) + e_4$$

and that $Be_4 = -e_1 + e_2 + e_3 = -e_1 + 2f_5 + 2f_4 - 2f_3 + 2e_4$, which shows that $Be_4 = -e_1 + 2e_4 + B(2f_4 + 2e_7 - 2e_5)$. Therefore we choose $\mathcal{G}_2 = \operatorname{span}\{e_1 - (2f_4 + e_7 - e_5), e_4 - (2f_4 + 2e_7 - 2e_5)\}$. Let us denote $g_1 = e_1 - (2f_4 + e_7 - e_5)$ and $g_2 = e_4 - (2f_4 + 2e_7 - 2e_5)$. The final step in the construction of $\mathcal{G}_1$ now follows. Compute that $Bg_1 = e_4$ and $Bg_2 = -e_1 + 2e_4$. We express e_1 and e_4 in terms of Be_5, Be_7, e_5, e_7, g_1 and g_2, and find that $e_1 = 2Be_7 + g - 1 + e_7 - e_5$ and $e_4 = 2Be_7 + 2e_7 - 2e_5 + g_2$. Thus we choose $\mathcal{G}_1 = \operatorname{span}\{g_1 - 2e_7, g_2 - 2e_7\}$. It must follow that $B[\mathcal{G}_1] \subset (\mathcal{G}_1 + \operatorname{Ker} P)$. It is easy to check that this is the case indeed. So $\mathcal{G}_\infty = \operatorname{span}\{g_1 - 2e_7, g_2 - 2e_7\}$.

Finally, we compute the desired basis for $\mathcal{F}_3$. Since $\mathcal{F}_3 = \mathcal{F}_2 \oplus \operatorname{span}\{f_5\}$ and $Bf_5 = 0$, we choose $e_{31} = f_5$. Choose $e_{21} = f_4$. Then $Be_{21} = P_{31} = e_{31}$. Because $\mathcal{F}_2 = \mathcal{F}_1 \oplus \operatorname{span}\{f_3, f_4\}$ and $B(f_3 + f_4) = 0$, we choose $e_{22} = f_3 + f_4$. Finally choose $e_{11} = e_7$ and $e_{12} = e_5 + e_7$. This finishes the construction of the basis of $\mathcal{F}_\kappa = \mathcal{F}_3$.

Proof of Theorem 1.1. Apply Proposition 1.3 to construct an independent set of vectors $\{e_{ij}\}_{i=1,j=1}^{\kappa_j+1\ ,q}$ and a subspace $\mathcal{G}_\infty$ that fulfill the conditions (i) to (iv) of that proposition. Let P' be the projection along $\operatorname{span}\{e_{11},\dots,e_{1q}\}$ onto the space

$$\mathcal{G}_\infty \oplus \operatorname{span}\{e_{ij}\}_{i=2,j=1}^{\kappa_j+1\ ,q} .$$

We write $B' = P'B$, and consider the block $(B'; P', I)$ on $\mathcal{X}$. The block $(B; P, I)$ is block similar to $(B'; P', I)$ with a block similarity of the second kind. We will show that $(B'; P', I)$ is a block similar to a direct sum of shifts of the first kind and the operator $J = B'|_{\mathcal{G}_\infty} : \mathcal{G}_\infty \to \mathcal{G}_\infty$. Put

$$\mathcal{X}_j = \operatorname{span}\{e_{1j},\dots,e_{\kappa_j+1\ j}\}.$$

Then $\mathcal{X}_j$ is regularly $(B'; P', I)$-invariant. Write $(B_j; P_j, I_j)$ for the regular restriction of $(B'; P', I)$ to $\mathcal{X}_j$. Remark that $(B_j; P_j, I_j)$ is block similar to a shift of the first kind with index κ_j. Clearly we get that

$$(B'; P', I) = (B_1; P_1, I_1) \oplus \cdots \oplus (B_q; P_q, I_q) \oplus (J; I_0, I_0),$$

where I_0 is the identity on $\mathcal{G}_\infty$. This proves the first part of the theorem.

It remains to show the uniqueness of the indices $\kappa_1, \dots, \kappa_q$ and the uniqueness of J up to similarity. Assume that $(B; P, I)$ is also similar to

$$(B''; P'', I) = (B_1'; P_1', I_1') \oplus \cdots \oplus (B_p'; P_p', I_p') \oplus (M; I_0', I_0'),$$

where $(B_i'; P_i', I_i')$ is a shift of the first kind with index ν_i. Assume $\nu_1 \geq \cdots \geq \nu_p \geq 0$. There exists a block similarity S of $(B'; P', I)$ and $(B''; P'', I)$. Put

$$\begin{aligned} &\mathcal{F}_0' = \mathcal{X},\ \mathcal{F}_1' = \operatorname{Ker} P',\ \mathcal{F}_j' = \mathcal{F}_1' + B'[\mathcal{F}_{j-1}'], \\ &\mathcal{F}_0'' = \mathcal{X},\ \mathcal{F}_1'' = \operatorname{Ker} P'',\ \mathcal{F}_j'' = \mathcal{F}_1' + B''[\mathcal{F}_{j-1}'']. \end{aligned} \tag{1.5}$$

Since S is a block similarity, we have that $S[\mathcal{F}_1'] = \mathcal{F}_1''$. Assume that $S[\mathcal{F}_{j-1}'] = \mathcal{F}_{j-1}''$ for some $j \geq 2$. We compute that

$$S\mathcal{F}_j' = S\mathcal{F}_1' + SB'\mathcal{F}_{j-1}' = \mathcal{F}_1'' + SB'\mathcal{F}_{j-1}'.$$

Because $(SB' - B''S)x \in \mathcal{F}_1''$, for each x, it follows that

$$\mathcal{F}_1'' + SB'\mathcal{F}_{j-1}' = \mathcal{F}_1'' + B''S\mathcal{F}_{j-1}' = \mathcal{F}_1'' + B''\mathcal{F}_{j-1}'' = \mathcal{F}_j''.$$

This shows that $S\mathcal{F}_j' = \mathcal{F}_j''$ for each j. Now we compute

$$\begin{aligned} \dim \mathcal{F}_j' &= \min\{\kappa_1 + 1, j\} + \cdots + \min\{\kappa_q + 1, j\}, \\ \dim \mathcal{F}_j'' &= \min\{\nu_1 + 1, j\} + \cdots + \min\{\nu_p + 1, j\}. \end{aligned}$$

In particular for $j = 1$ this proves that $p = q$. Remark that

$$\dim \mathcal{F}'_{j+1} - \dim \mathcal{F}'_j = \sum_{i=1}^{q} (\min\{\kappa_i + 1, j+1\} - \min\{\kappa_i + 1, j\}) = \#\{i \mid \kappa_i \geq j\},$$

$$\dim \mathcal{F}''_{j+1} - \dim \mathcal{F}''_j = \sum_{i=1}^{p} (\min\{\nu_i + 1, j+1\} - \min\{\nu_i + 1, j\}) = \#\{i \mid \nu_i \geq j\}.$$

Since both $(\kappa_i)_{i=1}^q$ and $(\nu_i)_{i=1}^q$ are descending sequences, this proves that $\kappa_i = \nu_i$, for $i = 1, \ldots, q$. Put

$$\kappa = \min\{j \mid \mathcal{F}'_j = \mathcal{F}'_{j+1}\} = \min\{j \mid \mathcal{F}''_j = \mathcal{F}''_{j+1}\}.$$

Consider the regular restrictions $(B'_0; P'_0, I')$ of $(B'; P', I)$ to $\mathcal{F}'_\kappa$ and $(B''_0; P''_0, I'')$ of $(B''; P'', I)$ to $\mathcal{F}''_\kappa$. These restrictions are block similar since each is the direct sum of shifts of the first kind with indices $\kappa_1, \ldots, \kappa_q$. In fact, the transformation $S|_{\mathcal{F}'_\kappa} : \mathcal{F}'_\kappa \to \mathcal{F}''_\kappa$ gives the block similarity. Write

$$(B'; P', I') = (B'_0; P'_0, I') \oplus (J; I_0, I_0),$$
$$(B''; P'', I) = (B''_0; P''_0, I'') \oplus (M; I'_0, I'_0).$$

Apply Theorem II.2.8 to obtain that $(J; I_0, I_0)$ is block similar to $(M; I'_0, I'_0)$. This shows that the operators J and M are similar. □

We continue with the example preceding the proof of Theorem 1.1. Recall that in this example $(B; P, I)$ is the block associated with the positioned submatrix (1.4). Here we construct the shifts of the first kind and the invariant polynomials of this block. We already have seen that the block $(B; P, I)$ has indices of the first kind 2 and 1. Choose P' to be the projection with

$$\operatorname{Im} P' = \operatorname{span}\{e_{21}, e_{31}, , e_{22}, g_1 - 2e_7, g_2 - 2e_7\}$$

and $\operatorname{Ker} P' = \operatorname{Ker} P$. One computes that the matrix of $P'B : \mathbb{C}^7 \to \operatorname{Im} P'$ with respect to the bases $e_{11}, e_{21}, e_{31}, e_{12}, e_{22}, g_1 - 2e_7, g_2 - 2e_7$ of $\mathbb{C}^7$ and $e_{21}, e_{31}, , e_{22}, g_1 - 2e_7, g_2 - 2e_7$ of $\operatorname{Im} P'$ is given by

$$\begin{pmatrix} 1 & 0 & 0 & 0 & 0 & 0 & 0 \\ 0 & 1 & 0 & 0 & 0 & 0 & 0 \\ 0 & 0 & 0 & 1 & 0 & 0 & 0 \\ 0 & 0 & 0 & 0 & 0 & 0 & -1 \\ 0 & 0 & 0 & 0 & 0 & 1 & 2 \end{pmatrix}.$$

So we see that the only non-constant invariant polynomial of the block is $(\lambda - 1)^2$.

Proof of Theorem 1.2. In Proposition 1.3 we constructed the independent set of vectors $\{e_{ij}\}_{i=1,\ j=1}^{\kappa,\ p_{i-1}} = \{e_{ij}\}_{i=1,j=1}^{\kappa_j+1\ ,q}$, where $p_j = \dim(\mathcal{F}_{j+1}/\mathcal{F}_j)$ and the relation between the sequences (κ_j) and (p_j) is given by (1.3). In the proof of Theorem 1.1 we showed that the numbers $\kappa_1 \geq \cdots \kappa_q$ are the indices of the first kind of the block $(B; P, I)$ indeed. This proves the first part of the theorem.

To prove the second part of the theorem we take P_1 to be the projection along $\operatorname{Ker} P$ onto $\operatorname{span}\{e_{ij}\}_{i=2,j=1}^{\kappa_j+1,q} \oplus \mathcal{G}_\infty$. Then, by definition, the invariant polynomials of $P_1 B|_{\mathcal{G}_\infty}$ are the invariant polynomials of the block. Next, let P_1' be any projection along $\operatorname{Ker} P$ with $\mathcal{G}_\infty \subset \operatorname{Im} P$. It is easy to check that $P_1' B|_{\mathcal{G}_\infty}$ is similar to $P_1 B|_{\mathcal{G}_\infty}$, the similarity transformation being $P_1|_{\mathcal{G}_\infty}$. □

Theorem 1.4. *The blocks $(B_1; P_1, I_1)$ on $\mathcal{X}_1$ and $(B_2; P_2, I_2)$ on $\mathcal{X}_2$ are block similar if and only if they have the same indices of the first kind and the same invariant polynomials.*

PROOF. Assume that the blocks $(B_1; P_1, I_1)$ and $(B_2; P_2, I_2)$ are block similar. Let $(B_1; P_1, I_1)$ be block similar to a direct sum of shifts of the first kind with indices $\kappa_1, \ldots, \kappa_q$ and an operator with invariant polynomials $p_1(\lambda), \ldots, p_n(\lambda)$. Then also $(B_2; P_2, I_2)$ is similar to this direct sum. This proves that the indices of the first kind of $(B_2; P_2, I_2)$ are $\kappa_1, \ldots, \kappa_q$ and the invariant polynomials are $p_1(\lambda), \ldots, p_n(\lambda)$. Conversely, assume that $(B_1; P_1, I_1)$ and $(B_2; P_2, I_2)$ have the same indices of the first kind and the same invariant polynomials. Then both $(B_1; P_1, I_1)$ and $(B_2; P_2, I_2)$ are block similar to the same direct sum of shifts of the first kind and an operator. So $(B_1; P_1, I_1)$ and $(B_2; P_2, I_2)$ are block similar. □

Let $(B; P, I)$ be a full width block on the space $\mathcal{X}$. Put $\mathcal{X}' = \operatorname{Im} P$. Let P' be a projection of $\mathcal{X}'$ with $\operatorname{Ker} P' = B[\operatorname{Ker} P] \subset \mathcal{X}'$, and write $B' = P'B|_{\mathcal{X}'} : \mathcal{X}' \to \operatorname{Im} P$. The block $(B'; P', I_{\mathcal{X}'})$ is called *a compression of* $(B; P, I)$ *to* $\operatorname{Im} P$. Note that different choices of P' lead to different compressions which are all block similar with a block similarity of the second kind. We give an example. Let $(B; P, I)$ be a shift of the first kind on $\mathbb{C}^n$ with $n \geq 2$. Thus $\operatorname{Ker} P = \operatorname{span}\{e_1\}$, $\operatorname{Im} P = \operatorname{span}\{e_2, \ldots, e_n\}$ and $Be_i = e_{i+1}$ for $i = 1, \ldots, n-1$, $Be_n = 0$. Then $B[\operatorname{Ker} P] = \operatorname{span}\{e_2\} \subset \mathcal{X}' = \operatorname{Im} P$. Let P' be the projection of $\mathcal{X}'$ along e_2 onto $\operatorname{span}\{e_3, \ldots, e_n\}$. Set $B'e_i = e_{i+1}$ for $i = 2, \ldots, n-1$ and $B'e_n = 0$. Then $(B'; P', I_{\mathcal{X}'})$ is a compression of $(B; P, I)$ to $\operatorname{Im} P$ and $(B'; P', I_{\mathcal{X}'})$ is block similar to a shift of the first kind with index $n-1$. The next theorem describes the general case.

Theorem 1.5. *Let $(B; P, I)$ be a block on the space $\mathcal{X}$ with indices of the first kind $\kappa_1 \geq \cdots \geq \kappa_p > 0 = \kappa_{p+1} = \cdots = \kappa_q$ and invariant polynomials $p_1, \ldots, p_n$. The compression of $(B; P, I)$ to $\operatorname{Im} P$ has invariant polynomials $p_1, \ldots, p_n$ and indices of the first kind $\kappa_1 - 1 \geq \cdots \geq \kappa_p - 1 \geq 0$.*

PROOF. We define the subspaces $\mathcal{F}_j$ by formula (1.1). The subspaces $\mathcal{F}_j'$ are defined in a similar way by

$$\mathcal{F}_0' = \{0\}, \ \mathcal{F}_1' = \operatorname{Ker} P', \ \mathcal{F}_j' = \operatorname{Ker} P' \oplus B'[\mathcal{F}_{j-1}'].$$

for $j = 1, \dots$. Remark that $\mathcal{F}_2 = \operatorname{Ker} P \oplus \mathcal{F}'_1$. Assume that we proved that $\mathcal{F}_j = \operatorname{Ker} P \oplus \mathcal{F}'_{j-1}$. We compute that $\mathcal{F}_{j+1} = \operatorname{Ker} P \oplus B[\operatorname{Ker} P] + B[\mathcal{F}'_{j-1}]$. Use the definition of $\operatorname{Ker} P'$ to obtain $\mathcal{F}_{j+1} = \operatorname{Ker} P \oplus \operatorname{Ker} P' + B[\mathcal{F}'_{j-1}]$. Now it follows from the definition of B' that $\mathcal{F}_{j+1} = \operatorname{Ker} P \oplus \operatorname{Ker} P' + B'[\mathcal{F}'_{j-1}]$. Again use the definition of $\mathcal{F}'_j$ to see that we may conclude that $\mathcal{F}_{j+1} = \operatorname{Ker} P \oplus \mathcal{F}'_j$. Now the relation between the indices is clear from Theorem 1.2.

The next step is to show that the invariant polynomials of $(B'; P', I_{\mathcal{X}'})$ are the invariant polynomials of $(B; P, I)$. We construct the subspace $\mathcal{G}'_\infty$ such that (1') $\mathcal{F}'_{\kappa-1} \oplus \mathcal{G}'_\infty = \mathcal{X}'$ and (2') $B'[\mathcal{G}'_\infty] \subset \mathcal{G}'_\infty \oplus \operatorname{Ker} P'$. Now choose P'_1 to be a projection with $\operatorname{Ker} P'_1 = \operatorname{Ker} P'$ and $\mathcal{G}'_\infty \subset \operatorname{Im} P'_1$. Then the invariant polynomials of $(B'; P', I_{\mathcal{X}'})$ are the invariant polynomials of $P'_1 B' : \mathcal{G}'_\infty \to \mathcal{G}'_\infty$. Remark that

$$\mathcal{X} = \operatorname{Ker} P \oplus \mathcal{X}' = \operatorname{Ker} P \oplus \mathcal{F}'_{\kappa-1} \oplus \mathcal{G}'_\infty = \mathcal{F}_\kappa \oplus \mathcal{G}'_\infty.$$

We do not necessarily have $B[\mathcal{G}'_\infty] \subset \mathcal{G}'_\infty \oplus \operatorname{Ker} P$. For $x \in \mathcal{G}'_\infty$ we compute

$$\begin{aligned} Bx &= (I - P'_1)Bx + P'_1 Bx = (I - P'_1)Bx + P'_1 P' Bx \\ &= (I - P'_1)Bx + P'_1 B' x. \end{aligned}$$

So we see that $Bx - (I - P'_1)Bx = P'_1 B' x \in \mathcal{G}'_\infty$. From the definition of P'_1 we know that $(I - P'_1)Bx \in B[\operatorname{Ker} P]$. So there exists a linear transformation $T : \mathcal{G}'_\infty \to \operatorname{Ker} P$ such that $(I - P'_1)B = BT$. For each $x \in \mathcal{G}'_\infty$ we see that

$$B(I - T)x \in \mathcal{G}'_\infty \subset (I - T)[\mathcal{G}'_\infty] + \operatorname{Ker} P.$$

We choose the subspace $\mathcal{G}_\infty = (I - T)[\mathcal{G}'_\infty]$ of $\mathcal{X}$. Then it follows from $\mathcal{X} = \mathcal{F}_\kappa \oplus \mathcal{G}'_\infty$ and the fact that $T : \mathcal{G}'_\infty \to \operatorname{Ker} P$ that $\mathcal{X} = \mathcal{F}_\kappa \oplus \mathcal{G}_\infty$. Also we constructed $\mathcal{G}_\infty$ such that $B[\mathcal{G}_\infty] \subset \mathcal{G}_\infty \oplus \operatorname{Ker} P$. From Theorem 1.2 we now conclude that the invariant polynomials of $(B; P, I)$ are the invariant polynomials of $P_1 B : \mathcal{G}_\infty \to \mathcal{G}_\infty$, where P_1 is a projection of $\mathcal{X}$ with $\operatorname{Ker} P = \operatorname{Ker} P_1$ and $\mathcal{G}_\infty \subset \operatorname{Im} P_1$. For $x \in \mathcal{G}'_\infty$ we see that

$$P(P_1 B)(I - T)x = B(I - T)x = P'_1 B' x$$

and

$$P(I - T) = I_{\mathcal{G}'_\infty} : \mathcal{G}'_\infty \to \mathcal{G}'_\infty.$$

This proves that $P_1 B : \mathcal{G}_\infty \to \mathcal{G}_\infty$ and $P'_1 B' : \mathcal{G}'_\infty \to \mathcal{G}'_\infty$ are similar. Therefore the invariant polynomials of $(B'; P', I_{\mathcal{X}'})$ and $(B; P, I)$ are the same □

Corollary 1.6. *The blocks $(B_1; P_1, I_1)$ on $\mathcal{X}_1$ and $(B_2; P_2, I_2)$ on $\mathcal{X}_2$ are block similar if and only if $\dim \mathcal{X}_1 = \dim \mathcal{X}_2$ and a compression $(B'_1; P'_1, I'_1)$ of $(B_1; P_1, I_1)$ to $\operatorname{Im} P_1$ is block similar to a compression $(B'_2; P'_2, I'_2)$ of $(B_2; P_2, I_2)$ to $\operatorname{Im} P_2$.*

PROOF. If $(B_1; P_1, I_1)$ and $(B_2; P_2, I_2)$ are block similar, then the indices of the first kind and the invariant polynomials of these blocks are equal. From Theorem 1.6 we obtain that the indices of the first kind and the invariant polynomials of $(B_1'; P_1', I_1')$ and $(B_2'; P_2', I_2')$ are equal. So $(B_1'; P_1', I_1')$ and $(B_2'; P_2', I_2')$ are block similar.

Now, conversely, assume that $\dim \mathcal{X}_1 = \dim \mathcal{X}_2$, and that the indices of the first kind and the invariant polynomials of $(B_1'; P_1', I_1')$ and $(B_2'; P_2', I_2')$ are equal. Then the invariant polynomials and the nonzero indices of the first kind of $(B_1; P_1, I_1)$ and $(B_2; P_2, I_2)$ are equal, according to Theorem 1.6. Since, for $i = 1, 2$, the total number of indices of the first kind of $(B_i; P_i, I_i)$ is equal to $\dim \mathcal{X}_i - \dim \operatorname{Im} P_i = \dim \mathcal{X}_i - \dim \mathcal{X}_i'$, the blocks $(B_1; P_1, I_1)$ and $(B_2; P_2, I_2)$ also have the same number of indices of the first kind that are equal to 0. So $(B_1; P_1, I_1)$ and $(B_2; P_2, I_2)$ have the same indices of the first kind and the same invariant polynomials. Apply Theorem 1.4 to conclude that $(B_1; P_1, I_1)$ and $(B_2; P_2, I_2)$ are block similar. □

V.2 Finite dimensional operator pencils

In Section III.2 we defined the notions of strict equivalence for matrix pencils and of direct sum of pencils. Here we will consider the meaning of these notions for a special class of linear operator pencils that is closely related to the class of full width blocks. We shall consider operator pencils of the form

$$L(\lambda) = \begin{pmatrix} \lambda I_{\mathcal{X}} - F & -G \end{pmatrix} = \lambda \begin{pmatrix} I_{\mathcal{X}} & 0 \end{pmatrix} - \begin{pmatrix} F & G \end{pmatrix}, \tag{2.1}$$

where $F : \mathcal{X} \to \mathcal{X}$ and $G : \mathcal{Y} \to \mathcal{X}$ are linear operators acting between finite dimensional spaces, and $I_{\mathcal{X}}$ is the identity on $\mathcal{X}$. A simple example of such a pencil is

$$\lambda U_\kappa - V_\kappa = \begin{pmatrix} \lambda & -1 & 0 & \cdots & 0 & 0 \\ 0 & \lambda & -1 & & 0 & 0 \\ 0 & 0 & \lambda & & 0 & 0 \\ \vdots & \vdots & & \ddots & & \\ 0 & 0 & 0 & & \lambda & -1 \end{pmatrix} : \mathbb{C}^{\kappa+1} = \mathbb{C}^\kappa \oplus \mathbb{C}^1 \to \mathbb{C}^\kappa \tag{2.2}$$

Note that in this case

$$U_\kappa = \begin{pmatrix} I_\kappa & 0 \end{pmatrix} : \mathbb{C}^\kappa \oplus \mathbb{C}^1 \to \mathbb{C}^\kappa, \quad V_\kappa = \begin{pmatrix} F & G \end{pmatrix} : \mathbb{C}^\kappa \oplus \mathbb{C}^1 \to \mathbb{C}^\kappa, \tag{2.3}$$

where $Fe_{i+1} = e_i$ for $i = 1, \ldots, \kappa - 1$ and $Fe_1 = 0$, and $G(1) = e_\kappa$.

We may also choose $\mathcal{Y} = \{0\}$. In this case we identify $\mathcal{X} \oplus \mathcal{Y}$ with $\mathcal{X}$. Thus then the pencil takes the form $\lambda I - F$. If $\mathcal{X} = \{0\}$, then the pencil reduces to the zero operator $O : \mathcal{Y} \to \{0\}$. A pencil of the form (2.2) is called an *elementary pencil with minimal column index* κ. The number κ is called the *minimal column index* of the pencil (2.2). We allow that $\kappa = 0$. One of the main results is that any operator pencil of the form (2.1) is strictly equivalent to a direct sum of pencils of the types described above. This is the contents of the next result.

Theorem 2.1. *Let $\mathcal{X}$ be an n-dimensional linear space, $\mathcal{Y}$ be an m-dimensional space, and let $L(\lambda)$ be an operator pencil given by (2.1). Then $L(\lambda)$ is strictly equivalent to a pencil $L'(\lambda)$ which is a direct sum of elementary pencils with minimal column indices $\kappa_1 \geq \kappa_2 \geq \cdots \geq \kappa_m$, and a pencil $\lambda I - J$.*

If in the pencil $L'(\lambda)$ the matrix J is in Jordan canonical form, then the pencil $L'(\lambda)$ is called the *Kronecker canonical form* of the pencil $L(\lambda)$. The numbers $\kappa_1 \geq \cdots \geq \kappa_m$, are called *the minimal column indices* of $L(\lambda)$. The invariant polynomials of $\lambda I - J$ are called *the invariant polynomials* of $L(\lambda)$.

The proof of this theorem will be based on a connection with full width blocks. Let $L(\lambda)$ be given by (2.1). We define

$$B = \begin{pmatrix} F & G \end{pmatrix} : \mathcal{X} \oplus \mathcal{Y} \to \mathcal{X}. \tag{2.4}$$

We identify $\mathcal{X}$ with the subspace $\mathcal{X} \oplus \{0\}$ of $\mathcal{X} \oplus \mathcal{Y}$. Put P to be the projection of $\mathcal{X} \oplus \mathcal{Y}$ onto $\mathcal{X} \oplus \{0\}$ and along $\{0\} \oplus \mathcal{Y}$. The block $(B; P, I)$ on $\mathcal{X} \oplus \mathcal{Y}$ will be called *the block associated to* $L(\lambda)$. Remark that the block associated to a direct sum of pencils is the direct sum of blocks associated with these pencils.

Consider the pencil (2.2). We view the operator V_κ defined in (2.3) as an operator $V_\kappa : \mathbb{C}^\kappa \oplus \mathbb{C}^1 \to \mathbb{C}^\kappa \oplus \{0\} \subset \mathbb{C}^\kappa \oplus \mathbb{C}^1$. Then $V_\kappa e_{i+1} = e_i$ for $i = 1, \ldots, \kappa$ and $V_\kappa e_1 = 0$. Here we identify $\mathbb{C}^\kappa \oplus \mathbb{C}^1$ with $\mathbb{C}^{\kappa+1}$ and $e_1, \ldots, e_{\kappa+1}$ is the standard basis in $\mathbb{C}^{\kappa+1}$. Let P be the projection of $\mathbb{C}^{\kappa+1}$ along $e_{\kappa+1}$ onto $\operatorname{span}\{e_1, \ldots, e_\kappa\}$. Then the block $(V_\kappa; P, I)$ is the block associated to $\lambda U_\kappa - V_\kappa$. It is a full width block. Moreover this block is block similar to a shift of the first kind. This one sees by considering the similarity

$$S_\kappa : \mathbb{C}^{\kappa+1} \to \mathbb{C}^{\kappa+1}, \quad S_\kappa e_i = e_{\kappa+2-i}. \tag{2.5}$$

Conversely, assume that the block $(T_\kappa; P', I)$ is a shift of the first kind on $\mathbb{C}^{\kappa+1}$. Then apply the block similarity S_κ defined above to this shift. We obtain the block $(V_\kappa; P, I)$, which is the block associated to the pencil (2.2).

In the proof of Theorem 2.1 the following result will play a key role.

Theorem 2.2. *Let $L_1(\lambda) = \begin{pmatrix} \lambda I_1 - F_1 & -G_1 \end{pmatrix} : \mathcal{X}_1 \oplus \mathcal{Y}_1 \to \mathcal{X}_1$ and $L_2(\lambda) = \begin{pmatrix} \lambda I_2 - F_2 & -G_2 \end{pmatrix} : \mathcal{X}_2 \oplus \mathcal{Y}_2 \to \mathcal{X}_2$. Then $L_1(\lambda)$ is strictly equivalent to $L_2(\lambda)$ if and only if the block on $\mathcal{X}_1 \oplus \mathcal{Y}_1$ associated with $L_1(\lambda)$ is block similar to the block on $\mathcal{X}_2 \oplus \mathcal{Y}_2$ associated to $L_2(\lambda)$.*

PROOF. Let $(B_i; P_i, I_i)$ denote the full width block associated with $L_i(\lambda)$, $i = 1, 2$. Assume that $L_1(\lambda)$ and $L_2(\lambda)$ are strictly equivalent. So there exist invertible linear operators $S : \mathcal{X}_1 \oplus \mathcal{Y}_1 \to \mathcal{X}_2 \oplus \mathcal{Y}_2$ and $T : \mathcal{X}_1 \to \mathcal{X}_2$ such that $TL_1(\lambda) = L_2(\lambda)S$. Partition S as follows:

$$S = \begin{pmatrix} S_{11} & S_{12} \\ S_{21} & S_{22} \end{pmatrix} : \mathcal{X}_1 \oplus \mathcal{Y}_1 \to \mathcal{X}_2 \oplus \mathcal{Y}_2.$$

Then, comparing the coefficients of λ in $TL_1(\lambda) = L_2(\lambda)S$, we see that $T\,(\,I_1 \quad 0\,) = (\,I_2 \quad 0\,)\,S$, which proves that $T = S_{11}$ and $S_{12} = 0$. It follows that $S\mathcal{Y}_1 = \mathcal{Y}_2$, i.e., $S[\operatorname{Ker} P_1] = \operatorname{Ker} P_2$. Next we use that $T\,(\,F_1 \quad G_1\,) = (\,F_2 \quad G_2\,)\,S$. In order to show that $P_2(SB_1 - B_2S) = 0$ we consider the matrix representation

$$\begin{pmatrix} I_2 & 0 \\ 0 & 0 \end{pmatrix} \left[\begin{pmatrix} T & 0 \\ S_{21} & S_{22} \end{pmatrix} \begin{pmatrix} F_1 & G_1 \\ 0 & 0 \end{pmatrix} - \begin{pmatrix} F_2 & G_2 \\ 0 & 0 \end{pmatrix} \begin{pmatrix} T & 0 \\ S_{21} & S_{22} \end{pmatrix} \right] =$$

$$\begin{pmatrix} T & 0 \\ 0 & 0 \end{pmatrix} \begin{pmatrix} F_1 & G_1 \\ 0 & 0 \end{pmatrix} - \begin{pmatrix} F_2 & G_2 \\ 0 & 0 \end{pmatrix} S = 0$$

This proves that the blocks $(B_1; P_1, I_1)$ and $(B_2; P_2, I_2)$ are block similar.

Conversely, assume that the blocks $(B_1; P_1, I_1)$ and $(B_2; P_2, I_2)$ are block similar. Thus there exists an invertible linear transformation $S : \mathcal{X}_1 \oplus \mathcal{Y}_1 \to \mathcal{X}_2 \oplus \mathcal{Y}_2$ such that $S[\operatorname{Ker} P_1] = \operatorname{Ker} P_2$ and $P_2(SB_1 - B_2S) = 0$. Then $S\mathcal{Y}_1 = \mathcal{Y}_2$ and therefore the block matrix representation of S is

$$S = \begin{pmatrix} T & 0 \\ S_{21} & S_{22} \end{pmatrix} : \mathcal{X}_1 \oplus \mathcal{Y}_1 \to \mathcal{X}_2 \oplus \mathcal{Y}_2.$$

So we see that $T\,(\,I_1 \quad 0\,) = (\,I_2 \quad 0\,)\,S$. This shows that the coefficient of λ in $TL_1(\lambda) - L_2(\lambda)S$ is equal to 0. Also the matrix representation of the equality $P_2(SB_1 - B_2S) = 0$ gives

$$\begin{pmatrix} I_2 & 0 \\ 0 & 0 \end{pmatrix} \left[\begin{pmatrix} T & 0 \\ S_{21} & S_{22} \end{pmatrix} \begin{pmatrix} F_1 & G_1 \\ 0 & 0 \end{pmatrix} - \begin{pmatrix} F_2 & G_2 \\ 0 & 0 \end{pmatrix} \begin{pmatrix} T & 0 \\ S_{21} & S_{22} \end{pmatrix} \right] = 0.$$

From the first row in this equality we read off that $T\,(\,F_1 \quad G_1\,) = (\,F_2 \quad G_2\,)\,S$. This shows that the constant term in $TL_1(\lambda) - L_2(\lambda)S$ is equal to 0. □

PROOF OF THEOREM 2.1. Let $(B; P, I)$ be the block associated to $L(\lambda)$. Then $(B; P, I)$ is a full width block, and hence block similar to a direct sum $(B'; P', I')$ of an operator J and shifts of the first kind with indices $\kappa_1 \geq \cdots \geq \kappa_p > 0 = \kappa_{p+1} = \cdots = \kappa_m$. Apply to each of the shifts the block similarity S_{κ_j} defined by (2.5). We get a block $(B''; P'', I'')$, which is block similar to $(B; P, I)$ and is associated to a direct sum $L'(\lambda)$ of the pencil $\lambda I - J$, pencils of the form (2.2) with minimal column indices $\kappa_1 \geq \cdots \geq \kappa_p > 0$ and a zero pencil $O_{m-p} : \mathbb{C}^{m-p} \to \{0\}$. Theorem 2.2 shows that $L(\lambda)$ and $L'(\lambda)$ are strictly equivalent. □

The next corollary results from the proof of Theorem 2.1.

Corollary 2.3. *The minimal column indices and the invariant polynomials of*

$$L(\lambda) = (\,\lambda I_{\mathcal{X}} - F \quad -G\,) : \mathcal{X} \to \mathcal{X} \oplus \mathcal{Y}$$

are the indices of the third kind and the invariant polynomials of the block

$$\left((F \quad G); \begin{pmatrix} I & 0 \\ 0 & 0 \end{pmatrix}, \begin{pmatrix} I & 0 \\ 0 & I \end{pmatrix}\right)$$

on $\mathcal{X} \oplus \mathcal{Y}$.

From Corollary 2.3 and Theorem 1.1 it follows that the minimal column indices and invariant polynomial are uniquely determined by the pencil. From Corollary 2.3 and Theorem 1.4 one deduces the next result.

Corollary 2.4. *The pencils*

$$L_1(\lambda) = (\,\lambda I_1 - F_1 \quad -G_1\,) : \mathcal{X}_1 \oplus \mathcal{Y}_1 \to \mathcal{X}_1,$$

and

$$L_2(\lambda) = (\,\lambda I_2 - F_2 \quad -G_2\,) : \mathcal{X}_2 \oplus \mathcal{Y}_2 \to \mathcal{X}_2,$$

are strictly equivalent if and only if they have the same minimal column indices and the same invariant polynomials.

We will say that the pencil $L(\lambda)$ has no invariant polynomials if the order of the operator J appearing in Theorem 2.1 is zero. The last theorem in this section characterizes the pencils without invariant polynomials.

Theorem 2.5. *Let $\mathcal{X}$ be an n-dimensional linear space, $\mathcal{Y}$ be an m-dimensional space, and let*

$$L(\lambda) = (\,\lambda I_{\mathcal{X}} - F \quad -G\,) : \mathcal{X} \oplus \mathcal{Y} \to \mathcal{X}$$

be an operator pencil. Then the following three statements are equivalent:

(1) *$L(\lambda)$ has no invariant polynomials;*

(2) *$\operatorname{rank}(\,G \quad FG \quad F^2G \quad \dots \quad F^{n-1}G\,) = n$;*

(3) *$\operatorname{rank} L(\lambda) = n$ for all λ.*

PROOF. First we prove that condition (1) is equivalent to condition (3). Apply Theorem 2.2 to obtain that the polynomial $L(\lambda)$ is strictly equivalent to a direct sum $L'(\lambda)$ of an $s \times s$ pencil $\lambda I - J$, a zero pencil and pencils of the form (2.2) with minimal column indices $\kappa_1 \geq \cdots \geq \kappa_p > 0$. Counting rows in this direct sum gives that $n = s + \sum_{i=1}^{p} \kappa_i$.

Now assume that $L(\lambda)$ has no invariant polynomials. This means $s = 0$. It follows that for each $\lambda \in \mathbb{C}$ the rank of $L'(\lambda)$ is equal to $\sum_{i=1}^{p} \kappa_i = n$. Hence $L(\lambda)$ has rank n for each $\lambda \in \mathbb{C}$.

Conversely, assume that $L(\lambda)$ has rank n for each $\lambda \in \mathbb{C}$. Then $L'(\lambda)$ has rank n for each $\lambda \in \mathbb{C}$. But, if $s > 0$, then $\operatorname{rank} L(\mu) < s + \sum_{i=1}^{p} \kappa_i$ if μ is an eigenvalue of $\lambda I - J$. So we conclude that $s = 0$. Remark that we also showed that $L(\lambda)$ has no invariant polynomials if and only if $\sum_{i=1}^{p} \kappa_i = n$, where $\kappa_1, \dots, \kappa_m$ are the minimal column indices of the pencil.

Next we show that condition (1) and condition (2) are equivalent. Put

$$B = (\,F \quad G\,) : \mathcal{X} \oplus \mathcal{Y} \to \mathcal{X}\,, \; P = \begin{pmatrix} I & 0 \\ 0 & 0 \end{pmatrix} : \mathcal{X} \oplus \mathcal{Y} \to \mathcal{X} \oplus \mathcal{Y}\,.$$

Then $(B; P, I)$ is a full width block on the space $\mathcal{X} \oplus \mathcal{Y}$ associated to $L(\lambda)$. The minimal column indices of the pencil $L(\lambda)$ are the indices of the first kind of the block $(B; P, I)$, and $L(\lambda)$ has no invariant polynomials if and only if $(B; P, I)$ has no invariant polynomials. Define $\mathcal{F}_k$ by (1.1) and p_k by (1.2). Let κ be such that $\mathcal{F}_{\kappa-1} \neq \mathcal{F}_\kappa = \mathcal{F}_{\kappa+1}$. From Theorem 1.2 it follows that $(B; P, I)$ has no invariant polynomials if and only if $\dim \mathcal{F}_\kappa = n + m$. The definitions of the space $\mathcal{F}_\kappa$ and of the operator B give that

$$\mathcal{F}_k = \mathcal{Y} \oplus G[\mathcal{Y}] + \cdots + F^{k-2}G[\mathcal{Y}]$$

for $k = 2, 3, \ldots$, and hence

$$\mathcal{F}_k = \mathcal{Y} \oplus \operatorname{Im} (G \quad FG \quad \ldots \quad F^{k-2}G) \ .$$

The Cayley-Hamilton theorem now guarantees that $\kappa \leq n + 1$. So $\mathcal{F}_\kappa = \mathcal{F}_{n+1}$. Moreover, $\dim \mathcal{F}_\kappa = n + m$ if and only if

$$\operatorname{rank}\begin{pmatrix} G & FG & F^2G & \ldots & F^{n-1}G \end{pmatrix} = n.$$

This shows that (1) and (2) are equivalent. □

V.3 Similarity of operators modulo a subspace

In this section we discuss a notion that plays in this chapter a role similar to that of the non-everywhere defined operators in Chapter III. Let $\mathcal{M}_1$ be a subspace of $\mathcal{X}_1$. The operator T_1 on $\mathcal{X}_1$ is called *similar modulo* $\mathcal{M}_1$ to the operator T_2 on $\mathcal{X}_2$ if there exists an invertible operator $S : \mathcal{X}_2 \to \mathcal{X}_1$ such that

$$(ST_2 - T_1S)x \in \mathcal{M}_1 \text{ for each } x \in \mathcal{X}_2.$$

The operator S is called a *similarity modulo* $\mathcal{M}_1$. Clearly T_1 is similar modulo $\mathcal{M}_1$ to T_2 if and only if T_2 is similar modulo $\mathcal{M}_2 = S^{-1}[\mathcal{M}_1]$ to T_1.

Let $\mathcal{M}$ be a subspace of $\mathcal{X}$ and T be a linear operator on $\mathcal{X}$. A subspace $\mathcal{N}$ is called *T-invariant modulo* $\mathcal{M}$ if $T[\mathcal{N}] \subset \mathcal{N} + \mathcal{M}$. The subspaces $\mathcal{X}_1$ and $\mathcal{X}_2$ are called *complementary subspaces modulo* $\mathcal{M}$ if $\mathcal{X} = \mathcal{X}_1 \oplus \mathcal{X}_2$ and $\mathcal{M} = (\mathcal{M} \cap \mathcal{X}_1) \oplus (\mathcal{M} \cap \mathcal{X}_2)$. The operator T is called *decomposable modulo* $\mathcal{M}$ if there exists non zero subspaces $\mathcal{X}_1$ and $\mathcal{X}_2$, which are T-invariant modulo $\mathcal{M}$ and complementary modulo $\mathcal{M}$.

We now give an example of an operator of which we will eventually show that it is *indecomposable* (i.e., not decomposable) modulo a given subspace $\mathcal{M}$. Let $e_1, \ldots, e_{\kappa+1}$ be the standard basis of $\mathbb{C}^{\kappa+1}$, $\mathcal{M} = \operatorname{span}\{e_1\}$, and define $T : \mathbb{C}^{\kappa+1} \to \mathbb{C}^{\kappa+1}$ by $Te_i = e_{i+1}$, for $i = 1, \ldots, n-1$, and $Te_{\kappa+1} = 0$. Then T is called a *shift of index* κ. Now let $f_1, \ldots, f_{\kappa+1}$ be a basis of the linear space $\mathcal{X}$, and let $\mathcal{M}_1 = \operatorname{span}\{f_1\}$. Let T_1 be a linear operator on $\mathcal{X}$ such that $T_1 f_1 - f_{i+1} \in \mathcal{M}_1$ for $i = 1, \ldots, \kappa$ and $T_1 f_{\kappa+1} \in \mathcal{M}_1$. Then T_1 is similar modulo $\mathcal{M}_1$ to the shift T.

To check this, define the invertible operator $S : \mathbb{C}^{\kappa+1} \to \mathcal{X}$ by putting $Se_i = f_i$ for $i = 1, \ldots, \kappa+1$. Then $(ST - T_1S)e_i \in \mathcal{M}_1$ for $i = 1, \ldots, \kappa+1$. Notice that $S[\mathcal{M}] = \mathcal{M}_1$. Hence T_1 is similar modulo $\mathcal{M}_1$ to the shift T.

If $\mathcal{M} = \{0\}$, then an operator $T : \mathcal{X} \to \mathcal{X}$ is indecomposable modulo $\mathcal{M}$ if and only if the Jordan normal form of T consists of one single Jordan block. We shall derive the next fundamental structure theorem.

Theorem 3.1. *Let $T : \mathcal{X} \to \mathcal{X}$ be a linear operator, and let $\mathcal{M}$ be a subspace of $\mathcal{X}$. Then there exist subspaces $\mathcal{X}_0, \ldots, \mathcal{X}_q$ of $\mathcal{X}$, which are T-invariant modulo $\mathcal{M}$, complementary modulo $\mathcal{M}$ and such that $\mathcal{M} \cap \mathcal{X}_0 = \{0\}$, and operators $T_j : \mathcal{X}_j \to \mathcal{X}_j$ such that for $j = 1, \ldots, q$ the operator T_j is similar modulo $\mathcal{M} \cap \mathcal{X}_j$ to a shift and T is similar modulo $\mathcal{M}$ to $T_0 \oplus \cdots \oplus T_q$*

PROOF. Let P be a projection of $\mathcal{X}$ with $\operatorname{Ker} P = \mathcal{M}$. Consider the block $(PT; P, I)$ on $\mathcal{X}$. We apply Proposition 1.3 to construct linearly independent vectors $\{e_{ij}\}_{i=1,j=1}^{\kappa_j+1\ ,q}$ and the subspace $\mathcal{G}_\infty$ with the properties (i), (ii), (iii) and (iv) stated in Proposition 1.3. Put $\mathcal{X}_0 = \mathcal{G}_\infty$ and $\mathcal{X}_j = \operatorname{span}\{e_{1j}, \ldots, e_{\kappa_j+1\ j}\}$. Then $\mathcal{X} = \mathcal{X}_0 \oplus \mathcal{X}_1 \oplus \cdots \oplus \mathcal{X}_q$, $\mathcal{M} = (\mathcal{M} \cap \mathcal{X}_1) \oplus \cdots \oplus (\mathcal{M} \cap \mathcal{X}_q)$ and $\mathcal{M} \cap \mathcal{X}_0 = \{0\}$, since $\mathcal{M} = \operatorname{span}\{e_{11}, \ldots, e_{1q}\}$. We put $T_je_{ij} = e_{i+1\ j}$ for $j = 1, \ldots, \kappa_j$ and $T_je_{\kappa_j+1\ j} = 0$. The operator T_j is similar modulo $\mathcal{M} \cap \mathcal{X}_j$ to a shift of index κ_j. For $j = 1, \ldots, q$ and for $i = 1, \ldots, \kappa_j$ we have that $(T - T_j)e_{ij} = Te_{ij} - e_{i+1\ j}$. We use property (iv), which states that $P(Te_{ij} - e_{i+1\ j}) = 0$, to see that $Te_{ij} - e_{i+1\ j} \in \operatorname{Ker} P$. This gives that $(T - T_j)e_{ij} \in \mathcal{M}$. For $i = \kappa_j + 1$ the operator T_j is defined such that $(T - T_j)e_{ij} = Te_{ij}$. Again from property (iv) we see that $PTe_{ij} = 0$, and therefore $(T - T_j)e_{ij} \in \mathcal{M}$ too if $i = \kappa_j + 1$. Now put P' to be the projection of $\mathcal{X}$ along $\mathcal{M}$ onto $\mathcal{X}_0 \oplus \operatorname{span}\{e_{ij}\}_{i=2,j=1}^{\kappa_j+1\ ,q}$. For $x_0 \in \mathcal{X}_0$ we define $T_0x_0 = P'Tx_0$. Since $\mathcal{M} = \operatorname{Ker} P'$, we see that $(T - T_0)x_0 \in \mathcal{M}$ for each $x_0 \in \mathcal{X}_0$. It now follows from property (i) that for each $x \in \mathcal{X}$

$$(T - [T_0 \oplus \cdots \oplus T_q])(x) \in \mathcal{M}.$$

So T is indeed similar modulo $\mathcal{M}$ to $T_0 \oplus \cdots \oplus T_q$. □

The ranks $\kappa_1 \geq \cdots \geq \kappa_q$ of the shifts $T_1, \ldots, T_q$ in Theorem 3.1 are called *indices of T modulo $\mathcal{M}$*, and the invariant polynomials of the operator T_0 in Theorem 3.1 are called *invariant polynomials of T modulo $\mathcal{M}$*. The operator T_0 will be called the *$\mathcal{M}$-free part of T*. We shall see that these notions do not depend on the particular choice of the operators $T_0, T_1, \ldots, T_q$ in Theorem 3.1.

Corollary 3.2. *Let $T : \mathcal{X} \to \mathcal{X}$ be an operator, $\mathcal{M}$ be a subspace of $\mathcal{X}$, and P be a projection of $\mathcal{X}$ with $\operatorname{Ker} P = \mathcal{M}$. Then a set of indices of T modulo $\mathcal{M}$ is equal to the set of indices of the first kind of the block $(PT; P, I)$, and a set of invariant polynomials of T modulo $\mathcal{M}$ is equal to the set invariant polynomials of $(PT; P, I)$.*

PROOF. Let T be similar modulo $\mathcal{M}$ to $T_0 \oplus \cdots \oplus T_q$ as in Theorem 3.1. Put $I_j = I_{\mathcal{X}_j}$ and write P_j for the projection of $\mathcal{X}_j$ along $\mathcal{M} \cap \mathcal{X}_j$ onto $\operatorname{Im} T_j$. Then

$(P_jT_j; P_j, I_j)$ is block similar to a shift of the first kind and the rank of T_j is the index of this shift. The block $(PT; P, I)$ is block similar to the direct sum

$$(T_0; P_0, I_0) \oplus (T_1; P_1, I_1) \oplus \cdots \oplus (T_q; P_q, I_q).$$

Now we compare the definitions of indices of T modulo $\mathcal{M}$ and of indices of the first kind of the block $(PT; P, I)$, and we compare the definitions of the invariant polynomials of $(PT; P, I)$ with those of the invariant polynomials of T modulo $\mathcal{M}$. The result is immediate from this comparison. □

The following result follows from Theorem 1.1 and Corollary 3.2. Note that this result implies that a shift of index κ is indecomposable.

Corollary 3.3. *Let $T : \mathcal{X} \to \mathcal{X}$ be an operator, $\mathcal{M}$ be a subspace of $\mathcal{X}$. The set of indices of T modulo $\mathcal{M}$ and the set of invariant polynomials of T modulo $\mathcal{M}$ are uniquely determined by T and $\mathcal{M}$.*

Theorem 3.4. *Let $T_1 : \mathcal{X}_1 \to \mathcal{X}_1$ and $T_2 : \mathcal{X}_2 \to \mathcal{X}_2$ be linear operators, and let $\mathcal{M}_1$ be a subspace of $\mathcal{X}_1$. Then S is a similarity modulo $\mathcal{M}_1$ of T_1 with T_2 if and only if the indices of T_1 modulo $\mathcal{M}_1$ are equal to the indices of T_2 modulo $S[\mathcal{M}_1]$ and the $\mathcal{M}_1$-free part of T_1 and the $S[\mathcal{M}_1]$-free part of T_2 are similar operators.*

PROOF. Put P_1 to be a projection of $\mathcal{X}_1$ with $\operatorname{Ker} P_1 = \mathcal{M}_1$, and put P_2 to be a projection of $\mathcal{X}_2$ with $\operatorname{Ker} P = S[\mathcal{M}_1] = \mathcal{M}_2$. Then S is a similarity modulo $\mathcal{M}_1$ of T_1 and T_2 if and only if S is a block similarity of the blocks $(PT_1; P, I)$ and $(PT_2; P, I)$. These blocks are block similar if and only if they share the same indices of the first kind and the same invariant polynomials. Remark that the indices of the first kind of $(P_iT_i; P_i, I)$ are the indices of T_i modulo $\mathcal{M}_i$, and the invariant polynomials of $(P_iT_i; P_i, I)$ are the invariant polynomials of the $\mathcal{M}_i$-free part of T_i, for $i = 1, 2$. So T_1 is similar modulo $\mathcal{M}_1$ to T_2 if and only if the the indices of T_1 modulo $\mathcal{M}_1$ are equal to the indices of T_2 modulo $\mathcal{M}_2$ and the $\mathcal{M}_1$-free part of T_1 and the $\mathcal{M}_2$-free part of T_2 have the same invariant polynomials. We finish the proof by the remark that the $\mathcal{M}_1$-free part of T_1 and the $\mathcal{M}_2$-free part of T_2 share the same invariant polynomials if and only if they are similar operators. □

V.4 Duality

In this section we use the notion of duality (introduced in Section II.4), and show that many of the results in this chapter may be derived by duality from the corresponding results in Chapter III.

In the case when the block $(B; P, I)$ is a full width block with underlying space $\mathcal{X}$, the dual block $(B^*; I^*, P^*)$ is a full length block with underlying space $\mathcal{X}^*$. The next proposition gives the duality relation between the shifts of the first and third kind.

Proposition 4.1. *The dual block of a shift of the first kind with index $n-1$ is block similar to a shift of the third kind with index $n-1$ with a block similarity of the first kind. The dual block of a shift of the third kind with index $n-1$ is block similar to a shift of the first kind with index $n-1$ with a block similarity of the first kind.*

PROOF. Let $(B;P,I)$ be a shift of the first kind with index $n-1$. Thus with respect to the standard basis $e_1,\dots,e_n$ of $\mathbb{C}^n$ we have $Pe_1=0$, $Pe_i=e_i$ for $i=2,\dots,n$, $Be_n=0$, and $Be_i=e_{i+1}$ for $i=1,\dots,n-1$. Let $e_1^*,\dots,e_n^*$ be the dual basis in $(\mathbb{C}^n)^*$. This means that $e_i^*(e_j)=\delta_{ij}$, where δ_{ij} is the Kronecker delta. We get that $P^*e_1^*=0$ and $P^*e_j^*=e_j^*$ if $j=2,\dots,n$. Let $e_j^*\in \operatorname{Im}P^*$. Then $B^*e_j^*(e_i)=e_j^*(Be_i)$, and thus $B^*e_j^*(e_i)=0$ if $i\neq j-1$ and $B^*e_j^*(e_i)=1$ if $i=j-1$. This proves that $B^*e_j^*=e_{j-1}^*$, for $j=2,\dots,n$. We construct the similarity of the first kind with a shift of the third kind. Let $S:\mathbb{C}^n\to(\mathbb{C}^n)$ be defined by $Se_i=e_{n+1-i}^*$, and let $(B_1;I,Q)$ be a shift of the third kind. Then it is clear that $SQ=P^*S$ and $SB_1=BS$. This proves that S is a similarity of the first kind of $(B_1;I,Q)$ and $(B^*;I^*,P^*)$.

The second statement can be proved in much the same way. □

A block $(B;I,I)$ on $\mathcal{X}$ can be identified with the operator B, and as such it has invariant polynomials. The dual of the block $(B;I,I)$ is the block $(B^*;I^*,I^*)$ on $\mathcal{X}^*$. If the matrix of B with respect to a basis $e_1,\dots,e_n$ of $\mathcal{X}$ is M, then the matrix of B^* with respect to the dual basis $e_1^*,\dots,e_n^*$ of $\mathcal{X}^*$ is the transposed M^{T} of the matrix M.

Using the results of Section II.4 and Proposition 4.1 one can see that by duality many results of Section III.1 carry over into the corresponding results in Section V.1 and vice versa. For instance, Theorem V.1.1 can be proved from Theorem III.1.1 by applying duality. The duality relation between the operator pencils of Section III.2 and those of Section V.2 is obvious. Since each operator pencil has a matrix representation, duality is just a matter of transposing matrices in this case.

We finish this section by considering the duality between simlarity of non-everywhere defined operators and similarity modulo a subspace. Assume that S is a similarity of the operators $T_1:\mathcal{X}\to\mathcal{X}$ and $T_2:\mathcal{X}\to\mathcal{X}$ modulo the subspace $\mathcal{M}$ of $\mathcal{X}$. Let $\mathcal{M}^\perp$ be the annihilator of $\mathcal{M}$, and consider the non-everywhere defined operators $T_1^*|_{\mathcal{M}^\perp}(\mathcal{X}^*\to\mathcal{X}^*)$ and $T_2^*|_{\mathcal{M}^\perp}(\mathcal{X}^*\to\mathcal{X}^*)$ both with domain of definition $\mathcal{M}^\perp$. We will prove that S^* is a similarity of these non-everywhere defined operators. First note that $S\mathcal{M}=\mathcal{M}$ and hence $S^*\mathcal{M}^\perp=\mathcal{M}^\perp$. Furthermore, we have that $(ST_1-T_2S)(x)\in\mathcal{M}$ for all $x\in\mathcal{X}$. Then $f((ST_1-T_2S)x)=0$ for each $f\in\mathcal{M}^\perp$ and each $x\in\mathcal{X}$. So $(T_1^*S^*-S^*T_2^*)(f)=0$ for each $f\in\mathcal{M}^\perp$. This proves that $T_1^*|_{\mathcal{M}^\perp}$ and $T_2^*|_{\mathcal{M}^\perp}$ are similar non-everywhere defined operators with domain $\mathcal{M}^\perp$. Conversely, if S^* is a similarity of $T_1^*|_{\mathcal{M}^\perp}(\mathcal{X}^*\to\mathcal{X}^*)$ and $T_2^*|_{\mathcal{M}^\perp}(\mathcal{M}^*\to\mathcal{X}^*)$, then S is a similarity of T_1 and T_2 modulo $\mathcal{M}$. To see this first note that $S^*\mathcal{M}^\perp=\mathcal{M}^\perp$ implies that $S\mathcal{M}=\mathcal{M}$. Next one has that

$(T_1^*S^* - S^*T_2^*)(f) = 0$ for each $f \in \mathcal{M}^\perp$. So $f((ST_1 - T_2S)x) = 0$ for each $f \in \mathcal{M}^\perp$ and each $x \in \mathcal{X}$, and therefore $(ST_1 - T_2S)(x) \in \mathcal{M}$ for all $x \in \mathcal{X}$. This proves that T_1 and T_2 are similar modulo $\mathcal{M}$.

V.5 The eigenvalue completion problem and related problems

The solution of the eigenvalue completion problem for full width blocks can be derived from the solution of the eigenvalue completion problem for full length blocks by using duality. The rough result on the eigenvalue completion problem (which describes only the location of the eigenvalues) can be deduced from the canonical form as it was done for the full length block in the beginning of Section IV.1.

The full solution of the eigenvalue completion problem for full width blocks is given by the next theorem.

Theorem 5.1. *Let $(B; P, I)$ be a full width block on the n dimensional space $\mathcal{X}$, with indices of the first kind $\kappa_1 \geq \cdots \geq \kappa_m \geq 0$ and invariant polynomials $p_n | \cdots | p_1$. Consider polynomials $q_n | \cdots | q_1$ with $\sum_{i=1}^n \deg q_i = n$. Then there exists a completion A of the block $(B; P, I)$ which has invariant polynomials $q_1, \dots, q_n$ if and only if*

(1) *$p_i | q_i$, for $i = 1, \dots, n$, and $q_{i+m} | p_i$, for $i = 1, \dots, n-m$;*

(2) *$\sum_{i=1}^j \deg s_i \geq \sum_{i=1}^j (\kappa_i + 1)$ for $j = 1, \dots, m$, with equality holding for $j = m$.*

The polynomials s_j are defined as $s_j = t_j / t_{j+1}$, for $j = 1, \dots, m$, where

$$t_j = \prod_{i=j}^{n} \text{l.c.m.}(p_{i-j+1}, q_i), \quad j = 1, \dots, m+1.$$

Proof. The operator A is a completion of the block $(B; P, I)$ means that $B = PA : \mathcal{X} \to \operatorname{Im} P$. The operator A' is a completion of $(B; I', P')$ if $B' = A'P'|_{\operatorname{Im} P'}$. Therefore we see that A is a completion of $(B; P, I)$ if and only if A' is a completion of $(B; I', P')$. So $(B; P, I)$ has a completion with invariant polynomials $q_1, \dots, q_n$ if and only if $(B; I', P')$ has a completion with invariant polynomials $q_1, \dots, q_n$. If the block $(B; P, I)$ is similar to a direct sum of shifts of the first kind with indices $\kappa_1 \geq \cdots \geq \kappa_m$ and an operator with invariant polynomials $p_n | \cdots | p_1$, then the block $(B; I', P')$ is similar to a direct sum of shifts of the third kind with indices $\kappa_1 \geq \cdots \geq \kappa_m$ and an operator with invariant polynomials $p_n | \cdots | p_1$ (see Lemma 4.1). We apply Theorem IV.1.1 to the block $(B; I', P')$ and find necessary and sufficient conditions for the existence of a completion A' with invariant polynomials $q_1, \dots, q_n$ in terms of the indices $\kappa_1 \geq \cdots \geq \kappa_m$ and invariant polynomials $p_n | \cdots | p_1$. So the necessary and sufficient conditions for the block $(B; P, I)$ to have a completion A with invariant polynomials $q_1, \dots, q_n$ are the conditions stated in Theorem IV.1.1. □

Next, we consider the eigenvalue restriction problem.

Theorem 5.2. *Let $q_n|\cdots|q_1$ be the invariant polynomials of the linear operator $A : \mathcal{X} \to \mathcal{X}$. Let $\mu_1 \geq \cdots \geq \mu_m \geq 0$ be integers and $p_n|\cdots|p_1$ be polynomials. Then A has a block of full width with indices of the third kind $\mu_1 \geq \cdots \geq \mu_m$ and invariant polynomials $p_1, \ldots, p_n$ if and only if*

(0) $\sum_{i=1}^{m}(\mu_i + 1) + \sum_{i=1}^{n} \deg p_i = n$;

(1) $p_i|q_i$, *for* $i = 1, \ldots, n$, *and* $q_{i+m}|p_i$, *for* $i = 1, \ldots, n-m$;

(2) $\sum_{i=1}^{j} \deg s_i \geq \sum_{i=1}^{j}(\mu_i+1)$ *for* $j = 1, \ldots, m$, *with equality holding for* $j = m$.

The polynomials s_j are defined as $s_j = t_j/t_{j+1}$, for $j = 1, \ldots, m$, where

$$t_j = \prod_{i=j}^{n} \text{l.c.m.}(p_{i-j+1}, q_j), \quad j = 1, \ldots, m+1.$$

We omit a detailed proof of this theorem. There are two simple ways to obtain a proof. The first is to translate the proof of Theorem IV.1.3 from the full length case to the full width case. The second is to use Theorem IV.1.3 and to apply duality.

By transposing we obtain the next result from Theorem IV.5.1.

Theorem 5.3. *Let F be an $n \times n$ matrix with invariant polynomials $q_n|\cdots|q_1$. There exists a $k \times n$ matrix G of rank m such that the pencil*

$$L(\lambda) = \lambda(\,I \quad 0\,) - (\,F \quad G\,)$$

has invariant polynomials $p_n|\cdots|p_1$ and minimal column indices $\kappa_1 \geq \cdots \geq \kappa_m > 0 = \kappa_{m+1} = \cdots = \kappa_k$ if and only if

(1) $p_i|q_i$, *for* $i = 1, \ldots, n$, *and* $q_{i+m}|p_i$, *for* $i = 1, \ldots, n-m$;

(2) $\sum_{i=1}^{j} \deg s_i \geq \sum_{i=1}^{j} \kappa_i$ *for* $j = 1, \ldots, m$, *with equality holding for* $j = m$.

The polynomials s_j are defined as $s_j = t_j/t_{j+1}$, for $j = 1, \ldots, m$, where

$$t_j = \prod_{i=j}^{n} \text{l.c.m.}(p_{i-j+1}, q_i) \quad j = 1, \ldots, m+1.$$

We shall also treat a question that does not appear in the full length case. Let the $n \times (n+k)$ polynomial $L_0(\lambda) = \lambda(\,I \quad 0\,) - (\,F_0 \quad G_0\,)$ be given. Let $q_n|\cdots|q_1$ be polynomials such that $\sum_{i=1}^{n} \deg q_i = n$. Under what conditions does there exist a matrix $n \times (n+k)$ polynomial $L(\lambda) = \lambda(\,I \quad 0\,) - (\,F \quad G\,)$ that is strictly equivalent to $L_0(\lambda)$ and such that $q_n|\cdots|q_1$ are the invariant polynomials of F? Before we give the answer to this question, we write out the relations between F, the equivalence transformations, F_0 and G_0. From the special form of $L_0(\lambda)$ it is clear that

$$L(\lambda) = T^{-1}L_0(\lambda)\begin{pmatrix} T & 0 \\ U & V \end{pmatrix}.$$

So we see that

$$F = T^{-1}(F_0 + G_0UT^{-1})T, \quad G = T^{-1}G_0V.$$

Thus the question is: under what conditions does there exist matrices F, G and U, and invertible matrices T and V such (5.1) is fulfilled and $q_n|\cdots|q_1$ are the invariant polynomials of F? We will return to this type of questions in Chapter IX, which deals with the relations with linear systems. The next result is the answer to the question.

Theorem 5.4. *Let $L_0(\lambda) = \lambda(\,I \quad 0\,) - (\,F_0 \quad G_0\,)$ be an $n \times (n+k)$ matrix pencil with invariant polynomials $p_n|\cdots|p_1$ and minimal column indices $\kappa_1 \geq \cdots \geq \kappa_m > 0 = \kappa_{m+1} = \cdots = \kappa_k$. There exists an $n \times (n+k)$ matrix polynomial $L(\lambda) = \lambda(\,I \quad 0\,) - (\,F \quad G\,)$ that is strictly equivalent to $L_0(\lambda)$ and such that $q_n|\cdots|q_1$ are the invariant polynomials of F if and only if*

(1) *$p_i|q_i$, for $i = 1, \dots, n$, and $q_{i+m}|p_i$, for $i = 1, \dots, n-m$;*

(2) *$\sum_{i=1}^{j} \deg s_i \geq \sum_{i=1}^{j} \kappa_i$ for $j = 1, \dots, m$, with equality holding for $j = m$.*

The polynomials s_j are defined as $s_j = t_j/t_{j+1}$, for $j = 1, \dots, m$, where

$$t_j = \prod_{i=j}^{n} \text{l.c.m.}(p_{i-j+1}, q_i) \quad j = 1, \dots, m+1.$$

PROOF. We take F such that F has invariant polynomials $q_1, \dots, q_n$. According to Theorem 5.2 there exists an operator G such that $L(\lambda) = \lambda(\,I \quad 0\,) - (\,F \quad G\,)$ has invariant polynomials $p_1, \dots, p_n$ and minimal column indices $\kappa_1 \geq \cdots \geq \kappa_m > 0 = \kappa_{m+1} = \cdots = \kappa_k$ if and only (1) and (2) hold true. Now for $L(\lambda)$ to have these invariant polynomials and minimal column indices means that $L(\lambda)$ is strictly equivalent to $L_0(\lambda)$. So there exists a matrix polynomial $L(\lambda)$ with the desired properties if and only (1) and (2) hold true. □

V.6 A matrix equation

In this section we treat the dual of the problem considered in Section IV.5. The main result could easily be obtained by using duality and the results of Chapter IV. We choose however to give an independent proof of this main result. Remember that for a monic polynomial $p(\lambda)$ of degree n such that $p(0) \neq 0$ we define $p^{\#}(\lambda) = \lambda^n p(\lambda^{-1})/p(0)$.

Theorem 6.1. *Let* $A : \mathbb{C}^n \to \mathbb{C}^n$ *and* $B : \mathbb{C}^m \to \mathbb{C}^n$ *be linear operators such that the pencil* $(\,\lambda I - A \quad -B\,)$ *has invariant polynomials* $p_1, \ldots, p_n$ *and minimal column indices* $\kappa_1 \geq \cdots \geq \kappa_r > 0 = \kappa_{r+1} = \cdots = \kappa_m$. *Let* $q_1, \ldots, q_n$ *be monic polynomials such that* $q_n | \cdots | q_1$ *and* $\sum_{i=1}^{n} \deg q_i = n$. *Then there exists a transformation* $T : \mathbb{C}^n \to \mathbb{C}^n$ *with invariant polynomials* $q_1, \ldots, q_n$ *and an operator* F *such that*

$$AT + BF = I$$

if and only if

(0) $p_1(0) \neq 0$;

(1) $p_i^{\#} | q_i$ *for* $i = 1, \ldots, n$, *and* $q_{i+r} | p_i^{\#}$ *for* $i = 1, \ldots, n-r$;

(2) $\sum_{i=1}^{j} \deg s_i \geq \sum_{i=1}^{j} \kappa_i$ *for* $j = 1, \ldots, r$, *with equality holding for* $j = r$.

The polynomials s_j *are defined as* $s_j = t_j / t_{j+1}$, *for* $j = 1, \ldots, r$, *with*

$$t_j = \prod_{i=j}^{n} \text{l.c.m.}(p_{i-j+1}^{\#}, q_i) \quad j = 1, \ldots, m+1.$$

PROOF. The proof will consist of five steps.

STEP 1. Let P be any projection such that $\operatorname{Ker} P = \operatorname{Im} B$. Let $T : \mathbb{C}^n \to \mathbb{C}^n$ be a linear operator. There exists a linear operator $F : \mathbb{C}^m \to \mathbb{C}^n$ such that $AT + BF = I$ if and only if $PAT = P$. To see this first assume that $AT + BF = I$. Apply P to the left of this equality and (since $PB = 0$) one gets that $PAT = P$. Conversely, assume that $PAT = P$. Then an elementary calculation gives that $AT + (I - P)(I - AT) = I$. Since $\operatorname{Im}(I - P)(I - AT) \subset \operatorname{Ker} P = \operatorname{Im} B$, we see that there exists an operator F such that $(I - P)(I - AT) = BF$. So we get $AT + BF = I$.

STEP 2. Consider the block $(PA; P, I)$, where P is such that $\operatorname{Ker} P = \operatorname{Im} B$. There is still much freedom in the choice of $\operatorname{Im} P$. We will use that freedom later on in the proof. For the moment we deal with any P. First we express the indices of the first kind of the block $(PA; P, I)$ and its invariant polynomials in terms of the minimal column indices $\kappa_1 \geq \cdots \geq \kappa_m$ and the invariant polynomials of the pencil $(\,\lambda I - A \quad -B\,)$, respectively. It follows easily from Corollary 2.3 and Theorem 1.5 that the indices of the block $(PA; P, I)$ are $\kappa_1 - 1, \ldots, \kappa_m - 1$, and its invariant polynomials are the invariant polynomials of the pencil $(\,\lambda I - A \quad -B\,)$.

STEP 3. In this step we choose a special projection P. We start with any projection P_0 such that $\operatorname{Ker} P_0 = \operatorname{Im} B$. From Proposition 1.3 we know that there exists an independent set of vectors $\{e_{ij}\}_{i=1,j=1}^{\kappa_j\;,m}$ and a subspace $\mathcal{G}_\infty$ of $\mathcal{X}$ such that:

(i) $\mathcal{X} = \operatorname{span}\{e_{ij}\}_{i=1,j=1}^{\kappa_j,m} \oplus \mathcal{G}_\infty$;

(ii) $\operatorname{Ker} P_0 = \operatorname{span}\{e_{11}, \ldots, e_{1m}\}$;

(iii) $P_0A[\mathcal{G}_\infty] \subset \mathcal{G}_\infty \oplus \operatorname{Ker} P_0$;

(iv) $P_0Ae_{ij} = P_0e_{i+1\ j}$, for $i = 1, \dots, \kappa_j - 1$, and $P_0Ae_{\kappa_j\ j} = 0$, for $j = 1, \dots, m$.

Now choose

$$\operatorname{Im} P = \operatorname{span}\{e_{ij}\}_{i=2,j=1}^{\kappa_j,m} \oplus \mathcal{G}_\infty$$

and, of course, $\operatorname{Ker} P = \operatorname{Ker} P_0$. Then $PA[\mathcal{G}_\infty] \subset \mathcal{G}_\infty$ and $PAe_{ij} = e_{i+1\ j}$ for $i = 1, \dots, \kappa_j - 1$, $PAe_{\kappa_j j} = 0$ for $j = 1, \dots, m$.

STEP 4. In this step we answer the question: when does there exist a T such that $PAT = P$. Since $P[\mathcal{G}_\infty] = \mathcal{G}_\infty$ and $PA[\mathcal{G}_\infty] \subset \mathcal{G}_\infty$, the operator T has to fulfill that $(PA)|_{\mathcal{G}_\infty} T|_{\mathcal{G}_\infty} = I_{\mathcal{G}_\infty}$. This means that a necessary condition is that $(PA)|_{\mathcal{G}_\infty}$ is invertible, or equivalently the operator $(PA)|_{\mathcal{G}_\infty}$ has no eigenvalue equal to zero. In other words the minimal polynomial of $(PA)|_{\mathcal{G}_\infty}$ does not vanish at zero. This is condition (1). On the other hand, if we define $T|_{\mathcal{G}_\infty} = ((PA)|_{\mathcal{G}_\infty})^{-1}$ and $Te_{i+1\ j} = e_{ij}$ for $i = 1, \dots, \kappa_j - 1$, $Te_{1j} = 0$ for $j = 1, \dots, m$, then we see that $PAT = P$. This proves that condition (0) is necessary and sufficient for the existence of a pair (T, F) such that $AT + BF = I$. Moreover the only freedom in the choice of the operator T is that we might add to it any operator that has its image in the subspace $\operatorname{Ker} PA$.

STEP 5. Define the projection P_1 by specifying $\operatorname{Ker} P_1 = \operatorname{Ker} PA$ and

$$\operatorname{Im} P_1 = \operatorname{span}\{e_{ij}\}_{i=1,j=1}^{\kappa_j-1,m} \oplus \mathcal{G}_\infty.$$

So $\operatorname{Ker} P_1 = \operatorname{span}\{e_{\kappa_1,1}, \dots, e_{\kappa_m,m}\}$. In the previous step we saw that $PAT_1 = P$ if and only if $T_1x - Tx \in \operatorname{Ker} PA$ for each $x \in \mathbb{C}^n$. This means that $P_1T_1 = P_1T$. So we see that an operator T_1 is such that $PAT_1 = P$ if and only if T_1 is a completion of the block $(P_1T; P_1, I)$, where T is the operator which we constructed in step 4. From the special form of T we see that the indices of the first kind of this block are $\kappa_1 - 1, \dots, \kappa_m - 1$ and the invariant polynomials are the invariant polynomials of $((PA)|_{\mathcal{G}_\infty})^{-1}$ supplemented with some constant polynomials equal to 1. So the invariant polynomials of $(P_1T; P_1, I)$ are $p_1^\#, \dots, p_n^\#$. We apply Theorem 5.1 to see that a completion of $(P_1T; P_1, I)$ with invariant polynomials $q_1, \dots, q_n$ exists if and only if the conditions (1) and (2) are fulfilled. □

Notes

The results in the first three sections may be viewed as duals of the results in Chapter III. Here the proofs are given directly, but they could be obtained by using duality arguments as is done for the results in Section 5 (see also Section 4). The theorems about full length blocks in Section 1 and their applications in Section 2 are improved versions of the analogous results in Gohberg-Kaashoek-Van Schagen [1]. The results in Section 3 come also from Gohberg-Kaashoek-Van Schagen [1]. Theorem 5.1 is taken from Zaballa [1] and Theorem 5.2 comes from Gohberg-Kaashoek-Van Schagen [5]. Theorem 6.1 is dual to a result of Gohberg-Kaashoek-Ran [1].

Chapter VI
Principal Blocks

This chapter contains the solution of the eigenvalue completion problem and the associated restriction problem for principal blocks. The first section gives the block similarity invariants for principal blocks.

VI.1 Structure theorem for principal blocks

Although this section is about principal blocks we start with a block of a somewhat more general type. Let $(B; P, Q)$ be a block on the linear space $\mathcal{X}$ with $\mathcal{X} = \operatorname{Ker} P \oplus \operatorname{Im} Q$. It is clear that a principal block has this property. We define P_1 the projection along $\operatorname{Ker} P$ onto $\operatorname{Im} Q$. Put $B_1 = P_1 B : \operatorname{Im} P_1 \to \operatorname{Im} P_1$. Then $(B_1; P_1, P_1)$ is similar to $(B; P, Q)$ with a similarity of the second kind, as we have seen in Section I.4. Conversely, if $(B; P, Q)$ on $\mathcal{X}$ is block similar to the principal block $(B_1; P_1, P_1)$ on $\mathcal{X}$, then there exists an invertible linear transformation such that $S[\operatorname{Ker} P] = \operatorname{Ker} P_1$ and $S[\operatorname{Im} Q] = \operatorname{Im} P_1$. Thus we see that $\mathcal{X} = \operatorname{Ker} P \oplus \operatorname{Im} Q$. If we characterize the similarity class of $(B_1; P_1, P_1)$, then we also have characterized the similarity class of $(B; P, Q)$.

Theorem 1.1. *The principal blocks $(B_1; P_1, P_1)$ on $\mathcal{X}_1$ and $(B_2; P_2, P_2)$ on $\mathcal{X}_2$ are block similar if and only if the operators $B_1 : \operatorname{Im} P_1 \to \operatorname{Im} P_1$ and $B_2 : \operatorname{Im} P_2 \to \operatorname{Im} P_2$ are similar operators and* $\dim \operatorname{Ker} P_1 = \dim \operatorname{Ker} P_2$.

PROOF. Assume that $S : \mathcal{X}_1 \to \mathcal{X}_2$ is a block similarity of $(B_1; P_1, P_1)$ and $(B_2; P_2, P_2)$. So $S[\operatorname{Ker} P_1] = \operatorname{Ker} P_2$, $S[\operatorname{Im} P_1] = \operatorname{Im} P_2$ and $(SB_1 - B_2 S)x \in \operatorname{Ker} P_2$ for each $x \in \operatorname{Im} P_1$. However $SB_1 x \in \operatorname{Im} P_2$ and $B_2 S x \in \operatorname{Im} P_2$, and thus $(SB_1 - B_2 S)x \in (\operatorname{Ker} P_2 \cap \operatorname{Im} P_2)$. So $(SB_1 - B_2 S)x = 0$ for each $x \in \operatorname{Im} P_1$. We see that $\dim \operatorname{Ker} P_1 = \dim \operatorname{Ker} P_2$, and that $S|_{\operatorname{Im} P_1}$ gives a similarity of the operators $B_1 : \operatorname{Im} P_1 \to \operatorname{Im} P_1$ and $B_2 : \operatorname{Im} P_2 \to \operatorname{Im} P_2$.

Conversely, assume that $\dim \operatorname{Ker} P_1 = \dim \operatorname{Ker} P_2$ and that $S_1 : \operatorname{Im} P_1 \to \operatorname{Im} P_2$ gives a similarity of $B_1 : \operatorname{Im} P_1 \to \operatorname{Im} P_1$ and $B_2 : \operatorname{Im} P_2 \to \operatorname{Im} P_2$. Then there exists an invertible operator $S_2 : \operatorname{Ker} P_1 \to \operatorname{Ker} P_2$. Define $S : \mathcal{X}_1 \to \mathcal{X}_2$ by $S = S_1 \oplus S_2 : \operatorname{Im} P_1 \oplus \operatorname{Ker} P_1 \to \operatorname{Im} P_2 \oplus \operatorname{Ker} P_2$. Then $S[\operatorname{Ker} P_1] = \operatorname{Ker} P_2$, $S[\operatorname{Im} P_1] = \operatorname{Im} P_2$ and $(SB_1 - B_2 S)x = 0$ for each $x \in \operatorname{Im} P_1$. This proves that S is a block similarity of $(B_1; P_1, P_1)$ and $(B_2; P_2, P_2)$. □

In Theorem 1.1 we have seen that the similarity class of the block $(B; P, P)$ is determined by the similarity class of the operator B and the number $\dim \operatorname{Ker} P$. So the invariant polynomials of B and the number $\dim \operatorname{Ker} P$ determine the similarity class of $(B; P, P)$ uniquely.

VI.2 The eigenvalue completion problem for principal blocks

We recall that $A : \mathcal{X} \to \mathcal{X}$ is a completion of the block $(B; P, P)$ on $\mathcal{X}$ if $PA|_{\operatorname{Im} P} = B$. The solution of the eigenvalue completion problem is given in the next result.

Theorem 2.1. *Let $(B; P, P)$ be a principal block on the n-dimensional space $\mathcal{X}$, and let $p_{n-m}|\cdots|p_1$ be the invariant polynomials of $(B; P, P)$. Here $m = \dim \operatorname{Ker} P$. The polynomials $q_n|\cdots|q_1$ are the invariant polynomials of a completion of $(B; P, P)$ if and only if*

(a) $\sum_{i=1}^{n} \deg q_i = n$,

(b) $p_{i+m}|q_i$ *for* $i = 1, \dots, n-2m$, *and* $q_{i+m}|p_i$ *for* $i = 1, \dots, n-m$.

Before we present the proof of this theorem, we formulate a result that will be handy both here and in Section VIII.2.

Lemma 2.2. *Let the sequence $\mu_1 \geq \cdots \geq \mu_m$ be such that $\sum_{i=1}^{m} \mu_i = \gamma$ and $\mu_1 - \mu_m \leq 1$, and let $\nu_1 \geq \cdots \geq \nu_m$ be a sequence with $\sum_{i=1}^{m} \nu_i = \gamma$. Then $\sum_{i=1}^{j} \nu_i \geq \sum_{i=1}^{j} \mu_i$, for $j = 1, \dots, m$.*

PROOF. Assume that $\sum_{i=1}^{\alpha} \nu_i \geq \sum_{i=1}^{\alpha} \mu_i$ for $\alpha = 1, \dots, j-1$, where $j = 1$ is allowed, and $\sum_{i=1}^{j} \nu_i < \sum_{i=1}^{j} \mu_i$. It follows that $\nu_j < \mu_j$ and, since $\mu_j \leq \mu_{j+\beta} + 1$ for $\beta = 1, \dots, m-j$, also $\nu_j \leq \mu_{j+\beta}$ for $\beta = 1, \dots, m-j$. So $\nu_{j+\beta} \leq \mu_{j+\beta}$ for $\beta = 1, \dots, m-j$. We obtain that

$$\sum_{i=1}^{m} \nu_i = \sum_{i=1}^{j} \nu_i + \sum_{i=j+1}^{m} \nu_i < \sum_{i=1}^{j} \mu_i + \sum_{i=j+1}^{m} \mu_i = \gamma,$$

which contradicts $\sum_{i=1}^{m} \nu_i = \gamma$. □

PROOF OF THEOREM 2.1. First assume that $A : \mathcal{X} \to \mathcal{X}$ is a completion of the block $(B; P, P)$ with invariant polynomials $q_1, \dots, q_n$. Then $(A|_{\operatorname{Im} P}; I, P)$ is a block restriction of A. Let $r_n|\cdots|r_1$ be the invariant polynomials of $(A|_{\operatorname{Im} P}; I, P)$. From Theorem IV.1.3 we obtain that

$$r_i|q_i, \quad q_{i+m}|r_i, \quad i = 1, \dots, n-m. \tag{2.1}$$

Recall that $B = PA|_{\operatorname{Im} P}$. Consider the operator pencil

$$L(\lambda) = \lambda \begin{pmatrix} I_{\operatorname{Im} P} \\ 0 \end{pmatrix} - \begin{pmatrix} B \\ (I-P)A|_{\operatorname{Im} P} \end{pmatrix} : \operatorname{Im} P \to \operatorname{Im} P \oplus \operatorname{Ker} P.$$

This pencil has invariant polynomials $r_1, \dots, r_n$. We apply Theorem IV.5.1 to obtain that

$$r_i|p_i, \ (i = 1, \dots, n-m), \quad p_{i+m}|r_i, \ (i = 1, \dots, n-2m). \tag{2.2}$$

Combining (2.1) and (2.2) we obtain that $p_{i+m}|q_i$ for $i = 1, \dots, n-2m$ and $q_{i+m}|p_i$ for $i = 1, \dots, n-m$.

Conversely, assume that the polynomials $q_n|q_{n-1}|\cdots|q_1$ are given such that (a) and (b) in the theorem hold. Put $r_i = \text{g.c.d.}(p_i, q_i)$ for $i = 1, \dots, n-m$. Then $r_i|q_i$ and $r_i|p_i$ for $i = 1, \dots, n-m$. Moreover, $p_{i+m}|p_i$ and $p_{i+m}|q_i$ imply that $p_{i+m}|r_i$ for $i = 1, \dots, n-2m$. Put $\gamma = n - m - \sum_{i=1}^{n-m} \deg r_i$. We define the sequence $\mu_1 \geq \cdots \geq \mu_m$ such that $\sum_{i=1}^{m} \mu_i = \gamma$ and $\mu_1 - \mu_m \leq 1$. This determines the sequence $\mu_1, \dots, \mu_m$ uniquely. In fact, if $\gamma = m\alpha + \beta$ with $0 \leq \beta < m$, then $\mu_1 = \cdots = \mu_\beta = \alpha + 1$ and $\mu_{\beta+1} = \cdots = \mu_m = \alpha$.

The next step is to apply Theorem IV.5.1 to see that there exists an operator $C : \operatorname{Im} P \to \operatorname{Ker} P$ such that the pencil

$$L(\lambda) = \lambda \begin{pmatrix} I_{\operatorname{Im} P} \\ 0 \end{pmatrix} - \begin{pmatrix} B \\ C \end{pmatrix} : \operatorname{Im} P \to \operatorname{Im} P \oplus \operatorname{Ker} P,$$

has invariant polynomials $r_1, \dots, r_{n-m}$ and minimal row indices $\mu_1, \dots, \mu_m$. The condition (1) of Theorem IV.5.1 has already been checked for the case when there is no minimal row index equal to 0, i.e., the case when $\gamma \geq m$. If $\gamma < m$, we see that $\mu_{\gamma+1} = \cdots = \mu_1 = 0$, and we then have to prove that $p_{i+\gamma}|r_i$ for $i = 1, \dots, n-m-\gamma$. Fix a value of i, and let us prove this division relation. Remark that $\sum_{j=1}^{n-m} \deg p_j = n - m$ and $r_j|p_j$. So

$$\gamma = \sum_{j=1}^{n-m} \deg p_j - \sum_{j=1}^{n-m} \deg r_j = \sum_{j=1}^{n-m} \deg \left(\frac{p_j}{r_j} \right).$$

It follows that for at most γ values of j the quotient p_j/r_j is non-constant. So for at least one value of j, with $i \leq j \leq i+\gamma$, we see that $p_j|r_j$. Therefore $p_{i+\gamma}|p_j|r_j|r_i$. This proves that condition (1) in Theorem IV.5.1 is fulfilled also if $\gamma < m$. Now we put $\nu_i = \deg s_i$, where the polynomials s_i are those appearing in condition (2) of Theorem IV.5.1. From Proposition IV.4.3 we see that $\nu_1 \geq \cdots \geq \nu_m$ and from the definition of s_i one computes that $\sum_{i=1}^{m} \nu_i = \gamma$. From Lemma 2.2 and the special choice of the sequence $\mu_1, \dots, \mu_m$ it follows that condition (2) of Theorem IV.5.1 is fulfilled.

Next we apply Corollary III.2.3 to conclude that the full length block

$$\left(\begin{pmatrix} B \\ C \end{pmatrix} ; I_{\mathcal{X}}, P \right)$$

has invariant polynomials $r_1, \dots, r_{n-m}$ and indices of the third kind $\mu_1, \dots, \mu_m$. In order to show the existence of a completion of this block with invariant polynomials $q_1, \dots, q_n$, we check the conditions (1) and (2) of Theorem IV.1.1. To verify condition (1) remark that $q_{i+m}|q_i$ and $q_{i+m}|p_i$, and thus $q_{i+m}|r_i$, for $i = 1, \dots, n-m$.

Condition (2) is again automatically fulfilled, due to the special choice of the sequence of the indices of the third kind $\mu_1, \dots, \mu_m$. So there exists an operator $A : \mathcal{X} \to \mathcal{X}$ with invariant polynomials $q_1, \dots, q_n$ and such that

$$A|_{\operatorname{Im} P} = \begin{pmatrix} B \\ C \end{pmatrix} : \operatorname{Im} P \to \operatorname{Im} P \oplus \operatorname{Ker} P.$$

In particular we see that $PA|_{\operatorname{Im} P} = B$. This proves that a completion of $(B; P, P)$ with invariant polynomials $q_1, \dots, q_n$ exists. □

We end this section with two examples. First, assume that our principal block is the principal block associated with the $(n-1) \times (n-1)$ positioned submatrix

$$A_{\mathcal{P}} = \begin{pmatrix} 0 & 0 & \cdots & 0 & 0 & ? \\ 1 & 0 & & 0 & 0 & ? \\ & \vdots & \ddots & & \vdots & \vdots \\ 0 & 0 & & 1 & 0 & ? \\ ? & ? & \cdots & ? & ? & ? \end{pmatrix}.$$

Thus $\operatorname{Im} P = \operatorname{span}\{e_1, \dots, e_{n-1}\} \subset \mathbb{C}^n$ and $Be_i = e_{i+1}$ for $i = 1, \dots, n-2$, $Be_{n-1} = 0$. The invariant polynomials of the block $(B; P, P)$ are $p_1(\lambda) = \lambda^{n-1}$ and $p_2(\lambda) = \cdots = p_{n-1}(\lambda) = 1$. We see that for a completion A the invariant polynomials $q_n | \cdots | q_1$ have to be such that $q_2 | p_1$, and hence q_2 is a power of λ, and $q_3 | p_2$ which gives that $q_3(\lambda) = 1$. On the other hand, $p_{i+1} | q_i$ does not give any restriction. We find that A can have any eigenvalue provided that the geometric multiplicity of an eigenvalue μ is at most 1 if $\mu \neq 0$ and at most 2 if $\mu = 0$.

In the second example we consider the positioned submatrix

$$A_{\mathcal{P}} = \begin{pmatrix} \alpha_1 & 0 & \cdots & 0 & ? \\ 0 & \alpha_2 & & 0 & ? \\ \vdots & & \ddots & & \vdots \\ 0 & 0 & & \alpha_{n-1} & ? \\ ? & ? & \cdots & ? & ? \end{pmatrix}, \quad \alpha_i \neq \alpha_j \quad \text{if} \quad i \neq j.$$

Then $p_1(\lambda) = \prod_{i=1}^{n-1} (\lambda - \alpha_i)$, $p_i(\lambda) = 1$ if $i = 2, \dots, n-1$. Again $p_{i+1} | q_i$ is always fulfilled. Furthermore, q_1 is free to choose and $q_2 | \text{g.c.d.}(p_1, q_1)$, $q_3(\lambda) = 1$. So any eigenvalue may occur, but only the eigenvalues $\mu \in \{\alpha_1, \dots, \alpha_{n-1}\}$ may have geometric multiplicity 2. All other eigenvalues have geometric multiplicity 1. Moreover, an eigenvalue $\mu \in \{\alpha_1, \dots, \alpha_{n-1}\}$ can have at most two Jordan blocks of which the smallest has (in case there are indeed two) the order 1.

VI.3 The eigenvalue restriction problem for principal blocks

The next result gives the solution of the eigenvalue restriction problem for principal blocks.

Theorem 3.1. *Let A be an operator on the n-dimensional space $\mathcal{X}$ with invariant polynomials $q_n|\cdots|q_1$. Let $p_{n-m}|\cdots|p_1$ be polynomials such that $\sum_{i=1}^{n-m}\deg p_i = n-m$. Then A has a principal block on $\mathcal{X}$ with invariant polynomials $p_1,\dots,p_{n-m}$ if and only if $p_{i+m}|q_i$ for $i=1,\dots,n-2m$ and $q_{i+m}|p_i$ for $i=1,\dots,n-m$.*

PROOF. First assume that $p_{i+m}|q_i$ for $i=1,\dots,n-2m$ and $q_{i+m}|p_i$ for $i=1,\dots,n-m$. Let $(B_0;P_0,P_0)$ be a block on the space $\mathcal{X}_0$ with invariant polynomials $p_1,\dots,p_{n-m}$. According to Theorem 2.1 there exists a completion A_0 of $(B_0;P_0,P_0)$ with invariant polynomials $q_n|\cdots|q_1$. Thus there exists linear operator $S:\mathcal{X}\to\mathcal{X}_0$ such that $A=S^{-1}A_0S$. Put $P=S^{-1}P_0S^{-1}$. Then $S[\operatorname{Im}P]=\operatorname{Im}P_0$, and hence $B=(S|_{\operatorname{Im}P})^{-1}B_0S|_{\operatorname{Im}P}$ is well defined. Moreover the operator S is a block similarity of the first kind of $(B_0;P_0,P_0)$ and $(B;P,P)$. Thus $(B;P,P)$ has invariant polynomials $p_1,\dots,p_{n-m}$. Furthermore, one computes for $x\in\operatorname{Im}P$ that

$$\begin{aligned}Bx =&(S|_{\operatorname{Im}P})^{-1}B_0S|_{\operatorname{Im}P}x = (S|_{\operatorname{Im}P})^{-1}P_0A_0|_{\operatorname{Im}P_0}S|_{\operatorname{Im}P}x\\ =&(S|_{\operatorname{Im}P})^{-1}SPS^{-1}SAS^{-1}S|_{\operatorname{Im}P}x = (S|_{\operatorname{Im}P})^{-1}SPAx = PAx.\end{aligned}$$

Thus $(B;P,P)$ is a restriction of A.

Conversely, assume that $(B;P,P)$ is a restriction of A with invariant polynomials $p_1,\dots,p_{n-m}$. Then A is an extension of $(B;P,P)$ with invariant polynomials $q_n|\cdots|q_1$. Hence Theorem 2.1 gives that $p_{i+m}|q_i$ for $i=1,\dots,n-2m$ and $q_{i+m}|p_i$ for $i=1,\dots,n-m$. □

Remark that in the first part of the proof, after the construction of A_0 and $(B_0;P_0,P_0)$, we could simply apply Theorem I.6.2 and conclude that A has a restriction with invariant polynomials $p_1,\dots,p_{n-m}$. The argument given in the proof is just a specification of the proof of Theorem I.6.2 to the present situation.

Notes

Theorem 2.2 is due to Thompson [1] and De Sà [1]. Another proof appears in Zaballa [2]. Here, as in Zaballa [2], we prove the theorem by first extending the principal block to a full length block and using next the solution of the eigenvalue completion problem for full length blocks from Chapter IV. A novelty in our proof is the use of the eigenvalue restriction problem for full length blocks in the first step of the proof. The result in Section VI.3 is probably new.

Chapter VII
General Blocks

This chapter presents the general theory of block similarity for arbitrary (finite dimensional) operator blocks, including the block similarity invariants and the corresponding canonical form. The connection with Kronecker's theorem about the canonical form of matrix pencils under strict equivalence is discussed. As another application we derive the canonical form under similarity of a non-everywhere defined operator modulo a subspace. For such operators the eigenvalue completion problem is reformulated as a lifting problem.

VII.1 Block similarity invariants, completion and restriction problems

In this section we present the main structure theorems for general blocks modulo block similarity. We discuss their implications for the eigenvalue completion problem and for the eigenvalue restriction problem. Recall that we introduced shifts of the first, second and third kind in Section II.3. The next result contains as special cases Theorems III.1.1, V.1.1 and VI.1.1.

Theorem 1.1. *An operator block $(B; P, Q)$ is block similar to a direct sum of an operator J, and a direct sum of shifts of the first kind with indices $\kappa_1 \geq \cdots \geq \kappa_p \geq 0$, shifts of the second kind with indices $\nu_1 \geq \cdots \geq \nu_r \geq 0$, and shifts of the third kind with indices $\mu_1 \geq \cdots \geq \mu_q \geq 0$. Furthermore, the indices $\kappa_1 \geq \cdots \geq \kappa_p$, $\nu_1 \geq \cdots \geq \nu_r$, and $\mu_1 \geq \cdots \geq \mu_q$, and the invariant polynomials of the operator J are uniquely determined by the block $(B; P, Q)$.*

Theorem 1.1 will be proved in Section VII.4. The numbers $\kappa_1, \ldots, \kappa_p$ will be called *the indices of the first kind of* $(B; P, Q)$, the numbers $\nu_1, \ldots, \nu_r$ will be called *the indices of the second kind of* $(B; P, Q)$ and the numbers $\mu_1, \ldots, \mu_q$ will be called *the indices of the third kind of* $(B; P, Q)$. The invariant polynomials of the operator J will be called *the invariant polynomials of* $(B; P, Q)$. As before we extend the sequence of invariant polynomials of $(B; P, Q)$ with constant polynomials equal to 1 in order to get n invariant polynomials for $(B; P, Q)$, where n is the dimension of the underlying space $\mathcal{X}$. Note that these definitions are consistent with the definitions that were given earlier in Sections III.1 and V.1.

Any direct sum of the type described in Theorem 1.1 is called *a canonical form* for the block $(B; P, Q)$ in Theorem 1.1. The block associated with the following positioned submatrix (empty places denoting zero entries) is an example of such a canonical form

$$B = \begin{pmatrix}
? & ? & ? & ? & ? & ? & ? & ? & ? & ? & ? & ? \\
1 & 0 & 0 & & ? & & & & ? & & & \\
0 & 1 & 0 & & ? & & & & ? & & & \\
? & ? & ? & ? & ? & ? & ? & ? & ? & ? & ? & ? \\
 & & & 1 & ? & & & & ? & & & \\
 & & & & ? & 0 & 0 & 0 & ? & & & \\
 & & & & ? & 1 & 0 & 0 & ? & & & \\
 & & & & ? & 0 & 1 & 0 & ? & & & \\
 & & & & ? & 0 & 0 & 1 & ? & & & \\
 & & & & ? & & & & ? & 1 & 1 & 0 \\
 & & & & ? & & & & ? & 0 & 1 & 0 \\
 & & & & ? & & & & ? & 0 & 0 & 1
\end{pmatrix}.$$

Notice that the block associated with B is the direct sum of a shift of the first kind with index 2, a shift of the second kind with index 1, a shift of the third kind with index 3, and a Jordan part with invariant polynomials $p_1(\lambda) = (\lambda - 1)^2$, $p_2(\lambda) = \lambda - 1$, $p_j(\lambda) = 1$ for $j = 3, \dots, 12$.

The next theorem characterizes the block similarity classes in terms of the indices and invariant polynomials of its members.

Theorem 1.2. *The blocks $(B; P, Q)$ on $\mathcal{X}$ and $(B'; P', Q')$ on $\mathcal{X}'$ are block similar if and only if they have the same indices of the first, second, and third kind and the same invariant polynomials.*

PROOF. Assume that $(B; P, Q)$ is block similar to $(B'; P', Q')$. Let $(B; P, Q)$ be block similar to the direct sum $(B''; P'', Q'')$ of an operator and of shifts of the first, second, and third kind. Then also $(B'; P', Q')$ is block similar to $(B''; P'', Q'')$. The indices of the first, second and third kind of both $(B; P, Q)$ and $(B'; P', Q')$ are defined to be the indices of the shifts of the first, second and third kind occurring in the direct sum $(B''; P'', Q'')$. The invariant polynomials of $(B; P, Q)$ and $(B'; P', Q')$ are defined to be the invariant polynomials of the operator occurring in the direct sum $(B''; P'', Q'')$. So $(B; P, Q)$ and $(B'; P', Q')$ share the same indices and invariant polynomials.

Conversely, assume that $(B; P, Q)$ and $(B'; P', Q')$ have the same indices of the first, second and third kind and the same invariant polynomials. Then $(B; P, Q)$ and $(B'; P', Q')$ are each similar to a direct sum of an operator with these invariant polynomials and of shifts with these indices. Remark that two shifts of the same kind are block similar if they have the same index, and two operators are similar if they have the same invariant polynomials. We finish the proof with the remark that two direct sums of block similar blocks are block similar. □

The eigenvalue completion problem (Problem A or A' in Section I.6) asks to describe the eigenvalues and their multiplicities of the completions of all the blocks in the block similarity class of a block. Note that giving the eigenvalues and multiplicities of an operator is equivalent to giving the invariant polynomials. On the other hand the block similarity class of a block is fully determined by the indices (of first, second and third kind) and the invariant polynomials of the block. So, in other words, to solve the eigenvalue completion problem means to describe the conditions that the indices and invariant polynomials of a block impose on the invariant polynomials of a completion of the block. In Theorems IV.1.1 and V.5.1 the solution of the eigenvalue completion problems for full length and full width blocks are given in this form. We consider the problem for general blocks in the last section. In this chapter we will also describe how the solution of the eigenvalue completion problem for principal block, which was given in Theorem VI.2.1, fits into this general frame work. In fact, we will show that indices appear in Theorem VI.2.1 in a disguised form.

The eigenvalue restriction problem (Problem B in Section I.6) asks to give the indices and invariant polynomials of blocks that can appear as a block of an operator with a certain given set of invariant polynomials. For full width and full length blocks we showed that the solution can be derived from the solution of the eigenvalue completion problem.

VII.2 Structure theorems and canonical form

We will give formulas for the indices of the first, second and third kind of a block $(B; P, Q)$ on $\mathcal{X}$ in terms of the associated definition spaces $\mathcal{D}_i$ and iterated images $\mathcal{F}_i$.

Recall that the i-th definition space $\mathcal{D}_i$ (see Section III.1) is defined by

$$\mathcal{D}_0 = \mathcal{X}, \quad \mathcal{D}_1 = \operatorname{Im} Q, \quad \mathcal{D}_i = \{x \in \mathcal{D}_{i-1} \mid Bx \in \mathcal{D}_{i-1} + \operatorname{Ker} P\}, \ i = 2, 3, \dots . \tag{2.1}$$

Furthermore, the residual subspace $\mathcal{D}_\infty$ is by definition given by $\mathcal{D}_\infty = \mathcal{D}_\mu$, where μ is such that $\mathcal{D}_{\mu-1} \neq \mathcal{D}_\mu = \mathcal{D}_{\mu+1}$. As we have seen in Section III.1, the space $\mathcal{D}_\infty$ may also be characterized as the largest $(B; P, Q)$-invariant subspace of $\operatorname{Im} Q$.

The i-th iterated image $\mathcal{F}_i$ is introduced in Section V.1 by the formulas

$$\mathcal{F}_0 = \{0\}, \quad \mathcal{F}_1 = \operatorname{Ker} P, \quad \mathcal{F}_{i+1} = B[\mathcal{F}_i \cap \operatorname{Im} Q] \oplus \operatorname{Ker} P, \ i = 1, 2, \dots . \tag{2.2}$$

The final iterated image $\mathcal{F}_\infty$ is by definition given by $\mathcal{F}_\infty = \mathcal{F}_\kappa$, where κ is such that $\mathcal{F}_{\kappa-1} \neq \mathcal{F}_\kappa = \mathcal{F}_{\kappa+1}$. This space is precisely the smallest $(B; P, Q)$-block-invariant subspace that contains $\operatorname{Ker} P$.

This next theorem contains as special cases Theorems III.1.2 and V.1.2.

Theorem 2.1. *Let $(B; P, Q)$ be a block on the linear space $\mathcal{X}$ with indices of the first kind $\kappa_1 \geq \cdots \geq \kappa_p$, indices of the second kind $\nu_1 \geq \cdots \geq \nu_r$ and indices of the third kind $\mu_1 \geq \cdots \geq \mu_q$. Let $\mathcal{D}_i$ be the i-th definition space, $\mathcal{D}_\infty$ the* *residual space, $\mathcal{F}_i$ the i-th iterated image, and $\mathcal{F}_\infty$ the final iterated image, all corresponding to $(B; P, Q)$. Put*

$$\begin{aligned}
\alpha_i &= \dim \mathcal{F}_{i+1} - \dim \mathcal{F}_i, \\
\beta_i &= \dim \mathcal{F}_i - \dim(\mathcal{F}_i \cap \operatorname{Im} Q), \\
\gamma_i &= \dim \operatorname{Ker} P - \dim(\mathcal{D}_i \cap \operatorname{Ker} P), \\
\delta_i &= \dim \mathcal{D}_i - \dim \mathcal{D}_{i+1}, \\
\beta_\infty &= \max\{\beta_i \mid i = 1, 2, \dots\} = \dim \mathcal{F}_\infty - \dim(\mathcal{F}_\infty \cap \operatorname{Im} Q), \\
\gamma_\infty &= \max\{\gamma_i \mid i = 1, 2, \dots\} = \dim \operatorname{Ker} P - \dim(\mathcal{D}_\infty \cap \operatorname{Ker} P).
\end{aligned}$$

Then

$$r = \beta_\infty = \gamma_\infty, \quad p = \alpha_0 - r, \quad q = \delta_0 - r,$$

and

$$\begin{aligned}
\kappa_i &= \#\{k \geq 1 \mid \alpha_k - (\beta_\infty - \beta_k) \geq i\}, \\
\nu_i &= \#\{k \geq 1 \mid \beta_\infty - \beta_k \geq i\} = \#\{k \geq 1 \mid \gamma_\infty - \gamma_k \geq i\}, \\
\mu_i &= \#\{k \geq 1 \mid \delta_k - (\gamma_\infty - \gamma_k) \geq i\}.
\end{aligned}$$

The proof of Theorem 2.1 will be given in Section VII.4. As an example we apply Theorem 2.1 to a principal block $(B; P, P)$. First, we note that $\mathcal{D}_0 = \mathcal{X}$ and $\mathcal{D}_i = \operatorname{Im} P$ for $i = 1, 2, \dots$. To see this recall that $\mathcal{D}_1 = \operatorname{Im} P$ by definition Furthermore, if we assume that $\mathcal{D}_{i-1} = \operatorname{Im} P$, then

$$\mathcal{D}_i = \{x \in \operatorname{Im} P \mid Bx \in \operatorname{Im} P + \operatorname{Ker} P\} = \operatorname{Im} P.$$

Also $\mathcal{D}_\infty = \operatorname{Im} P$. Next, we remark that $\mathcal{F}_0 = \{0\}$ and $\mathcal{F}_i = \operatorname{Ker} P$ for $i = 1, 2, \dots$. To show this recall that $\mathcal{F}_1 = \operatorname{Ker} P$ by definition. So, assume that $\mathcal{F}_{i-1} = \operatorname{Ker} P$. Then $\mathcal{F}_i = B[\operatorname{Ker} P \cap \operatorname{Im} P] + \operatorname{Ker} P = \operatorname{Ker} P$. Also $\mathcal{F}_\infty = \operatorname{Ker} P$. Now compute the numbers α_i, β_i, γ_i and δ_i. One obtains that $\alpha_0 = \dim \operatorname{Ker} P$, $\alpha_i = 0$ for $i \geq 1$, $\beta_i = \dim \operatorname{Ker} P$ for $i \geq 1$, $\gamma_i = \dim \operatorname{Ker} P$ for $i \geq 1$, $\delta_0 = \dim \operatorname{Ker} P$ and $\delta_i = 0$ for $i \geq 1$. It follows that $\beta_\infty = \gamma_\infty = \dim \operatorname{Ker} P$. So $p = 0$, $q = 0$, and $r = \dim \operatorname{Ker} P$. Hence there are no indices of the first and third kind, and $r = \dim \operatorname{Ker} P$ indices of the second kind. The i-th index of the second kind is given by $\nu_i = \#\{k \geq 1 \mid \beta_\infty - \beta_k \geq i\} = 0$. The first conclusion is that for a principal block we only have to know the number of indices of the second kind and its invariant polynomials to determine its similarity class. This results explains the appearance of the number $\dim \operatorname{Ker} P$ in Theorem VI.1.1. From the above computations we may also conclude that a block is similar a principal block if and only if there are no indices of the first and third kind and the indices of the second kind are all equal to zero. In this case the number of indices of the second kind is equal to the number $\dim \operatorname{Ker} P$ of the principal block.

The next result (which contains Propositions III.1.3 and V.1.3 as special cases) describes the properties of a special basis for a block; the proof of Theorem 1.1 will be based on this result.

Proposition 2.2. *Let $(B; P, Q)$ be an operator block on the space $\mathcal{X}$. Then there exist a subspace $\mathcal{E}_\infty$ and numbers $\kappa_1 \geq \cdots \geq \kappa_p \geq 0$, $\nu_1 \geq \cdots \geq \nu_r \geq 0$, and $\mu_1 \geq \cdots \geq \mu_q \geq 0$, and independent sets of of linearly independent vectors*

$$\{e_{1j}, \ldots, e_{\kappa_j+1\ j} \mid j = 1, \ldots, p\},$$
$$\{f_{1j}, \ldots, f_{\nu_j+1\ j} \mid j = 1, \ldots, r\},$$
$$\{g_{1j}, \ldots, g_{\mu_j+1\ j} \mid j = 1, \ldots, q\},$$

such that

(i) $\mathcal{X} = \left(\operatorname{span}\{e_{ij}\}_{i=1,j=1}^{\kappa_j+1\ ,p}\right) \oplus \left(\operatorname{span}\{f_{ij}\}_{i=1,j=1}^{\nu_j+1\ ,r}\right) \oplus \left(\operatorname{span}\{g_{ij}\}_{i=1,j=1}^{\mu_j+1\ ,q}\right) \oplus \mathcal{E}_\infty$;

(ii) $\operatorname{Ker} P = \operatorname{span}\{e_{11}, \ldots, e_{1p}, f_{11}, \ldots, f_{1r}\}$;

(iii) $\operatorname{Im} Q = \left(\operatorname{span}\{e_{ij}\}_{i=1,j=1}^{\kappa_j+1\ ,p}\right) \oplus \left(\operatorname{span}\{f_{ij}\}_{i=1,j=1}^{\nu_j\ ,r}\right) \oplus \left(\operatorname{span}\{g_{ij}\}_{i=1,j=1}^{\mu_j\ ,q}\right) \oplus \mathcal{E}_\infty$;

(iv) $B[\mathcal{E}_\infty] \subset \mathcal{E}_\infty \oplus \operatorname{Ker} P$;

(v) $Be_{ij} = Pe_{i+1\ j}$ *for* $i = 1, \ldots, \kappa_j$, *and* $Be_{\kappa_j+1\ j} = 0$ *for* $j = 1, \ldots, p$;
$Bf_{ij} = Pf_{i+1\ j}$ *for* $i = 1, \ldots, \nu_j$ *and* $j = 1, \ldots, r$;
$Bg_{ij} = Pg_{i+1\ j}$ *for* $i = 1, \ldots, \mu_j$ *and* $j = 1, \ldots, q$.

The proof of this proposition will be given in the next section.

VII.3 Proof of Proposition 2.2

The proof consists of several steps. As before $\mathcal{F}_\infty$ denotes final iterated image of $(B; P, Q)$.

STEP 1. In the first step we construct a $(B; P, Q)$-block-invariant subspace $\mathcal{G}_\infty$ of $\mathcal{X}$ that is (P, Q)-complementary to $\mathcal{F}_\infty$.

Let $\mathcal{F}_i$ be given by (2.2), and let $\mathcal{F}_\rho = \mathcal{F}_\infty$. We choose subspaces $\mathcal{G}_\rho$ and $\mathcal{G}'$ in $\mathcal{X}$ such that

$$\mathcal{X} = \mathcal{F}_\infty \oplus \mathcal{G}_\rho \oplus \mathcal{G}', \quad \operatorname{Im} Q = (\mathcal{F}_\infty \cap \operatorname{Im} Q) \oplus \mathcal{G}_\rho.$$

Then we have for $j = \rho$ that

(1) $\mathcal{X} = \mathcal{F}_\infty \oplus \mathcal{G}_j \oplus \mathcal{G}'$ and $\operatorname{Im} Q = (\mathcal{F}_\infty \cap \operatorname{Im} Q) \oplus \mathcal{G}_j$;

(2) $B[\mathcal{G}_j] \subset \mathcal{F}_j \oplus \mathcal{G}_j \oplus \mathcal{G}'$.

We want to find a subspace $\mathcal{G}_1$ that fulfills the conditions (1) and (2) for $j = 1$. So assume that we have $\mathcal{G}_j$ that fulfills the conditions (1) and (2) for some j with $2 \leq j \leq \rho$. We construct $\mathcal{G}_{j-1}$. For $y \in \mathcal{G}_j$ one has $By = f_1 + Bf_{j-1} + y_j + y'$ with $f_1 \in \mathcal{F}_1$, $f_{j-1} \in \mathcal{F}_{j-1} \cap \operatorname{Im} Q$, $y_j \in \mathcal{G}_j$ and $y' \in \mathcal{G}'$. It follows that there exist linear transformations $T_j : \mathcal{G}_j \to \mathcal{F}_1$, $S_j : \mathcal{G}_j \to \mathcal{F}_{j-1} \cap \operatorname{Im} Q$, $R_j : \mathcal{G}_j \to \mathcal{G}_j$, and $U_j : \mathcal{G}_j \to \mathcal{G}'$ such that $By = (T_j + BS_j + R_j + U_j)y$ for each $y \in \mathcal{G}_j$.

Then $B(I-S_j)y = (T_j + S_jR_j)y + (I-S_j)R_jy + U_jy$ for each $y \in \mathcal{G}_j$. We put $\mathcal{G}_{j-1} = (I-S_j)\mathcal{G}_j$. Since $S_j[\mathcal{G}_j] \subset \mathcal{F}_{j-1} \cap \operatorname{Im} Q$ and $\mathcal{F}_{j-1} \subset \mathcal{F}_\infty$, condition (1) now holds for $j-1$ replacing j. Furthermore it follows that $B[\mathcal{G}_{j-1}] \subset \mathcal{F}_{j-1} \oplus \mathcal{G}_{j-1} \oplus \mathcal{G}'$. Thus, by induction, there exists a subspace $\mathcal{G}_1$ satisfying (1) and (2) for $j=1$.

Put $\mathcal{G}_\infty = \mathcal{G}_1 \oplus \mathcal{G}'$. Then the first equality in (1) shows that $\mathcal{X} = \mathcal{F}_\infty \oplus \mathcal{G}_\infty$. From the second equality of (1) we see that $\mathcal{G}_1 \subset \operatorname{Im} Q$, and therefore $\mathcal{G}_1 \subset \mathcal{G}_\infty \cap \operatorname{Im} Q$. Now

$$\operatorname{Im} Q = (\mathcal{F}_\infty \cap \operatorname{Im} Q) \oplus \mathcal{G}_1 \subset (\mathcal{F}_\infty \cap \operatorname{Im} Q) \oplus (\mathcal{G}_\infty \cap \operatorname{Im} Q) \subset \operatorname{Im} Q,$$

and thus $\mathcal{G}_1 = \mathcal{G}_\infty \cap \operatorname{Im} Q$ and $\operatorname{Im} Q = (\mathcal{F}_\infty \cap \operatorname{Im} Q) \oplus (\mathcal{G}_\infty \cap \operatorname{Im} Q)$. Since $\operatorname{Ker} P \subset \mathcal{F}_\infty$, we see that $\mathcal{G}_\infty \cap \operatorname{Ker} P = \{0\}$, and thus

$$\operatorname{Ker} P = (\mathcal{F}_\infty \cap \operatorname{Ker} P) \oplus (\mathcal{G}_\infty \cap \operatorname{Ker} P).$$

This proves that $\mathcal{G}_\infty$ is (P,Q)-complementary to $\mathcal{F}_\infty$. From condition (2) and the fact that $\mathcal{G}_1 = \mathcal{G}_\infty \cap \operatorname{Im} Q$ it is clear that $\mathcal{G}_\infty$ is $(B;P,Q)$-block-invariant. So the subspaces $\mathcal{F}_\infty$ and $\mathcal{G}_\infty$ form a pair of (P,Q)-complementary and $(B;P,Q)$-block-invariant subspaces. Hence $\mathcal{F}_\infty$ and $\mathcal{G}_\infty$ are a pair of decomposing subspaces. According to Theorem II.2.5 there exists a block $(P'B;P',Q')$ with $\operatorname{Im} Q' = \operatorname{Im} Q$ and $\operatorname{Ker} P' = \operatorname{Ker} P$, and such that $(P'B;P',Q')$ is the direct sum of its regular restrictions to $\mathcal{F}_\infty$ and $\mathcal{G}_\infty$. Let $(B_1;P_1,Q_1)$ be the regular restriction of $(P'B;P',Q')$ to $\mathcal{F}_\infty$, and let $(B_2;P_2,Q_2)$ be the regular restriction of $(P'B;P',Q')$ to $\mathcal{G}_\infty$.

STEP 2. Next, we construct the linearly independent vectors $\{g_{ij}\}_{i=1,j=1}^{\mu_j+1\ ,q}$ and the subspace $\mathcal{E}_\infty$. Since $\operatorname{Ker} P \cap \mathcal{G}_\infty = \{0\}$, we know that $P_2 = I_{\mathcal{G}_\infty}$. Therefore $(B_2;P_2,Q_2)$ is a full length block. Let $\mathcal{E}_\infty$ be the residual space of the block $(B_2;P_2,Q_2)$. We apply Proposition III.1.3 to the block $(B_2;P_2,Q_2)$ to obtain a set of vectors $\{g_{ij}\}_{i=1,j=1}^{\mu_j+1\ ,q}$ such that

$$\mathcal{G}_\infty = \operatorname{span}\{g_{ij}\}_{i=1,j=1}^{\mu_j+1,q} \oplus \mathcal{E}_\infty;$$

$$\operatorname{Im} Q_2 = \operatorname{span}\{g_{ij}\}_{i=1,j=1}^{\mu_j,q} \oplus \mathcal{E}_\infty;$$

$$B_2 g_{ij} = g_{i+1\ j}, \text{ for } i=1,\ldots,\mu_j \text{ and } j=1,\ldots,q.$$

Since $B = PP'B$, we have that $Bx = PB'x = PB_2x$, for each $x \in \mathcal{G}_\infty$. Hence

$$Bg_{ij} = Pg_{i+1\ j}, \text{ for } i=1,\ldots,\mu_j \text{ and } j=1,\ldots,q;$$

$$B[\mathcal{E}_\infty] \subset \mathcal{E}_\infty \oplus \operatorname{Ker} P.$$

STEP 3. In this step we treat the block $(B_1;P_1,Q_1)$ and prepare the construction of the independent sets of linearly independent vectors $\{e_{ij}\}_{i=1,j=1}^{\kappa_j+1\ ,p}$ and $\{f_{ij}\}_{i=1,j=1}^{\nu_j+1\ ,r}$. Remark that $\operatorname{Ker} P = \operatorname{Ker} P_1$, $\operatorname{Im} Q_1 = \mathcal{F}_\infty \cap \operatorname{Im} Q$, and that $B_1x = P_1Bx$ for each $x \in \operatorname{Im} Q_1$. From these equalities it follows that $\mathcal{F}_1 = \operatorname{Ker} P_1$ and $\mathcal{F}_{j+1} = \mathcal{F}_1 \oplus B_1[\mathcal{F}_j \cap \operatorname{Im} Q_1]$, for $j=1,\ldots,\rho$, with $\mathcal{F}_\rho = \mathcal{F}_\infty$. Then $B_1 : \mathcal{F}_{i-1} \cap \operatorname{Im} Q_1 \to \mathcal{F}_i$ induces a quotient operator

$$[B_1]_i : (\mathcal{F}_{i-1} \cap \operatorname{Im} Q_1)/(\mathcal{F}_{i-2} \cap \operatorname{Im} Q_1) \to \mathcal{F}_i/\mathcal{F}_{i-1}, i=2,3,\ldots,\rho.$$

Denote the class of g in $(\mathcal{F}_{i-1}\cap \operatorname{Im} Q_1)/(\mathcal{F}_i \cap \operatorname{Im} Q_1)$ by $[g]_{i-1}$. and the class of f in $\mathcal{F}_i/\mathcal{F}_{i-1}$ by $[f]_i$. Assume that $f\in\mathcal{F}_i$ and $f\notin\mathcal{F}_{i-1}$. Then $f=f_0+B_1g$ with $f_0\in\operatorname{Ker}P_1$ and $g\in\mathcal{F}_{i-1}\cap\operatorname{Im}Q_1$. Thus $P_1f=P_1B_1g=B_1g$. Since $(I-P_1)f\in\operatorname{Ker}P_1\subset\mathcal{F}_{i-1}$, it follows that

$$[f]_i=[P_1f]_i=[B_1g]_i=[B_1]_i[g]_{i-1}.$$

So $[B_1]_i$ is surjective. On the other hand $(\mathcal{F}_{i-1}\cap\operatorname{Im}Q_1)/(\mathcal{F}_{i-2}\cap\operatorname{Im}Q_1)$ is a subspace of $\mathcal{F}_{i-1}/\mathcal{F}_{i-2}$. Assume that $g\in\mathcal{F}_{i-1}\cap\operatorname{Im}Q_1$ and $[B_1]_i[g]_{i-1}=0$. Then $B_1g\in\mathcal{F}_{i-1}$, and thus $B_1g=f_0+B_1g_1$ with $f_0\in\operatorname{Ker}P_1$ and $g_1\in\mathcal{F}_{i-2}\cap\operatorname{Im}Q_1$. Put $h=g-g_1$. Then $[h]_{i-1}=[g]_{i-1}$ and $B_1h=P_1B_1(g-g_1)=P_1f_0=0$. Thus if $[g]_{i-1}\in\operatorname{Ker}[B]_i$, then there is a representative of $[g]_{i-1}$ which is in $\operatorname{Ker}B_1$.

We start with the construction of the linearly independent vectors $\{e_{ij}\}_{i=1,j=1}^{\kappa_j+1,p}$ and $\{f_{ij}\}_{i=1,j=1}^{\nu_j+1\ ,r}$. Let α_i and β_i be as defined in the theorem, and put

$$\sigma_i=\alpha_{i-1}+\beta_{i-1}-\beta_i=\dim(\mathcal{F}_i\cap\operatorname{Im}Q_1)-\dim(\mathcal{F}_{i-1}\cap\operatorname{Im}Q_1).$$

Write $\tau_i=\alpha_i-\sigma_i$. Then τ_i is the codimension of $(\mathcal{F}_{i-1}\cap\operatorname{Im}Q_1)/(\mathcal{F}_{i-2}\cap\operatorname{Im}Q_1)$ in $\mathcal{F}_{i-1}/\mathcal{F}_{i-2}$. Remark that $\mathcal{F}_{\rho+1}/\mathcal{F}_\rho=\{0\}$. Choose a basis $[e_{\rho1}],\dots,[e_{\rho\sigma_\rho}]$ of $(\mathcal{F}_\rho\cap\operatorname{Im}Q_1)/(\mathcal{F}_{\rho-1}\cap\operatorname{Im}Q_1)$ with $B_1e_{\rho i}=0$. Extend this basis with $[f_{\rho1},\dots,[f_{\rho\tau_\rho}]$ to a basis

$$[e_{\rho1}],\dots,[e_{\rho\sigma_\rho}],[f_{\rho1},\dots,[f_{\rho\tau_\rho}]$$

of $\mathcal{F}_\rho/\mathcal{F}_{\rho-1}$.

Assume that we have found a basis

$$[e_{i1}]_i,\dots,[e_{i\sigma_i}]_i,[f_{i1}]_i,\dots,[f_{i\tau_i}]_i$$

of $\mathcal{F}_i/\mathcal{F}_{i-1}$. Since $[B_1]_i$ is surjective there is an independent set of vectors

$$[e_{i-1\ 1}]_{i-1},\dots,[e_{i-1\ \sigma_i}]_{i-1},[f_{i-1\ 1}]_{i-1},\dots,[f_{i-1\ \tau_i}]_{i-1} \tag{3.1}$$

in $(\mathcal{F}_{i-1}\cap\operatorname{Im}Q_1)/(\mathcal{F}_{i-2}\cap\operatorname{Im}Q_1)$ such that $B_1e_{i-1\ j}=P_1e_{ij}$ for $j=1,\dots,\sigma_i$ and $B_1f_{i-1\ j}=P_1f_{ij}$ for $j=1,\dots,\tau_i$. Hence

$$[B_1]_i[e_{i-1\ j}]_{i-1}=[e_{ij}]_i,\quad [B_1]_i[f_{i-1\ j}]_{i-1}=[f_{ij}]_i.$$

Now extend the set (3.1) to a basis of $(\mathcal{F}_{i-1}\cap\operatorname{Im}Q_1)/(\mathcal{F}_{i-2}\cap\operatorname{Im}Q_1)$ by adding a basis

$$[e_{i-1\ \sigma_i+1}]_{i-1},\dots,[e_{i-1\ \sigma_{i-1}}]_{i-1}$$

of $\operatorname{Ker}[B_1]_i$. We may choose $e_{i-1\ j}$ such that for $j=\sigma_i+1,\dots,\sigma_{i-1}$ we have $B_1e_{i-1\ j}=0$. Next we extend the basis

$$[e_{i-1\ 1}]_{i-1},\dots,[e_{i-1\ \sigma_{i-1}}]_{i-1},[f_{i-1\ 1}]_{i-1},\dots,[f_{i-1\ \tau_i}]_{i-1}$$

of $(\mathcal{F}_{i-1} \cap \operatorname{Im} Q_1)/(\mathcal{F}_{i-2} \cap \operatorname{Im} Q_1)$ with vectors $[f_{i-1\ \tau_i+1}]_{i-1}, \dots, [f_{i-1\ \tau_{i-1}}]_{i-1}$ to a basis

$$[e_{i-1\ 1}]_{i-1}, \dots, [e_{i-1\ \sigma_{i-1}}]_{i-1}, [f_{i-1\ 1}]_{i-1}, \dots, [f_{i-1\ \tau_{i-1}}]_{i-1}$$

of $\mathcal{F}_{i-1}/\mathcal{F}_{i-2}$.

We have constructed linearly independent vectors

$$\{e_{ij}\}_{i=1,j=1}^{\rho,\sigma_i} \bigcup \{f_{ij}\}_{i=1,j=1}^{\rho,\tau_i} \tag{3.2}$$

in $\mathcal{F}_\rho = \mathcal{F}_\infty$ such that

$$\mathcal{F}_i = \mathcal{F}_{i-1} \oplus \operatorname{span}\{e_{i1}, \dots, e_{i\sigma_i}, f_{i1}, \dots, f_{i\tau_i}\}.$$

Since $\mathcal{F}_0 = \{0\}$, the set (3.2) is a basis of $\mathcal{F}_\infty$. Also

$B_1 e_{ij} = P_1 e_{i+1\ j}$ for $j = 1, \dots, \sigma_{i+1}$, and $B_1 e_{i\ j} = 0$ for $j = \sigma_{i+1}+1, \dots, \sigma_i$, and $i = 1, \dots, \rho$;

$B_1 f_{ij} = P_1 f_{i+1\ j}$ for $j = 1, \dots, \tau_i$ and $i = 1, \dots, \rho$;

Recall that $Bx = PB_1x$ and $Px = PP_1x$ for each $x \in \mathcal{F}_\infty$.

$Be_{ij} = Pe_{i+1\ j}$ for $j = 1, \dots, \sigma_{i+1}$, and $Be_{i\ j} = 0$ for $j = \sigma_{i+1}+1, \dots, \sigma_i$, and $i = 1, \dots, \rho$;

$Bf_{ij} = Pf_{i+1\ j}$ for $j = 1, \dots, \tau_i$ and $i = 1, \dots, \rho$;

STEP 4. In principle the construction is now finished. We have to check that the sets of vectors that we found fulfill the requirements. First we check condition (i). Remark that $i = \kappa_j + 1$ if and only if $\sigma_{i+1} < j \le \sigma_i$. Hence the sequence $\{\kappa_j + 1\}$ is the dual sequence of the sequence $\{\sigma_i\}$. Analogously, we get that $\{\mu_j + 1\}$ is the dual sequence of $\{\tau_i\}$. From the definition of dual sequences it follows that $\{e_{ij}\}_{i=1,j=1}^{\rho,\sigma_i}$ can be rewritten as $\{e_{ij}\}_{i=1,j=1}^{\kappa_j+1\ ,p}$ and $\{f_{ij}\}_{i=1,j=1}^{\rho,\tau_i}$ as $\{f_{ij}\}_{i=1,j=1}^{\nu_j+1\ ,r}$. Furthermore, we get $p = \sigma_1$ and $r = \tau_1$. For condition (ii) we just note that $\mathcal{F}_1 = \operatorname{Ker} P_1 = \operatorname{Ker} P$. Condition (iii) follows from $\operatorname{Im} Q = \operatorname{Im} Q_1 \oplus \operatorname{Im} Q_2$ and $\operatorname{Im} Q_2 = \left(\operatorname{span}\{g_{ij}\}_{i=1,j=1}^{\mu_j\ ,q}\right) \oplus \mathcal{E}_\infty$, and $\operatorname{Im} Q_1 = \left(\operatorname{span}\{e_{ij}\}_{i=1,j=1}^{\kappa_j+1\ ,p}\right) \oplus \left(\operatorname{span}\{f_{ij}\}_{i=1,j=1}^{\nu_j\ ,r}\right)$. To see this last equality just note that

$$\mathcal{F}_i \cap \operatorname{Im} Q_1 = (\mathcal{F}_{i-1} \cap \operatorname{Im} Q_1) \oplus \operatorname{span}\{e_{i-1\ 1}, \dots, e_{i-1\ \sigma_{i-1}}, f_{i-1\ 1}, \dots, f_{i-1\ \tau_i}\}.$$

So $\operatorname{Im} Q_1 = \operatorname{span}\left(\{e_{ij}\}_{i=1,j=1}^{\rho,\sigma_i} \bigcup \{f_{ij}\}_{i=1,j=1}^{\rho,\tau_{i+1}}\right)$. To obtain (iii), it remains to use the dual sequences of $\{\sigma_i\}$ and $\{\tau_{i+1}\}$, to reorder this basis in same way as we rewrote the basis of $\mathcal{F}_\infty$. Finally we remark that all the properties in (iv) and (v) are already checked. $\square$

VII.4 Proof of Theorems 1.1 and 2.1

First we state and prove a lemma which contains the first part of Theorem 1.1. The lemma shows that there are indices and invariant polynomials associated with a block, although at this stage we do not yet know that these items are uniquely determined by the block.

Lemma 4.1. *An operator block $(B; P, Q)$ is block similar, with a block similarity of the second kind, to a direct sum of shifts of the first kind, shifts of the second kind, shifts of the third kind, and an operator J.*

PROOF. We apply Theorem 1.3 to obtain the independent sets of linearly independent vectors

$$\{e_{ij}\}_{i=1,j=1}^{\kappa_j+1\ ,p},\ \{f_{ij}\}_{i=1,j=1}^{\nu_j+1\ ,r},\ \{g_{ij}\}_{i=1,j=1}^{\mu_j+1\ ,q}$$

and a subspace $\mathcal{E}_\infty$ of $\mathcal{X}$, with the properties (i), (ii), (iii), (iv) and (v). Put $\mathcal{X}_{1j} = \operatorname{span}\{e_{1j}, \dots, e_{\kappa_j+1\ j}\}$ for $j = 1, \dots, p$, $\mathcal{X}_{2j} = \operatorname{span}\{f_{1j}, \dots, f_{\nu+1\ j}\}$ for $j = 1, \dots, r$, and $\mathcal{X}_{3j} = \operatorname{span}\{g_{1j}, \dots, g_{\mu+1\ j}\}$ for $j = 1, \dots, q$. Then

$$\mathcal{X} = \left(\bigoplus_{j=1}^{p} \mathcal{X}_{1j}\right) \oplus \left(\bigoplus_{j=1}^{r} \mathcal{X}_{2j}\right) \oplus \left(\bigoplus_{j=1}^{q} \mathcal{X}_{3j}\right) \oplus \mathcal{E}_\infty.$$

The properties (i), (ii) and (iii) give us that the set of subspaces

$$\{\mathcal{X}_{11}, \dots, \mathcal{X}_{1p}, \mathcal{X}_{21}, \dots, \mathcal{X}_{2r}, \mathcal{X}_{31}, \dots, \mathcal{X}_{3q}, \mathcal{E}_\infty\}$$

is a set of (P, Q)-complementary subspaces of $\mathcal{X}$. From the properties (iv) and (v) we see that each member of this set is $(B; P, Q)$-block-invariant. Let P_{1j} be the projection of $\mathcal{X}_{1j}$ with $\operatorname{Ker} P_{1j} = \operatorname{span}\{e_{1j}\}$ and $\operatorname{Im} P_{1j} = \operatorname{span}\{e_{2j}, \dots, e_{\kappa_j+1\ j}\}$, and set $Q_{1j} = I_{\mathcal{X}_{1j}}$. We define $B_{1j} : \mathcal{X}_{1j} \to \operatorname{Im} P_{1j}$ by putting $B_{1j} e_{ij} = e_{i+1\ j}$ for $i = 1, \dots, \kappa_j$ and $B_{1j} e_{\kappa_j+1\ j} = 0$. Then the block $(B_{1j}; P_{1j}, Q_{1j})$ is block similar to a shift of the first kind with index κ_j. Next, define P_{2j} to be the projection of $\mathcal{X}_{2j}$ with $\operatorname{Im} P_{2j} = \operatorname{span}\{f_{2j}, \dots, f_{\nu+1\ j}\}$ and $\operatorname{Ker} P_{2j} = \operatorname{span}\{f_{1j}\}$, and let Q_{2j} be the projection of $\mathcal{X}_{2j}$ with $\operatorname{Im} Q_{2j} = \operatorname{span}\{f_{1j}, \dots, f_{\nu_j j}\}$ and $\operatorname{Ker} Q_{2j} = \operatorname{span}\{f_{\nu_j+1\ j}\}$. We define the operator $B_{2j} : (\mathcal{X}_{2j} \cap \operatorname{Im} Q_{2j}) \to \operatorname{Im} P_{2j}$ by putting $B_{2j} f_{ij} = e_{i+1\ j}$ for $i = 1, \dots, \nu_j$ and $B_{1j} f_{\nu_j+1\ j} = 0$. Then the block $(B_{2j}; P_{2j}, Q_{2j})$ is block similar to a shift of the second kind with index ν_j. On the space $\mathcal{X}_{3j}$ we define $P_{3j} = I_{\mathcal{X}_{3j}}$ and Q_{3j} to be the projection of $\mathcal{X}_{3j}$ with $\operatorname{Im} Q_{3j} = \operatorname{span}\{g_{1j}, \dots, g_{\mu_j j}\}$ and $\operatorname{Ker} Q_{3j} = \operatorname{span}\{g_{\mu_j+1\ j}\}$. The operator $B_{3j} : (\mathcal{X}_{3j} \cap \operatorname{Im} Q_{3j}) \to \mathcal{X}_{3j}$ we define by $B_{3j} g_{ij} = e_{i+1\ j}$ for $i = 1, \dots, \mu_j$ and $B_{3j} f_{\mu_j+1\ j} = 0$. Then the block $(B_{3j}; P_{3j}, Q_{3j})$ is block similar to a shift of the third kind with index μ_j and $Bx = PB_{3j}x$ for each $x \in \mathcal{X}_{3j} \cap \operatorname{Im} Q$.

Next we define a projection P' of $\mathcal{X}$ by

$$P' = \left(\bigoplus_{j=1}^{p} P_{1j}\right) \oplus \left(\bigoplus_{j=1}^{r} P_{2j}\right) \oplus \left(\bigoplus_{j=1}^{q} P_{3j}\right) \oplus I_{\mathcal{E}_\infty}$$

and a projection Q' of $\mathcal{X}$ by

$$Q' = \left(\bigoplus_{j=1}^{p} Q_{1j}\right) \oplus \left(\bigoplus_{j=1}^{r} Q_{2j}\right) \oplus \left(\bigoplus_{j=1}^{q} Q_{3j}\right) \oplus I_{\mathcal{E}_\infty}.$$

Then $\operatorname{Ker} P = \operatorname{Ker} P'$ and $\operatorname{Im} Q = \operatorname{Im} Q'$. We define $B_\infty : \mathcal{E}_\infty \to \mathcal{E}_\infty$ by $B_\infty x = P'Bx$ for each $x \in \mathcal{E}_\infty$. We obtained that

$$(P'B; P', Q') = \left(\bigoplus_{j=1}^{p} (B_{1j}; P_{1j}, Q_{1j}\right) \oplus \left(\bigoplus_{j=1}^{r} (B_{2j}; P_{2j}, Q_{2j}\right) \oplus \left(\bigoplus_{j=1}^{q} (B_{3j}; P_{3j}, Q_{3j}\right) \oplus (B_\infty; I_{\mathcal{E}_\infty}, I_{\mathcal{E}_\infty}).$$

Finally we see that the blocks $(B; P, Q)$ and $(P'B; P', Q')$ are block similar with a similarity of the second kind, which finishes the proof. □

The next result shows that a block similarity carries over the definition spaces $\mathcal{D}_i$ and the iterated images $\mathcal{F}_i$ of the one block into the corresponding spaces of the other block.

Lemma 4.2. *Let $(B; P, Q)$ be a block with underlying space $\mathcal{X}$ and $(B'; P', Q')$ be a block on $\mathcal{X}'$. Let S be a block similarity of $(B; P, Q)$ and $(B'; P', Q')$. Put*

$$\begin{aligned} &\mathcal{D}_0 = \mathcal{X},\ \mathcal{D}_1 = \operatorname{Im} Q,\ \mathcal{D}_i = \{x \in \mathcal{D}_{i-1} \mid Bx \in \mathcal{D}_{i-1} + \operatorname{Ker} P\},\ i = 2, 3, \dots, \\ &\mathcal{D}'_0 = \mathcal{X}',\ \mathcal{D}'_1 = \operatorname{Im} Q',\ \mathcal{D}'_i = \{x \in \mathcal{D}'_{i-1} \mid B'x \in \mathcal{D}'_{i-1} + \operatorname{Ker} P'\},\ i = 2, 3, \dots, \end{aligned} \tag{4.1}$$

and

$$\begin{aligned} &\mathcal{F}_0 = \{0\},\ \mathcal{F}_1 = \operatorname{Ker} P,\ \mathcal{F}_i = \mathcal{F}_1 \oplus B[\mathcal{F}_{i-1} \cap \operatorname{Im} Q],\ i = 2, 3, \dots, \\ &\mathcal{F}'_0 = \{0\},\ \mathcal{F}'_1 = \operatorname{Ker} P',\ \mathcal{F}'_i = \mathcal{F}'_1 \oplus B'[\mathcal{F}'_{i-1} \cap \operatorname{Im} Q'],\ i = 2, 3, \dots . \end{aligned} \tag{4.2}$$

Then $S[\mathcal{D}_i] = \mathcal{D}'_i$ and $S[\mathcal{F}_i] = \mathcal{F}'_i$ for $i = 0, 1, \dots$.

Proof. Clearly $S[\mathcal{D}_0] = \mathcal{D}'_0$ and $S[\mathcal{D}_1] = \mathcal{D}'_1$. Assume that $S[\mathcal{D}_{i-1}] = \mathcal{D}'_{i-1}$. Let $x \in \mathcal{D}_i$. Thus $x \in \mathcal{D}_{i-1}$ and $Bx \in \mathcal{D}_{i-1} + \operatorname{Ker} P$. Then $Sx \in \mathcal{D}'_{i-1}$ and $B'Sx = SBx - (I-P)'BSx \in \mathcal{D}'_{i-1} + \operatorname{Ker} P'$. This proves that $S[\mathcal{D}_i] \subset \mathcal{D}'_i$.

Applying this result to S^{-1}, we get $S^{-1}[\mathcal{D}_i'] \subset \mathcal{D}_i$, and we see that $S[\mathcal{D}_i] = \mathcal{D}_i'$. This proves that $S[\mathcal{D}_i] = \mathcal{D}_i'$, for $i = 0, 1, \dots$.

To prove that $S[\mathcal{F}_i] = \mathcal{F}_i'$, for $i = 0, 1, \dots$, we start with remarking that $S[\operatorname{Ker} P] = \operatorname{Ker} P'$. Assume that $S[\mathcal{F}_{i-1}] = \mathcal{F}_{i-1}'$, and take $x_i \in \mathcal{F}_i$. Then $x_i = x_1 + Bx_{i-1}$ with $x_1 \in \operatorname{Ker} P$ and $x_{i-1} \in \mathcal{F}_{i-1}$. We compute

$$Sx_i = Sx_1 + (I - P')SBx_{i-1} + B'Sx_{i-1},$$

and see that $Sx_i \in \operatorname{Ker} P' \oplus B'[\mathcal{F}_{i-1}'] = \mathcal{F}_i'$. This proves that $S[\mathcal{F}_i] \subset \mathcal{F}_i'$. Applying this result to S^{-1}, we get $S^{-1}[\mathcal{F}_i'] \subset \mathcal{F}_i$, and thus $S[\mathcal{F}_i] = \mathcal{F}_i'$. □

PROOF OF THEOREM 2.1. From Lemma 4.1 we know that there exists a direct sum $(B'; P', Q')$ of shifts of the first kind with indices $\kappa_1 \geq \dots \geq \kappa_p \geq 0$, shifts of the second kind with indices $\nu_1 \geq \dots \geq \nu_r \geq 0$, shifts of the third kind with indices $\mu_1 \geq \dots \geq \mu_q \geq 0$, and a block $(B_0'; I_0', I_0')$ such that $(B; P, Q)$ is block similar to $(B'; P', Q')$. Let S be the block similarity transformation of $(B; P, Q)$ and $(B'; P', Q')$. We define the subspaces $\mathcal{D}_i'$ and $\mathcal{F}_i'$ by the formulas (4.1) and (4.2), respectively. From Lemma 4.2 we see that $S[\mathcal{D}_i] = \mathcal{D}_i'$ and $S[\mathcal{F}_i] = \mathcal{F}_i'$, for $i = 0, 1, \dots$. Also it is clear that $S[\mathcal{D}_i \cap \operatorname{Ker} P] = \mathcal{D}_i' \cap \operatorname{Ker} P'$ and $S[\mathcal{F}_i \cap \operatorname{Im} Q] = \mathcal{F}_i' \cap \operatorname{Im} Q'$, for $i = 0, 1, \dots$. Therefore we can compute the numbers α_i, β_i, γ_i and δ_i by using $(B'; P', Q')$ in stead of $(B; P, Q)$. We will express these numbers in terms of the indices $\kappa_1, \dots, \kappa_p$, $\nu_1, \dots, \nu_r$ and $\mu_1, \dots, \mu_q$.

Assume that $(B; P, Q)$ is a shift of the first kind with index κ. Since this implies that $Q = I$, we obtain that $\mathcal{D}_i = \mathcal{X}$ and $\dim \operatorname{Ker} P - \dim(\mathcal{D}_i \cap \operatorname{Ker} P) = 0$ for all i. Furthermore, we see that $\dim \mathcal{F}_i = \min\{i, \kappa + 1\}$ and $\dim \mathcal{F}_i - \dim(\mathcal{F}_i \cap \operatorname{Im} Q) = 0$, for all i.

Assume that $(B; P, Q)$ is a shift of the second kind with index ν. We compute that $\dim \mathcal{X} - \dim \mathcal{D}_i = \min\{i, \nu + 1\}$ and $\dim \operatorname{Ker} P - \dim(\mathcal{D}_i \cap \operatorname{Ker} P) = 0$ if $i \leq \nu$ and $\dim \operatorname{Ker} P - \dim(\mathcal{D}_i \cap \operatorname{Ker} P) = 1$ if $i \geq \nu + 1$. Also it is easy to see that $\dim \mathcal{F}_i = \min\{i, \nu + 1\}$ and $\dim \mathcal{F}_i - \dim(\mathcal{F}_i \cap \operatorname{Im} Q) = 0$ if $i \leq \nu$, and $\dim \mathcal{F}_i - \dim(\mathcal{F}_i \cap \operatorname{Im} Q) = 1$ if $i \geq \nu + 1$.

Assume that $(B; P, Q)$ is a shift of the third kind with index μ. Then $\operatorname{Ker} P = \{0\}$, and thus $\mathcal{F}_i = \{0\}$, and $\dim \mathcal{F}_i - \dim(\mathcal{F}_i \cap \operatorname{Im} Q) = 0$ for all i. One computes that $\dim \mathcal{X} - \dim \mathcal{D}_i = \min\{i, \mu + 1\}$ and that $\dim \operatorname{Ker} P - \dim(\mathcal{D}_i \cap \operatorname{Ker} P) = 0$ for all i.

For the block $(B_0'; I_0', I_0')$ one computes that $\mathcal{F}_i = \{0\}$ and $\mathcal{D}_i = \mathcal{X}$ for all i. So also $\dim \operatorname{Ker} P - \dim(\mathcal{D}_i \cap \operatorname{Ker} P) = 0$ and $\dim \mathcal{F}_i - \dim(\mathcal{F}_i \cap \operatorname{Im} Q) = 0$ for all i.

For a direct sum the i-th definition space $\mathcal{D}_i$ is the direct sum of the subspaces $\mathcal{D}_i$ corresponding to the components, and the i-th iterated image $\mathcal{F}_i$ is the direct sum of the subspaces $\mathcal{F}_i$ corresponding to the components. So we compute that

$$\dim \mathcal{F}_i = \sum_{j=1}^{p} \min\{i, \kappa_j + 1\} + \sum_{j=1}^{r} \min\{i, \nu_j + 1\}, \tag{4.3}$$

$$\dim \mathcal{F}_i - \dim(\mathcal{F}_i \cap \operatorname{Im} Q) = \dim \operatorname{Ker} P - \dim(\mathcal{D}_i \cap \operatorname{Ker} P) = \#\{j \mid i \geq \nu_j + 1\}, \quad (4.4)$$

and

$$\dim \mathcal{X} - \dim \mathcal{D}_i = \sum_{j=1}^{r} \min\{i, \nu_j + 1\} + \sum_{j=1}^{q} \min\{i, \mu_j + 1\}. \quad (4.5)$$

From (4.3) one sees that

$$\alpha_i = \dim \mathcal{F}_{i+1} - \dim \mathcal{F}_i = \#\{j \mid \kappa_j \geq i\} + \#\{j \mid \nu_j \geq i\}$$

and from (4.5) that

$$\delta_j = \dim \mathcal{D}_i - \dim \mathcal{D}_{i+1} = \#\{j \mid \nu_j \geq i\} + \#\{j \mid \mu_j \geq i\}.$$

Observe that formula (4.4) shows that $\beta_i = \gamma_i$, and that the sequences $(\beta_i)_i$ and $(\gamma_i)_i$ are non-increasing and bounded. So the maximal numbers in these sequences are β_∞ and γ_∞. Then

$$\beta_\infty - \beta_i = \gamma_\infty - \gamma_i = \#\{j \mid \nu_j \geq i\}.$$

It follows that

$$\begin{aligned} &\alpha_i - (\beta_\infty - \beta_i) = \#\{j \mid \kappa_j \geq i\}, \\ &\beta_\infty - \beta_i = \gamma_\infty - \gamma_i = \#\{j \mid \nu_j \geq i\}, \\ &\delta_i - (\gamma_\infty - \gamma_i) = \#\{j \mid \mu_j \geq i\}. \end{aligned} \quad (4.6)$$

We proved that the sequences $\{\alpha_i - (\beta_\infty - \beta_i)\}$ and $\{\kappa_j\}$ are dual index sequences, which gives the desired formula for κ_i (see Proposition III.4.1). Similarly, the sequences $\{\beta_\infty - \beta_i\}$ and $\{\nu_j\}$, and $\{\delta_i - (\gamma_\infty - \gamma_i)\}$ and $\{\mu_j\}$ are pairs of dual index sequences. This gives the formulas for ν_i and μ_i. □

PROOF OF THEOREM 1.1. We apply Lemma 4.1 and see that the block $(B; P, Q)$ is block similar to a direct sum of shifts of the first, second and third kind and a block $(B_0; I_0, I_0)$. From the formulas for the indices, given in Theorem 2.1, we see that the indices of the first, second and third kind are uniquely determined by the block $(B; P, Q)$. It remains to show that the invariant polynomials are uniquely determined by $(B; P, Q)$.

Let S be a block similarity of the blocks $(B; P, Q)$ and $(B'; P', Q')$ which we both suppose to be a direct sum of shifts of the first, second and third kind and an operator on a subspace. Let $\mathcal{F}_i$ and $\mathcal{F}'_i$ be given by (4.2). Then $S[\mathcal{F}_i] = \mathcal{F}'_i$ for each i, and hence $S[\mathcal{F}_\infty] = \mathcal{F}'_\infty$. Decompose $(B; P, Q)$ as $(B; P, Q) = (B_1; P_1, Q_1) \oplus (B_2; P_2, Q_2)$, where $(B_1; P_1, Q_1)$ is the direct sum of the shifts of the first and second kind, and $(B_2; P_2, Q_2)$ is the direct sum of the shifts of the third kind and the block $(B_0; I_0, I_0)$. Similarly decompose $(B'; P', Q')$ as $(B'; P', Q') = (B'_1; P'_1, Q'_1) \oplus (B'_2; P'_2, Q'_2)$. The underlying space of $(B_1; P_1, Q_1)$ is $\mathcal{F}_\infty$ and the underlying space of $(B'_1; P'_1, Q'_1)$ is $\mathcal{F}'_\infty$. We apply Theorem II.2.6 to obtain that the blocks $(B_2; P_2, Q_2)$ and $(B'_2; P'_2, Q'_2)$ are block similar. These two blocks are full length blocks. Hence it follows from Theorem III.1.1 that the invariant polynomials of $(B_2; P_2, Q_2)$ and $(B_2'; P_2', Q_2')$ are equal. Since the invariant polynomials of $(B; P, Q)$ coincide with the invariant polynomials of $(B_2; P_2, Q_2)$ and the invariant polynomials of $(B'; P', Q')$ are the invariant polynomials of $(B_2'; P_2', Q_2')$, the invariant polynomials of a block are uniquely determined by the block. □

VII.5 Finite dimensional operator pencils

In this section we derive Kronecker's theorem about the canonical form of operator pencils under strict equivalence as a corollary of Theorems 1.1 and 1.2. Furthermore we shall establish the relation of the minimal row and minimal column indices with the indices of the third and first kind, respectively.

Recall from Section III.2 that two operator pencils $\lambda G_1 - H_1 : \mathcal{X}_1 \to \mathcal{Y}_1$ and $\lambda G_2 - H_2 : \mathcal{X}_2 \to \mathcal{Y}_2$ are called strictly equivalent if there exist invertible operators $S : \mathcal{X}_1 \to \mathcal{X}_2$ and $T : \mathcal{Y}_1 \to \mathcal{Y}_2$ such that $\lambda G_2 - H_2 = T(\lambda G_1 - H_1)S^{-1}$. Apart from the elementary pencils with minimal row index (defined in Section III.2) and the elementary pencils with minimal column index (defined in Section V.2) we will need here a third special kind of elementary pencil, namely

$$\lambda W_\nu - Z_\nu = \begin{pmatrix} -1 & \lambda & 0 & \cdots & 0 & 0 \\ 0 & -1 & \lambda & & 0 & 0 \\ 0 & 0 & -1 & & 0 & 0 \\ \vdots & \vdots & & \ddots & & \vdots \\ 0 & 0 & 0 & & -1 & \lambda \\ 0 & 0 & 0 & \cdots & 0 & -1 \end{pmatrix} : \mathbf{C}^\nu \to \mathbf{C}^\nu. \tag{5.1}$$

The pencil $\lambda W_\nu - Z_\nu$ is called an *elementary pencil based at infinity* and the number ν is called its *partial multiplicity at infinity*. We allow the number ν to be 1. A pencil of the form $\lambda I - J$, where J is a Jordan matrix, is called a *Jordan pencil*. The next result is the main theorem of this section.

Theorem 5.1. *A linear finite dimensional operator pencil is strictly equivalent to a direct sum of p elementary pencils with minimal row index, r elementary pencils based at infinity, q elementary pencils with minimal column index, and a Jordan pencil $\lambda I_{\mathcal{X}_0} - B_0$. Moreover, the numbers p, q, r, the corresponding minimal row indices $\kappa_1 \geq \cdots \geq \kappa_p \geq 0$, the corresponding partial multiplicities $\nu_1 \geq \cdots \geq \nu_r \geq 1$ at ∞, the corresponding minimal column indices $\mu_1 \geq \cdots \mu_q \geq 0$, and the invariant polynomials of B_0 are uniquely determined by the pencil.*

The numbers $\kappa_1 \geq \cdots \geq \kappa_p \geq 0$, occuring in Theorem 5.1, will be called the *minimal row indices* of the pencil, the numbers $\nu_1 \geq \cdots \geq \nu_r \geq 1$ will be called *the partial multiplicities at* ∞ of the pencil, the numbers $\mu_1 \geq \cdots \mu_q \geq 0$ will be called the *minimal column indices* of the pencil, and the invariant polynomials of $\lambda I_{\mathcal{X}_0} - B_0$ will be called the *the invariant polynomials* of the pencil. These definitions of the indices, multiplicities and invariant polynomials coincide with the classical definitions as is immediate from the the direct sum that appears in Theorem 5.1.

The proof of Theorem 5.1 will be based on several auxiliary results. The first of these results shows that each pencil is strictly equivalent to an operator pencil of a special type.

Lemma 5.2. *Let $\lambda G - H : \mathcal{X} \to \mathcal{Y}$ be an operator pencil. Then there exist spaces $\mathcal{X}_1$, $\mathcal{X}_2$ and $\mathcal{X}_3$, and an operator pencil*

$$\lambda J - B = \lambda \begin{pmatrix} 0 & I \\ 0 & 0 \end{pmatrix} - \begin{pmatrix} B_{11} & B_{12} \\ B_{21} & B_{22} \end{pmatrix} : \mathcal{X}_1 \oplus \mathcal{X}_2 \to \mathcal{X}_2 \oplus \mathcal{X}_3, \tag{5.2}$$

which is strictly equivalent to $\lambda G - H$.

PROOF. Choose $\mathcal{X}_1 = \operatorname{Ker} G$, $\mathcal{X}_2 = \operatorname{Im} G$, and $\mathcal{X}_3$ to be a complement to $\mathcal{X}_2$ in $\mathcal{Y}$. Then there exists an operator $S : \mathcal{X}_1 \oplus \mathcal{X}_2 \to \mathcal{X}$ such that

$$GS = \begin{pmatrix} 0 & I \\ 0 & 0 \end{pmatrix} : \mathcal{X}_1 \oplus \mathcal{X}_2 \to \mathcal{X}_2 \oplus \mathcal{X}_3 \,.$$

So the pencil $(\lambda G - H)S$ has the desired form and is strictly equivalent to $\lambda G - H$. □

Let $\lambda G - H : \mathcal{X} \to \mathcal{Y}$ be an operator pencil, and let $\lambda J - B$, given by (5.2), be strictly equivalent to $\lambda G - H$. Put $\mathcal{X}_0 = \mathcal{X}_1 \oplus \mathcal{X}_2 \oplus \mathcal{X}_3$, and let P and Q be projections of $\mathcal{X}_0$ such that $\operatorname{Ker} P = \mathcal{X}_1$ and $\operatorname{Im} Q = \mathcal{X}_1 \oplus \mathcal{X}_2$. The block $(B; P, Q)$ will be called *a block associated with the pencil* $\lambda G - H$. Of course a block associated with the pencil $\lambda G - H$ is far from unique. The following result shows that the block similarity class of a block associated with a pencil is uniquely determined by the pencil.

Proposition 5.3. *Let $\lambda G - H : \mathcal{X} \to \mathcal{Y}$ and $\lambda G' - H' : \mathcal{X}' \to \mathcal{Y}'$ be operator pencils, and let $(B; P, Q)$ be a block associated with $\lambda G - H$ and $(B'; P', Q')$ be a block associated with $\lambda G' - H'$. Then $\lambda G - H$ and $\lambda G' - H'$ are strictly equivalent if and only if $(B; P, Q)$ and $(B'; P', Q')$ are block similar.*

PROOF. Let $\lambda J - B : \mathcal{X}_1 \oplus \mathcal{X}_2 \to \mathcal{X}_2 \oplus \mathcal{X}_3$ be the pencil of the form (5.2) that is strictly equivalent to $\lambda G - H$ and is used to construct the block $(B; P, Q)$ associated with $\lambda G - H$. Similarly assume that $\lambda J' - B' : \mathcal{X}_1' \oplus \mathcal{X}_2' \to \mathcal{X}_2' \oplus \mathcal{X}_3'$ is the pencil of the form (5.2) that is strictly equivalent to $\lambda G' - H'$ and is used to construct the block $(B'; P', Q')$ associated with $\lambda G' - H'$.

Assume that $\lambda G - H$ and $\lambda G' - H'$ are strictly equivalent. Then $\lambda J - B$ and $\lambda J' - B'$ are strictly equivalent. So there exist $T = (T_{ij})_{i,j=1}^2 : \mathcal{X}_1 \oplus \mathcal{X}_2 \to \mathcal{X}_1' \oplus \mathcal{X}_2'$ and $U = (U_{ij})_{i,j=1}^2 : \mathcal{X}_2 \oplus \mathcal{X}_3 \to \mathcal{X}_2' \oplus \mathcal{X}_3'$ which are invertible and such that

$(\lambda J - B)T = U(\lambda J' - B')$. From the equality $JT = UJ'$ one obtains that $T_{21} = 0$, $U_{21} = 0$ and $T_{22} = U_{11}$. Put

$$S = \begin{pmatrix} T_{11} & T_{12} & 0 \\ 0 & T_{22} & U_{12} \\ 0 & 0 & U_{22} \end{pmatrix} : \mathcal{X}_1 \oplus \mathcal{X}_2 \oplus \mathcal{X}_3 \to \mathcal{X}_1' \oplus \mathcal{X}_2' \oplus \mathcal{X}_3' .$$

Then $S[\operatorname{Ker} P] = \operatorname{Ker} P'$, $S[\operatorname{Im} Q] = \operatorname{Im} Q'$ and $P(SBx - B'Sx) = 0$ for each $x \in \operatorname{Im} Q$. The last equality is easily checked from the block matrix partitionings of P, S, B and B'.

Conversely, assume that the blocks $(B; P, Q)$ on $\mathcal{X}$ and $(B'; P', Q')$ are block similar. Then the block similarity S is such that $S[\operatorname{Ker} P] = \operatorname{Ker} P'$ and $S[\operatorname{Im} Q] = \operatorname{Im} Q'$, and hence S has the following block matrix partitioning:

$$S = \begin{pmatrix} S_{11} & S_{12} & S_{13} \\ 0 & S_{22} & S_{23} \\ 0 & 0 & S_{33} \end{pmatrix} : \mathcal{X}_1 \oplus \mathcal{X}_2 \oplus \mathcal{X}_3 \to \mathcal{X}_1' \oplus \mathcal{X}_2' \oplus \mathcal{X}_3'.$$

Then one puts

$$T = \begin{pmatrix} S_{11} & S_{12} \\ 0 & S_{22} \end{pmatrix}, \quad U = \begin{pmatrix} S_{22} & S_{23} \\ 0 & S_{33} \end{pmatrix}.$$

It is easy to check that $JT = UJ'$. Since $P(SBx - B'Sx) = 0$ for each $x \in \operatorname{Im} Q$, it follows that $BT = UB'$. So $\lambda J - B$ is strictly equivalent to $\lambda J' - B'$. Thus we see that $\lambda G - H$ is strictly equivalent to $\lambda G' - H'$. □

The question arises whether or not each block is block similar to a block associated with a pencil. To answer this question, first remark that if $(B; P, Q)$ is a block associated with the pencil $\lambda G - H$, then $\operatorname{Ker} P \subset \operatorname{Im} Q$. So assume that $(B; P, Q)$ is a block with $\operatorname{Ker} P \subset \operatorname{Im} Q$. We will show that there exist a block $(B_0; P_0, Q_0)$, which is block similar to $(B; P, Q)$ with a block similarity of the second kind, and a pencil $\lambda G - H$ such that $(B_0; P_0, Q_0)$ is a block associated with the pencil $\lambda G - H$. The pencil $\lambda G - H$ will be called *a pencil associated with the block* $(B; P, Q)$. Let us construct such a block $(B_0; P_0, Q_0)$. First we choose P_0 such that $\operatorname{Ker} P_0 = \operatorname{Ker} P$ and $\operatorname{Im} Q = \operatorname{Ker} P_0 \oplus (\operatorname{Im} P_0 \cap \operatorname{Im} Q)$. Next we choose Q_0 such that $\operatorname{Im} Q_0 = \operatorname{Im} Q$ and $\operatorname{Ker} Q_0 \subset \operatorname{Im} P_0$. Then $\operatorname{Im} P_0 = (\operatorname{Im} P_0 \cap \operatorname{Im} Q_0) \oplus \operatorname{Ker} Q_0$. We put $B_0 = P_0 B : \operatorname{Im} Q_0 \to \operatorname{Im} P_0$. The block $(B_0; P_0, Q_0)$ is block similar to $(B; P, Q)$ with a block similarity of the second kind. Put $\mathcal{X}_1 = \operatorname{Ker} P_0$, $\mathcal{X}_2 = \operatorname{Im} P_0 \cap \operatorname{Im} Q_0$, and $\mathcal{X}_3 = \operatorname{Ker} Q_0$. We define $H = B_0 : \mathcal{X}_1 \oplus \mathcal{X}_2 \to \mathcal{X}_2 \oplus \mathcal{X}_3$ and $G : \mathcal{X}_1 \oplus \mathcal{X}_2 \to \mathcal{X}_2 \oplus \mathcal{X}_3$ by $\operatorname{Ker} G = \mathcal{X}_1$ and $Gx = x$ for each $x \in \mathcal{X}_2$. Then $(B_0; P_0, Q_0)$ is a block associated with the pencil $\lambda G - H$. So $\lambda G - H$ is a pencil associated with the block $(B; P, Q)$. Remark that $\lambda G - H$ is of type (5.2).

Proposition 5.4. *Let $(B; P, Q)$ be a block on $\mathcal{X}$ with $\operatorname{Ker} P \subset \operatorname{Im} Q$, and let $(B'; P', Q')$ be a block on $\mathcal{X}'$ with $\operatorname{Ker} P' \subset \operatorname{Im} Q'$. Let $\lambda G - H$ be a pencil associated with the block $(B; P, Q)$, and $\lambda G' - H'$ be a pencil associated with the block*

$(B';P',Q')$ Then $(B;P,Q)$ is block similar to $(B';P',Q')$ if and only if $\lambda G - H$ is strictly equivalent to $\lambda G' - H'$.

PROOF. There exist a block $(B_0;P_0,Q_0)$ that is block similar to $(B;P,Q)$, and a block $(B_0';P_0',Q_0')$ block similar to $(B';P',Q')$, such that $(B_0;P_0,Q_0)$ is a block associated with the pencil $\lambda G - H$ and $(B_0';P_0',Q_0')$ is a block associated with the pencil $\lambda G' - H'$. Proposition 5.3 gives us that $\lambda G - H$ is strictly equivalent to $\lambda G' - H'$ if and only if $(B_0;P_0,Q_0)$ is block similar to $(B_0';P_0',Q_0')$. Therefore $\lambda G - H$ is strictly equivalent to $\lambda G' - H'$ if and only if $(B;P,Q)$ is block similar to $(B';P',Q')$. □

We consider the pencils

$$\lambda J - B = \lambda \begin{pmatrix} 0 & I \\ 0 & 0 \end{pmatrix} - \begin{pmatrix} B_{11} & B_{12} \\ B_{21} & B_{22} \end{pmatrix} : \mathcal{X}_1 \oplus \mathcal{X}_2 \to \mathcal{X}_2 \oplus \mathcal{X}_3 \tag{5.3}$$

and

$$\lambda J' - B' = \lambda \begin{pmatrix} 0 & I \\ 0 & 0 \end{pmatrix} - \begin{pmatrix} B_{11}' & B_{12}' \\ B_{21}' & B_{22}' \end{pmatrix} : \mathcal{X}_1' \oplus \mathcal{X}_2' \to \mathcal{X}_2' \oplus \mathcal{X}_3'. \tag{5.4}$$

By definition the direct sum of these pencils is the pencil

$$\lambda(J \oplus J') - (B \oplus B') : (\mathcal{X}_1 \oplus \mathcal{X}_2) \oplus (\mathcal{X}_1' \oplus \mathcal{X}_2') \to (\mathcal{X}_2 \oplus \mathcal{X}_3) \oplus (\mathcal{X}_2' \oplus \mathcal{X}_3').$$

By a reordering of the spaces we get that this pencil is strictly equivalent to a pencil

$$\lambda J_0 - B_0 : (\mathcal{X}_1 \oplus \mathcal{X}_1') \oplus (\mathcal{X}_2 \oplus \mathcal{X}_2') \to (\mathcal{X}_2 \oplus \mathcal{X}_2') \oplus (\mathcal{X}_3 \oplus \mathcal{X}_3'),$$

which is of the type (5.2). Put $\mathcal{X} = \mathcal{X}_1 \oplus \mathcal{X}_2 \oplus \mathcal{X}_3$, $\mathcal{X}' = \mathcal{X}_1' \oplus \mathcal{X}_2' \oplus \mathcal{X}_3'$ and $\mathcal{X}_0 = (\mathcal{X}_1 \oplus \mathcal{X}_2) \oplus (\mathcal{X}_1' \oplus \mathcal{X}_2') \oplus (\mathcal{X}_2' \oplus \mathcal{X}_3')$. Define the projections P, P', P_0, Q, Q' and Q_0 by

$$\begin{aligned} \operatorname{Ker} P &= \mathcal{X}_1, & \operatorname{Im} P &= \mathcal{X}_2 \oplus \mathcal{X}_3, \\ \operatorname{Ker} P' &= \mathcal{X}_1', & \operatorname{Im} P' &= \mathcal{X}_2' \oplus \mathcal{X}_3', \\ \operatorname{Ker} P_0 &= \mathcal{X}_1 \oplus \mathcal{X}_1', & \operatorname{Im} P_0 &= (\mathcal{X}_2 \oplus \mathcal{X}_2') \oplus (\mathcal{X}_3 \oplus \mathcal{X}_3'), \\ \operatorname{Ker} Q &= \mathcal{X}_3, & \operatorname{Im} Q &= \mathcal{X}_1 \oplus \mathcal{X}_2, \\ \operatorname{Ker} Q' &= \mathcal{X}_3', & \operatorname{Im} Q' &= \mathcal{X}_1' \oplus \mathcal{X}_2', \\ \operatorname{Ker} Q_0 &= \mathcal{X}_3 \oplus \mathcal{X}_3', & \operatorname{Im} Q_0 &= (\mathcal{X}_1 \oplus \mathcal{X}_1') \oplus (\mathcal{X}_2 \oplus \mathcal{X}_2'). \end{aligned}$$

Then clearly the block $(B_0;P_0,Q_0)$ is block similar to the direct sum of the blocks $(B;P,Q)$ and $(B';P',Q')$. So a block associated with the direct sum of two pencils is block similar to the direct sum of blocks associated with each of the pencils separately. In the same way we see that a pencil associated with a direct sum of two blocks is strictly equivalent to the direct sum of blocks associated with the terms in the direct sum.

Let $(B;P,I)$ be a shift of the first kind with index κ. In this case $Q = I$, and $\operatorname{Ker} P \subset \operatorname{Im} Q$. So we may use the construction preceding Proposition 5.4 to obtain a pencil $\lambda I - B$ associated with $(B;P,I)$. Let $e_1, \dots, e_{\kappa+1}$ be the standard basis in $\mathbb{C}^n$. Then $\operatorname{Ker} P = \operatorname{span}\{e_1\}$, $\operatorname{Im} P = \operatorname{span}\{e_2, \dots, e_{\kappa+1}\}$, and $Be_i = e_{i+1}$ for $i = 1, \dots, \kappa$. Put $\mathcal{X}_1 = \operatorname{Ker} P$, $\mathcal{X}_2 = \operatorname{Im} P$, and $\mathcal{X}_3 = \{0\}$. With respect to the bases e_1 of $\mathcal{X}_1$ and $e_2, \dots, e_{n+1}$ of $\mathcal{X}_2$ the associated pencil $\lambda J - B : \mathcal{X}_1 \oplus \mathcal{X}_2 \to \mathcal{X}_2 \oplus \mathcal{X}_3$ has the $\kappa \times (\kappa+1)$ matrix representation

$$\lambda \begin{pmatrix} 0 & 1 & & \\ \vdots & & \ddots & \\ 0 & & & 1 \end{pmatrix} - \begin{pmatrix} 1 & & & 0 \\ & \ddots & & \vdots \\ & & 1 & 0 \end{pmatrix}. \tag{5.5}$$

This pencil is strictly equivalent to the pencil (2.2) in Section V.2. To see this one has to apply to the left of the pencil (5.5) the permutation transformation T on $\mathbb{C}^\kappa$ given by $Te_i = e_{\kappa+1-i}$ and to the right the permutation transformation S on $\mathbb{C}^{\kappa+1}$ with $Se_i = e_{\kappa+2-i}$. From Section V.2 we know that the number κ is the minimal row index of this pencil.

Let $(B;P,Q)$ be a shift of the second kind with index $\nu \geq 1$. Then there is a basis $e_1, \dots, e_{\nu+1}$ of $\mathcal{X}$ such that $\operatorname{Ker} P = \operatorname{span}\{e_1\}$, $\operatorname{Im} P = \operatorname{span}\{e_2, \dots, e_{\nu+1}\}$, $\operatorname{Im} Q = \operatorname{span}\{e_1, \dots, e_\nu\}$, $\operatorname{Ker} Q = \operatorname{span}\{e_{\nu+1}\}$ and $Be_i = e_{i+1}$, for $i = 1, \dots, \nu$. Since $\operatorname{Ker} P \subset \operatorname{Im} Q$, we may again use the construction preceding Proposition 5.4 to obtain a pencil $\lambda I - B$ associated with $(B;P,I)$. We put $\mathcal{X}_1 = \operatorname{span}\{e_1\}$, $\mathcal{X}_2 = \operatorname{span}\{e_1, \dots, e_\nu\}$ and $\mathcal{X}_3 = \operatorname{span}\{e_{\nu+1}\}$. Then the associated pencil $\lambda J - B : \mathcal{X}_1 \oplus \mathcal{X}_2 \to \mathcal{X}_2 \oplus \mathcal{X}_3$ has with respect to the bases e_1 of $\mathcal{X}_1$, $e_2, \dots, e_\nu$ of $\mathcal{X}_2$, and $e_{\nu+1}$ of $\mathcal{X}_3$ the $\nu \times \nu$ matrix representation (5.1)

For a shift $(B;I,Q)$ of the third kind with index μ we follow the same procedure as for shifts of the first and second kind. We end up with a $(\mu+1) \times \mu$ matrix pencil

$$\lambda \begin{pmatrix} 1 & & \\ & \ddots & \\ & & 1 \\ 0 & \dots & 0 \end{pmatrix} - \begin{pmatrix} 0 & \dots & 0 \\ 1 & & \\ & \ddots & \\ & & 1 \end{pmatrix}, \tag{5.6}$$

which is associated with the block $(B;I,Q)$. Note that the pencil (5.6) has the form (2.2) of Section III.2. According to Section III.2 the number μ is the minimal column index of this pencil.

For a block (B,I,I) of course the pencil $\lambda I - B$ is a pencil associated with the block.

Now we are ready to prove Theorem 5.1.

Proof of Theorem 5.1. Let $\lambda G - H : \mathcal{X} \to \mathcal{Y}$ be an operator pencil and let $(B;P,Q)$ be an associated block. Then $(B;P,Q)$ is block similar to a direct sum $(B_1;P_1,Q_1)$ of shifts of the first kind with indices $\kappa_1 \geq \dots \geq \kappa_p \geq 0$, shifts of the second kind with indices $\nu_1 \geq \dots \geq \nu_r \geq 0$, shifts of the third kind with indices $\mu_1 \geq \cdots \mu_q \geq 0$ and a block $(B_0; I_{\mathcal{X}_0}, I_{\mathcal{X}_0})$. Let us show that $\nu_r \geq 1$. Since $\operatorname{Ker} P \subset \operatorname{Im} Q$, the number $\dim \operatorname{Ker} P - \dim(\operatorname{Ker} P \cap \operatorname{Im} Q)$ is equal to 0. Therefore the number β_1, appearing in Theorem 2.1, is equal to 0. So we see that $\nu_r = \#\{k \geq 1 \mid r - \beta_k \geq r\} \geq 1$. With the block $(B_1; P_1, Q_1)$ is associated a direct sum $\lambda G_1 - H_1$ of pencils of the type (5.5) with minimal row indices $\kappa_1 \geq \cdots \geq \kappa_p \geq 0$, the type (5.1) with partial multiplicities $\nu_1 \geq \cdots \geq \nu_r \geq 1$ at ∞, the type (5.6) with minimal column indices $\mu_1 \geq \cdots \geq \mu_q \geq 0$, and a pencil $\lambda I_{\mathcal{X}_0} - B_0$. So the block $(B_1; P_1, Q_1)$ is block similar to a block $(B_2; P_2, Q_2)$ associated to $\lambda G_1 - H_1$. Since $(B_2; P_2, Q_2)$ is block similar to $(B; P, Q)$, it follows that $\lambda G - H$ is strictly equivalent to $\lambda G_1 - H_1$. This proves the first part of the theorem.

To prove the second part, assume that we have two direct sums of pencils of types (5.5), (5.1), (5.6) and a pencil of type $\lambda I - B_0$. Assume that these direct sums are strictly equivalent. We want to prove that for these two direct sums the minimal row indices, the multiplicities at ∞, the minimal column indices and the invariant polynomials of the pencils of type $\lambda I - B_0$ coincide. Remark that with each of the two direct sums there is an associated block, which is a direct sum of shifts of the first, second and third kind and a (I, I)-block (i.e., a (P, Q)-block with $P = Q = I$). Now these two blocks are block similar. So the indices of the first, second and third kind and the invariant polynomials of the (I, I)-block are the same for these two direct sums of blocks. Since the minimal row indices, the multiplicities at ∞ and the minimal column indices of a pencil are equal to the indices of the first, second and third kind, respectively, of an associated block, the minimal row indices, the multiplicities at ∞ and the minimal column indices of the two direct sums of pencils coincide. Moreover, since the invariant polynomials of the block $(B_0; I, I)$ are the invariant polynomials of the pencil $\lambda I - B_0$, the invariant polynomials of the pencils of type $\lambda I - B_0$ in the two direct sums are the same. □

From the proof of Theorem 5.1 we read off the following result.

Corollary 5.5. *Let $\lambda G - H$ be an operator pencil, and let $(B; P, Q)$ be an associated block. Then the minimal row indices of $\lambda G - H$ are the indices of the first kind of $(B; P, Q)$, the multiplicities at ∞ of $\lambda G - H$ are the indices of the second kind of $(B; P, Q)$, the minimal column indices of $\lambda G - H$ are the indices of the third kind of $(B; P, Q)$, and the invariant polynomials of $\lambda G - H$ are the invariant polynomials of $(B; P, Q)$.*

Our next goal is to translate the formulas for the indices in Theorem 2.1 to the setting of operator pencils. We restrict ourselves to pencils of the type (5.2). To avoid too many subscripts in the notation we write

$$\lambda J - G = \lambda \begin{pmatrix} 0 & I \\ 0 & 0 \end{pmatrix} - \begin{pmatrix} B & A \\ D & C \end{pmatrix} : \mathcal{U} \oplus \mathcal{X} \to \mathcal{X} \oplus \mathcal{Y}.$$

Before we formulate the result we introduce some more notation. Put

$$s_0 = D, \; s_i = CA^{i-1}B, \; (i = 1, 2, \ldots), \; s_i = 0, \; (i < 0),$$

and consider the block lower triangular Toeplitz matrix $T_k = (s_{i-j})_{i,j=1}^k$. Also, consider the block column

$$(C,A)_k = \operatorname{col}\left(CA^{j-1}\right)_{j=1}^k$$

and the block row

$$(A,B)_k = \operatorname{row}\left(A^{j-1}B\right)_{j=1}^k .$$

Furthermore, we set $S_k = \operatorname{col}(s_j)_{j=1}^k$ and $\tilde{S}_k = \operatorname{row}(s_j)_{j=1}^k$.

Theorem 5.6. *Let the operator pencil*

$$\lambda J - G = \lambda \begin{pmatrix} 0 & I \\ 0 & 0 \end{pmatrix} - \begin{pmatrix} B & A \\ D & C \end{pmatrix} : \mathcal{U} \oplus \mathcal{X} \to \mathcal{X} \oplus \mathcal{Y}$$

have minimal row indices $\kappa_1 \geq \cdots \geq \kappa_p \geq 0$, partial multiplicities $\nu_1 \geq \cdots \geq \nu_r \geq 1$ at ∞, and minimal column indices $\mu_1 \geq \cdots \mu_q \geq 0$. With $(A,B)_k$, $(C,A)_k$, T_k, S_k and $\tilde{S}_k$ defined as above, we define the spaces $\mathcal{E}_1 = \mathcal{U} \oplus \mathcal{X}$,

$$\mathcal{E}_2 = \{\begin{pmatrix} u \\ x \end{pmatrix} \in \mathcal{U} \oplus \mathcal{X} \mid Du + Cx = 0\}$$

and for $k = 3, 4, \ldots$

$$\mathcal{E}_k = \{\begin{pmatrix} u \\ x \end{pmatrix} \in \mathcal{U} \oplus \mathcal{X} \mid Du + Cx = 0,\ S_{k-2}u + (C,A)_{k-2}Ax \in \operatorname{Im} T_{k-2}\},$$

and the subspaces $\mathcal{G}_k$ of $\mathcal{X} \oplus \mathcal{Y}$ by $\mathcal{G}_1 = \{0\}$ and for $k = 2, 3, \ldots$

$$\mathcal{G}_k = \operatorname{Im}\begin{pmatrix} (A,B)_{k-1} \\ \tilde{S}_{k-1} \end{pmatrix}.$$

Put

$$\begin{aligned} \alpha_i &= \dim \mathcal{G}_{i+1} - \dim \mathcal{G}_i, \\ \beta_i &= \dim \mathcal{G}_i - \dim(\mathcal{G}_i \cap \mathcal{X}), \quad \beta_\infty = \max\{\beta_i \mid i = 1,2,\ldots\}, \\ \gamma_i &= \dim \mathcal{U} - \dim(\mathcal{E}_i \cap \mathcal{U}), \quad \gamma_\infty = \max\{\gamma_i \mid i = 1,2,\ldots\}, \\ \delta_i &= \dim \mathcal{E}_i - \dim \mathcal{E}_{i+1}. \end{aligned}$$

Then

$$r = \beta_\infty = \gamma_\infty,\ p = \alpha_1 - r,\ q = \delta_1 - r,$$

and

$$\begin{aligned} \kappa_i &= \#\{k \geq 1 \mid \alpha_k - (\beta_\infty - \beta_k) \geq i\}, \\ \nu_i &= \#\{k \geq 1 \mid \beta_\infty - \beta_k \geq i\} = \#\{k \geq 1 \mid \gamma_\infty - \gamma_k \geq i\}, \\ \mu_i &= \#\{k \geq 1 \mid \delta_k - (\gamma_\infty - \gamma_k) \geq i\}. \end{aligned} \tag{5.7}$$

PROOF. This result is based on Theorem 2.1. We relate the spaces $\mathcal{D}_i$ and $\mathcal{F}_i$, given in Theorem 2.1 to the spaces $\mathcal{E}_i$ and $\mathcal{G}_i$ defined above. With $\operatorname{Im} Q = \mathcal{U} \oplus \mathcal{X}$ and $\operatorname{Ker} P = \mathcal{U}$ it is straight forward to check that $\mathcal{D}_i = \mathcal{E}_i$ and that $\mathcal{F}_i = \mathcal{U} \oplus \mathcal{G}_i$. This gives that the numbers α_i, β_i, γ_i and δ_i as defined in Theorem 2.1 coincide with the corresponding numbers in the statement of the present theorem. So the indices of the block $(G; P, Q)$ are given by (5.7). We apply Corollary 5.5 to see that in fact (5.7) gives formulas for the minimal row and column indices and the multiplicities at ∞ of the pencil. □

VII.6 Non-everywhere defined operators modulo a subspace

In this section we combine the notion of non-everywhere defined operator (Section III.3) and similarity modulo a subspace (Section V.3). Let $T_1(\mathcal{X}_1 \to \mathcal{X}_1)$ and $T_2(\mathcal{X}_2 \to \mathcal{X}_2)$ be non-everywhere defined operators with domain of definition $\mathcal{D}(T_1)$ and $\mathcal{D}(T_2)$, respectively. Let $\mathcal{M}_1$ be a linear subspace of $\mathcal{X}_1$. The invertible operator $S : \mathcal{X}_2 \to \mathcal{X}_1$ is called *a similarity modulo* $\mathcal{M}_1$ of $T_1(\mathcal{X}_1 \to \mathcal{X}_1)$ and $T_2(\mathcal{X}_2 \to \mathcal{X}_2)$ if

(i) $S[\mathcal{D}(T_2)] = \mathcal{D}(T_1)$

(ii) $(T_1 S - S T_2)x \in \mathcal{M}_1$ for each $x \in \mathcal{D}(T_2)$.

If S is a similarity modulo $\mathcal{M}_1$ of $T_1(\mathcal{X}_1 \to \mathcal{X}_1)$ and $T_2(\mathcal{X}_2 \to \mathcal{X}_2)$, then S^{-1} is a similarity modulo $S^{-1}[\mathcal{M}_1]$ of $T_2(\mathcal{X}_2 \to \mathcal{X}_2)$ and $T_1(\mathcal{X}_1 \to \mathcal{X}_1)$.

Let $T(\mathcal{X} \to \mathcal{X})$ be a non-everywhere defined operator with domain of definition $\mathcal{D}(T)$, and let $\mathcal{M} \subset \mathcal{X}$ be a subspace. A subspace $\mathcal{N}$ of $\mathcal{X}$ is called *T-invariant modulo $\mathcal{M}$* if $T[\mathcal{N} \cap \mathcal{D}(T)] \subset \mathcal{N} + \mathcal{M}$. The operator T is called *decomposable modulo $\mathcal{M}$* if there exist non zero subspaces $\mathcal{X}_1$ and $\mathcal{X}_2$, which are T-invariant modulo $\mathcal{M}$ and such that

$$\mathcal{X} = \mathcal{X}_1 \oplus \mathcal{X}_2,\ \mathcal{D}(T) = (\mathcal{X}_1 \cap \mathcal{D}(T)) \oplus (\mathcal{X}_2 \cap \mathcal{D}(T)),\ \mathcal{M} = (\mathcal{X}_1 \cap \mathcal{M}) \oplus (\mathcal{X}_2 \cap \mathcal{M}). \quad (6.1)$$

We now consider four classes of (non-everywhere defined) operators that are indecomposable (i.e., not decomposable) modulo a subspace. The first of these classes is the class of the non-everywhere defined shifts with the subspace $\mathcal{M} = \{0\}$. This class was considered in Section III.3. For the second class we take again the non-everywhere defined shift $T(\mathbb{C}^{\kappa+1} \to \mathbb{C}^{\kappa+1})$ but now with $\mathcal{M} = \operatorname{span}\{e_1\}$, where $e_1, \dots, e_{\kappa+1}$ is the standard basis of $\mathbb{C}^{\kappa+1}$. We will show that $T(\mathbb{C}^{\kappa+1} \to \mathbb{C}^{\kappa+1})$ is indecomposable modulo $\mathcal{M}$. For the third class we take the shift $T : \mathbb{C}^{\kappa+1} \to \mathbb{C}^{\kappa+1}$ given by $Te_i = e_{i+1}$, $i = 1, \dots, \kappa$, $Te_{\kappa+1} = 0$ and $\mathcal{M} = \operatorname{span}\{e_1\}$. This T is indecomposable modulo $\mathcal{M}$ as was shown in Section V.3. Finally, the indecomposable operators $T : \mathcal{X} \to \mathcal{X}$ modulo $\{0\}$ are just those operators that have a matrix representation that consists of a single Jordan block. Such an indecomposable operator will be called an *elementary Jordan operator*. The next theorem is the main result of this section.

Theorem 6.1. *Let $T(\mathcal{X} \to \mathcal{X})$ be a non-everywhere defined operator, and let $\mathcal{M}$ be a subspace of $\mathcal{X}$. Then T is similar modulo $\mathcal{M}$ to a direct sum of shifts and elementary Jordan operators.*

Note that for the shifts in Theorem 6.1 all three cases that we described above can occur. The three sets of ranks of these shifts are called *the three sets of indices of $T(\mathcal{X} \to \mathcal{X})$ modulo $\mathcal{M}$*. The direct sum of the elementary Jordan operators is called the *Jordan part of T modulo $\mathcal{M}$*.

Theorem 6.2. *Let $T_1(\mathcal{X}_1 \to \mathcal{X}_1)$ and $T_2(\mathcal{X}_2 \to \mathcal{X}_2)$ be non-everywhere defined operators, and let $\mathcal{M}_1$ be a subspace of $\mathcal{X}_1$ and $\mathcal{M}_2$ be a subspace of $\mathcal{X}_2$. There exists a similarity $S : \mathcal{X}_2 \to \mathcal{X}_1$ modulo $\mathcal{M}_1$ of $T_1(\mathcal{X}_1 \to \mathcal{X}_1)$ and $T_2(\mathcal{X}_2 \to \mathcal{X}_2)$ with $\mathcal{M}_1 = S[\mathcal{M}_2]$ if and only if the indices of $T_1(\mathcal{X}_1 \to \mathcal{X}_1)$ modulo $\mathcal{M}_1$ are equal to the indices of $T_2(\mathcal{X}_2 \to \mathcal{X}_2)$ modulo $\mathcal{M}_2$ and the Jordan part of $T_1(\mathcal{X}_1 \to \mathcal{X}_1)$ modulo $\mathcal{M}_1$ is similar to the Jordan part of $T_2(\mathcal{X}_2 \to \mathcal{X}_2)$ modulo $\mathcal{M}_2$.*

The proofs will be based on a translation of results on blocks to the present situation. This is done as follows. First we assign to a subspace $\mathcal{M}$ and a non-everywhere defined operator $T(\mathcal{X} \to \mathcal{X})$ with domain of definition $\mathcal{D}(T)$ a block $(B; P, Q)$. The projection P is such that $\operatorname{Ker} P = \mathcal{M}$, the projection Q has $\operatorname{Im} Q = \mathcal{D}(T)$ and $B = PT : \operatorname{Im} Q \to \operatorname{Im} P$. Any block constructed this way is called a *block associated with* $T(\mathcal{X} \to \mathcal{X})$ and $\mathcal{M}$. Clearly, $(B; P, Q)$ is not uniquely determined by $\mathcal{M}$ and $T(\mathcal{X} \to \mathcal{X})$ because $(B; P, Q)$ depends on the choice of $\operatorname{Im} P$ and $\operatorname{Ker} Q$. However, different choices of $\operatorname{Im} P$ and $\operatorname{Ker} Q$ lead to block similar associated blocks. Conversely, if $(B; P, Q)$ is given, then $\mathcal{M} = \operatorname{Ker} P$, $\mathcal{D}(T) = \operatorname{Im} Q$ and $T = B$ lead to a subspace $\mathcal{M}$ and a non-everywhere defined operator $T(\mathcal{X} \to \mathcal{X})$ such that $(B; P, Q)$ is a block associated with $T(\mathcal{X} \to \mathcal{X})$ and $\mathcal{M}$.

From the definitions of block similarity and similarity modulo a subspace of non-everywhere defined operators it is clear that two non-everywhere defined operators are similar modulo a subspace if and only if the associated blocks are block similar. Also it is immediate from the definitions that a subspace is $T(\mathcal{X} \to \mathcal{X})$-invariant modulo $\mathcal{M}$ if and only if the subspace is block-invariant for the block associated with $T(\mathcal{X} \to \mathcal{X})$ and $\mathcal{M}$. A non-everywhere defined operator is decomposable modulo a subspace if and only if an associated block is decomposable. This one deduces from an inspection of the definition of decomposable modulo a subspace and the definition of decomposable block as it was given in Section II.2.

Above (in the third paragraph of this section) we considered three ways to combine a (non-everywhere defined) shift and a subspace. The first type already appeared in Chapter III, where it was shown that with this first type of non-everywhere defined operators one may associate a block which is a shift of the third kind. The third type we also met before. In Chapter V we saw that the shift of the first kind is a block associated with this third combination of a shift and a subspace. The second type did not yet appear. Let $T(\mathbb{C}^{\kappa+1} \to \mathbb{C}^{\kappa+1})$ be the non-everywhere defined shift and $\mathcal{M} = \operatorname{span}\{e_1\}$. Put P to be the projection of $\mathbb{C}^{\kappa+1}$

along $\mathcal{M}$ onto span$\{e_2, \ldots, e_{\kappa+1}$. Choose Q the projection along span$\{e_{\kappa+1}\}$) onto span$\{e_1, \ldots, e_\kappa\}$. The associated block $(PT; P, Q)$ is then a shift of the second kind with index κ.

With the translation above available, Theorems 6.1 and 6.2 are immediate corollaries of Theorems 1.1 and 1.2, respectively.

We conclude this section with the formulation of the eigenvalue completion problem for non-everywhere defined operators modulo a subspace. The operator $A : \mathcal{X} \to \mathcal{X}$ is called a *completion modulo* $\mathcal{M}$ *of* $T(\mathcal{X} \to \mathcal{X})$ if $A|_{\mathcal{D}(T)}(\mathcal{X} \to \mathcal{X})$ is similar modulo $\mathcal{M}$ to $T(\mathcal{X} \to \mathcal{X})$. The eigenvalue completion problem is then to the problem of describing the eigenvalues and their multiplicities for all the completions modulo $\mathcal{M}$ of a given non-everywhere defined operator. The eigenvalue restriction problem is the problem of describing all classes of non-everywhere defined operators, similar modulo a subspace $\mathcal{M}$, that contain a non-everywhere defined operator which has the operator A as its completion.

VII.7 Duality of operator blocks

In Section I.4 we introduced the dual block of a block $(B; P, Q)$. We also proved Lemma V.4.1, which states that the dual block of a shift of the first kind is block similar to a shift of the third kind, and that the dual block of a shift of the third kind is block similar to a shift of the first kind. The next result fits shifts of the second kind into this pattern.

Lemma 7.1. *The dual of a shift of the second kind with index* μ *is block similar to a shift of the second kind with index* μ.

PROOF. Let $(B; P, Q)$ be a shift of the second kind with index $n-1 = \mu$. Thus with respect to the standard basis $e_1, \ldots, e_n$ of $\mathbb{C}^n$ we have $Pe_1 = 0$, $Pe_i = e_i$ for $i = 2, \ldots, n$, $Qe_n = 0$, $Qe_i = e_i$ for $i = 1, \ldots, n-1$, and $Be_i = e_{i+1}$ for $i = 1, \ldots, n-1$. Define a completion A of $(B; P, Q)$ by setting $Ae_n = 0$ and $Ae_i = Be_i$ for $i = 1, \ldots, n-1$. Let $e_1^*, \ldots, e_n^*$ be the dual basis in $(\mathbb{C}^n)^*$. This means that $e_i^*(e_j) = \delta_{ij}$, where δ_{ij} is the Kronecker delta. We get that $P^*e_1^* = 0$ and $P^*e_j^* = e_j^*$ if $j = 2, \ldots, n$, and that $Q^*e_n^* = 0$ and $Q^*e_j^* = e_j^*$ if $j = 1, \ldots, n-1$. Let $e_j^* \in \operatorname{Im} P^*$. Then $A^*e_j^*(e_i) = e_j^*(Ae_i)$, and thus $A^*e_j^*(e_i) = 0$ if $i \neq j-1$, and $A^*e_j^*(e_i) = 1$ if $i = j-1$. This proves that $A^*e_j^* = e_{j-1}^*$ for $j = 2, \ldots, n$ and $A^*e_1^* = 0$. We take $B^* = A^*|_{\operatorname{Im} P^*}$. Next, we construct the similarity of the first kind of $(B^*; Q^*, P^*)$ with $(B; P, Q)$. Let $S : \mathbb{C}^n \to (\mathbb{C}^n)^*$ be defined by $Se_i = e_{n+1-i}^*$. Then it is clear that $SQ = P^*S$, $SP = Q^*S$ and $(SB - B^*S)e_i = 0$ for $i = 1, \ldots, n-1$, i.e., for $e_i \in \operatorname{Im} Q$. This proves that S is a block similarity of the first kind of $(B^*; Q^*, P^*)$ with $(B; P, Q)$. □

The next theorem describes the behaviour of the block similarity invariants under duality.

Theorem 7.3. *The block* $(B; P, Q)$ *has indices of the first kind* $\kappa_1, \ldots, \kappa_p$, *indices of the second kind* $\mu_1, \ldots, \mu_q$, *indices of the third kind* $\nu_1, \ldots, \nu_r$ *and invariant* *polynomials $p_1, \dots, p_n$ if and only if the dual block $(B^*; Q^*, P^*)$ has indices of the first kind $\nu_1, \dots, \nu_r$, indices of the second kind $\mu_1, \dots, \mu_q$, indices of the third kind $\kappa_1, \dots, \kappa_p$ and invariant polynomials $p_1, \dots, p_n$.*

PROOF. The block $(B; P, Q)$ has indices of the first kind $\kappa_1, \dots, \kappa_p$, indices of the second kind $\mu_1, \dots, \mu_q$, indices of the third kind $\nu_1, \dots, \nu_r$ and invariant polynomials $p_1, \dots, p_n$ means that the block $(B; P, Q)$ is block similar to a direct sum $(B_0; P_0, Q_0)$ of shifts of the first kind with indices $\kappa_1, \dots, \kappa_p$, shifts of the second kind with indices $\mu_1, \dots, \mu_q$, shifts of the third kind with indices $\nu_1, \dots, \nu_r$ and a matrix with invariant polynomials $p_1, \dots, p_n$. According to Proposition II.4.1, this is equivalent to $(B^*; Q^*, P^*)$ being block similar to $(B_0^*; Q_0^*, P_0^*)$. According to Proposition II.4.2, the dual block $(B_0^*; Q_0^*, P_0^*)$ of $(B_0; P_0, Q_0)$ is a direct sum of the duals of the components of $(B_0; P_0, Q_0)$. Therefore $(B_0^*; Q_0^*, P_0^*)$ is a direct sum of shifts of the third kind with indices $\kappa_1, \dots, \kappa_p$, shifts of the second kind with indices $\mu_1, \dots, \mu_q$, shifts of the first kind with indices $\nu_1, \dots, \nu_r$, and a matrix with invariant polynomials $p_1, \dots, p_n$. So $(B^*; Q^*, P^*)$ has indices of the first kind $\nu_1, \dots, \nu_r$, indices of the second kind $\mu_1, \dots, \mu_q$, indices of the third kind $\kappa_1, \dots, \kappa_p$, and invariant polynomials $p_1, \dots, p_n$. This finishes the proof. □

VII.8 The eigenvalue completion problem

The following result on the eigenvalue completion problem can be derived easily from the canonical form of a general block.

Proposition 8.1. *Let $(B; P, Q)$ be a block on the n dimensional space $\mathcal{X}$, and let $p_n | \cdots | p_1$ be the invariant polynomials of $(B; P, Q)$. Put $k = \sum_{i=1}^n \deg p_i$. Then*

(i) *for any $(n-k)$-tuple of complex numbers $\lambda_1, \dots, \lambda_{n-k}$ there exists a completion A of $(B; P, Q)$ that has $\lambda_1, \dots, \lambda_{n-k}$ as its eigenvalues;*

(ii) *the set of eigenvalues common to all completions of $(B; P, Q)$ is a subset of the zeros of p_1.*

PROOF. In view of Theorem I.6.1 and Theorem 1.1 we may assume without loss of generality that $(B; P, Q)$ is the direct sum of shifts of the first, second and third kind, and a Jordan matrix. Let $\kappa_1 \geq \cdots \geq \kappa_p \geq 0$ be the indices of the first kind, $\nu_1 \geq \cdots \geq \nu_r \geq 0$ be the indices of the second kind, and $\mu_1 \geq \cdots \geq \mu_q \geq 0$ be the indices of the third kind of $(B; P, Q)$. Then

$$n - k = \sum_{i=1}^{p} (\kappa_i + 1) + \sum_{i=1}^{r} (\nu_i + 1) + \sum_{i=1}^{q} (\mu_i + 1).$$

Assign to a shift S with index α precisely $\alpha + 1$ of the numbers $\lambda_1, \dots, \lambda_{n-k}$, say $\lambda_1, \dots, \lambda_{\alpha+1}$. Do this in such a way that each of the numbers λ_i is assigned to only one of the shifts in the canonical form of $(B; P, Q)$. Now complete the shift S to a companion matrix with eigenvalues $\lambda_1, \dots, \lambda_{\alpha+1}$. In this way the direct sum of the shifts is completed to a matrix with eigenvalues $\lambda_1, \dots, \lambda_{n-k}$. This proves (i).

In order to prove (ii), note that all the completions constructed in (i) have precisely the eigenvalues of the Jordan part of $(B; P, Q)$ in common. Therefore the eigenvalues shared by all completions of $(B; P, Q)$ form a subset of this set, i.e., a subset of the zeros of p_1, which is the minimal polynomial of the Jordan part. □

Note that there may not be any eigenvalue common to all the completions of a block. To see this consider the block associated with the positioned principal submatrix

$$\begin{pmatrix} 1 & ? \\ ? & ? \end{pmatrix}.$$

Next, we give without a proof a modified version of a recent result of Cabral and Silva [1] about the eigenvalue completion problem for general blocks. In order to formulate the result we need to consider the homogeneous invariant polynomials of a block. Recall that a polynomial $\tilde{p}(\lambda, \mu)$ is called *homogeneous of degree m* if $\tilde{p}(\alpha\lambda, \alpha\mu) = \alpha^m \tilde{p}(\lambda, \mu)$.

Let $(B; P, Q)$ be a block with indices of the second kind $\nu_1 \geq \cdots \nu_r \geq 0$, and invariant polynomials $p_n | \cdots | p_1$. Put

$$\begin{aligned} \tilde{p}_i(\lambda, \mu) &= \mu^{\nu_i + \deg p_i} p_i\left(\frac{\lambda}{\mu}\right), \quad i = 1, \ldots, r, \\ \tilde{p}_i(\lambda, \mu) &= \mu^{\deg p_i} p_i\left(\frac{\lambda}{\mu}\right), \quad i = r+1, \ldots, n. \end{aligned} \tag{8.1}$$

It is clear that $\tilde{p}_i(\alpha\lambda, \alpha\mu) = \alpha^{\nu_i + \deg p_i} \tilde{p}_i(\lambda, \mu)$ for $i = 1, \ldots, r$ and $\tilde{p}_i(\alpha\lambda, \alpha\mu) = \alpha^{\deg p_i} \tilde{p}_i(\lambda, \mu)$ for $i = r+1, \ldots, n$. Hence the polynomials $\tilde{p}_i$ are homogeneous. The polynomials $\tilde{p}_1, \ldots, \tilde{p}_n$ are called the *homogeneous invariant polynomials* of the block $(B; P, Q)$. If there are no indices of the second kind, then $\tilde{p}_i(1, 0) \neq 0$ for all i. In this case $\tilde{p}_i(1, 0)$ is the coefficient of the term with the highest degree in the (non-homogeneous) invariant polynomial p_i of the block. In particular this occurs when $P = I$ or $Q = I$. In general, the indices of the second kind can be recovered from the number r of indices of the second kind and the homogeneous invariant polynomials. In fact, one only has to know the number ℓ of indices of the second kind that are equal to 0. Indeed, if $\tilde{p}_i(\lambda, \mu) = \mu^{\nu_i} \bar{p}_i(\lambda, \mu)$, with $\nu_i > 0$ and $\bar{p}_i(1, 0) \neq 0$, then ν_i is the i-th index of the second kind and $\bar{p}_i(\lambda, 1)$ is the i-th invariant polynomial. In this way we find all the nonzero indices of the second kind and to find the remaining zero indices of the second kind we only have to know additionally the number ℓ.

For homogeneous polynomials f and g, we write $f|g$ if $g(\lambda, \mu) = f(\lambda, \mu)h(\lambda, \mu)$ for some homogeneous polynomial h. It follows that the homogeneous invariant polynomials defined by (8.1) satisfy $\tilde{p}_n | \cdots | \tilde{p}_1$.

Theorem 8.2. *Let $(B; P, Q)$ be a block on the n dimensional space $\mathcal{X}$. Let $\kappa_1 \geq \cdots \geq \kappa_p \geq 0$ be its indices of the first kind, $\mu_1 \geq \cdots \geq \mu_q \geq 0$ be its indices of* *the third kind, r be the number of its indices of the second kind, and $\tilde{p}_n|\cdots|\tilde{p}_1$ be its homogeneous invariant polynomials. Consider homogeneous polynomials $q_n|\cdots|q_1$ with $\sum_{i=1}^n \deg q_i = n$ and $q_i(1,0) \neq 0$ for $i = 1,\ldots,n$. Then there exists a completion A of the block $(B;P,Q)$ which has $q_1,\ldots,q_n$ as its homogeneous invariant polynomials if and only if there exist homogeneous polynomials $r_n|\cdots|r_1$ such that*

(1a) $\tilde{p}_j|r_j$ *for* $j = 1,\ldots,n$, *and* $r_j|\tilde{p}_{j-p}$ *for* $j = p+1,\ldots,n$;

(1b) $q_{j+r+q}|r_j$ *for* $j = 1,\ldots,m$, *with* $m = \dim \operatorname{Im} P$, *and*
$r_j|q_{j-r}$ *for* $j = r+1,\ldots,n-r$;

(2) $\sum_{i=1}^j \deg \frac{s_i}{s_{i+1}} \geq \sum_{i=1}^j(\kappa_i+1)$ *for* $j = 1,\ldots,p$, *with equality holding for* $j = p$;

here the polynomials s_j are defined by

$$s_j = \prod_{i=j}^n \text{l.c.m.}(\tilde{p}_{i-j+1}, r_i), \quad j = 1,\ldots,p+1;$$

(3) $n-r \geq \deg t_1$ *and* $(n-r-\deg t_2) + \sum_{i=2}^j \deg \frac{t_i}{t_{i+1}} \geq \sum_{i=1}^j(\mu_i+1)$ *for* $j = 1,\ldots,q$, *with equality holding for* $j = q$;

here the polynomials t_j are defined by

$$t_j = \prod_{i=r+j}^n \text{l.c.m.}(r_{i-r-j+1}, q_i), \quad j = 1,\ldots,q+1.$$

The above theorem is a modification of Theorem 2 in Cabral-Silva [1]. To prove it one has to use additionally Proposition III.4.4. In fact, one needs Proposition III.4.4 to deduce that s_i/s_{i+1} and t_i/t_{i+1} in the conditions (2) and (3) are sequences of polynomials with decreasing degrees (cf., the proof of Proposition IV.4.3). The method of proof of Theorem 2 in Cabral-Silva [1] consists of a reduction of the general case to the simpler cases appearing in Chapters IV, V, and VI. The difficulty in applying Theorem 8.2 lies in the fact that there are no rules or algorithms for finding the polynomials r_i, and hence, in general, one has to guess the right candidates for these polynomials. However, in the cases of full length, full width, and principal blocks the situation is much simpler, because in these cases the choice of the polynomials r_i is obvious. Indeed, for example, if the block $(B;P,Q)$ is a full length block, then the numbers p and r appearing in Theorem 8.2 are both zero, and hence in this case condition (1a) tells us that r_i must be equal to p_i. In this way one can deduce Theorem IV.1.1 from Theorem 8.2.

Notes

The main results of this chapter are taken from Gohberg-Kaashoek-Van Schagen [1]. Theorem 5.1 is the original Kronecker theorem (see Kronecker [1]; also Gantmacher [5], Chapter XII) about the canonical form of matrix pencils under strict equivalence. The recent result of Cabral-Silva [1] presented in Section 8 makes an important step toward the solution of the eigenvalue completion problem for general blocks.

Chapter VIII
Off-diagonal Blocks

In the first section of this chapter the general block similarity theorem is specified for off-diagonal blocks. By using the canonical form for off-diagonal blocks, the eigenvalue completion problem for these blocks is reduced to the corresponding problem for full length blocks and subsequentially solved. The associated restriction problem is treated in the second section.

VIII.1 Structure theorems for off-diagonal blocks

In Section VII.1 we gave the structure theorems for general blocks. In this section we specify these theorems for off-diagonal blocks. Recall that $(B; P, Q)$ is an off-diagonal block if $\operatorname{Im} Q \subset \operatorname{Ker} P$.

Theorem 1.1. *Let $(B; P, Q)$ be an off-diagonal block on the n-dimensional space $\mathcal{X}$. Put*

$$p = \operatorname{rank} Q - \operatorname{rank} B, \quad q = \operatorname{rank} P - \operatorname{rank} B,$$

$$r_0 = n - \operatorname{rank} P - \operatorname{rank} Q, \quad r_1 = \operatorname{rank} B.$$

Then $(B; P, Q)$ is block similar to the direct sum of p shifts of the first kind with index 0, q shifts of the third kind with index 0, r_0 shifts of the second kind with index 0 and r_1 shifts of the second kind with index 1. In other words, $(B; P, Q)$ has p indices of the first kind, all equal to 0, q indices of the third kind, all equal to 0, r_0 indices of the second kind equal to 0, and r_1 indices of the second kind each equal to 1.

PROOF. We apply Theorem VII.1.3. First we compute the spaces

$$\mathcal{D}_0 = \mathcal{X}, \ \mathcal{D}_1 = \operatorname{Im} Q, \ \mathcal{D}_i = \{x \in \mathcal{D}_{i-1} \mid Bx \in \mathcal{D}_{i-1} + \operatorname{Ker} P\}, \ i = 2, 3, \ldots .$$

Since $\operatorname{Im} Q \subset \operatorname{Ker} P$ and $\operatorname{Im} B \subset \operatorname{Im} P$, it is clear that $\mathcal{D}_2 = \operatorname{Ker} B$. Now use that $\operatorname{Ker} B \subset \operatorname{Im} Q \subset \operatorname{Ker} P$ to see that $\mathcal{D}_3 = \mathcal{D}_2$. So it follows that $\mathcal{D}_i = \operatorname{Ker} B$ for $i = 2, 3, \ldots$. We see that the numbers $\delta_i = \dim \mathcal{D}_i - \dim \mathcal{D}_{i+1}$ are given by

$$\delta_0 = n - \operatorname{rank} Q, \ \delta_1 = \operatorname{rank} B, \ \delta_i = 0, \ (i = 2, 3, \ldots).$$

The numbers $\gamma_i = \dim \operatorname{Ker} P - \dim(\mathcal{D}_i \cap \operatorname{Ker} P)$ are given by

$$\gamma_1 = n - \operatorname{rank} P - \operatorname{rank} Q, \ \gamma_i = n - \operatorname{rank} P - (\operatorname{rank} Q - \operatorname{rank} B) \quad (i = 2, 3, \ldots).$$

Next we determine the spaces:

$$\mathcal{F}_0 = \{0\},\ \mathcal{F}_1 = \operatorname{Ker} P,\ \mathcal{F}_{i+1} = \mathcal{F}_1 \oplus B[\mathcal{F}_i \cap \operatorname{Im} Q], \quad i = 1, 2, \dots .$$

We see that $\mathcal{F}_2 = \operatorname{Ker} P + B[\operatorname{Im} Q]$. So $\mathcal{F}_2 \cap \operatorname{Im} Q = \operatorname{Im} Q$ and thus $\mathcal{F}_3 = \mathcal{F}_2$. This implies that $\mathcal{F}_i = \operatorname{Ker} P + B[\operatorname{Im} Q]$ for $i = 2, 3, \dots$. Thus the numbers $\alpha_i = \dim \mathcal{F}_{i+1} - \dim \mathcal{F}_i$ are given by

$$\alpha_0 = n - \operatorname{rank} P, \quad \alpha_1 = \operatorname{rank} B, \quad \alpha_i = 0 \quad (i = 2, 3, \dots).$$

The results now follow from the formulas for the numbers of indices of the three different kinds appearing in Theorem VII.2.1. □

Theorem 1.2. *The off-diagonal blocks $(B_1; P_1, Q_1)$ on $\mathcal{X}_1$ and $(B_2; P_2, Q_2)$ on $\mathcal{X}_2$ are block similar if and only if $\dim \mathcal{X}_1 = \dim \mathcal{X}_2$, $\operatorname{rank} P_1 = \operatorname{rank} P_2$, $\operatorname{rank} Q_1 = \operatorname{rank} Q_2$, and $\operatorname{rank} B_1 = \operatorname{rank} B_2$.*

PROOF. Assume that the off-diagonal blocks $(B_1; P_1, Q_1)$ and $(B_2; P_2, Q_2)$ are block similar. Then the number of indices of the second kind that are equal to 1 is the same for $(B_1; P_1, Q_1)$ and $(B_2; P_2, Q_2)$. So we see that $\operatorname{rank} B_1 = \operatorname{rank} B_2$. The numbers of indices of the first kind are the same. Thus $\operatorname{rank} Q_1 - \operatorname{rank} B_1 = \operatorname{rank} Q_2 - \operatorname{rank} B_2$. This proves that $\operatorname{rank} Q_1 = \operatorname{rank} Q_2$. Finally comparing the numbers of indices of the third kind gives that $\operatorname{rank} P_1 - \operatorname{rank} B_1 = \operatorname{rank} P_2 - \operatorname{rank} B_2$ and thus $\operatorname{rank} P_1 = \operatorname{rank} P_2$. Assume that $\dim \mathcal{X}_1 = \dim \mathcal{X}_2 = n$, $\operatorname{rank} P_1 = \operatorname{rank} P_2 = \ell$, $\operatorname{rank} Q_1 = \operatorname{rank} Q_2 = m$ and $\operatorname{rank} B_1 = \operatorname{rank} B_2 = k$. Then $(B_i; P_i, Q_i)$ is similar to a direct sum of $m - k$ shifts of the first kind with index 0, $\ell - k$ shifts of the third kind with index 0, $n - \ell - m$ shifts of the second kind with index 0 and k shifts of the second kind with index 1, for $i = 1, 2$. Therefore $(B_1; P_1, Q_1)$ and $(B_2; P_2, Q_2)$ are block similar. □

The results in this section have natural analogues for finite dimensional operator pencils and non-everywhere defined operators modulo a subspace. We omit the details.

VIII.2 The eigenvalue completion and restriction problems

Assume that $(B; P, Q)$ is an off-diagonal block on the space $\mathcal{X}$. Recall that the operator $A : \mathcal{X} \to \mathcal{X}$ is a completion of $(B; P, Q)$ if $B = PA|_{\operatorname{Im} Q}$. The eigenvalue completion problem (Problem A or A' in Section I.6) is to determine which sets of polynomials can appear as the invariant polynomials of a completion of $(B; P, Q)$. The next result is the solution of this problem.

Theorem 2.1. *Let $(B; P, Q)$ be an off-diagonal block on the n-dimensional space $\mathcal{X}$. Then the polynomials $q_n | \cdots | q_1$, with $\sum_{i=1}^{n} \deg q_i = n$, are the invariant polynomials of a completion of $(B; P, Q)$ if and only if $\deg q_{n-k+1} = \cdots = \deg q_n = 0$, where $k = \operatorname{rank} B$. Equivalently, A is a completion of $(B; P, Q)$ if and only if each eigenvalue of A has a geometric multiplicity less than or equal to $n - \operatorname{rank} B$.*

PROOF. Since $(B; P, Q)$ is off-diagonal, we have $\operatorname{Im} Q \subset \operatorname{Ker} P$, and thus $PI|_{\operatorname{Im} Q}x = 0$ for each $x \in \operatorname{Im} Q$, where $I : \mathcal{X} \to \mathcal{X}$ is the identity operator. So, if A is a completion of $(B; P, Q)$, then $B = P(-\lambda I + A)|_{\operatorname{Im} Q}$ for each value of λ. Therefore $\operatorname{rank}(-\lambda I + A) \geq \operatorname{rank} B$, and thus for any λ the geometric multiplicity $\dim \operatorname{Ker}(-\lambda I + A)$ is less than $n - \operatorname{rank} B$.

Conversely, assume that $q_n | \cdots | q_1$, with $\sum_{i=1}^n \deg q_i = n$, are polynomials such that $\deg q_{n-k+1} = \cdots = \deg q_n = 0$, where $k = \operatorname{rank} B$. Take A_0 to be any operator with invariant polynomials $q_1, \ldots, q_n$. In two steps we construct a restriction of A_0 that is block similar to $(B; P, Q)$. In the first step we construct a full length block $(B_0; I, Q_0)$ which has specially choosen indices of the third kind and invariant polynomials. In the second step a projection P_0 is constructed such that the block $\big(P_0(I - Q_0)B_0; P_0, Q_0\big)$ is block similar to $(B; P, Q)$.

Put $m = \dim \operatorname{Ker} Q$. First, we prove that $\ell := n - m - k - \sum_{i=1}^{n-m} \deg q_{i+m} \geq 0$. It is sufficient to see that $\sum_{i=1}^m \deg q_i \geq m + k$. Assume that $\sum_{i=1}^m \deg q_i < m + k$. Then $\deg q_m \leq 1$ because the sequence $(\deg q_i)_{i=1}^n$ is descending and $k \leq m$. So $\sum_{i=1}^{n-m} \deg q_{i+m} \leq n - m - k$ because each term is at most 1 and k terms are equal to 0. Thus $\sum_{i=1}^n \deg q_i < n$, which contradicts the assumption on the sequence $(\deg q_i)_{i=1}^n$. We conclude that $\ell \geq 0$.

Now choose t such that

$$\ell = \sum_{i=1}^{t} (\deg q_i - \deg q_{i+m}) + \alpha,$$

with $0 \leq \alpha < \deg q_{t+1} - \deg q_{t+m+1}$. We put $r_i = q_i / q_{i+m}$ for $i = 1, \ldots, t$, $r_i = 1$ for $i = t+2, \ldots, n-m$, and r_{t+1} is a divisor of degree α of q_{t+1}/q_{t+m+1}. Put $p_i = r_i q_{i+m}$ for $i = 1, \ldots, n-m$, and $p_i = 1$ for $i = n-m+1, \ldots, n$. Put $\mu_1 = \cdots \mu_k = 1$ and $\mu_{k+1} = \cdots = \mu_m = 0$. Then

(0) $\sum_{i=1}^m (\mu_i + 1) + \sum_{i=1}^n \deg p_i = n$;

(1) $p_i | q_i$ for $i = 1, \ldots, n$, and $q_{i+m} | p_i$ for $i = 1, \ldots, n-m$;

Next, we use Proposition IV.4.3, Lemma VI.2.2 and the special choice of the numbers $\mu_1, \ldots, \mu_m$ to see that

(2) $\sum_{i=1}^j \deg s_i \geq \sum_{i=1}^j (\mu_i + 1)$ for $j = 1, \ldots, m$, with equality holding for $j = m$,

where the polynomials s_j are defined as $s_j = t_j / t_{j+1}$ for $j = 1, \ldots, m$, with

$$t_j = \prod_{i=j}^{n} \text{l.c.m.}(p_{i-j+1}, q_j), \quad j = 1, \ldots, m+1.$$

We apply Theorem IV.1.1 to construct a block $(B_0; I, Q_0)$ with invariant polynomials $p_1, \ldots, p_n$ and indices of the third kind $\mu_1, \ldots, \mu_m$.

The number of strictly positive indices of the block $(B_0; I, Q_0)$ is equal to $\dim \operatorname{Im} Q_0 - \dim\{x \in \operatorname{Im} Q_0 \mid B_0 x \in \operatorname{Im} Q_0\}$. This follows from Theorem III.1.2. So one finds that k is equal to the rank of the operator $(I - Q_0)B_0 : \operatorname{Im} Q_0 \to \operatorname{Ker} Q_0$. Choose a projection P_0 such that

$$\operatorname{Im} Q_0 \subset \operatorname{Ker} P_0, \quad \operatorname{Im}(I - Q_0)B_0 \subset \operatorname{Im} P_0, \quad \operatorname{rank} P_0 = \operatorname{rank} P.$$

This can be done, since $\operatorname{rank} P \geq k = \operatorname{rank}(I - Q_0)B_0$ and

$$\dim \operatorname{Ker} P_0 = \dim \operatorname{Ker} P \geq \dim \operatorname{Im} Q = \dim \operatorname{Im} Q_0.$$

Consider the block $(P_0(I - Q_0)B_0; P_0, Q_0)$. The operator A_0 is completion of $(P_0(I - Q_0)B_0; P_0, Q_0)$ with invariant polynomials $q_1, \ldots, q_n$. Since

$$\operatorname{rank} P_0(I - Q_0)B_0 = \operatorname{rank} B, \quad \operatorname{rank} P_0 = \operatorname{rank} P, \quad \operatorname{rank} Q_0 = \operatorname{rank} Q,$$

this block is an off-diagonal block, which is block similar to $(B; P, Q)$. Let S be the block similarity. Then we obtain from Theorem I.6.1 that $A = S^{-1}A_0 S$ is a completion of $(B; P, Q)$ with the desired invariant polynomials. □

We conclude this section with the eigenvalue restriction problem (Problem B in Section I.6) for off-diagonal blocks. So given $A : \mathcal{X} \to \mathcal{X}$, determine all the block similarity classes of off-diagonal blocks of A. The solution is given in the next theorem.

Theorem 2.2. *Let $A : \mathcal{X} \to \mathcal{X}$ and natural numbers k, ℓ and m be given such that $k \leq \ell$, $k \leq m$ and $\ell \leq \dim \mathcal{X} - m$. Then there exists an off-diagonal block $(B; P, Q)$ with $\operatorname{rank} P = m$, $\operatorname{rank} Q = \ell$, $\operatorname{rank} B = k$ and $B = PA|_{\operatorname{Im} Q} : \operatorname{Im} Q \to \operatorname{Im} P$ if and only if no eigenvalue of A has geometric multiplicity more than $\dim \mathcal{X} - k$.*

PROOF. First let the off-diagonal block $(B; P, Q)$ be such that $\operatorname{rank} P = m$, $\operatorname{rank} Q = \ell$, $\operatorname{rank} B = k$ and $B = PA|_{\operatorname{Im} Q} : \operatorname{Im} Q \to \operatorname{Im} P$. Then A is a completion of $(B; P, Q)$, and according to Theorem 2.1 this implies that no eigenvalue of A has geometric multiplicity exceeding $\dim \mathcal{X} - k$.

Conversely, assume $(B_0; P_0, Q_0)$ is an off-diagonal block with $\operatorname{rank} P_0 = m$, $\operatorname{rank} Q_0 = \ell$, $\operatorname{rank} B_0 = k$ and that A has at most $\dim \mathcal{X} - k$ non-constant invariant polynomials. By Theorem 2.1 there exists a completion A_0 of $(B_0; P_0, Q_0)$ that has the same invariant polynomials as A. So A_0 is similar to A. Let S be the similarity, i.e., $A = SA_0S^{-1}$. Take $P = SP_0S^{-1}$, $Q = SQ_0S^{-1}$ and $B = PA|_{\operatorname{Im} Q} : \operatorname{Im} Q \to \operatorname{Im} P$. Then S is a block similarity of $(B_0; P_0, Q_0)$ and $(B; P, Q)$. So indeed $(B; P, Q)$ is a restriction of A and $\operatorname{rank} P = m$, $\operatorname{rank} Q = l$, and $\operatorname{rank} B = k$. □

Notes

The main results in this chapter are taken from Gohberg-Kaashoek-Van Schagen [4]. Independently Theorem 2.1 was also proved by Silva [3] and Zaballa [5].

Chapter IX
Connections with Linear Systems

This chapter gives an introduction to the theory of discrete time-invariant linear input/output systems. Included is the analysis of observability and controllability of systems, theorems about minimal realization, the eigenvalue assignment theorem and state feedback, and theorems on output stabilization and output injection. The connection with operator blocks and block similarity theory is one of the main features of this chapter. For example, observability indices and controllability indices are identified as block similarity invariants of a certain kind, state feedback is viewed as block similarity, and the eigenvalue assignment problem is treated as an eigenvalue completion problem. These connections are exploited throughout the chapter.

IX.1 Linear input/output systems and transfer functions

Consider the system

$$\Sigma \quad \begin{cases} x_{k+1} = Ax_k + Bu_k, & \text{for } k \geq 0, \\ y_k = Cx_k + Du_k, & \text{for } k \geq 0, \\ x_0 = x, \end{cases} \tag{1.1}$$

where $A : \mathcal{X} \to \mathcal{X}$, $B : \mathcal{U} \to \mathcal{X}$, $C : \mathcal{X} \to \mathcal{Y}$ and $D : \mathcal{U} \to \mathcal{Y}$ are linear transformations of the n-dimensional space $\mathcal{X}$, the m-dimensional space $\mathcal{U}$ and the r-dimensional space $\mathcal{Y}$. So the vector u_k, which is referred to as the *input at time* k, is a vector from $\mathcal{U}$, the *input space*, and the vector y_k, which is called the *output at time* k, is in the space $\mathcal{Y}$, the *output space*. The vector x_k is referred to as the *state* of the system at time k, and the space $\mathcal{X}$ is called the *state space*. The vector x is called *the initial state* of the system. Usually we consider the state of the system as unknown. The *state space dimension* n may be much larger than the numbers m and r. The system Σ can be symbolically represented as in figure 1.

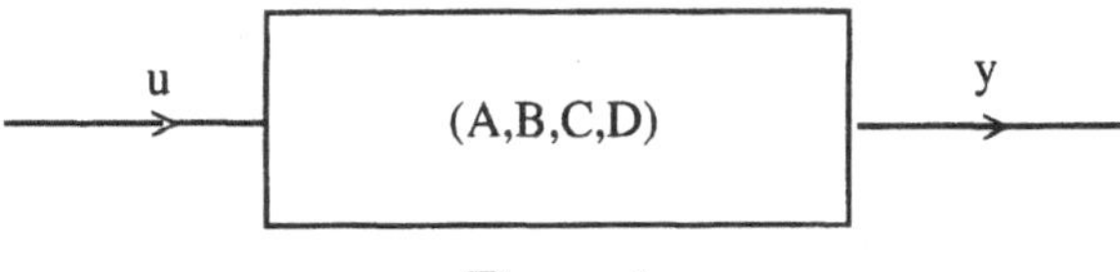

Figure 1

Instead of (1.1) we shall use the shorthand notation $\Sigma = (A, B, C, D)$. The system Σ is assumed to be causal, i.e., the state at time k depends only on previous states and previous inputs. One computes from the equations (1.1) that

$$x_k = A^k x + \sum_{j=0}^{k-1} A^{k-1-j} B u_j, \quad k \geq 0, \tag{1.2}$$

and

$$y_k = CA^k x + \sum_{j=0}^{k-1} CA^{k-1-j} B u_j + D u_k, \quad k \geq 0. \tag{1.3}$$

Formula (1.3) describes the output sequence in terms of the sequence of input vectors and the initial state. If in (1.3) the initial state is taken to be 0, then the resulting map is called the *input-output map* of the system Σ. In other words, the input-output map of the system Σ is given by

$$T_\Sigma\left((u_k)_{k=0}^\infty\right) = (y_k)_{k=0}^\infty, \quad y_k = Du_k + \sum_{j=0}^{k-1} CA^{k-1-j} B u_j, \quad k \geq 0.$$

Consider the system Σ with the initial value $x = 0$. We encode the sequences of inputs, outputs and states also in a different way. Put

$$x(z) = \sum_{k=1}^\infty x_k z^{-k}, \quad y(z) = \sum_{k=1}^\infty y_k z^{-k}, \quad u(z) = \sum_{k=1}^\infty u_k z^{-k}.$$

Here z is a (formal) complex parameter. The transformation of a sequence $(x_k)_{k=0}^\infty$ to the formal power series $x(z)$ is called the *z-transform*. This transformation clearly is bijective. Using the z-transform and assuming the initial state to be zero, we can rewrite (1.1) as

$$\begin{cases} zx(z) = Ax(z) + Bu(z), \\ y(z) = Cx(z) + Du(z). \end{cases} \tag{1.4}$$

One refers to the previous equations as the frequency domain representation of Σ. We solve $x(z)$ from the first of these two equations and get $x(z) = (zI - A)^{-1}Bu(z)$. Substituting the result in the second, we obtain that

$$y(z) = [D + C(zI - A)^{-1}B]u(z).$$

The function $W(z) = D + C(zI - A)^{-1}B$ is called the *transfer function* of the system Σ. It has the property that from the input series $u(z)$ one computes the output series $y(z)$ as $y(z) = W(z)u(z)$. Observe that the transfer function is a rational matrix function which is analytic at infinity and its value at infinity is precisely the feed through coefficient D.

Two systems have the same input-output map if and only if they have the same transfer function. To prove this first assume that two systems have the same transfer function. So with the same $u(z)$ they generate the same $y(z)$ and hence with the same input sequence $(u_k)_{k=0}^{\infty}$ they give the same output sequence $(y_k)_{k=0}^{\infty}$, which proves that the input-output maps are equal. Conversely, assume that the input-output maps are the same. From a given $u(z)$, and hence a given input sequence $(u_k)_{k=0}^{\infty}$, the two systems give the same output sequence $(y_k)_{k=0}^{\infty}$, and therefore the same $y(z)$. Thus with the two transfer functions W_1 and W_2, we have for each $u(z)$ that $y(z) = W_1(z)u(z) = W_2(z)u(z)$. This implies that $W_1 = W_2$.

A rational $r \times m$ matrix function which is analytic at infinity is called *proper*. Let $W(z)$ be a proper rational $r \times m$ matrix function with value D at infinity. The system Σ is called a *realization* of $W(z)$ if $W(z)$ is the transfer function of Σ. In other words,

$$W(z) = D + C(zI - A)^{-1}B. \tag{1.5}$$

In this case we also call the quadruple (A, B, C, D) a *realization* of $W(z)$.

Theorem 1.1. *Let W be a $n \times m$ rational matrix function. Then W admits a realization if and only if W is proper.*

Proof. First assume that $\Sigma = (A, B, C, D)$ is a realization of W. Then W is given by (1.5) and hence $\lim_{z\to\infty} W(z)$ exists. Thus W is proper.

Conversely, assume that W is proper. Put $D = \lim_{z\to\infty} W(z)$. Assume that the poles of W are $z_1, \dots, z_k$, and let

$$W_i(z) = \sum_{j=1}^{m_i} (z - z_i)^{-j} L_{ij}$$

be the Laurent principal part of W at z_i. So $W(z) = D + \sum_{i=1}^{k} W_i(z)$. With

$$C_i = \begin{pmatrix} I_n & 0 & \cdots & 0 \end{pmatrix},$$

$$A_i = \begin{pmatrix} z_iI_n & I_n & & & \\ 0 & z_iI_n & \cdot & & \\ & \cdot & \ddots & \cdot & \\ & & \cdot & z_iI_n & I_n \\ & & & 0 & z_iI_n \end{pmatrix}, \quad B_i = \begin{pmatrix} L_{i1} \\ \cdot \\ \cdot \\ \cdot \\ L_{i\,m_i} \end{pmatrix}$$

matrices of sizes $n \times m_in$, $m_in \times m_in$ and $m_in \times m$, respectively, we have that $W_i(z) = C_i(zI_{m_in} - A_i)^{-1}B_i$. Put

$$C = \begin{pmatrix} C_1 & \cdots & C_k \end{pmatrix}, \quad A = \begin{pmatrix} A_1 & 0 & & & \\ 0 & A_2 & \cdot & & \\ & \cdot & \ddots & \cdot & \\ & & \cdot & A_{k-1} & 0 \\ & & & 0 & A_k \end{pmatrix}, \quad B = \begin{pmatrix} B_1 \\ \cdot \\ \cdot \\ \cdot \\ B_k \end{pmatrix}.$$

Then $W(z) = D + C(zI - A)^{-1}B$, and thus we have constructed a realization of W. □

Note that a realization of the function W is not unique. We will describe the freedom in the choice of the realization in Section 4.

IX.2 Blocks and controllability

Let the system Σ be as in (1.1), and let P be any projection of $\mathcal{X}$ along the subspace $\operatorname{Im} B$. We refer to the block $(PA; P, I)$ as a *full width block associated with* Σ. The relevance of this block for the system is explained by the next result. Recall that the k-times iterated images $\mathcal{F}_k$ have been introduced in Section V.1.

Proposition 2.1. *Let $\mathcal{F}_k$ be the k-times iterated image of the full width block $(PA; P, I)$ associated with the system Σ. Then $\mathcal{F}_k$ is the subspace of the state space consisting of those states x which can be reached at time k from the zero state using k subsequent inputs, i.e.,*

$$\mathcal{F}_k = \left\{ \sum_{i=1}^{k} A^{i-1} B u_i \mid u_1, \ldots, u_k \in \mathcal{U} \right\}. \tag{2.1}$$

PROOF. Recall that by definition, see Section V.1,

$$\mathcal{F}_1 = \operatorname{Ker} P, \quad \mathcal{F}_{k+1} = \operatorname{Ker} P \oplus PA[\mathcal{F}_k] \quad (k = 1, 2, \ldots).$$

So

$$\mathcal{F}_k = \operatorname{Ker} P + PA[\operatorname{Ker} P] + \cdots + (PA)^{k-1}[\operatorname{Ker} P] \quad (k = 1, 2, \ldots).$$

We compute, using $\operatorname{Ker} P = \operatorname{Im} B$, that

$$\begin{aligned} \mathcal{F}_k &= \operatorname{Im} B + PA[\operatorname{Im} B] + \cdots + (PA)^{k-1}[\operatorname{Im} B] \\ &= \operatorname{Im} B + A[\operatorname{Im} B] + \cdots + A^{k-1}[\operatorname{Im} B] \\ &= \operatorname{Im}\begin{pmatrix} B & AB & \cdots & A^{k-1}B \end{pmatrix}. \end{aligned} \tag{2.2}$$

So for a vector $x \in \mathcal{F}_k$ there exists a sequence of inputs $u_1, \ldots, u_k$ such that $x = \sum_{i=1}^{k} A^{i-1} B u_i$. □

As we know from Section V.1 there exists a number κ such that $\mathcal{F}_\kappa = \mathcal{F}_{\kappa+1}$ and $\mathcal{F}_j \neq \mathcal{F}_{j+1}$ if $j < \kappa$. The space $\mathcal{F}_\kappa$ was called the final iterated image of the block $(PA; P, I)$. From (2.2) one sees that $\mathcal{F}_\kappa$ is the smallest A-invariant subspace of $\mathcal{X}$ that contains $\operatorname{Im} B$. It follows from the Cayley-Hamilton theorem that $\kappa \leq n - 1$. Thus $\mathcal{F}_\kappa = \operatorname{Im}\begin{pmatrix} B & AB & \cdots & A^{n-1}B \end{pmatrix}$. We denote this subspace by

$$\operatorname{Im}(A|B) = \operatorname{Im}\begin{pmatrix} B & AB & \cdots & A^{n-1}B \end{pmatrix},$$

and we call it *the image of the pair* (A, B). Of special interest is the case when $\operatorname{Im}(A|B) = \mathcal{X}$. By Proposition 2.1 the latter means for the system that each state $x \in \mathcal{X}$ can be reached from the zero state with κ subsequent inputs. Therefore the system is called *controllable* if $\operatorname{Im}(A|B) = \mathcal{X}$. Sometimes also the term *reachable* is used in stead of controllable. Since in the notion of controllability the output does not play a role, we also call the *pair* (A, B) *controllable* if $\operatorname{Im}(A|B) = \mathcal{X}$.

Remark that the freedom of choice in $\operatorname{Im} P$ does not affect the spaces $\mathcal{F}_k$. Indeed, if S is a block similarity of the second kind of $(PA; P, I)$ and $(P'A; P', I)$, then the corresponding iterated images are equal (see Lemma VII.4.2).

In the system Σ not every state can be reached from the zero state with the same number of inputs. For instance, vectors in the space $\operatorname{Im} B$ can be reached with only one input, and vectors in the subspace $\operatorname{Im}(B\ AB\ \ldots\ A^{j-1}B)$ can be reached with j subsequent inputs. We consider the numbers

$$p_k = \dim \operatorname{Im}(B\ AB\ \ldots\ A^{k-1}B) - \dim \operatorname{Im}(B\ AB\ \ldots\ A^{k-2}B), \tag{2.3}$$

for $k \geq 2$ and $p_1 = \dim \operatorname{Im} B$. They represent the gains in dimension of the reachable subspace which occur by adding one extra input vector. Clearly, these numbers contain important information about the system. It is common to store the information contained in the sequence $p_1, p_2, \ldots$ into its dual sequence of indices, which is defined by

$$\kappa_j = \#\{k \mid p_k \geq j\} \quad j \geq 1. \tag{2.4}$$

The numbers $\kappa_1, \ldots, \kappa_m$, where m is the dimension of the input space $\mathcal{U}$, are called the *controllability indices of the system* Σ. Remark that they do not depend on the second equation of Σ. Therefore they are often referred to as the *controllability indices of the pair* (A, B). There exists a close relation between the controllability indices of the pair (A, B) and the indices of the first kind of a full width block associated with the system Σ. This relation will be described in the next result.

Theorem 2.2. *Let $A : \mathcal{X} \to \mathcal{X}$ and $B : \mathcal{U} \to \mathcal{X}$ be linear operators, and let $\kappa_1 \geq \kappa_2 \geq \cdots \geq \kappa_p > 0 = \kappa_{p+1} = \cdots = \kappa_m$ be natural numbers, where $m = \dim \mathcal{U}$. Then the following statements are equivalent:*

(1) *$\kappa_1, \ldots, \kappa_m$ are the controllability indices of the pair (A, B);*

(2) *$\kappa_1 - 1, \ldots, \kappa_p - 1$ are the indices of the first kind of the block $(P'A; P', I_{\mathcal{X}})$, where P' is any projection of $\mathcal{X}$ with $\operatorname{Ker} P' = \operatorname{Im} B$;*

(3) *$\kappa_1, \ldots, \kappa_m$ are the indices of the first kind of the block $(Z; P, I)$ given by*

$$P = \begin{pmatrix} I_{\mathcal{X}} & 0 \\ 0 & 0 \end{pmatrix} : \mathcal{X} \times \mathcal{U} \to \mathcal{X} \times \mathcal{U}\ ,$$

$$Z = \begin{pmatrix} A & B \end{pmatrix} : \mathcal{X} \times \mathcal{U} \to \operatorname{Im} P = \mathcal{X} \times \{0\}\ ;$$

(4) *$\kappa_1, \ldots, \kappa_m$ are the minimal column indices of the operator pencil*

$$\lambda\begin{pmatrix} I_{\mathcal{X}} & 0 \end{pmatrix} - \begin{pmatrix} A & B \end{pmatrix} : \mathcal{X} \times \mathcal{U} \to \mathcal{X}.$$

Proof. Let $p_1, p_2, \dots$ be given by (2.3). Put

$$B_k = \begin{pmatrix} B & AB & \cdots & A^k B \end{pmatrix}.$$

Then

$$\begin{aligned} p_k &= \operatorname{rank}\begin{pmatrix} B & AB_{k-2} \end{pmatrix} - \operatorname{rank} B_{k-2} \leq \\ &\operatorname{rank} B + \operatorname{rank}(AB_{k-2}) - \operatorname{rank} B_{k-2} \leq \operatorname{rank} B = p_1. \end{aligned}$$

So for $p = p_1 = \dim \operatorname{Im} B$ formula (2.4) yields that $\kappa_{p+i} = 0$ for $i \geq 1$. We now prove the equivalence of (1) and (2). Let $\mathcal{F}_k$ be the k-th iterated image of the block $(P'A; P', I_{\mathcal{X}})$. Put $r_k = \dim \mathcal{F}_{k+1} - \dim \mathcal{F}_k$ for $k = 1, 2, \dots$, and $r_0 = \dim \mathcal{F}_1$. Then the indices of the first kind of the block $(P'A; P', I_{\mathcal{X}})$, say $\kappa_1' \geq \kappa_2' \geq \cdots$, form the dual sequence of the sequence $r_1 \geq r_2 \geq \cdots$. From formulas (2.2) and (2.3) one sees that $p_k = r_{k-1}$ for $k \geq 1$. So

$$\kappa_j = \#\{k \geq 1 \mid p_k \geq j\} = \#\{k \geq 1 \mid r_{k-1} \geq j\} = \#\{k \geq 0 \mid r_k \geq j\}.$$

Since $r_0 \geq r_j$ for all $j \geq 1$, we see that $\kappa_j = \kappa_j' + 1$ for $j = 1, \dots, r_0$. With the remark that $p = p_1 = r_0$ this part of the proof is finished.

The equivalence of (2) and (3) easily follows from Theorem V.1.5.

The equivalence of (3) and (4) one gets from Corollary V.2.3. □

The next theorem gives a list of characterization of the notion of controllability.

Theorem 2.3. *Let $A : \mathcal{X} \to \mathcal{X}$ and $B : \mathcal{U} \to \mathcal{X}$ be linear operators, and let $\kappa_1, \dots, \kappa_m$ be the controllability indices of the pair (A, B). Then the following statements are equivalent:*

(1) *the pair (A, B) is controllable;*

(2) $\sum_{i=1}^m \kappa_i = n$, *where $n = \dim \mathcal{X}$;*

(3) $\operatorname{Im}\begin{pmatrix} B & AB & \cdots & A^{n-1}B \end{pmatrix} = \mathcal{X}$, *with $n = \dim \mathcal{X}$;*

(4) *the block $(P'A; P', I_{\mathcal{X}})$, where P' is any projection of $\mathcal{X}$ with* $\operatorname{Ker} P' = \operatorname{Im} B$, *has no invariant polynomials;*

(5) *the block $(Z; P, I)$ given by*

$$P = \begin{pmatrix} I_{\mathcal{X}} & 0 \\ 0 & 0 \end{pmatrix} : \mathcal{X} \times \mathcal{U} \to \mathcal{X} \times \mathcal{U},$$

$$Z = \begin{pmatrix} A & B \end{pmatrix} : \mathcal{X} \times \mathcal{U} \to \operatorname{Im} P = \mathcal{X} \times \{0\},$$

has no invariant polynomials;

(6) *the operator pencil*

$$\lambda\begin{pmatrix} I_{\mathcal{X}} & 0 \end{pmatrix} - \begin{pmatrix} A & B \end{pmatrix} : \mathcal{X} \times \mathcal{U} \to \mathcal{X}.$$

has no invariant polynomials;

(7) $\operatorname{rank}\left(\lambda\begin{pmatrix} I_{\mathcal{X}} & 0 \end{pmatrix} - \begin{pmatrix} A & B \end{pmatrix}\right) = \dim \mathcal{X}$ *for each value of λ.*

PROOF. First we prove the equivalence of (1) and (2). Put $p = m$, if $\kappa_m > 0$, or else choose p such that $\kappa_p > 0$ and $\kappa_{p+1} = 0$. Then $\kappa_1 - 1, \dots, \kappa_p - 1$ are the indices of the first kind of the block $(PA; P, I)$ with P a projection of $\mathcal{X}$ with $\operatorname{Ker} P = \operatorname{Im} B$. Remark that $p = \dim \operatorname{Ker} P$. We know that the dimension of the iterated image $\mathcal{F}_\kappa$ of $(PA; P, I)$ is equal to the sum of the indices of the first kind plus the dimension of $\operatorname{Ker} P$, i.e., $\dim \mathcal{F}_\kappa = p + \sum_{i=1}^p (\kappa_i - 1)$. So we see that $\mathcal{F}_\kappa = \mathcal{X}$ if and only if $n = p + \sum_{i=1}^p (\kappa_i - 1) = \sum_{i=1}^p \kappa_i = \sum_{i=1}^m \kappa_i$. Finally recall that the pair is controllable if and only if $\mathcal{F}_\kappa = \mathcal{X}$.

Next use (2.2) to see that (1) and (3) are equivalent. From Theorem V1.2 we see that $\mathcal{F}_\kappa = \mathcal{X}$ is equivalent to the requirement that the block $(P'A; P', I_{\mathcal{X}})$ has no invariant polynomials. This proves that (1) and (4) are equivalent. Remark that it follows from Theorem V.1.5 that the sets of invariant polynomials of the blocks $(P'A; P', I_{\mathcal{X}})$ and $(Z; P, I)$ coincide. This proves that (4) and (5) are equivalent. Finally Theorem V.2.5 gives that (3), (6) and (7) are equivalent. □

In Chapter V we proved that for a full width block there exists a special basis such that the matrix representation of the block with respect to this basis has a simple form. Moreover, it is easy to read off the indices of the first kind from this representation. We apply this to a full width block associated with Σ and describe the result in the next theorem. We will find special bases for $\mathcal{X}$ and $\mathcal{U}$ and matrix representations of A and B with respect to these bases. In these matrix representations simple types of matrices will appear. First, we describe these types. The first type will be an $\alpha \times \alpha$ matrix

$$\begin{pmatrix} * & * & * & \cdots & \cdots & * & * \\ 1 & 0 & 0 & \cdots & \cdots & 0 & 0 \\ 0 & 1 & 0 & \cdots & \cdots & 0 & 0 \\ \vdots & & \cdot & \ddots & & \vdots & \vdots \\ \vdots & & & \ddots & \ddots & \vdots & \vdots \\ 0 & 0 & 0 & & \cdot & 0 & 0 \\ 0 & 0 & 0 & & & 1 & 0 \end{pmatrix}, \tag{2.5}$$

where $*$ denotes an unspecified entry. The second type will be an $\alpha \times \beta$ matrix

$$\begin{pmatrix} * & * & \cdots & * & * \\ 0 & 0 & \cdots & 0 & 0 \\ \vdots & \vdots & & \vdots & \vdots \\ 0 & 0 & \cdots & 0 & 0 \end{pmatrix}, \tag{2.6}$$

and the third type will be a simple one column matrix

$$\begin{pmatrix} 1 \\ 0 \\ \vdots \\ 0 \end{pmatrix}. \tag{2.7}$$

Theorem 2.4. *Let $A : \mathcal{X} \to \mathcal{X}$ and $B : \mathcal{U} \to \mathcal{X}$ be linear operators. Let $\kappa_1 \geq \cdots \geq \kappa_p > 0 = \kappa_{p+1} = \cdots = \kappa_m$ be the controllability indices of the pair (A, B). Then there exists a basis $\{e_{ij}\}_{i=1,j=1}^{\kappa_j, p} \cup \{f_i\}_{i=1}^{s}$ of $\mathcal{X}$ and a basis $\{g_i\}_{i=1}^{m}$ of $\mathcal{U}$ such that with respect to these bases the matrices of A and B are given as*

$$A = \left(A_{ij}\right)_{i,j=1}^{p+1}, \quad B = \left(B_{ij}\right)_{i=1,j=1}^{p+1,m},$$

where

(1) *$A_{p+1\ p+1}$ is just a matrix of size $s \times s$;*

(2) *$A_{p+1\ j} = 0$ for $j = 1, \dots, p$;*

(3) *for $i = 1, \dots, p$ the $\kappa_i \times \kappa_i$ matrix A_{ii} will be of type (2.5), i.e., has all its entries equal to 0 except for the entries in the diagonal just below the main diagonal, which are all equal to 1, and the entries in the first row;*

(4) *for $i = 1, \dots, p$, $j = 1, \dots, p+1$ and $i \neq j$ the matrix A_{ij} will be of type (2.6), i.e., has all its entries equal to 0 except for those in the first row;*

(5) *the $\kappa_i \times 1$ matrix $B_{ij} = 0$ for $i = 1, \dots, p$ and $i \neq j$;*

(6) *the $s \times 1$ matrix $B_{p+1\ j} = 0$ for $j = 1, \dots, m$;*

(7) *the $\kappa_i \times 1$ matrix B_{ii} will be of type (2.7), i.e., has an entry 1 on the first place and entries 0 elsewhere, for $i = 1, \dots, p$.*

PROOF. Let P be a projection of $\mathcal{X}$ with $\operatorname{Ker} P = \operatorname{Im} B$. We consider the block $(PA; P, I)$. Remark that $\kappa_1 - 1, \dots, \kappa_p - 1$ are the indices of the first kind of $(PA; P, I)$. Apply Proposition V.1.3. to obtain an independent set of vectors $\{e_{ij}\}_{i=1,j=1}^{\kappa_j,\ p}$ and a subspace $\mathcal{G}_\infty$. Let $f_1, \dots, f_s$ be a basis of $\mathcal{G}_\infty$. Then

$$e_{11}, \dots, e_{\kappa_1 1}, \dots, e_{1p}, \dots, e_{\kappa_p, p}, f_1, \dots, f_s$$

is a basis of $\mathcal{X}$. Note that $PAe_{ij} - Pe_{i+1\ j} = 0$ for $i = 1, \dots, \kappa_j - 1$, and $PAe_{\kappa_j j} = 0$. Hence $Ae_{ij} - e_{i+1\ j} \in \operatorname{Ker} P = \operatorname{span}\{e_{11}, \dots, e_{1p}\}$ for $i = 1, \dots, \kappa_j - 1$, and $Ae_{\kappa_j j} \in \operatorname{span}\{e_{11}, \dots, e_{1p}\}$. This proves the properties (1), (2), (3) and (4). Next recall that $\operatorname{Im} B = \operatorname{Ker} P = \operatorname{span}\{e_{11}, \dots, e_{1p}\}$. Choose $g_1, \dots, g_p$ such that $Bg_i = e_{1i}$. Then $g_1, \dots, g_p$ is an independent set in $\mathcal{U}$. Extend this set with $g_{p+1}, \dots, g_m$ in $\operatorname{Ker} B$ to a basis of $\mathcal{U}$. With this choice of $g_1, \dots, g_m$ we also have (5), (6) and (7). □

The invariant polynomials $p_1, \dots, p_s$ of the matrix $A_{p+1\ p+1}$ appearing in item (1) of Theorem 2.4 are called *the invariant polynomials* of the pair (A, B). Since they are the invariant polynomials of the block $(Z; P, I)$, described in Theorem 2.2, they are uniquely determined by the block. Remark that the order of the matrix $A_{p+1\ p+1}$ is equal to $n - \sum_{j=1}^{m} \kappa_j$. From Theorem 2.3 it follows that in case the pair (A, B) is controllable the order of the matrix $A_{p+1\ p+1}$ is equal to 0. So in addition to the statements of Theorem 2.3 we have also that (A, B) is controllable if and only if there are no invariant polynomials for (A, B).

The invariant polynomials of the pair (A, B) can be characterized also in an other way.

Corollary 2.5. *Let $A : \mathcal{X} \to \mathcal{X}$ and $B : \mathcal{U} \to \mathcal{X}$ be linear operators. Decompose $\mathcal{X}$ as $\mathrm{Im}(A|B) \oplus \mathcal{X}_2$, where $\mathrm{Im}(A|B)$ is the image of the pair (A, B). With respect to this decomposition write*

$$A = \begin{pmatrix} A_{11} & A_{12} \\ 0 & A_{22} \end{pmatrix}, \quad B = \begin{pmatrix} B_1 \\ 0 \end{pmatrix}.$$

Then the non-constant invariant polynomials of the pair (A, B) are the non-constant invariant polynomials of A_{22} and the controllability indices of the pair (A, B) are the controllability indices of the pair (A_{11}, B_1). Furthermore, the invariant polynomials of (A, B) coincide with the invariant polynomials of the pencil $(\,\lambda I_{\mathcal{X}} - A \quad -B\,)$.

PROOF. To prove the equality of the controllability indices of the pairs (A, B) and (A_{11}, B_1) it is sufficient to note that

$$\dim \mathrm{Im}(\,B \quad AB \quad \ldots \quad A^k B\,) = \dim \mathrm{Im}(\,B_1 \quad A_{11}B_1 \quad \ldots \quad A_{11}^k B_1\,)$$

for each value of k. To prove the statement on the invariant polynomials we show that the matrix A_{22} is similar to the matrix $A_{p+1\ p+1}$, as it appears in Theorem 2.4. Remark that for the basis $\{e_{ij}\}_{i=1,j=1}^{\kappa_j\ ,p} \cup \{f_i\}_{i=1}^{s}$ in Theorem 2.4 one gets that $\mathrm{Im}(A|B) = \mathrm{span}\{e_{ij}\}_{i=1,j=1}^{\kappa_j\ ,p}$. This follows from the matrix representation of the pair (A, B). Put $\mathcal{X}_2' = \mathrm{span}\{f_i\}_{i=1}^{s}$. With respect to the decomposition $\mathcal{X} = \mathrm{Im}(A|B) \oplus \mathcal{X}_2'$, the matrix representation of A is

$$A = \begin{pmatrix} A_{11} & A_{12}' \\ 0 & A_{p+1\ p+1} \end{pmatrix}.$$

It is easy to see that this implies that the matrices A_{22} and $A_{p+1\ p+1}$ are similar, and hence they have the same invariant polynomials. We already noted above that the invariant polynomials of the pair (A, B) are the invariant polynomials of the block $(Z; P, I)$, described in Theorem 2.2. The last statement in the corollary now follows from Corollary V.2.3. □

The next corollary specializes Theorem 2.4 for the case of a controllable pair.

Corollary 2.6. *Let $A : \mathcal{X} \to \mathcal{X}$ and $B : \mathcal{U} \to \mathcal{X}$ be linear operators. Assume that the pair (A, B) is controllable. Let $\kappa_1 \geq \cdots \geq \kappa_p > 0 = \kappa_{p+1} = \cdots = \kappa_m$ be the controllability indices of the pair (A, B). Then there exists a basis $\{e_{ij}\}_{i=1,j=1}^{\kappa_j\ ,p}$ of $\mathcal{X}$ and a basis $\{g_i\}_{i=1}^{m}$ of $\mathcal{U}$ such that with respect to these bases the matrices of A and B are given as*

$$A = \left(A_{ij}\right)_{i,j=1}^{p}, \quad B = \left(B_{ij}\right)_{i=1,j=1}^{p,m},$$

where

(1) *for* $i = 1, \dots, p$ *the* $\kappa_i \times \kappa_i$ *matrix* A_{ii} *will be of type (2.5), i.e., has all its entries equal to 0 except for the entries in the diagonal just below the main diagonal, which are all equal to 1, and the entries in the first row;*

(2) *for* $i, j = 1, \dots, p$ *and* $i \neq j$ *the matrix* A_{ij} *will be of type (2.6), i.e., has all its entries equal to 0 except for those in the first row;*

(3) *the* $\kappa_i \times 1$ *matrix* $B_{ij} = 0$ *for* $i = 1, \dots, p$ *and* $i \neq j$;

(4) *the* $\kappa_i \times 1$ *matrix* B_{ii} *will be of type (2.7), i.e., has an entry 1 on the first place and entries 0 elsewhere, for* $i = 1, \dots, p$.

PROOF. First note that according to Theorem 2.3 we have $\dim \mathcal{X} = \sum_{i=1}^{p} \kappa_i$, and hence the number s appearing in Theorem 2.4 is equal to 0. Now apply Theorem 2.4 to get a basis $\{e_{ij}\}_{i=1,j=1}^{\kappa_j, p}$ of $\mathcal{X}$ and a basis $\{g_i\}_{i=1}^{m}$ of $\mathcal{U}$ such that with respect to these bases the matrices of A and B are given as $A = \big(A_{ij}\big)_{i,j=1}^{p}$ and $B = \big(B_{ij}\big)_{i=1,j=1}^{p,m}$. The matrices A_{ij} and B_{ij} have the properties (3), (4), (5), and (7) of Theorem 2.4, which are the same as the properties (1) – (4) in this corollary. □

IX.3 Blocks and observability

We consider the system $\Sigma = (A, B, C, D)$ given by (1.1). Recall that $A : \mathcal{X} \to \mathcal{X}$, $B : \mathcal{U} \to \mathcal{X}$, $C : \mathcal{X} \to \mathcal{Y}$ and $D : \mathcal{U} \to \mathcal{Y}$ are linear transformations of the n-dimensional space $\mathcal{X}$, the m-dimensional space $\mathcal{U}$ and the r-dimensional space $\mathcal{Y}$. In this section we associate with the system a full length block. Let Q be a projection of $\mathcal{X}$ onto $\operatorname{Ker} C$. Then $(A|_{\operatorname{Im} Q}; I, Q)$ is called *a full length block associated with* Σ. The next proposition describes the k-th definition space $\mathcal{D}_k$ (see Section III.1) of this block.

Proposition 3.1. *Let* $\mathcal{D}_k$ *be the* k*-th definition space of the full length block* $(A|_{\operatorname{Im} Q}; I, Q)$ *associated with the system* Σ. *Then* $\mathcal{D}_k$ *is the subspace consisting of the initial states* x *which generate* k *subsequent zero outputs if the system gets* k *subsequent zero inputs, i.e.,*

$$\mathcal{D}_k = \{x \in \mathcal{X} \mid CA^{j-1}x = 0, j = 1, \dots, k\} = \bigcap_{j=1}^{k} \operatorname{Ker} CA^{j-1}. \tag{3.1}$$

PROOF. Recall that, by definition,

$$\mathcal{D}_1 = \operatorname{Im} Q, \ \mathcal{D}_j = \{x \in \mathcal{D}_{j-1} \mid Ax \in \mathcal{D}_{j-1}\}, \ j = 2, 3, \dots .$$

So $\mathcal{D}_1 = \operatorname{Ker} C$, and $\mathcal{D}_2$ consists of those $x \in \operatorname{Ker} C$ satisfying $Ax \in \operatorname{Ker} C$. In other words, $\mathcal{D}_2 = \operatorname{Ker} C \cap \operatorname{Ker} CA$. Put $\mathcal{S}_k = \{x \in \mathcal{X} \mid CA^{j-1}x = 0, j = 1, \dots, k\}$. Then $\mathcal{S}_k = \bigcap_{j=1}^{k} \operatorname{Ker} CA^{j-1}$. We have that $\mathcal{D}_1 = \mathcal{S}_1$ and $\mathcal{D}_2 = \mathcal{S}_2$. Assume that $\mathcal{D}_k = \mathcal{S}_k$. Then $\mathcal{D}_{k+1} = \{x \in \mathcal{S}_k \mid Ax \in \mathcal{S}_k\}$. So we see that $x \in \operatorname{Ker} C$ and $Ax \in \bigcap_{j=1}^{k} \operatorname{Ker} CA^{j-1}$. The second inclusion means that $x \in \bigcap_{j=1}^{k} \operatorname{Ker} CA^{j}$. Thus $x \in \bigcap_{j=1}^{k+1} \operatorname{Ker} CA^{j-1}$. We have proved formula (3.1). □

There exists a number μ such that $\mathcal{D}_\mu = \mathcal{D}_{\mu+1}$ and $\mathcal{D}_j \neq \mathcal{D}_{j+1}$ if $j < \mu$. This space $\mathcal{D}_\mu$ was identified as the largest $A|_{\operatorname{Im} Q}$-invariant subspace in $\operatorname{Im} Q$. Since (by formula (3.1)) it depends on the pair (C, A) only, we denote this space as $\operatorname{Ker}(C|A)$, and we call this space the *kernel of the pair* (C, A). A special situation occurs if $\operatorname{Ker}(C|A) = \{0\}$. According to Proposition 3.1 this means that the only initial state that generates μ zero outputs, given zero inputs, is the zero state. So two different initial states generate different output sequences provided that the same sequence of inputs is applied to the system. Therefore the system is called *observable* if $\operatorname{Ker}(C|A) = \{0\}$. Since the inputs and the matrices B and D do not play a role in the definition of $\operatorname{Ker}(C|A)$, the pair (C, A) is also called *observable.* Remark that the freedom of choice in $\operatorname{Ker} Q$ does not affect the spaces $\mathcal{D}_k$. Indeed, if S is a block similarity of the second kind of $(A|_{\operatorname{Im} Q}; I, Q)$ and $(A'|_{\operatorname{Im} Q'}; I, Q')$, then the corresponding definition spaces are equal (see Lemma VII.4.2).

Next, we shall define the observability indices. Recall that the dimension of $\mathcal{X}$ is n and the dimension of $\mathcal{Y}$ is r. The state vectors that differ by a vector $x \in \operatorname{Ker} C$ can not be distinguished by looking at one output only since the difference between the first outputs is then 0 by definition. However, some of these state vectors may not be in $\operatorname{Ker} C \cap \operatorname{Ker} CA$, and can therefore be distinguished by considering the differences between two subsequent outputs, Cx and CAx. So, if k outputs are zero, then the subspace $\bigcap_{i=0}^{k-1} \operatorname{Ker} CA^i$ is the space of all possible initial states. In other words, two states x_1 and x_2 generate the same first k outputs with the same k inputs if and only if $x_1 - x_2 \in \bigcap_{i=0}^{k-1} \operatorname{Ker} CA^i$. Therefore we consider the numbers

$$q_k = \dim \bigcap_{i=0}^{k-2} \operatorname{Ker} CA^i - \dim \bigcap_{i=0}^{k-1} \operatorname{Ker} CA^i, \tag{3.2}$$

for $k = 2, 3, \ldots$, and $q_1 = n - \dim \operatorname{Ker} C$. The number q_k represents the difference made in the determination of the initial state by observing an additional k-th output. The information contained in the sequence $q_1, q_2, \ldots$ is represented by its dual sequence

$$\nu_j = \#\{k \mid q_k \geq j\}, \quad j \geq 1.$$

The numbers $\nu_1, \ldots, \nu_r$ are called the *observability indices* of the pair (C, A).

Theorem 3.2. *Let $A : \mathcal{X} \to \mathcal{X}$ and $C : \mathcal{X} \to \mathcal{Y}$ be linear operators, and let $\nu_1 \geq \nu_2 \geq \cdots \geq \nu_p > 0 = \nu_{p+1} = \cdots = \nu_r$ be natural numbers, where $r = \dim \mathcal{Y}$. Then the following statements are equivalent:*

(1) *$\nu_1, \ldots, \nu_r$ are the observability indices of the pair (C, A);*

(2) *$\nu_1 - 1, \ldots, \nu_p - 1$ are the indices of the third kind of the block $(A|_{\operatorname{Im} Q'}; I, Q')$, where Q' is any projection of $\mathcal{X}$ with $\operatorname{Im} Q' = \operatorname{Ker} C$;*

(3) *$\nu_1, \ldots, \nu_r$ are the indices of the first kind of the block $(Z; I, Q)$ given by*

$$Q = \begin{pmatrix} I_{\mathcal{X}} & 0 \\ 0 & 0 \end{pmatrix} : \mathcal{X} \times \mathcal{Y} \to \mathcal{X} \times \mathcal{Y}, \quad Z = \begin{pmatrix} A \\ C \end{pmatrix} : \operatorname{Im} Q \to \mathcal{X} \times \mathcal{Y};$$

(4) $\nu_1, \ldots, \nu_r$ *are the minimal column indices of the operator pencil*

$$\lambda \begin{pmatrix} I \\ 0 \end{pmatrix} - \begin{pmatrix} A \\ C \end{pmatrix} : \mathcal{X} \to \mathcal{X} \times \mathcal{Y}.$$

PROOF. Consider q_j given by formula (3.2). First we prove that $q_1 \geq q_k$ for all k. Remark that $q_1 = n - \dim \operatorname{Ker} C$, and therefore $q_1 = \operatorname{rank} C$. Furthermore

$$q_k = \operatorname{rank} \operatorname{col}\big(CA^{j-1}\big)_{j=1}^{k-1} - \operatorname{rank} \operatorname{col}\big(CA^{j-1}\big)_{j=1}^{k-2}.$$

Put $T = \operatorname{col}\big(CA^{j-1}\big)_{j=1}^{k-2}$. Then we see that

$$q_k = \operatorname{rank} \begin{pmatrix} C \\ TA \end{pmatrix} - \operatorname{rank} T \leq \operatorname{rank} TA + \operatorname{rank} C - \operatorname{rank} T \leq \operatorname{rank} C = q_1.$$

Next we prove the equivalence of (1) and (2). Put $r_k = \dim \mathcal{D}_k - \dim \mathcal{D}_{k+1}$, with $\mathcal{D}_0 = \mathcal{X}$, $\mathcal{D}_1 = \operatorname{Im} Q' = \operatorname{Ker} C$ and $\mathcal{D}_j$ given by (3.1) for $j \geq 2$. Proposition 3.1 gives that $q_k = \dim \mathcal{D}_{k-2} - \dim \mathcal{D}_{k-1}$. So we see that $q_k = r_{k-1}$ for $k = 1, 2, \ldots$. The indices of the third kind, $\mu_1 \geq \mu_2 \geq \cdots$ of the block $(A|_{\operatorname{Im} Q'}; I, Q')$ are the dual numbers of the sequence $r_1 \geq r_2 \geq \cdots$, i.e., $\mu_j = \#\{k \geq 1 \mid r_k \geq j\}$. Use $q_1 \geq q_j$ for all j to see that

$$\mu_j = \#\{k \geq 1 \mid q_{k+1} \geq j\} = \#\{k \geq 1 \mid q_k \geq j\} - 1 = \nu_j - 1.$$

This proves the equivalence of (1) and (2).

To prove the equivalence of (2) and (3), one applies Theorem III.1.5 to the block $(Z; I, Q)$.

The equivalence of (3) and (4) is proved in Corollary III.2.3. □

The next result gives characterizations of the notion of observability.

Theorem 3.3. *Let $A : \mathcal{X} \to \mathcal{X}$ and $C : \mathcal{X} \to \mathcal{Y}$ be linear operators, and let $\nu_1, \ldots, \nu_r$ be the observability indices of the pair (C, A). Then the following statements are equivalent:*

(1) *the pair (C, A) is observable;*

(2) $\sum_{i=1}^{r} \nu_i = n$, *where $n = \dim \mathcal{X}$;*

(3) $\bigcap_{j=1}^{n} \operatorname{Ker} CA^{j-1} = \{0\}$, *with $n = \dim \mathcal{X}$;*

(4) *the block $(A|_{\operatorname{Im} Q'}; I, Q')$, where Q' is any projection of $\mathcal{X}$ with $\operatorname{Im} Q' = \operatorname{Ker} C$, has no invariant polynomials;*

(5) *the block $(Z; I, Q)$ on $\mathcal{X} \times \mathcal{Y}$, given by*

$$Q = \begin{pmatrix} I_{\mathcal{X}} & 0 \\ 0 & 0 \end{pmatrix} : \mathcal{X} \times \mathcal{Y} \to \mathcal{X} \times \mathcal{Y}, \quad Z = \begin{pmatrix} A \\ C \end{pmatrix} : \operatorname{Im} Q \to \mathcal{X} \times \mathcal{Y};$$

has no invariant polynomials;

(6) the operator pencil

$$\lambda \begin{pmatrix} I \\ 0 \end{pmatrix} - \begin{pmatrix} A \\ C \end{pmatrix} : \mathcal{X} \to \mathcal{X} \times \mathcal{Y}.$$

has no invariant polynomials;

(7) rank $\begin{pmatrix} \lambda I - A \\ -C \end{pmatrix} = \dim \mathcal{X}$ for each value of λ.

PROOF. Put $p = r$, if $\nu_r > 0$, or else choose p such that $\nu_p > 0$ and $\nu_{p+1} = 0$. Then $\nu_1 - 1, \nu_2 - 1, \ldots, \nu_p - 1$ are the indices of the third kind of the block $(A|_{\operatorname{Im} Q'}; I, Q')$, where Q' is a projection of $\mathcal{X}$ with $\operatorname{Im} Q' = \operatorname{Ker} C$. We know that $p = \dim \operatorname{Ker} Q'$. The codimension of $\operatorname{Ker}(C|A)$, the largest $(A|_{\operatorname{Im} Q'})$-invariant subspace of $\operatorname{Im} Q'$, is equal to the sum of the indices of the third kind and $\dim \operatorname{Ker} Q'$. So $n - \dim \operatorname{Ker}(C|A) = p + \sum_{i=1}^{p} (\nu_i - 1)$. We see that $\operatorname{Ker}(C|A) = \{0\}$ if and only if $n = p + \sum_{i=1}^{p} (\nu_i - 1) = \sum_{i=1}^{p} \nu_i$. Since the pair is observable if and only if $\operatorname{Ker}(C|A) = \{0\}$, we proved that (1) and (2) are equivalent.

The pair (C, A) is observable if and only if $\operatorname{Ker}(C|A) = \bigcap_{j=1}^{\mu} \operatorname{Ker} CA^{j-1} = \{0\}$ for some number μ. By the Cayley-Hamilton theorem this is equivalent to $\bigcap_{j=1}^{n} \operatorname{Ker} CA^{j-1} = \{0\}$. This proves that (1) and (3) are equivalent. Recall that, by Theorem III.1.2, the invariant polynomials of $(A|_{\operatorname{Im} Q'}; I, Q')$ are the invariant polynomials of the operator $A|_{\operatorname{Ker}(C|A)}$. So $\operatorname{Ker}(C|A) = \{0\}$ if and only if the the block $(A|_{\operatorname{Im} Q'}; I, Q')$ has no non constant invariant polynomials. This shows the equivalence of (1) and (4).

According to Theorem III.1.5 the sets of invariant polynomials of the blocks $(A|_{\operatorname{Im} Q'}; I, Q')$ and $(Z; I, Q)$ are the same. This proves the equivalence of (4) and (5). From Corollary III.2.3 we see that (5) and (6) are equivalent, and finally Theorem III.2.5 gives that (6) holds if and only if (7) does. □

The next aim is to construct special bases, for which the matrix representation of the pair (C, A) has a simple form. To do this we consider the pair (A^T, C^T), and we apply to this pair the results for pairs of the type (A, B) of the previous section. Here X^T denotes the transposed of the matrix X. We get a special matrix representation which we translate back by again transposing. One has also to perform an extra reordering of the obtained bases to get the result presented in the next theorem. Before we give the result we introduce some matrices, that will appear in the matrix representation of (C, A). The first type is an $\alpha \times \alpha$ matrix

$$\begin{pmatrix} 0 & 0 & 0 & \cdots & \cdots & 0 & * \\ 1 & 0 & 0 & \cdots & \cdots & 0 & * \\ 0 & 1 & 0 & \cdots & \cdots & 0 & * \\ \vdots & & \cdot & \ddots & & \vdots & \vdots \\ \vdots & & & \ddots & \ddots & \vdots & \vdots \\ 0 & 0 & 0 & & \cdot & 0 & * \\ 0 & 0 & 0 & & & 1 & * \end{pmatrix}, \tag{3.3}$$

where $*$ denotes an unspecified entry. The second type will be an $\alpha \times \beta$ matrix

$$\begin{pmatrix} 0 & 0 & \cdots & 0 & * \\ 0 & 0 & \cdots & 0 & * \\ \vdots & \vdots & & \vdots & \vdots \\ 0 & 0 & \cdots & 0 & * \end{pmatrix}, \tag{3.4}$$

and the third type will be a simple row matrix

$$\begin{pmatrix} 0 & 0 & \cdots & 1 \end{pmatrix}. \tag{3.5}$$

Theorem 3.4. *Let $A : \mathcal{X} \to \mathcal{X}$ and $C : \mathcal{X} \to \mathcal{Y}$ be linear transformations. Let $\nu_1 \geq \cdots \geq \nu_p > 0 = \nu_{p+1} = \cdots = \nu_r$ be the observability indices of the pair (C, A). Then there exists a basis $\{e_{ij}\}_{i=1,j=1}^{\nu_j,p} \cup \{f_i\}_{i=1}^{s}$ of $\mathcal{X}$ and a basis $\{g_i\}_{i=1}^{r}$ of $\mathcal{Y}$ such that with respect to these bases the matrices of A and C are given as*

$$A = (A_{ij})_{i,j=1}^{p+1}, \; C = (C_{ij})_{i=1,j=1}^{r,p+1},$$

where

1. *$A_{p+1\ p+1}$ is an $s \times s$ matrix;*
2. *$A_{i\ p+1} = 0$ for $i = 1, \ldots, p$;*
3. *for $i = 1, \ldots, p$ the $\nu_i \times \nu_i$ matrix A_{ii} is of type (3.3), i.e., has all its entries equal to 0 except for the entries in the diagonal just below the main diagonal, which are all equal to 1, and the entries in the last column;*
4. *for $i = 1, \ldots, p+1$, $j = 1, \ldots, p$ and $i \neq j$ the matrix A_{ij} is of type (3.4), i.e., has all entries equal to 0 except for those in the last column;*
5. *the $1 \times \nu_i$ matrix $C_{ij} = 0$ for $i = 1, \ldots, p$ and $i \neq j$;*
6. *the $1 \times s$ matrix $C_{i\ p+1} = 0$ for $i = 1, \ldots, r$;*
7. *the $1 \times \nu_i$ matrix C_{ii} is of type (3.5), i.e., has an entry 1 on the last place and entries 0 elsewhere, for $i = 1, \ldots, p$.*

PROOF. Apply Theorem 2.4 to the pair (A^T, C^T), and use an additional reordering of the obtained bases to get the desired result. □

The invariant polynomials $p_1, \ldots, p_s$ of the matrix $A_{p+1\ p+1}$ appearing in item (1) of Theorem 3.4 are called *the invariant polynomials* of the pair (C, A). They are the invariant polynomials of the block

$$\left(\begin{pmatrix} A \\ C \end{pmatrix}; I_{\mathcal{X} \oplus \mathcal{Y}}, \begin{pmatrix} I & 0 \\ 0 & 0 \end{pmatrix} \right).$$

This one sees from the matrix representation. We specify Theorem 3.4 for the case when the pair (C, A) is observable.

Corollary 3.5. *Let $A : \mathcal{X} \to \mathcal{X}$ and $C : \mathcal{X} \to \mathcal{Y}$ be linear transformations such that the pair (C, A) is observable. Let $\nu_1 \geq \cdots \geq \nu_p > 0 = \nu_{p+1} = \cdots = \nu_r$ be the observability indices of the observable pair (C, A). Then there exists a basis $\{e_{ij}\}_{i=1,j=1}^{\nu_j,p}$ of $\mathcal{X}$ and a basis $\{g_i\}_{i=1}^{r}$ of $\mathcal{Y}$ such that with respect to these bases the matrices of A and C are given as*

$$A = (A_{ij})_{i,j=1}^{p}, \quad C = (C_{ij})_{i=1,j=1}^{r,p},$$

where

(1) *for $i = 1, \ldots, p$ the $\nu_i \times \nu_i$ matrix A_{ii} is of type (3.3), i.e., has all its entries equal to 0 except for the entries in the diagonal just below the main diagonal, which are all equal to 1, and the entries in the last column;*

(2) *for $i, j = 1, \ldots, p$ and $i \neq j$ the matrix A_{ij} is of type (3.4), i.e., has all entries equal to 0 except for those in the last column;*

(3) *the $1 \times \nu_i$ matrix $C_{ij} = 0$ for $i = 1, \ldots, p$ and $i \neq j$;*

(4) *the $1 \times \nu_i$ matrix C_{ii} is of type (3.5), i.e., has an entry 1 on the last place and entries 0 elsewhere, for $i = 1, \ldots, p$.*

PROOF. We take the basis obtained in Theorem 3.4. Since the pair (C, A) is observable, we conclude from Theorem 3.3 that $\sum_{i=1}^{r} \nu_i = \dim \mathcal{X}$. This gives that the number s, occurring in Theorem 3.4 is equal to zero. therefore only the items (3), (4), (5) and (7) in Theorem 3.4 have non void contents. These items are restated here as (1), (2), (3) and (4). □

IX.4 Minimal systems

The material in this section is not directly related to operator blocks. However, the notion of minimality is an important tool in the sequel. Consider the input-output map of the system

$$\Sigma \quad \begin{cases} x_{k+1} = Ax_k + Bu_k, & \text{for } k \geq 0, \\ y_k = Cx_k + Du_k, & \text{for } k \geq 0. \end{cases} \tag{4.1}$$

In stead of (4.1) we use the shorthand notation $\Sigma = (A, B, C, D)$.

There are two basic operations which change Σ into a system with the same input-output map. The first is similarity. A system $(S^{-1}AS, S^{-1}B, CS, D)$, with $S : \mathcal{X}_1 \to \mathcal{X}$ invertible, is called *similar* to Σ. Two similar systems have the same input-output map. Next put

$$A' = \begin{pmatrix} * & * & * \\ 0 & A & * \\ 0 & 0 & * \end{pmatrix}, \quad B' = \begin{pmatrix} * \\ B \\ 0 \end{pmatrix}, \quad C' = \begin{pmatrix} 0 & C & * \end{pmatrix}, \quad D' = D,$$

where $*$ stands for unspecified entries. The system (A', B', C', D') is called a *dilation* of Σ. It is easy to see that (A', B', C', D') and Σ have the same input-output map.

A system is called *minimal* if the dimension of its state space is minimal among all systems that have the same input-output map. The two basic operations mentioned above describe all the freedom one has in constructing systems with the same input-output map. This statement follows from the next three theorems.

Theorem 4.1. *A system is minimal if and only if it is observable and controllable.*

Theorem 4.2. *Two minimal systems with the same input-output map are similar.*

Theorem 4.3. *Each system is a dilation of a minimal system.*

Theorem 4.4. *Two systems have the same input-output map if and only if they are dilations of similar systems.*

For the proofs of these theorems we need two propositions.

Proposition 4.5. *Each system is a dilation of an observable and controllable system. Indeed, for the system (A, B, C, D) with state space $\mathcal{X}$ of dimension n, put*

$$\mathcal{D}_\infty = \bigcap_{i=0}^{n-1} \operatorname{Ker}(CA^i), \quad \mathcal{F}_\infty = \operatorname{Im}\begin{pmatrix} B & AB & \cdots & A^{n-1}B \end{pmatrix}$$

and write $\mathcal{X} = \mathcal{X}_1 \oplus \mathcal{X}_2 \oplus \mathcal{X}_3$ with $\mathcal{X}_1 = \mathcal{F}_\infty \cap \mathcal{D}_\infty$, with $\mathcal{X}_2$ such that $\mathcal{F}_\infty = \mathcal{X}_1 \oplus \mathcal{X}_2$, and $\mathcal{X}_3$ such that $\mathcal{X} = \mathcal{F}_\infty \oplus \mathcal{X}_3$. With respect to this decomposition the operator matrices of A, B and C take the form

$$A = \begin{pmatrix} * & * & * \\ 0 & A_0 & * \\ 0 & 0 & * \end{pmatrix}, \quad B = \begin{pmatrix} * \\ B_0 \\ 0 \end{pmatrix}, \quad C = \begin{pmatrix} 0 & C_0 & * \end{pmatrix}, \tag{4.2}$$

and the system (A_0, B_0, C_0, D) is observable and controllable and has the same input-output map as (A, B, C, D).

PROOF. First remark that the Cayley-Hamilton theorem gives that both $\mathcal{F}_\infty$ and $\mathcal{D}_\infty$ are invariant subspaces for the operator A. This proves that the matrix representation of A is as given in (4.2). Moreover $\mathcal{X}_1 \subset \operatorname{Ker} C$ and $\operatorname{Im} B \subset \mathcal{X}_1 \oplus \mathcal{X}_2$. which proves that the operator matrices of B and C are given by (4.2). Thus (A, B, C, D) is a dilation of the system (A_0, B_0, C_0, D). We now show that (C_0, A_0) is observable. Assume that $x_2 \in \mathcal{X}_2$ is such that $C_0 A_0^j x_2 = 0$ for each $j \geq 0$. Then $CA^j x_2 = 0$ for each j. This proves that $x_2 = 0$. Next let $x_2 \in \mathcal{X}_2$. Since $x_2 \in \mathcal{F}_\infty$, it is clear that there exist y_j $(j = 1, \ldots, n)$ such that $x_2 = By_1 + ABy_2 + \cdots + A^{n-1}By_n$. Set $y_j = y_{j1} + y_{j2} + y_{j3}$ with $y_{ji} \in \mathcal{X}_i$ $(i = 1, 2, 3)$. Then we see that $x_2 = B_0 y_{12} + A_0 B_0 y_{22} + \cdots + A_0^{n-1} B_0 y_{n2}$. This proves that the pair (A_0, B_0) is controllable. Finally remark that $CA^j B = C_0 A_0^j B_0$ for $j \geq 0$. □

Proposition 4.6. *Let (A_1,B_1,C_1,D_1) and (A_2,B_2,C_2,D_2) have the same input-output map. If (A_1,B_1) and (A_2,B_2) are controllable, and (C_1,A_1) and (C_2,A_2) are observable, then $D_2=D_1$ and the systems (A_1, B_1, C_1, D_1) and (A_2, B_2, C_2, D_2) are similar. Moreover the similarity is uniquely determined by (A_1,B_1,C_1,D_1) and (A_2,B_2,C_2,D_2).*

PROOF. We have to prove that there exists an invertible linear operator S such that

$$C_2=C_1S, \quad A_2=S^{-1}A_1S, \quad B_2=S^{-1}B_1. \tag{4.3}$$

Since (A_1,B_1,C_1,D_1) and (A_2,B_2,C_2,D_2) have of the same input-output map, we see that $D_2=D_1$ and $C_1A_1^jB_1=C_2A_2^jB_2$ for $j=0,1,2,\dots$. Put, for $k=1,2$,

$$\Omega_k=\operatorname{col}(C_kA_k^j)_{j=0}^{p-1}, \quad \Delta_k=\begin{pmatrix} B_k & A_kB_k & \dots & A_k^{p-1}B_k\end{pmatrix},$$

where p is any number larger than the degrees of the minimal polynomials of A_1 and A_2. It follows from $C_1A_1^jB_1=C_2A_2^jB_2$ that $\Omega_1\Delta_1=\Omega_2\Delta_2$. Note that $\operatorname{Ker}\Omega_k=\{0\}$ for $k=1,2$. So Ω_k has a left inverse Ω_k^{-L} such that $\Omega_k^{-L}\Omega_k=I_k$. Also $\operatorname{Im}\Delta_k=\mathcal{X}_k$ for $k=1,2$. So Δ_k has a right inverse Δ_k^{-R} such that $\Delta_k\Delta_k^{-R}=I_k$. Now we multiply the equality $\Omega_1\Delta_1=\Omega_2\Delta_2$ from the left with Ω_2^{-L} and from the right with Δ_1^{-R}. We put $S=\Omega_2^{-L}\Omega_1=\Delta_2\Delta_1^{-R}$. Then $S:\mathcal{X}_1\to\mathcal{X}_2$ is invertible. Indeed $\Omega_1^{-L}\Omega_2S=I_1$ and $S\Delta_1\Delta_2^{-R}=I_2$. So the dimensions n_1 and n_2 of $\mathcal{X}_1$ and $\mathcal{X}_2$, respectively, must be the same.

Let us check the equalities (4.3). Write

$$\Omega_2A_2\Delta_2=\Omega_1A_1\Delta_1=\Omega_1\Delta_1\Delta_1^{-R}A_1\Delta_1=\Omega_2\Delta_2\Delta_1^{-R}A_1\Delta_1\ .$$

Premultiply by Ω_2^{-L} and postmultiply by Δ_1^{-R} to get $A_2S=SA_1$. Now $SB_1=\Omega_2^{-L}\Omega_1B_1=\Omega_2^{-L}\Omega_2B_2=B_2$ and $C_2S=C_2\Delta_2\Delta_1^{-R}=C_1\Delta_1\Delta_1^{-R}=C_1$.

Finally, assume that the invertible transformation T fulfills

$$C_2=C_1T, \quad A_2=T^{-1}A_1T, \quad B_2=T^{-1}B_1$$

Then $\Omega_2=\Omega_1T$, and we already had $\Omega_2=\Omega_1S$. So $\Omega_1(T-S)=0$. Since $\operatorname{Ker}\Omega_1=\{0\}$, this shows that $T-S=0$, and hence the uniqueness of the similarity is proved. □

PROOF OF THEOREM 4.1. Assume that the system (A,B,C,D) is minimal. Apply Proposition 4.5 to (A,B,C,D). We get a system (A_0,B_0,C_0,D) with the same input-output map and state space $\mathcal{X}_2$. Since (A,B,C,D) is minimal, $\dim\mathcal{X}\le\dim\mathcal{X}_2$. On the other hand $\mathcal{X}_2\subset\mathcal{X}$, and this proves that $\mathcal{X}_2=\mathcal{X}$. Thus $(A_0,B_0,C_0,D)=(A,B,C,D)$, which proves that (A,B,C,D) is observable and controllable.

Next assume that the system (A,B,C,D) is controllable and observable. There exists a minimal system with the same input-output map. This system we denote as (A_1,B_1,C_1,D_1). Then (A_1,B_1,C_1,D_1) is controllable and observable, as we proved above. Now apply Proposition 4.6 to see that $A_1=SAS^{-1}$ for some invertible S. This proves that the orders of the matrices of A and A_1 are equal. So (A,B,C,D) is a minimal system. □

Theorems 4.2 and 4.3 now follow from Theorem 4.1 and Propositions 4.5 and 4.6. To prove Theorem 4.4 we first note that if two systems are dilations of similar systems they share the same input output map. Conversely, suppose that two systems have the same input-output map. Each of them is a dilation of a minimal system. These minimal systems have the same input-output map. Thus these minimal systems are similar.

In the statements of the theorems we could replace the input-output map by the transfer function, since two systems have the same transfer function if and only if they have the same input-output map.

IX.5 Feedback and block similarity

We consider the following system

$$x_{k+1} = Ax_k + Bu_k, \quad x_0 = x, \quad k \geq 0, \tag{5.1}$$

with $A : \mathcal{X} \to \mathcal{X}$ a linear operator on the n-dimensional linear space $\mathcal{X}$ and $B : \mathcal{U} \to \mathcal{X}$ a linear operator from the m-dimensional linear space $\mathcal{U}$. For a moment let us assume that the inputs u_k are 0 for $k = 0, 1, \ldots$. Then the k-th state of the system x_k is equal to $A^k x$, for $k = 0, 1, \ldots$. It is natural to require that in the case when inputs are not present that the system goes to rest when time goes to infinity, i.e., $A^k x \to 0$ for any x if $k \to \infty$. The system (5.1) is called *stable* if $\lim_{k\to\infty} A^k x = 0$ for each state $x \in \mathcal{X}$. Equivalently, (5.1) is stable if A has all its eigenvalues in the open unit disk. If a system is not stable, a common way to stabilize the system is to apply feedback. For the system (5.1) this means that one replaces the input u_k by $u_k + Fx_k$ for some operator F. This type of feedback is called *state feedback* and F is called a *(state) feedback operator.* One can express this with a block diagram like in figure 2.

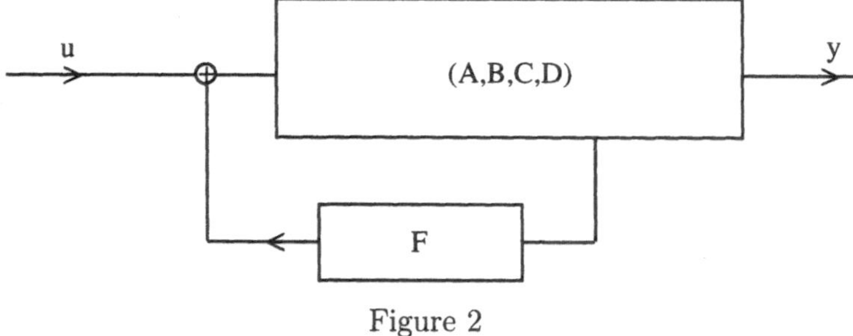

Figure 2

The resulting system is

$$x_{k+1} = (A + BF)x_k + Bu_k, \quad x_0 = x, \quad k \geq 0, \tag{5.2}$$

with $F : \mathcal{X} \to \mathcal{U}$. So, the system (5.1) is called *stabilizable (by state feedback)* if there exists a state feedback F such that the system (5.2) is stable. An important question is whether or not a certain given system is stabilizable by state feedback. Corollary 6.5 gives the answer. The question leads to the problem to describe the properties of the operators $A+BF$, with $A:\mathcal{X}\to\mathcal{X}$ and $B:\mathcal{U}\to\mathcal{X}$ given linear transformations and $F:\mathcal{X}\to\mathcal{U}$ a free parameter.

The pair (A',B') with $A':\mathcal{X}'\to\mathcal{X}'$ and $B':\mathcal{U}'\to\mathcal{X}'$ is called *feedback equivalent* to (A,B) if there exist invertible linear transformations $S:\mathcal{X}'\to\mathcal{X}$, $T:\mathcal{U}'\to\mathcal{U}$, and a feedback operator $F:\mathcal{X}\to\mathcal{U}$ such that

$$A'=S^{-1}(A+BF)S,\quad B'=S^{-1}BT. \tag{5.3}$$

It is straightforward to check that feedback equivalence is an equivalence relation indeed. This is also clear from the next theorem.

Theorem 5.1. *Let $A:\mathcal{X}\to\mathcal{X}$, $A':\mathcal{X}'\to\mathcal{X}'$, $B:\mathcal{U}\to\mathcal{X}$ and $B':\mathcal{U}'\to\mathcal{X}'$ be operators. The following statements are equivalent:*

(1) *the pairs (A,B) and (A',B') are feedback equivalent;*

(2) *the blocks $(Z;P,I)$ and $(Z';P',I')$ with*

$$P=\begin{pmatrix} I_{\mathcal{X}} & 0\\ 0 & 0\end{pmatrix}:\mathcal{X}\times\mathcal{U}\to\mathcal{X}\times\mathcal{U},\quad P'=\begin{pmatrix} I_{\mathcal{X}'} & 0\\ 0 & 0\end{pmatrix}:\mathcal{X}'\times\mathcal{U}'\to\mathcal{X}'\times\mathcal{U}',$$

$$Z=\begin{pmatrix} A & B\end{pmatrix}:\mathcal{X}\times\mathcal{U}\to\operatorname{Im}P,\quad Z'=\begin{pmatrix} A' & B'\end{pmatrix}:\mathcal{X}'\times\mathcal{U}'\to\operatorname{Im}P',$$

are block similar;

(3) *the blocks $(PA;P,I_{\mathcal{X}})$ on $\mathcal{X}$ and $(P'A';P',I_{\mathcal{X}'})$ on $\mathcal{X}'$ with P (P') a projection of $\mathcal{X}$ $(\mathcal{X}')$ with $\operatorname{Ker}P=\operatorname{Im}B$ $(\operatorname{Ker}P'=\operatorname{Im}B')$ are block similar;*

(4) *the pairs (A,B) and (A',B') have the same controllability indices and the same invariant polynomials;*

(5) *The pencils $\lambda\begin{pmatrix} I_{\mathcal{X}} & 0\end{pmatrix}-\begin{pmatrix} A & B\end{pmatrix}$ and $\lambda\begin{pmatrix} I'_{\mathcal{X}} & 0\end{pmatrix}-\begin{pmatrix} A' & B'\end{pmatrix}$ are strictly equivalent.*

Proof. First we prove the equivalence of the statements (1) and (2). The pairs (A,B) and (A',B') are feedback equivalent if and only if there exist invertible linear transformations $S:\mathcal{X}'\to\mathcal{X}$ and $T:\mathcal{U}'\to\mathcal{U}$ and a feedback operator $F:\mathcal{X}\to\mathcal{U}$ such that (5.3) holds. The equations (5.3) can be written in the following equivalent form:

$$S\begin{pmatrix} A' & B'\end{pmatrix}-\begin{pmatrix} A & B\end{pmatrix}\begin{pmatrix} S & 0\\ FS^{-1} & T\end{pmatrix}=\begin{pmatrix} 0 & 0\end{pmatrix}.$$

This means that the blocks $(Z;P,I)$ and $(Z';P',I')$ are block similar.

The blocks $(Z;P,I)$ and $(Z';P',I')$ are block similar if and only if they have the same indices of the first kind and the same invariant polynomials. From Theorem 2.2 it follows that the indices of the first kind of the block $(Z;P,I)$ $((Z';P',I'))$

are the controllability indices of the pair (A, B) $((A', B'))$. The invariant polynomials of $(Z; P, I)$ $((Z'; P', I'))$ are the invariant polynomials of $(\,\lambda I - A \quad -B\,)$ $((\,\lambda I - A' \quad -B'\,))$ (see Corollary V.2.3), which by Corollary 2.5 are the invariant polynomials of (A, B) $((A', B'))$. This together shows the equivalence of (2) and (4). Corollary V.1.6 proves that (2) and (3) are equivalent, and finally Corollary V.2.3 gives the equivalence of (2) and (5). □

The next theorem shows that each equivalence class of feedback equivalent pairs contains a representative with a simple matrix representation.

Theorem 5.2. *Let $\mathcal{X}$ be an n-dimensional and $\mathcal{U}$ an m-dimensional linear space, and let $A : \mathcal{X} \to \mathcal{X}$ and $B : \mathcal{U} \to \mathcal{X}$ be linear operators. Let $\kappa_1 \geq \cdots \geq \kappa_q > 0 = \kappa_{q+1} = \cdots = \kappa_m$ be the controllability indices of the pair (A, B). Then there exist a basis $\{e_{ij}\}_{i=1,j=1}^{\kappa_j\ ,q} \cup \{f_i\}_{i=1}^{s}$ of $\mathcal{X}$, a basis $\{g_i\}_{i=1}^{m}$ of $\mathcal{U}$ and an operator $F : \mathcal{X} \to \mathcal{U}$ such that the matrix of $A + BF$ with respect to this basis of $\mathcal{X}$ is*

$$J_{\kappa_1} \oplus \cdots \oplus J_{\kappa_q} \oplus M, \tag{5.4}$$

where J_{κ_i} is the $\kappa_i \times \kappa_i$ matrix given by

$$J_{\kappa_i} = \begin{pmatrix} 0 & & & & & \\ 1 & 0 & & & & \\ & 1 & \ddots & & & \\ & & \ddots & \ddots & & \\ & & & 1 & 0 & \\ & & & & 1 & 0 \end{pmatrix}, \tag{5.5}$$

and M is a $s \times s$ matrix. Furthermore, with respect to these bases the matrix of B has zero entries except for those in the positions $(1,1)$, $(\kappa_1 + 1, 2)$, ..., $(\kappa_1 + \cdots + \kappa_{q-1} + 1, q)$ which entries are equal to 1.

PROOF. We apply Theorem 2.3 to obtain a suitable matrix representation of both A and B. So we take the basis $\{e_{ij}\}_{i=1,j=1}^{\kappa_j\ ,p} \cup \{f_i\}_{i=1}^{s}$ of $\mathcal{X}$ and a basis $\{g_i\}_{i=1}^{m}$ of $\mathcal{U}$ with the properties mentioned in Theorem 2.3. We choose the operator F such that its matrix with respect to the basis $\{e_{ij}\}_{i=1,j=1}^{\kappa_j\ ,p} \cup \{f_i\}_{i=1}^{s}$ of $\mathcal{X}$ and the basis $\{g_i\}_{i=1}^{m}$ of $\mathcal{U}$, is given as

$$F = \left(F_{ij}\right)_{i=1,j=1}^{m,p+1},$$

where

(1) the $1 \times \kappa_j$ matrix F_{ij} is equal to the first row of $-A_{ij}$ for $i = 1, \ldots, p$ and $j = 1, \ldots, p$;

(2) the $1 \times s$ matrix $F_{i\ p+1} = -A_{i\ p+1}$ for $i = 1, \ldots, p$;

(3) the matrices $F_{ij} = 0$, for $i = p+1, \ldots, m$ and $j = 1, \ldots, p+1$.

So the i-th row of the matrix of F contains the row of the matrix of $-A$ that has the same row number as the i-th nonzero row of B. This i-th nonzero row of B contains only one nonzero entry which is equal to 1 and is positioned in column i. The $n \times n$ matrix $-BF$ consists of zeros except for the rows $1, \kappa_1 + 1, \dots, \kappa_1 + \cdots + \kappa_{q-1} + 1$, which contain the entries of the row with the same row number of the matrix of A. Therefore the matrix of $A + BF$ with respect to the basis $\{e_{ij}\}_{i=1,j=1}^{\kappa_j,q} \cup \{f_i\}_{i=1}^{s}$ is a direct sum of an $s \times s$ matrix and matrices J_{κ_i} given by (5.5) for $i = 1, \dots, q$. □

The matrix (5.4) together with the matrix representation of B is called the *Brunovsky canonical form* of the pair (A, B). Note that up to a reordering of the matrices J_{κ_i} in (4.3) and a similarity transformation on M the Brunovsky canonical form of (A, B) is uniquely determined by the feedback equivalence class of (A, B).

IX.6 Eigenvalue assignment and eigenvalue completion

Throughout this section we assume that $A : \mathcal{X} \to \mathcal{X}$ is a linear operator on the n-dimensional linear space $\mathcal{X}$ and $B : \mathcal{U} \to \mathcal{X}$ a linear operator from the m-dimensional linear space $\mathcal{U}$ to $\mathcal{X}$. Recall that if we want to stabilize (A, B) by state feedback, we have to find F such that $A + BF$ has all its eigenvalues in the open unit disk. A somewhat more general problem is to require that all the eigenvalues of $A + BF$ are in a certain region in the complex plane. The latter problem is called the *eigenvalue assignment problem.* Also the more restrictive versions of this problem which asks only to prescribe the characteristic polynomial or even the invariant polynomials of $A + BF$ can be considered. In this section we present the solutions to all these problems.

First we link the eigenvalue assignment problem to the eigenvalue completion problem of a full width block.

Theorem 6.1. *Let $A : \mathcal{X} \to \mathcal{X}$ is a linear operator on the n-dimensional linear space $\mathcal{X}$ and $B : \mathcal{U} \to \mathcal{X}$ a linear operator from the m-dimensional linear space $\mathcal{U}$ to $\mathcal{X}$. Let P be a projection of $\mathcal{X}$ with $\operatorname{Ker} P = \operatorname{Im} B$. An operator A_0 on $\mathcal{X}$ can be written as $A_0 = A + BF$ if and only if A_0 is a completion of the full width block $(PA; P, I)$.*

PROOF. If $A_0 = A + BF$, then $PA_0 = PA$, and thus A_0 is a completion of $(PA; P, I)$. Conversely, if $PA_0 = PA$, then $\operatorname{Im}(A_0 - A) \subset \operatorname{Ker} P = \operatorname{Im} B$, and thus $A_0 - A = BF$ for some $F : \mathcal{X} \to \mathcal{U}$. □

Theorem 6.2. *Let the pair (A, B) have controllability indices $\kappa_1 \geq \kappa_2 \geq \cdots \geq \kappa_k > 0 = \kappa_{k+1} = \cdots = \kappa_m$ and invariant polynomials $p_n | \cdots | p_1$. Let $q_1, \dots, q_n$ be polynomials such that $q_{i+1} | q_i$ for $i = 1, \dots, n-1$. Then there exists an operator F such that $A + BF$ has invariant polynomials $q_1, \dots, q_n$ if and only if*

(1) *$p_i | q_i$ for $i = 1, \dots, n$, and $q_{i+k} | p_i$ for $i = 1, \dots, n-k$;*

(2) $\sum_{i=1}^{j} \deg s_i \geq \sum_{i=1}^{j} \kappa_i$ *for* $j = 1, \ldots, k$, *with equality holding for* $j = k$. *The polynomials* s_j *are defined as* $s_j = t_j / t_{j+1}$ *for* $j = 1, \ldots, k$, *where*

$$t_j = \prod_{i=j}^{n} \text{l.c.m.}(p_{i-j+1}, q_i), \quad j = 1, \ldots, k+1.$$

PROOF. Let P be a projection of $\mathcal{X}$ with $\operatorname{Ker} P = \operatorname{Im} B$. Then it follows from Theorem 6.1 that there exists an operator F such that $A + BF$ has invariant polynomials $q_1, \ldots, q_n$ if and only if there is a completion of $(PA; P, I)$ with invariant polynomials $q_1, \ldots, q_n$. Theorem V.5.1 gives necessary and sufficient conditions for the existence of such a completion in terms of the indices of the first kind and the invariant polynomials of the block $(PA; P, I)$. It only remains to translate these conditions into the controllability indices and invariant polynomials of the pair (A, B). Now, note that Theorem 2.2 gives the key to translate these conditions from the the indices of the first kind into the controllability indices. Moreover, by Corollary 2.5, the invariant polynomials of the pair (A, B) are the invariant polynomials of the pencil $\begin{pmatrix} \lambda I_{\mathcal{X}} - A & -B \end{pmatrix}$. These in turn are the invariant polynomials of the block $(PA; P, I)$ as one easily derives from Corollary V.2.3 and Theorem V.1.5. □

The next corollary specializes Theorem 6.2 for the case when the pair (A, B) is controllable.

Corollary 6.3. *Let the pair* (A, B) *be controllable with controllability indices* $\kappa_1 \geq \kappa_2 \geq \cdots \geq \kappa_k > 0 = \kappa_{k+1} = \cdots = \kappa_m$. *Let* $q_1, \ldots, q_n$ *be polynomials such that* $q_{i+1} | q_i$ *for* $i = 1, \ldots, n-1$. *Then there exists an operator* F *such that* $A + BF$ *has invariant polynomials* $q_1, \ldots, q_n$ *if and only if*

$$\sum_{i=1}^{j} \deg q_i \geq \sum_{i=1}^{j} \kappa_i, \tag{6.1}$$

for $j = 1, \ldots, k$ *and with equality for* $j = k$.

PROOF. To prove this one has to remark that a controllable pair (A, B) has only constant invariant polynomials. So condition (1) in Theorem 6.2 is void. In condition (2) one gets that the $s_j = q_j$. So condition (2) translates into the condition (6.1). □

The weaker version of the problem, where one only requires that the eigenvalues are in a prescribed domain, is solved by the next result.

Corollary 6.4. *Let the pair* (A, B) *and the non empty set* $\Omega \subset \mathbb{C}$ *be given. Then there exists an operator* F *such that* $A + BF$ *has all its eigenvalues in* Ω *if and only if the first invariant polynomial of* (A, B) *has all its zeros in* Ω.

PROOF. Let λ_0 be a zero of the first invariant polynomial p_1 of the pair (A, B). Theorem 6.2 gives that $p_1 | q_1$, where q_1 is the minimal polynomial of $A + BF$

independent of our choice of F. So λ_0 is a zero of q_1, and therefore λ_0 is an eigenvalue of $A+BF$, whatever we choose for F. The proves that if there exists an operator F such that $A+BF$ has all its eigenvalues in Ω, then each zero of p_1 must be in Ω.

To prove the converse, we again apply Theorem 6.2. Let $r = \deg p_1 \cdots p_n$. Let $\kappa_1 \geq \cdots \geq \kappa_k > 0 = \kappa_{k+1} = \cdots = \kappa_m$ be the controllability indices of the pair (A, B). Then $n - r = \kappa_1 + \cdots + \kappa_k$. Choose the number $\alpha \in \Omega$. Put $q_1(\lambda) = (\lambda - \alpha)^{n-r} p_1(\lambda)$ and $q_i = p_i$, for $i = 2, \ldots, n$. Then the conditions (1) and (2) of Theorem 6.2 are fulfilled. For (1) this is obvious. Next, we compute the polynomials s_j in condition (2) of Theorem 6.2. The polynomials t_j in (2) are

$$t_1 = q_1 p_2 \cdots p_n, \quad t_j = p_1 p_2 \cdots p_{n+1-j} \quad j = 2, \ldots, k+1.$$

Hence $s_1 = (\lambda - \alpha)^{n-r} p_n$ and $s_j = p_{n+1-j}$ for $j = 2, \ldots, k$. Now $\sum_{k=1}^{n} p_j = r = n - (\kappa_1 + \cdots + \kappa_k) \leq n - k$. This proves that $p_{n+1-j} = 1$ for $j = 1, \ldots, k$. Thus we get $s_1 = (\lambda - \alpha)^{n-r}$ and $s_j = 1$ for $j = 2, \ldots, k$. So (2) is fulfilled. We conclude that there exists a matrix F such that the invariant polynomials of $A + BF$ are $q_1, \ldots, q_n$. Since q_1 has all its zeros in Ω and q_1 is the minimal polynomial of $A + BF$, we see that $A + BF$ has all its eigenvalues in Ω. □

By specializing Corollary 6.4 for the case when Ω is the open unit disk we finally solve the problem when a system (5.1) is stabilizable.

Corollary 6.5. *The system*

$$x_{k+1} = Ax_k + Bu_k, \quad x_0 = x, \quad k \geq 0,$$

is stabilizable by state feedback if and only if the invariant polynomials of the pair (A, B) have all their zeros inside the open unit disk.

IX.7 Assignment of controllability indices and eigenvalue restriction

In this section we treat the following control problem. Let A be a given $n \times n$ matrix, and the numbers m and k with $k \leq m$ be given. The question is to describe all possible sets of natural numbers that can appear as controllability indices of a pair (A, B), where B is an $n \times m$ matrix of rank k. The solution is given by the next theorem.

Theorem 7.1. *Let A be an $n \times n$ matrix with invariant polynomials $q_n | \cdots | q_1$, let $\kappa_1 \geq \cdots \geq \kappa_k > 0$ be natural numbers, and let $p_n | \cdots | p_1$ be polynomials. Then there exists a $n \times m$ matrix B of rank k such that $\kappa_1, \ldots, \kappa_k$ are the positive controllability indices of (A, B) and $p_1, \ldots, p_n$ are the invariant polynomials of (A, B) if and only if*

(1) *$p_i | q_i$ for $i = 1, \ldots, n$, and $q_{i+k} | p_i$ for $i = 1, \ldots, n - k$;*

(2) *$\sum_{i=1}^{j} \deg s_i \geq \sum_{i=1}^{j} \kappa_i$ for $j = 1, \ldots, k$, with equality holding for $j = k$.*

The polynomials s_j are defined as $s_j = t_j/t_{j+1}$, for $j = 1, \dots, k$, where

$$t_j = \prod_{i=j}^{n} \text{l.c.m.}(p_{i-j+1}, q_i), \quad j = 1, \dots, k+1.$$

PROOF. Recall that the invariant polynomials of the pair (A, B) are the invariant polynomials of the pencil $\begin{pmatrix} \lambda I - A & -B \end{pmatrix}$ (Corollary 2.5) and that the controllability indices of (A, B) are the minimal column indices of this pencil (Theorem 2.4). With this observation the theorem is just a translation of Theorem V.5.3. □

In the next corollary we specify Theorem 7.1 for the special case when all the polynomials $p_1, \dots, p_n$ are constant (and hence the pair (A, B) is controllable).

Corollary 7.2. *Let A be an $n \times n$ matrix with invariant polynomials $q_n | \cdots | q_1$ and let $\kappa_1 \geq \cdots \geq \kappa_k > 0$ be natural numbers. Then there exists a $n \times m$ matrix B of rank k $(m \geq k)$ such that $\kappa_1, \dots, \kappa_k$ are the positive controllability indices of the controllable pair (A, B) if and only if*

(1) $q_{k+1} = 1$,

(2) $\sum_{i=1}^{j} \deg q_i \geq \sum_{i=1}^{j} \kappa_i$ *for $j = 1, \dots, k$, with equality holding for $j = k$.*

PROOF. Note that (A, B) is controllable if and only if the invariant polynomials of (A, B) are all constant. In that case the polynomials s_j appearing in Theorem 7.1 are easily shown to be $s_j = q_j$. Hence condition (2) in Theorem 7.1 translates to condition (2) above. Moreover (1) in Theorem 7.1 simplifies to (1) above. □

We end this section with the remark that similar results about the assignment of observability indices can be formulated and proved. We omit the details.

IX.8 (A, B)-invariant subspaces

Consider the following system

$$x_{k+1} = Ax_k + Bu_k, \quad x_0 = x, \quad k \geq 0,$$

where $A : \mathcal{X} \to \mathcal{X}$ and $B : \mathcal{U} \to \mathcal{X}$ are linear operators acting between finite dimensional spaces. We consider subspaces $\mathcal{M}$ of $\mathcal{X}$ with the property that for each $x \in \mathcal{M}$ there exists an input u such that $x' = Ax + Bu \in \mathcal{M}$. In other words, for each initial state x_0 there is a sequence of inputs that keep the consecutive states in the space $\mathcal{M}$. Subspaces with this property appear in several important applications; they are called (A, B)–invariant. Thus, by definition, $\mathcal{M}$ is (A, B)-*invariant* if

$$A[\mathcal{M}] \subset \mathcal{M} + \operatorname{Im} B.$$

If there exists a feedback F such that $(A+BF)[\mathcal{M}] \subset \mathcal{M}$ one sees that for each $x_0 \in \mathcal{M}$ the vector $x_1 = Ax_0 + B(-Fx_0)$ is again in $\mathcal{M}$. This means that a sequence of inputs that keeps the sequence of states inside the subspace $\mathcal{M}$ can be obtained by state feedback. This looks to be a more restricted notion than (A, B)-invariance. The next theorem shows that this is not the case. This theorem also relates the notions of (A, B)-invariance and block invariance.

Theorem 8.1. *Let $\mathcal{X}$ and $\mathcal{U}$ be finite dimensional linear spaces and $A : \mathcal{X} \to \mathcal{X}$ and $B : \mathcal{U} \to \mathcal{X}$ be linear operators. Let P be a projection of the space $\mathcal{X}$ with $\operatorname{Ker} P = \operatorname{Im} B$. For the subspace $\mathcal{M}$ of $\mathcal{X}$ the following statements are equivalent:*

(a) *$\mathcal{M}$ is (A, B)-invariant;*

(b) *$\mathcal{M}$ is $(PA; P, I)$-invariant;*

(c) *there exists an operator $F : \mathcal{X} \to \mathcal{U}$ such that $\mathcal{M}$ is $(A+BF)$-invariant.*

PROOF. The equivalence of the properties (a) and (b) is a direct consequence of the definitions. Indeed, $\mathcal{M}$ is (A, B)-invariant if and only if $A[\mathcal{M}] \subset \mathcal{M} + \operatorname{Im} B$. This is the same as $A[\mathcal{M}] \subset \mathcal{M} + \operatorname{Ker} P$, and therefore also equivalent to $PA[\mathcal{M}] \subset \mathcal{M} + \operatorname{Ker} P$. The last formula gives that $\mathcal{M}$ is $(PA; P, I)$-invariant. One checks that property (c) implies property (a) by noting that for each $x \in \mathcal{M}$ it follows from $(A+BF)x \in \mathcal{M}$ that $Ax \in \mathcal{M} + \operatorname{Im} B$. Finally, we check that property (b) implies property (c). First, we apply Lemma II.1.2 to get a block $(P'A; P', I)$ such that $\operatorname{Ker} P' = \operatorname{Ker} P$ and $P'A[\mathcal{M}] \subset \mathcal{M}$. Since $\operatorname{Im}(I - P')A \subset \operatorname{Ker} P' = \operatorname{Im} B$, there exists an operator $F : \mathcal{X} \to \mathcal{U}$ such that $(I - P')A = BF$. So we get that $P'A = A + BF$, and since $P'A[\mathcal{M}] \subset \mathcal{M}$, it follows that $(A+BF)[\mathcal{M}] \subset \mathcal{M}$. $\square$

IX.9 Output stabilization by state feedback

Consider the system

$$\Sigma \begin{cases} x_{k+1} = Ax_k + Bu_k, & \text{for } k \geq 0, \\ y_k = Cx_k, & \text{for } k \geq 0, \\ x = x_0, \end{cases} \tag{9.1}$$

where $A : \mathcal{X} \to \mathcal{X}$, $B : \mathcal{U} \to \mathcal{X}$ and $C : \mathcal{X} \to \mathcal{Y}$ are linear transformations acting between finite dimensional spaces. We say that Σ is *output stable* if $\lim_{k\to\infty} y_k = 0$ for any initial state x_0 and with inputs $u_k = 0$ for each k. In this case, $y_k = CA^k x_0$, $k \geq 0$, and hence Σ is *output stable* if and only if

$$\lim_{k\to\infty} CA^k = 0. \tag{9.2}$$

In this section we deal with the following problem. Given the system Σ in (9.1), can we find a state feedback $F : \mathcal{X} \to \mathcal{U}$ such that the resulting system

$$\begin{cases} x_{k+1} = (A+BF)x_k + Bu_k, & \text{for } k \geq 0, \\ y_k = Cx_k, & \text{for } k \geq 0, \\ x_0 = x \end{cases}$$

is output stable. In other words, the problem is to find an operator $F : \mathcal{X} \to \mathcal{U}$ such that (9.2) holds for $A + BF$ in place of A. In this case we say that Σ is made output stable by state feedback.

To state the solution of this problem, we need the notion of generalized eigenspace. For $A : \mathcal{X} \to \mathcal{X}$ as above, the *generalized eigenspace* $\mathcal{N}_\lambda(A)$ of A at the point $\lambda \in \mathbf{C}$ is the space

$$\mathcal{N}_\lambda(A) = \operatorname{Ker}(A - \lambda I)^n,$$

where $n = \dim \mathcal{X}$. Also, recall that

$$\operatorname{Ker}(C|A) = \bigcap_{i=0}^{n-1} \operatorname{Ker} CA^i, \quad \operatorname{Im}(A|B) = \operatorname{Im}\begin{pmatrix} B & AB & \cdots & A^{n-1}B \end{pmatrix}.$$

The next theorem is the main result of this section.

Theorem 9.1. *The system* (9.1) *can be made output stable by a state feedback* $F : \mathcal{X} \to \mathcal{U}$ *if and only if*

$$\mathcal{N}_\lambda(A) \subset \operatorname{Im}(A|B) + \mathcal{M}, \quad |\lambda| \geq 1, \tag{9.3}$$

where $\mathcal{M}$ *is the maximal* (A, B)*-invariant subspace contained in* $\operatorname{Ker} C$. *In other words, there exists a* $\mathcal{F} : \mathcal{X} \to \mathcal{U}$ *such that* $\lim_{k\to\infty} C(A + BF)^k = 0$ *if and only if* (9.3) *holds.*

To prove Theorem 9.1 we have to make some preparations. The first proposition tells us when the system is output stable.

Proposition 9.2. *Let* $C : \mathcal{X} \to \mathcal{Y}$ *and* $A : \mathcal{X} \to \mathcal{X}$ *be operators acting between the* n*-dimensional space* $\mathcal{X}$ *and the* r*-dimensional space* $\mathcal{Y}$. *Then* $\lim_{k\to\infty} CA^k = 0$ *if and only if for each* $|\lambda| \geq 1$ *the generalized eigenspace* $\mathcal{N}_\lambda(A)$ *is contained in* $\operatorname{Ker}(C|A)$, *i.e.,*

$$\mathcal{N}_\lambda(A) \subset \operatorname{Ker}(C|A), \quad |\lambda| \geq 1. \tag{9.4}$$

PROOF. We apply Proposition 4.5 with $B = I$ to get a decomposition $\mathcal{X} = \operatorname{Ker}(C|A) \oplus \mathcal{X}_2$ such that

$$C = \begin{pmatrix} 0 & C_2 \end{pmatrix}, \quad A = \begin{pmatrix} A_{11} & A_{12} \\ 0 & A_{22} \end{pmatrix}.$$

The pair (C_2, A_{22}) is observable. Assume that λ is an eigenvalue of A with $|\lambda| \geq 1$ and that $\lim_{k\to\infty} CA^k = 0$. Let x be an eigenvector of A with $Ax = \lambda x$. Then $\lim_{k\to\infty} CA^{k+\ell}x = 0$, and thus $\lim_{k\to\infty} \lambda^k CA^\ell x = 0$. Since $|\lambda| \geq 1$, this proves that $CA^\ell x = 0$ for each value of ℓ. So $x \in \operatorname{Ker}(C|A)$, and λ is therefore not an eigenvalue of the operator A_{22}. This means that $A_{22} - \lambda I_2$ is invertible. If now $(A - \lambda I)^n x = 0$ for any $x = x_1 + x_2 \in \operatorname{Ker}(C|A) \oplus \mathcal{X}_2$, then it is clear that $(A_{22} - \lambda I_2)^n x_2 = 0$ and thus $x_2 = 0$, which proves that each generalized eigenvector $x \in \mathcal{N}_\lambda(A)$ is in $\operatorname{Ker}(C|A)$. We proved $\mathcal{N}_\lambda(A) \subset \operatorname{Ker}(C|A)$.

Conversely, assume that (9.4) holds. So, if λ is an eigenvalue of A_{22}, then $|\lambda| < 1$. This implies that $\lim_{k\to\infty} A_{22}^k x_2 = 0$ for each $x_2 \in \mathcal{X}_2$. Since $CA^k x = C_2 A_{22}^k x_2$ if $x = x_1 + x_2 \in \operatorname{Ker}(C|A) \oplus \mathcal{X}_2$, it follows that $\lim_{k\to\infty} CA^k x = \lim_{k\to\infty} C_2 A_{22}^k x_2 = 0$ for each x. So $\lim_{k\to\infty} CA^k = 0$. □

Proposition 9.2 shows that our output stabilization problem is to find an operator F such that for each λ with $|\lambda| \geq 1$ the generalized eigenspace $\mathcal{N}_\lambda(A+BF)$ is contained in the space $\operatorname{Ker}(C|A+BF)$. Since $\mathcal{N}_\lambda(A+BF)$ is $(A+BF)$-invariant and $\operatorname{Ker}(C|A+BF)$ is the largest $(A+BF)$-invariant subspace in $\operatorname{Ker} C$, it is sufficient to find F such that $\mathcal{N}_\lambda(A+BF) \subset \operatorname{Ker} C$ for every λ with $|\lambda| \geq 1$. To find such an F we treat a more general problem. Given $A : \mathcal{X} \to \mathcal{X}$, $B : \mathcal{U} \to \mathcal{X}$, a subspace $\mathcal{E}$ of $\mathcal{X}$, and a subset Ω in $\mathbb{C}$, find $F : \mathcal{X} \to \mathcal{U}$ such that $\mathcal{N}_\lambda(A+BF) \subset \mathcal{E}$ for each λ in Ω. Here the subset Ω plays the role of the exterior of the open unit disk, and $\mathcal{E}$ plays the role of the subspace $\operatorname{Ker} C$.

Before we proceed we state two lemmas about generalized eigenspaces, which will come handy in the sequel.

Lemma 9.3. *Let the n-dimensional space $\mathcal{X}$ have a direct sum decomposition $\mathcal{X} = \mathcal{X}_1 \oplus \mathcal{X}_2$. Assume that the subspace $\mathcal{X}_1$ is invariant for the operator*

$$A = \begin{pmatrix} A_{11} & A_{12} \\ 0 & A_{22} \end{pmatrix} : \mathcal{X}_1 \oplus \mathcal{X}_2 \to \mathcal{X}_1 \oplus \mathcal{X}_2.$$

Then, for $\lambda \in \mathbb{C}$, the generalized eigenspace $\mathcal{N}_\lambda(A_{22})$ consists of the second components of the vectors in the generalized eigenspace $\mathcal{N}_\lambda(A)$, i.e.,

$$\mathcal{N}_\lambda(A_{22}) = \{x_2 \mid \begin{pmatrix} x_1 \\ x_2 \end{pmatrix} \in \mathcal{N}_\lambda(A) \text{ for some } x_1\}.$$

PROOF. Choose $p(\mu) = (\mu - \lambda)^n$. Then the generalized eigenspace $\mathcal{N}_\lambda(A)$ is defined as $\mathcal{N}_\lambda(A) = \operatorname{Ker} p(A)$. For a vector $x \in \mathcal{X}$ the second component of $p(A)x$, which we denote as $(p(A)x)_2$, is equal to $p(A_{22})x_2$. So if $x \in \mathcal{N}_\lambda(A)$, then $p(A_{22})x_2 = 0$. This proves that

$$\{x_2 \mid \begin{pmatrix} x_1 \\ x_2 \end{pmatrix} \in \mathcal{N}_\lambda(A) \text{ for some } x_1\} \subset \mathcal{N}_\lambda(A_{22}).$$

To prove the converse inclusion we take $x_2 \in \mathcal{N}_\lambda(A_{22})$ and construct a vector u such that $u_2 = x_2$ and $u \in \mathcal{N}_\lambda(A)$. Choose any x such that its second component is x_2. Put $q(\mu) = \prod_{i=0}^r (\mu - \lambda_i)^n$, where $\lambda_1, \ldots, \lambda_r$ are the eigenvalues of A different from the eigenvalue λ. Since the polynomials $p(\mu)$ and $q(\mu)$ do not have a common zero, we can find polynomials $g(\mu)$ and $h(\mu)$ such that $g(\mu)p(\mu) + h(\mu)q(\mu) = 1$. Remark that $p(A)x \in \mathcal{X}_1$ since $(p(A)x)_2 = p(A_{22})x_2 = 0$. So we get that $g(A)p(A)x \in \mathcal{X}_1$. Since $p(\mu)q(\mu)$ is a multiple of the characteristic polynomial of A, we can apply the Cayley-Hamilton Theorem to deduce that $p(A)h(A)q(A)x = 0$. Choose $u = h(A)q(A)x$. Then $p(A)u = 0$ and thus $u \in \mathcal{N}_\lambda(A)$. Moreover $x = g(A)p(A)x + u$, and since $g(A)p(A)x \in \mathcal{X}_1$, it follows that $u - x \in \mathcal{X}_1$. This proves that $u_2 = x_2$. □

Lemma 9.4. *Let $B : \mathcal{U} \to \mathcal{X}$ and $A : \mathcal{X} \to \mathcal{X}$ be linear operators acting between the finite dimensional spaces $\mathcal{X}$ and $\mathcal{U}$, and let $\Omega \subset \mathbb{C}$. Then for each $\lambda \in \Omega$ the subspace $\mathcal{N}_\lambda(A)$ is contained in* $\operatorname{Im}(A|B)$ *if and only if there exists a linear transformation $F : \mathcal{X} \to \mathcal{U}$ such that $A + BF$ has no eigenvalues in Ω.*

PROOF. Recall that $\operatorname{Im}(A|B)$ is the smallest A-invariant subspace of $\mathcal{X}$ that contains $\operatorname{Im} B$, and that $\operatorname{Im}(A|B) = \operatorname{Im}\begin{pmatrix} B & AB & \dots & A^{n-1}B \end{pmatrix}$. Decompose $\mathcal{X}$ as $\mathcal{X} = \operatorname{Im}(A|B) \oplus \mathcal{X}_2$. With respect to this decomposition the matrices of A and B take the form

$$A = \begin{pmatrix} A_{11} & A_{12} \\ 0 & A_{22} \end{pmatrix}, \quad B = \begin{pmatrix} B_1 \\ 0 \end{pmatrix}.$$

According to Corollary 2.5 the invariant polynomials of the pair (A, B) are the invariant polynomials of A_{22}. Now $\mathcal{N}_\lambda(A) \subset \operatorname{Im}(A|B)$ if and only if λ is not an eigenvalue of A_{22}. So $\mathcal{N}_\lambda(A) \subset \operatorname{Im}(A|B)$ for each $\lambda \in \Omega$ if and only if the invariant polynomials of the pair (A, B) have no zero in Ω. By Corollary 6.4 this is equivalent to the existence of an operator $F : \mathcal{X} \to \mathcal{U}$ such that $A + BF$ has no eigenvalues in Ω. □

Theorem 9.5. *Let $A : \mathcal{X} \to \mathcal{X}$ and $B : \mathcal{U} \to \mathcal{X}$ be linear operators acting between finite dimensional spaces, let $\mathcal{E}$ be a subspace of $\mathcal{X}$, and let Ω be a subset of $\mathbb{C}$. Then there exists a linear transformation $F : \mathcal{X} \to \mathcal{U}$ such that $\mathcal{N}_\lambda(A + BF) \subset \mathcal{E}$ for every $\lambda \in \Omega$ if and only if*

$$\mathcal{N}_\lambda(A) \subset \operatorname{Im}(A|B) + \mathcal{M}, \quad \lambda \in \Omega, \tag{9.5}$$

where $\mathcal{M}$ is the maximal (A, B)-invariant subspace that is contained in $\mathcal{E}$.

PROOF. Assume that $\mathcal{N}_\lambda(A + BF) \subset \mathcal{E}$. Since $\mathcal{N}_\lambda(A + BF)$ is $(A + BF)$-invariant, the space $\mathcal{N}_\lambda(A + BF)$ is also (A, B)-invariant. Indeed,

$$A\mathcal{N}_\lambda(A+BF) \subset (A+BF)\mathcal{N}_\lambda(A+BF) + BF\mathcal{N}_\lambda(A+BF) \subset \mathcal{N}_\lambda(A+BF) + \operatorname{Im} B.$$

This proves that $\mathcal{N}_\lambda(A + BF) \subset \mathcal{M}$. Now write the matrix representations of A, B and F with respect to the decomposition $\mathcal{X} = \operatorname{Im}(A|B) \oplus \mathcal{X}_2$. One gets

$$A = \begin{pmatrix} A_{11} & A_{12} \\ 0 & A_{22} \end{pmatrix}, \ B = \begin{pmatrix} B_1 \\ 0 \end{pmatrix}, F = \begin{pmatrix} F_1 & F_2 \end{pmatrix}.$$

Therefore

$$A + BF = \begin{pmatrix} A_{11} + B_1 F_1 & A_{12} + B_1 F_2 \\ 0 & A_{22} \end{pmatrix}.$$

Apply Lemma 9.3 to see that

$$\{x_2 \mid \begin{pmatrix} x_1 \\ x_2 \end{pmatrix} \in \mathcal{N}_\lambda(A) \text{ for some } x_1\} = \mathcal{N}_\lambda(A_{22}) =$$

$$\{x_2 \mid \begin{pmatrix} x_1 \\ x_2 \end{pmatrix} \in \mathcal{N}_\lambda(A + BF) \text{ for some } x_1\}.$$

Since $\operatorname{Im}(A_{11}|B_1) = \operatorname{Im}(A|B)$, this means that $\mathcal{N}_\lambda(A) \subset \operatorname{Im}(A|B) + \mathcal{M}$ if and only if $\mathcal{N}_\lambda(A+BF) \subset \operatorname{Im}(A|B) + \mathcal{M}$. Use that $\mathcal{N}_\lambda(A+BF) \subset \mathcal{M}$ to obtain that $\mathcal{N}_\lambda(A) \subset \operatorname{Im}(A|B) + \mathcal{M}$.

Conversely, assume that (9.5) holds. We will construct an operator F such that $\mathcal{N}_\lambda(A+BF) \subset \mathcal{E}$ for every $\lambda \in \Omega$. Since $\mathcal{M}$ is (A,B)-invariant, there exists an $F_0 : \mathcal{X} \to \mathcal{U}$ such that $(A+BF_0)[\mathcal{M}] \subset \mathcal{M}$. The subspace $\mathcal{R}_1 = \operatorname{Im}(A|B) + \mathcal{M}$ is $(A+BF_0)$-invariant since both $\operatorname{Im}(A|B)$ and $\mathcal{M}$ are. It is also A-invariant since $\operatorname{Im}(A|B)$ is A-invariant and $A[\mathcal{M}] \subset \mathcal{M} + \operatorname{Im} B \subset \mathcal{M} + \operatorname{Im}(A|B)$. Now, let $\mathcal{R}_2$ be such that $\mathcal{X} = \mathcal{R}_1 \oplus \mathcal{R}_2$. With respect to this new decomposition we write

$$B = \begin{pmatrix} B_1 \\ 0 \end{pmatrix}, \quad F_0 = \begin{pmatrix} F_{01} & F_{02} \end{pmatrix}, \quad A = \begin{pmatrix} A_{11} & A_{12} \\ 0 & A_{22} \end{pmatrix},$$

$$A + BF_0 = \begin{pmatrix} A_{11} + B_1 F_{01} & A_{12} + B_1 F_{02} \\ 0 & A_{22} \end{pmatrix}.$$

We apply Lemma 9.3 to find that

$$\{x_2 \mid \begin{pmatrix} x_1 \\ x_2 \end{pmatrix} \in \mathcal{N}_\lambda(A) \text{ for some } x_1\} = \{x_2 \mid \begin{pmatrix} x_1 \\ x_2 \end{pmatrix} \in \mathcal{N}_\lambda(A + BF_0) \text{ for some } x_1\}.$$

Since $\mathcal{N}_\lambda(A) \subset \mathcal{R}_1$, one sees that $\mathcal{N}_\lambda(A+BF_0) \subset \mathcal{R}_1$.

The subspace $\mathcal{M}$ is the largest $(A+BF_0)$-invariant subspace of $\mathcal{E}$, because $\mathcal{M}$ is both $(A+BF_0)$-invariant and the largest (A,B)-invariant subspace of $\mathcal{E}$. Write $\overline{A} = A + BF_0$. Choose $\mathcal{M}_2$ such that $\mathcal{X} = \mathcal{M} \oplus \mathcal{M}_2$. For the rest of the proof all block matrix representations will be with respect to the latter decomposition. The block matrix representations of $\overline{A}$ and B are

$$\overline{A} = \begin{pmatrix} \overline{A}_{11} & \overline{A}_{12} \\ 0 & \overline{A}_{22} \end{pmatrix}, \quad B = \begin{pmatrix} \overline{B}_1 \\ \overline{B}_2 \end{pmatrix}.$$

We intend to construct an $F_1 = \begin{pmatrix} 0 & F_{12} \end{pmatrix} : \mathcal{X} \to \mathcal{U}$ such that $\mathcal{N}_\lambda(\overline{A}_{22} + \overline{B}_2 F_{12}) \cap \Omega = \{0\}$. Therefore we consider the subspace $\mathcal{N}_\lambda(\overline{A}_{22})$, for $\lambda \in \Omega$, and we first prove that

$$\mathcal{N}_\lambda(\overline{A}_{22}) \subset \operatorname{Im}\begin{pmatrix} \overline{B}_2 & \overline{A}_{22}\overline{B}_2 & \dots & \overline{A}_{22}^{\,n-1}\overline{B}_2 \end{pmatrix}.$$

We apply Lemma 9.3 to see that

$$\mathcal{N}_\lambda(\overline{A}_{22}) = \{x_2 \mid \begin{pmatrix} x_1 \\ x_2 \end{pmatrix} \in \mathcal{N}_\lambda(\overline{A}) \text{ for some } x_1\}.$$

Since $\mathcal{N}_\lambda(\overline{A}) \subset \operatorname{Im}(A|B) + \mathcal{M}$, we deduce that

$$\mathcal{N}_\lambda(\overline{A}_{22}) \subset \{x_2 \mid \begin{pmatrix} x_1 \\ x_2 \end{pmatrix} \in \operatorname{Im}(A|B) + \mathcal{M} \text{ for some } x_1\}.$$

Next, the fact that the first coordinate space is $\mathcal{M}$, gives that

$$\{x_2 \mid \begin{pmatrix} x_1 \\ x_2 \end{pmatrix} \in \operatorname{Im}(A|B) + \mathcal{M} \text{ for some } x_1 \in \mathcal{M}\} =$$
$$\{x_2 \mid \begin{pmatrix} x_1 \\ x_2 \end{pmatrix} \in \operatorname{Im}(A|B) \text{ for some } x_1 \in \mathcal{M}\}.$$

Recall that $\operatorname{Im}(A|B) = \operatorname{Im}\begin{pmatrix} B & AB & \dots & A^{n-1}B \end{pmatrix}$. So we see that

$$\{x_2 \mid \begin{pmatrix} x_1 \\ x_2 \end{pmatrix} \in \operatorname{Im}(A|B) \text{ for some } x_1\} = \operatorname{Im}\begin{pmatrix} \overline{B}_2 & \overline{A}_{22}\overline{B}_2 & \dots & \overline{A}_{22}^{n-1}\overline{B}_2 \end{pmatrix}.$$

This proves that

$$\mathcal{N}_\lambda(\overline{A}_{22}) \subset \operatorname{Im}\begin{pmatrix} \overline{B}_2 & \overline{A}_{22}\overline{B}_2 & \dots & \overline{A}_{22}^{n-1}\overline{B}_2 \end{pmatrix}.$$

We apply Lemma 9.4 to obtain an $F_{12} : \mathcal{M}_2 \to \mathcal{U}$ such that $\overline{A}_{22} + \overline{B}_2 F_{12}$ does not have any eigenvalue in Ω. Choose $F_1 = \begin{pmatrix} 0 & F_{12} \end{pmatrix}$. Then

$$\overline{A} + BF_1 = \begin{pmatrix} \overline{A}_{11} & \overline{A}_{12} + \overline{B}_1 F_{12} \\ 0 & \overline{A}_{22} + \overline{B}_2 F_{12} \end{pmatrix}.$$

Since $\mathcal{N}_\lambda(\overline{A}_{22} + \overline{B}_2 F_{12}) = \{0\}$, for any $\lambda \in \Omega$, Lemma 9.3 gives that $\mathcal{N}_\lambda(\overline{A} + BF_1) \subset \mathcal{M}$ for each $\lambda \in \Omega$. We choose $F = F_0 + F_1$ and we see that $\mathcal{N}_\lambda(A + BF) \subset \mathcal{M} \subset \mathcal{E}$ for each $\lambda \in \Omega$. □

PROOF OF THEOREM 9.1 Combine Proposition 9.2 and Theorem 9.4 (with Ω equal to the exterior of the open unit disc) in the way described in the paragraph following the proof of Proposition 9.2. □

IX.10 Output injection

We start this section with a brief discussion of a notion that is dual to the notion of state feedback. Consider the system

$$\begin{cases} x_{k+1} = Ax_k, & \text{for } k \geq 0, \\ y_k = Cx_k, & \text{for } k \geq 0, \\ x_0 = x, \end{cases}$$

where $A : \mathcal{X} \to \mathcal{X}$ and $C : \mathcal{X} \to \mathcal{Y}$ are linear transformations acting between finite dimensional spaces. Choose a linear transformation $G : \mathcal{Y} \to \mathcal{X}$. Such a transformation is called an *output injection*. The next picture illustrates this terminology.

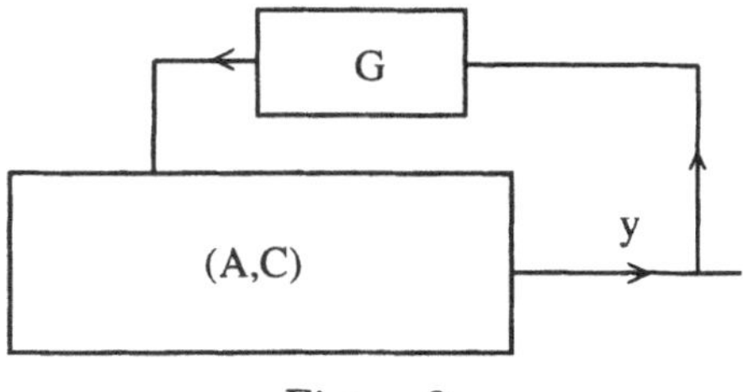

Figure 3

The resulting system is

$$\begin{cases} x_{k+1} = (A + GC)x_k, & \text{for } k \geq 0, \\ y_k = Cx_k, & \text{for } k \geq 0, \\ x_0 = x. \end{cases}$$

We define the pair (C', A'), with $A' : \mathcal{X}' \to \mathcal{X}'$ and $C' : \mathcal{X}' \to \mathcal{Y}'$, to be *output injection equivalent* to the pair (C, A) if there exist a linear transformation $G : \mathcal{Y} \to \mathcal{X}$ and invertible linear transformations $S : \mathcal{X}' \to \mathcal{X}$ and $T : \mathcal{Y}' \to \mathcal{Y}$ such that

$$C' = T^{-1}CS, \ A' = S^{-1}(A + GC)S. \tag{10.1}$$

The first result ties up this notion with block similarity.

Theorem 10.1. *Let $A : \mathcal{X} \to \mathcal{X}$, $A' : \mathcal{X}' \to \mathcal{X}'$, $C : \mathcal{X} \to \mathcal{Y}$ and $C' : \mathcal{X}' \to \mathcal{Y}'$ be linear operators acting between finite dimensional spaces. The following statements are equivalent:*

(1) *the pairs (C, A) and (C', A') are output injection equivalent;*

(2) *the blocks $(Z; I, Q)$ and $(Z'; I', Q')$ with*

$$Q = \begin{pmatrix} I & 0 \\ 0 & 0 \end{pmatrix} : \mathcal{X} \times \mathcal{Y} \to \mathcal{X} \times \mathcal{Y} \quad Q' = \begin{pmatrix} I & 0 \\ 0 & 0 \end{pmatrix} : \mathcal{X}' \times \mathcal{Y}' \to \mathcal{X}' \times \mathcal{Y}'$$

$$Z = \begin{pmatrix} A \\ C \end{pmatrix} : \operatorname{Im} Q \to \mathcal{X} \oplus \mathcal{Y} \quad , \qquad Z' = \begin{pmatrix} A' \\ C' \end{pmatrix} : \operatorname{Im} Q' \to \mathcal{X}' \oplus \mathcal{Y}' \quad ,$$

are block similar;

(3) *the blocks* $(A|_{\operatorname{Im} Q}; I_{\mathcal{X}}, Q)$ *on* $\mathcal{X}$ *and* $(A'|_{\operatorname{Im} Q'}; I_{\mathcal{X}'}, Q')$ *on* $\mathcal{X}'$ *with* Q (Q') *a projection of* $\mathcal{X}$ $(\mathcal{X}')$ *with* $\operatorname{Im} Q = \operatorname{Ker} C$ $(\operatorname{Im} Q' = \operatorname{Ker} C')$ *are block similar;*

(4) *the pairs* (C, A) *and* (C', A') *have the same observability indices and the same invariant polynomials;*

(5) *The pencils* $\lambda \begin{pmatrix} I_{\mathcal{X}} \\ 0 \end{pmatrix} - \begin{pmatrix} A \\ C \end{pmatrix}$ *and* $\lambda \begin{pmatrix} I_{\mathcal{X}'} \\ 0 \end{pmatrix} - \begin{pmatrix} A' \\ C' \end{pmatrix}$ *are strictly equivalent.*

PROOF. We prove that the statements (1) and (2) are equivalent. The pairs (C, A) and (C', A') are output injection equivalent if and only if there exist invertible linear transformations $S : \mathcal{X}' \to \mathcal{X}$ and $T : \mathcal{Y}' \to \mathcal{Y}$ and an operator $G : \mathcal{Y} \to \mathcal{X}$ such that (10.1) holds. These equalities are equivalent to the matrix equality

$$\begin{pmatrix} S & -GT \\ 0 & T \end{pmatrix} \begin{pmatrix} A' \\ C' \end{pmatrix} = \begin{pmatrix} A \\ C \end{pmatrix} S.$$

This means that the blocks $(Z; I, Q)$ and $(Z'; I', Q')$ are block similar.

To equivalence of the statements (2) and (3) follows from Corollary III.1.6. To see that (2) and (4) are equivalent we apply Theorem 3.2 to get that the controllability indices of the pair (C, A) $((C', A'))$ are the indices of the third kind of the block $(Z; I, Q)$ $\big((Z'; I', Q')\big)$. Furthermore, the matrix representation in Theorem 3.4 shows that the invariant polynomials of the pair (C, A) $\big((C', A')\big)$ and of the block $(Z; I, Q)$ $\big((Z'; I', Q')\big)$ coincide. Now Theorem III.1.4 states that the blocks are block similar if and only if the the indices of the third kind and the invariant polynomials are the same.

Recall that Corollary III.2.3 states that the minimal row indices and the invariant polynomials of a matrix pencil are the indices of the third kind and the invariant polynomials of the block that corresponds to the matrix pencil in the way described in (2) and (5). Theorem III.2.3 states that two pencils are strictly equivalent if and only if the minimal row indices and the invariant polynomials are the same for the two pencils. Theorem III.1.4 gives that two blocks are block similar if and only if they have the same indices of the third kind and the same invariant polynomials. So we see that (2) and (5) are equivalent. □

The next result shows that each equivalence class of output injection equivalent pairs contains a representative with a simple matrix representation.

Theorem 10.2. *Let* $\mathcal{X}$ *be a* n *dimensional and* $\mathcal{Y}$ *be a* r *dimensional linear space and* $A : \mathcal{X} \to \mathcal{X}$ *and* $C : \mathcal{X} \to \mathcal{Y}$ *be linear operators. Let* $\mu_1 \geq \cdots \geq \mu_p > 0 = \mu_{p+1} = \cdots = \mu_r$ *be the observability indices of the pair* (C, A). *Then there exist a basis* $\{f_{ij}\}_{i=1,j=1}^{\mu_j\ ,p} \cup \{h_i\}_{i=1}^{s}$ *of* $\mathcal{X}$, *a basis* $\{g_i\}_{i=1}^{m}$ *of* $\mathcal{Y}$, *and an operator* $G : \mathcal{Y} \to \mathcal{X}$ *such that the matrix of* $A + GC$ *with respect to this basis of* $\mathcal{X}$ *is*

$$J_{\mu_1} \oplus \cdots \oplus J_{\mu_p} \oplus M,$$

where J_{μ_i} is the $\mu_i \times \mu_i$ matrix given by (5.4) and M is an $s \times s$ matrix. Furthermore, with respect to these bases the matrix of C has zero entries except for those in the positions $(1, \mu_1)$, $(2, \mu_1 + \mu_2)$, ..., $(p, \mu_1 + \cdots + \mu_p)$ and these exceptional entries are 1.

PROOF. This theorem can be proved by applying Theorem 5.2 to the pair (A^T, C^T). We get an operator F and a matrix representation of $A^T + C^T F$ in the form $J_{\mu_1} \oplus \cdots \oplus J_{\mu_p} \oplus M$ and a simple matrix for C^T. So with $G = F^T$ we get a matrix $J^T_{\mu_1} \oplus \cdots \oplus J^T_{\mu_p} \oplus M$ for $A + GC$. A simple reordering of the bases gives the desired matrix representations of $A + GC$ and C. □

We may combine the notions of feedback equivalence and output injection. The triples (A, B, C) and (A', B', C') are said to be *equivalent under simultaneous state feedback and output injection* if there exist invertible operators $S : \mathcal{X}' \to \mathcal{X}$, $T : \mathcal{U}' \to \mathcal{U}$ and $V : \mathcal{Y}' \to \mathcal{Y}$ and operators $F : \mathcal{X} \to \mathcal{U}$ and $G : \mathcal{Y} \to \mathcal{X}$ such that

$$A' = S^{-1}(A + BF + GC)S, \quad C' = V^{-1}CS, \quad B' = S^{-1}BT. \tag{10.2}$$

It is easy to check that this is an equivalence relation. This fact is however also an easy consequence of the next result.

Theorem 10.3. *Let $\mathcal{X}$, $\mathcal{U}$, $\mathcal{Y}$, $\mathcal{X}'$, $\mathcal{U}'$ and $\mathcal{Y}'$ be finite dimensional linear spaces, and let $A : \mathcal{X} \to \mathcal{X}$, $A' : \mathcal{X}' \to \mathcal{X}'$, $B : \mathcal{U} \to \mathcal{X}$, $B' : \mathcal{U}' \to \mathcal{X}'$, $C : \mathcal{X} \to \mathcal{Y}$, and $C' : \mathcal{X}' \to \mathcal{Y}'$ be linear operators. The triples (A, B, C) and (A', B', C') are equivalent under simultaneous state feedback and output injection if and only if the pencils*

$$\lambda \begin{pmatrix} 0 & I_{\mathcal{X}} \\ 0 & 0 \end{pmatrix} - \begin{pmatrix} B & A \\ 0 & C \end{pmatrix} : \mathcal{U} \oplus \mathcal{X} \to \mathcal{X} \oplus \mathcal{Y}$$

and

$$\lambda \begin{pmatrix} 0 & I_{\mathcal{X}'} \\ 0 & 0 \end{pmatrix} - \begin{pmatrix} B' & A' \\ 0 & C' \end{pmatrix} : \mathcal{U}' \oplus \mathcal{X}' \to \mathcal{X}' \oplus \mathcal{Y}'$$

are strictly equivalent.

PROOF. Assume that (10.2) holds. Then

$$\begin{pmatrix} S & -GS \\ 0 & S \end{pmatrix} \left(\lambda \begin{pmatrix} 0 & I_{\mathcal{X}'} \\ 0 & 0 \end{pmatrix} - \begin{pmatrix} B' & A' \\ 0 & C' \end{pmatrix} \right) =$$

$$\left(\lambda \begin{pmatrix} 0 & I_{\mathcal{X}} \\ 0 & 0 \end{pmatrix} - \begin{pmatrix} B & A \\ 0 & C \end{pmatrix} \right) \begin{pmatrix} T & FS \\ 0 & S \end{pmatrix}.$$

This proves the strict equivalence of the pencils.

Conversely, assume that we have invertible transformations $V : \mathcal{U}' \oplus \mathcal{X}' \to \mathcal{U} \oplus \mathcal{X}$ and $W : \mathcal{X}' \oplus \mathcal{Y}' \to \mathcal{X} \oplus \mathcal{Y}$ such that

$$\begin{pmatrix} W_{11} & W_{12} \\ W_{21} & W_{22} \end{pmatrix} \left(\lambda \begin{pmatrix} 0 & I_{\mathcal{X}'} \\ 0 & 0 \end{pmatrix} - \begin{pmatrix} B' & A' \\ 0 & C' \end{pmatrix} \right) =$$

$$\left(\lambda \begin{pmatrix} 0 & I_{\mathcal{X}} \\ 0 & 0 \end{pmatrix} - \begin{pmatrix} B & A \\ 0 & C \end{pmatrix} \right) \begin{pmatrix} V_{11} & V_{12} \\ V_{21} & V_{22} \end{pmatrix}.$$

Comparing the first degree terms in λ we see that $W_{11} = V_{22}$, $W_{21} = 0$, and $V_{21} = 0$. Since both W and V are invertible, it follows that W_{11}, V_{11} and W_{22} are invertible. Put $S = W_{11}$, $T_{\mathcal{U}} = V_{11}$ and $T_{\mathcal{Y}} = W_{22}$. Furthermore, set $F = V_{12}S^{-1}$ and $G = -W_{12}T_{\mathcal{Y}}^{-1}$. One checks that (10.2) holds true. This proves that the triples (A, B, C) and (A', B', C') are equivalent under simultaneous state feedback and output injection. □

The problem of stabilization by output injection will not be considered here; its solution is dual to the one for stabilization by state feedback.

From the previous theorem and Theorem VII.5.1 it follows that there exists a canonical form for the discussed equivalence relation. This canonical form can be derived easily from the results of Section VII.5.

Notes

The results on linear input/output systems presented here are well-known and can be found in books (see, e.g., Kalman-Falb-Arbib [1], Kailath [1], Gohberg-Lancaster-Rodman [2]). The connections with operator blocks and the block similarity theory is taken from Gohberg-Kaashoek-Van Schagen [1]. The Brunovsky canonical form for a pair (A, B) in Section 5 is due to Brunovsky [1]. A link between the feedback equivalence and the full width eigenvalue completion problem was first made in Wimmer [1]. Theorem 6.1 which describes the connection between the eigenvalue completion problem and the eigenvalue assignment problem, was suggested in Gohberg-Kaashoek-Van Schagen [5]. Theorem 6.2 for controllable pairs (i.e., Corollary 6.3) is due to Rosenbrock [1] and in its full generality to Zaballa [2]. Sections 7–10 contain the standard material about feedback and its dual operation of output injection. Sections 7, 8 and 10 are mainly concerned with the transition from blocks to systems. Section 9 contains the main theorem about stabilization by state feedback from systems theory. The results in Section 7 on the assignment of controllability indices are taken from Gohberg-Kaashoek-Van Schagen [5].

Chapter X
Applications to Matrix Polynomials

This chapter contains two applications to the theory of regular matrix polynomials. The first is connected with the zero structure, and relates the behaviour of a matrix polynomial at infinity to block indices of a certain kind. The second deals with Wiener-Hopf factorization and identifies the factorization indices as block similarity invariants.

X.1 Preliminaries

In this section we bring together some of the main notions and results on matrix polynomials needed in this chapter. For the convenience of the reader an introduction to the local spectral theory of regular analytic matrix functions is presented in an appendix at the end of this book. Here we take as our starting point Theorem 5.1 in this appendix.

Let C be an $m \times n$ matrix and A be an $n \times n$ matrix. The pair (C, A) is called a *zero kernel pair* if $\bigcap_{i=0}^{n-1} \operatorname{Ker} CA^i = \{0\}$. In the previous chapter such a pair is called observable. Here and in the sequel we shall use the mathematical term "zero kernel pair". From Theorem III.2.5 we know that (C, A) is a zero kernel pair if and only if the matrix pencil

$$\begin{pmatrix} \lambda I - A \\ -C \end{pmatrix}$$

has full rank for each value of the complex parameter λ. Theorem III.2.5 and Corollary III.2.3 yield that (C, A) is a zero kernel pair if and only if the block

$$\left(\begin{pmatrix} A \\ C \end{pmatrix}; \begin{pmatrix} I_n & 0 \\ 0 & 0 \end{pmatrix}, I_{n+m} \right)$$

has no invariant polynomials.

A regular matrix polynomial is a square, say $m \times m$, matrix polynomial in one variable with a determinant that does not vanish identically. Let Γ be a contour in the complex plane with interior domain Ω. A pair (C, A) is a *(right) null pair with respect to* Ω for the regular matrix polynomial $L(\lambda) = \sum_{k=0}^{\ell} \lambda^k L_k$ if

(1) A has all its eigenvalues in Ω and the order of A is equal to the sum of the multiplicities of the zeros of $\det L(\lambda)$ in Ω;

(2) the pair (C, A) is a zero kernel pair;

(3) $L(\lambda)C(\lambda I - A)^{-1}$ is a matrix polynomial in λ.

According to Theorem 5.7 in the appendix, condition (3) is equivalent to the following condition:

(3') $L_0C + L_1CA + \cdots + L_\ell CA^\ell = 0$.

If $\det L$ has only one zero in Ω, μ say, then $A - \mu I$ is nilpotent, and the pair (C, A) is called a *right null pair of L at μ*. A (right) null pair with respect to the entire complex plane will simply be called a *(right) null pair* of the polynomial.

Dual to the notion of zero kernel pair there is the notion of full range pair. So, with A an $n \times n$ matrix and B an $n \times m$ matrix, the pair (A, B) is called a *full range pair* if $\operatorname{Im}\begin{pmatrix} B & AB & \dots & A^{n-1}B \end{pmatrix} = \mathbb{C}^n$. In the previous chapter such a pair is called controllable. Here and in the sequel we will use the mathematical term "full range pair". According to Theorem V.2.5 a pair (A, B) is full range if and only if the matrix pencil $\begin{pmatrix} (\lambda I - A) & -B \end{pmatrix}$ has full rank n for all values of λ. The full range property is also equivalent to the requirement that the block

$$\left(\begin{pmatrix} A & B \end{pmatrix}; I_{n+m}, \begin{pmatrix} I_n & 0 \\ 0 & 0 \end{pmatrix}\right)$$

has no invariant polynomials (see Corollary V.2.3).

Using the notion of full range pair we can give a different characterization of a null pair. The zero kernel pair (C, A) is a null pair with respect to Ω for the regular matrix polynomial L if and only if there exists a matrix B such that $L(\lambda)^{-1} - C(\lambda I - A)^{-1}B$ has an analytic continuation to Ω and the pair (A, B) is full range (see Appendix, Theorem 5.3). This characterization is very useful. It allows us to show that a null pair is unique up to similarity. That is, if (C', A') is also a null pair with respect to Ω of the matrix polynomial $L(\lambda)$, then there exists a unique invertible transformation S such that $C' = CS$ and $SA' = AS$ (see Appendix, Theorem 5.4).

Using the transposes of the matrices involved one easily obtains the definition and properties of a left null pair of the matrix polynomial $L(\lambda)$. So the pair (A, B), with A a $n \times n$ matrix and B a $n \times m$ matrix, is a *left null pair with respect to Ω* of the regular matrix polynomial $L(\lambda) = \sum_{k=0}^{\ell} \lambda^k L_k$ if

(i) A has all its eigenvalues in Ω and the order of A is equal to the sum of the multiplicities of the zeros of $\det L(\lambda)$ in Ω;

(ii) The pair (A, B) is a full range pair;

(iii) $(\lambda I - A)^{-1}BL(\lambda)$ is a matrix polynomial in λ.

Condition (iii) is equivalent to the following condition:

(iv) $BL_0 + ABL_1 + \cdots + A^\ell BL_\ell = 0$.

If μ is the only zero of $\det L$ in Ω, then (A, B) is said to be a *left null pair at μ*. A left null pair with respect to the entire complex plane will simply be called a *left null pair* of the polynomial. Furthermore, the full range pair (A, B) is a left null pair with respect to Ω if and only if there exists a matrix C such that $L(\lambda)^{-1} - C(\lambda I - A)^{-1}B$ has an analytic continuation to Ω and the pair (C, A) is a zero kernel pair (to see this use Theorem 5.3 of the Appendix and transpose).

Next we consider singularities at infinity. Let $L(\lambda)$ be a matrix polynomial, and consider the rational matrix function $L(\lambda^{-1})$. In general, both this function and its inverse can have a pole at $\lambda = 0$. *In the sequel we shall assume that* $L(\lambda^{-1})^{-1}$ *is analytic at* $\lambda = 0$. Under this additional assumption the order of 0 as a pole of $\det L(\lambda^{-1})$ is $\deg\det L(\lambda)$. Again using Theorem 5.1 of the Appendix as a starting point, we call the pair (T, G) a *left null pair of* $L(\lambda^{-1})^{-1}$ at zero if

(1) T is nilpotent and the order of T is equal to the order of 0 as a zero of $\det L(\lambda^{-1})^{-1}$;

(2) the pair (T, G) is a full range pair;

(3) there exists a matrix X such that $L(\lambda^{-1}) - X(\lambda I - T)^{-1}G$ is analytic at 0.

Let $\ell_1 \geq \ell_2 \geq \cdots \ell_k > 0$ be the sizes of the Jordan blocks in the Jordan canonical form of T. Put $\ell_{k+1} = \cdots = \ell_m = 0$. Then $\ell_1, \ldots, \ell_m$ are called *the partial pole multiplicities at infinity* of the polynomial L. These numbers do not depend on the particular choice of the pair (T, G) (cf., Appendix, Theorem 5.4). One can show that

$$L(\lambda) = E(\lambda)\begin{pmatrix} \lambda^{\ell_1} & & \\ & \ddots & \\ & & \lambda^{\ell_m} \end{pmatrix} F(\lambda),$$

where E and F are analytical and invertible at infinity, and the integers $\ell_1, \ldots, \ell_m$ are the partial pole multiplicities of L at infinity.

X.2 Matrix polynomials with prescribed zero structure

In this section we study an inverse problem. We assume that a zero kernel pair (C, A) is given with A an $n \times n$ matrix and C an $m \times n$ matrix. The aim is to construct a matrix polynomial $L(\lambda)$, which has (C, A) as a null pair. We want the polynomial to have some extra properties. First, for a number α, not an eigenvalue of A, we prescribe that $L(\alpha) = D$, where D is a given invertible matrix. Secondly, we want that $L(\lambda)^{-1}$ is analytic at infinity. If (C, A) is a null pair of such a polynomial $L(\lambda)$ and B is such that $L(\lambda)^{-1} - C(\lambda I - A)^{-1}B$ is a polynomial, then both terms are analytic at infinity, and therefore $L(\lambda)^{-1} - C(\lambda I - A)^{-1}B$ is a constant, which can be computed by simply substituting α for λ. Before we state the main theorem we derive an auxiliary result.

Lemma 2.1. *Let the regular matrix pencil $\lambda G - A$ and the matrices B and C be such that for each $\lambda \in \mathbb{C}$ the equality $x\begin{pmatrix}\lambda G - A & B\end{pmatrix} = 0$ implies $x = 0$ and*

$$\begin{pmatrix} \lambda G - A \\ C \end{pmatrix} y = 0$$

implies $y=0$. Then $\det(\mu G - A) = 0$ for some number μ if and only if μ is a pole for the rational matrix function $C(\lambda G - A)^{-1}B$.

PROOF. Let $\mu \in \mathbb{C}$. Since the matrices $\begin{pmatrix} \mu G - A & B \end{pmatrix}$ and

$$\begin{pmatrix} \mu G - A \\ C \end{pmatrix}$$

have full rank there exists matrices D, E, F and H such that the matrices

$$\begin{pmatrix} D & E \\ \lambda G - A & B \end{pmatrix}, \quad \begin{pmatrix} F & \lambda G - A \\ H & C \end{pmatrix}$$

are invertible for $\lambda = \mu$. This means that these matrices are invertible for λ in a neighbourhood of μ. So they have inverses that are analytic at the point μ. Next compute

$$\begin{pmatrix} F & \lambda G - A \\ H & C \end{pmatrix} \begin{pmatrix} I & 0 \\ 0 & (\lambda G - A)^{-1} \end{pmatrix} \begin{pmatrix} D & E \\ \lambda G - A & B \end{pmatrix}$$
$$= \begin{pmatrix} FD + \lambda G - A & FE + B \\ HD + C & HE + C(\lambda G - A)^{-1}B \end{pmatrix}.$$

We see that the middle term in the left hand side has a pole at μ if and only if the right hand side has a pole at μ. Since $(\lambda G - A)^{-1}$ has a pole at μ if and only if $\det(\mu G - A) = 0$, we proved the lemma. □

We present the main result in this section.

Theorem 2.2. *Let (C, A) be a zero kernel pair, with A an $n \times n$ matrix and C an $m \times n$ matrix, let α be a complex number, not an eigenvalue of A, and let D be an invertible matrix. The general form of a regular matrix polynomial $L(\lambda)$ satisfying:*

(i) *$L(\alpha) = D$;*

(ii) *(C, A) is a null pair of $L(\lambda)$;*

(iii) *$L(\lambda)^{-1}$ is analytic at infinity,*

is given by

$$L(\lambda) = D + (\lambda - \alpha)DC(\alpha I - A)^{-1}(I - \lambda T)^{-1}G. \tag{2.1}$$

Here (T, G) is any pair of matrices, with T an $n \times n$ nilpotent matrix and G an $n \times m$ matrix, and satisfying

$$TA + GC = I. \tag{2.2}$$

Moreover (2.1) gives a 1–1 correspondence between all such pairs (T, G) and regular matrix polynomials L satisfying (i), (ii) *and* (iii). *Finally, the partial pole multiplicities $\mu_1 \geq \cdots \geq \mu_m$ of L at infinity in (2.1) satisfy*

$$\sum_{j=1}^{k} \mu_j \geq \sum_{j=1}^{k} \kappa_j, \quad k = 1, \ldots, m, \tag{2.3}$$

where $\kappa_1 \geq \cdots \geq \kappa_m$ are the indices of the third kind of the block

$$\left(\begin{pmatrix} A \\ C \end{pmatrix}, \begin{pmatrix} I & 0 \\ 0 & 0 \end{pmatrix}, I\right)$$

on $\mathbb{C}^{n+m}$.

PROOF. Assume that the pair (T, G) satisfies (2.2). Note that (2.1) can be rewritten as

$$L(\lambda) = D + (\lambda - \alpha) \sum_{j=0}^{n-1} \lambda^j DC(\alpha I - A)^{-1} T^j G. \tag{2.4}$$

Hence (2.1) indeed defines a polynomial. Next we will show that

$$L(\lambda)^{-1} = D^{-1} + (\lambda - \alpha) C (A - \lambda I)^{-1} (I - \alpha T)^{-1} G D^{-1}. \tag{2.5}$$

To see this we multiply the right hand side of (2.1) on the left by the right hand side of (2.5), and get

$$\begin{aligned} &I + (\lambda - \alpha) C(\alpha I - A)^{-1}(I - \lambda T)^{-1} G + (\lambda - \alpha) C(A - \lambda I)^{-1}(I - \alpha T)^{-1} G + \\ &(\lambda - \alpha)^2 C(A - \lambda I)^{-1}(I - \alpha T)^{-1} G C(\alpha I - A)^{-1}(I - \lambda T)^{-1} G. \end{aligned} \tag{2.6}$$

We consider the second, third and fourth term of (2.6) leaving out the common factors $(\lambda - \alpha)C$ on the left and G on the right. One sees that

$$\begin{aligned} &(\alpha I - A)^{-1}(I - \lambda T)^{-1} + (A - \lambda I)^{-1}(I - \alpha T)^{-1} + \\ &(\lambda - \alpha)(A - \lambda I)^{-1}(I - \alpha T)^{-1} G C(\alpha I - A)^{-1}(I - \lambda T)^{-1} = \\ &(A - \lambda I)^{-1}(I - \alpha T)^{-1} \\ &\big((I - \alpha T)(A - \lambda I) + (I - \lambda T)(\alpha I - A) + (\lambda - \alpha) G C\big) \\ &(\alpha I - A)^{-1}(I - \lambda T)^{-1}. \end{aligned}$$

The middle factor in the right hand side of this equality reduces to

$$(\alpha - \lambda)(I - TA - GC),$$

and hence this factor is zero because of (2.2). This proves that the formulas (2.5) and (2.1) represent inverses of each other. It is obvious that $L(\alpha) = D$. Let us check that $L(\lambda)^{-1}$ is analytic at infinity. Rewrite (2.5) as

$$L(\lambda)^{-1} = D^{-1} + (\lambda - \alpha)C(A - \lambda I)^{-1}(I - \alpha T)^{-1}GD^{-1} =$$

$$D^{-1} - C(I - \alpha T)^{-1}GD^{-1} + C(A - \lambda I)^{-1}(A - \alpha I)(I - \alpha T)^{-1}GD^{-1}.$$

This last expression clearly shows that $L(\lambda)^{-1}$ is analytic at infinity.

Next we will show that the pair (C, A) is a null pair of $L(\lambda)$. Now (C, A) is a null pair of $L(\lambda)$ if there exists an operator E such that $L(\lambda)^{-1} - C(\lambda I - A)^{-1}E$ is analytic on $\mathbb{C}$ and (A, E) is a full range pair. With $E = (A - \alpha I)(I - \alpha T)^{-1}GD^{-1}$ this property is fulfilled provided that we show that

$$\big(A, (A - \alpha I)(I - \alpha T)^{-1}GD^{-1}\big) \tag{2.7}$$

is a full range pair. Assume that the row vector y is such that

$$yA = \lambda y, \quad y(A - \alpha I)(I - \alpha T)^{-1}GD^{-1} = 0. \tag{2.8}$$

Put $x = y(A - \alpha I)$, and note that (2.8) is equivalent to $xA = \lambda x$ and $x(I - \alpha T)^{-1}G = 0$. Use (2.2) and $x(I - \alpha T)^{-1}G = 0$ to compute

$$x(I - \alpha T)^{-1}TA = x(I - \alpha T)^{-1} = x + x(I - \alpha T)^{-1}T\alpha.$$

So $x = x(I - \alpha T)^{-1}T(A - \alpha I)$ or, equivalently, since α is not an eigenvalue of A, $x(A - \alpha I)^{-1} = x(I - \alpha T)^{-1}T$. Now use that $xA = \lambda A$ to see that $(\lambda - \alpha)^{-1}x = x(I - \alpha T)^{-1}T$. It follows that $(\lambda - \alpha)^{-n}x = x(I - \alpha T)^{-n}T^n = 0$ because T is nilpotent. This proves that $x = 0$, and thus $y = 0$. We conclude that the pair (2.8) is a full range pair, and hence (C, A) is a null pair of $L(\lambda)$.

Conversely, assume that (i), (ii) and (iii) are satisfied. There exists B and X such that

$$L(\lambda)^{-1} = X + C(\lambda I - A)^{-1}B. \tag{2.9}$$

We may assume that the pair (A, B) is full range. We rewrite (2.9) as

$$\begin{aligned} L(\lambda)^{-1} &= X - C(A - \alpha I)^{-1}B + (\lambda - \alpha)C(\lambda I - A)^{-1}B \\ &= D^{-1} + (\lambda - \alpha)C(A - \alpha I)^{-1}(\lambda I - A)^{-1}B. \end{aligned} \tag{2.10}$$

Then the polynomial $L(\lambda)$ can be represented as

$$L(\lambda) = D - (\lambda - \alpha)DC(A - \alpha I)^{-1}(\lambda H^\times - A^\times)^{-1}BD, \tag{2.11}$$

where $H^\times = I + BDC(A-\alpha I)^{-1}$ and $A^\times = A + \alpha BDC(A-\alpha I)^{-1}$. To check this formula multiply the right hand side of (2.11) on the left by the right hand side of (2.10). This gives

$$\begin{aligned}
I-&(\lambda-\alpha)C(A-\alpha I)^{-1}(\lambda I - A)^{-1}BD+\\
&+(\lambda-\alpha)C(A-\alpha I)^{-1}(\lambda H^\times - A^\times)^{-1}BD+\\
&-(\lambda-\alpha)C(A-\alpha I)^{-1}(\lambda I - A)^{-1}(\lambda-\alpha)BDC(A-\alpha I)^{-1}(\lambda H^\times - A^\times)^{-1}BD\\
=&I+(\lambda-\alpha)C(A-\alpha I)^{-1}ZBD,
\end{aligned}$$

with

$$\begin{aligned}
Z =&(\lambda I - A)^{-1} - (\lambda H^\times - A^\times)^{-1}+\\
&-(\lambda I - A)^{-1}(\lambda-\alpha)BDC(A-\alpha I)^{-1}(\lambda H^\times - A^\times)^{-1}\\
=&(\lambda I - A)^{-1} - (\lambda H^\times - A^\times)^{-1}+\\
&-(\lambda I - A)^{-1}(\lambda H^\times - A^\times - \lambda I + A)(\lambda H^\times - A^\times)^{-1}\\
=&0
\end{aligned}$$

Let us show that

$$\begin{pmatrix} \lambda H^\times - A^\times \\ C(A-\alpha I)^{-1} \end{pmatrix} y = 0$$

implies $y=0$. One computes that $0 = (\lambda H^\times - A^\times)y = (\lambda I - A)y$, and therefore

$$(\lambda I - A)(A-\alpha I)^{-1}y = 0.$$

Since also $C(A-\alpha I)^{-1}y = 0$, the fact that (C,A) is a zero kernel pair gives that $(A-\alpha I)^{-1}y = 0$, which in turn implies that $y=0$. Also, if $x\begin{pmatrix} \lambda H^\times - A^\times & B \end{pmatrix} = 0$ then $x=0$. Indeed, from $xB=0$ one obtains that $x(\lambda H^\times - A^\times) = x(\lambda I - A)$. So we see that $xB=0$ and $x(\lambda I - A) = 0$. Since (A,B) is a full range pair, these equalities imply that $x=0$. Hence we may apply Lemma 2.1 to show that $DC(A-\alpha I)^{-1}(\lambda H^\times - A^\times)^{-1}BD$ has a pole at every eigenvalue of $\lambda H^\times - A^\times$. Since $L(\lambda)$, as given by (2.11), has no pole at all, we see that $\lambda H^\times - A^\times$ has no eigenvalue different from α. However $\alpha H^\times - A^\times = \alpha I - A$ and α is not an eigenvalue of A. We conclude that the pencil $\lambda H^\times - A^\times$ has no eigenvalue in the finite complex plane. This shows that $A^\times$ is invertible and $H^\times$ is nilpotent. Put $T = (A^\times)^{-1}H^\times$ and $G = -(A^\times)^{-1}BD$. Then

$$L(\lambda) = D + (\lambda-\alpha)DC(\alpha I - A)^{-1}(I-\lambda T)^{-1}G.$$

as required in (2.1). Furthermore

$$TA + GC = (A^\times)^{-1}H^\times A + (A^\times)^{-1}BDC = (A^\times)^{-1}(H^\times A + BDC)$$

and

$$\begin{aligned}H^{\times}A + BDC &= A + BDC(A-\alpha I)^{-1}A + BDC\\ &= A + BDC\big((A-\alpha I)^{-1}(A-\alpha I) - \alpha I(A-\alpha I)^{-1}\big)\\ &= A + \alpha BDC(A-\alpha I)^{-1}\\ &= A^{\times}.\end{aligned}$$

So indeed $TA + GC = I$.

Let us prove the 1–1 correspondence between the polynomials and the pairs. Assume that $L(\lambda)$ is given by (2.1) for pairs (T_1, G_1) and (T_2, G_2) which both satisfy (2.2). We get that $L(\lambda)^{-1}$ is given by (2.5) for the pairs (T_1, G_1) and (T_2, G_2). Rewrite this (as it has been done above) as

$$L(\lambda)^{-1} = D^{-1} - C(I-\alpha T_i)^{-1}G_iD^{-1} + C(A-\lambda I)^{-1}(A-\alpha I)(I-\alpha T_i)^{-1}G_iD^{-1}$$

for $i = 1, 2$. Above we checked that these realizations are minimal (use Theorem IX.4.1). According to Theorem IX.4.2 there exists a unique invertible S such that $CS = C$, $AS = SA$. So $S = I$, and we get that

$$(A-\alpha I)(I-\alpha T_1)^{-1}G_1D^{-1} = (A-\alpha I)(I-\alpha T_2)^{-1}G_2D^{-1}$$

and thus $(I-\alpha T_1)^{-1}G_1 = (I-\alpha T_2)^{-1}G_2$. Use that

$$-(I-\alpha T_i)A + \alpha G_iC = \alpha I - A,$$

for $i = 1, 2$. Hence

$$-A + (I-\alpha T_i)^{-1}\alpha G_iC = (I-\alpha T_i)^{-1}(\alpha I - A).$$

So $(I-\alpha T_1)^{-1}(\alpha I - A) = (I-\alpha T_2)^{-1}(\alpha I - A)$. It follows that $T_1 = T_2$, and hence also $G_1 = G_2$, i.e., the pair is (T, G) is unique.

It remains to prove (2.3). To find T and G we have to solve (2.2). Conditions for solvability of this equation, or rather its transposed form, are given in Theorem V.6.1. The numbers $\kappa_1 \geq \cdots \geq \kappa_m$ are the minimal row indices of the pencil

$$\begin{pmatrix}\lambda I - A\\ -C\end{pmatrix}$$

and hence the minimal column indices of $\big((\lambda I - A^T) \quad -C^T\big)$. Since the pair (C, A) is a zero kernel pair, this pencil has no invariant polynomials. Hence, by Corollary III.2.3, the block

$$\left(\begin{pmatrix}A\\ C\end{pmatrix}, \begin{pmatrix}I & 0\\ 0 & 0\end{pmatrix}, I\right)$$

has no invariant polynomials either. Therefore the conditions for solvability of (2.2) reduce to

$$\sum_{j=1}^{k} \deg q_j \geq \sum_{j=1}^{k} \kappa_j, \quad k = 1, \ldots, m \quad ,$$

where $q_1, \ldots, q_m$ are the invariant polynomials of T. Since T is nilpotent, we get that $\deg q_i = \ell_i$, with ℓ_i the size of the i–th Jordan block in the Jordan canonical form of T at 0.

To complete the proof it remains to prove that the numbers ℓ_i are the partial pole multiplicities of L at infinity. For this purpose we show that (T, G) is a left null pair of $L(\lambda^{-1})^{-1}$ at 0. First, we check condition (1) in the definition of left null pair. Since $L(\lambda^{-1})^{-1}$ is analytic at 0, the order p of $\lambda = 0$ as a zero of $\det L(\lambda^{-1})^{-1}$ is equal to the order of $\lambda = 0$ as a pole of $\det L(\lambda^{-1})$. Thus p is equal to the degree of $\det L(\lambda)$. Because (C, A) is a null pair of $L(\lambda)$, we conclude that p is equal to the order of A and thus also to the order of T. So condition (1) is satisfied. Next, let us prove that the pair (T, G) is a full range pair. Assume that $x^T T = \lambda T$ and $x^T G = 0$ for an eigenvalue λ of T. We have to show that $x^T = 0$. Now 0 is the only eigenvalue of T, and hence it suffices to compute

$$x^T = x^T I = x^T (TA + GC) = x^T TA + x^T GC = 0.$$

We conclude that (T, G) is a full range pair. Finally, we compute, using (2.1) and $(1 - \lambda\alpha)I = I - \alpha T - \alpha(\lambda I - T)$, that

$$\begin{aligned} L(\lambda^{-1}) &= D + (1 - \lambda\alpha) DC(\alpha I - A)^{-1}(\lambda I - T)^{-1} G \\ &= D - \alpha DC(\alpha I - A)^{-1} G + DC(\alpha I - A)^{-1}(I - \alpha T)(I - \lambda T)^{-1} G \\ &= D + DC(\alpha I - A)^{-1} \alpha G + X(I - \lambda T)^{-1} G \end{aligned}$$

with $X = DC(\alpha I - A)^{-1}(I - \alpha T)$. So also the third condition is fulfilled. □

X.3 Wiener-Hopf factorization and indices

We begin by recalling the definition of Wiener-Hopf factorization. Throughout this section Γ is a closed rectifiable curve in the complex plane with a bounded inner domain Ω, and, for simplicity, we shall assume that $0 \in \Omega$. Let $L_1(\lambda)$ and $L_2(\lambda)$ be regular $m \times m$ matrix polynomials. We call $L_1(\lambda)$ and $L_2(\lambda)$ *left Wiener-Hopf equivalent* with respect to Γ if $L_1(\lambda)$ and $L_2(\lambda)$ are invertible for each $\lambda \in \Gamma$ and

$$L_2(\lambda) = E_-(\lambda) L_1(\lambda) E_+(\lambda), \quad \lambda \in \Gamma, \tag{3.1}$$

where $E_+(\lambda)$ is holomorphic on Ω and $E_-(\lambda)$ is holomorphic on $\mathbb{C}_\infty \setminus \overline{\Omega}$, both are continuous up to the boundary Γ, for each $\lambda \in \overline{\Omega}$ the operator $E_+(\lambda)$ is invertible, and for each $\lambda \in \mathbb{C}_\infty \setminus \Omega$ the operator $E_-(\lambda)$ is invertible.

We call (3.1) a *left Wiener-Hopf factorization* of $L_2(\lambda)$ if in the right side of (3.1) the middle factor $L_1(\lambda) = D(\lambda) = \operatorname{diag}(\lambda^{\nu_i})_{i=1}^m$, where $\nu_1 \geq \cdots \geq \nu_m \geq 0$ are integers. The numbers $\nu_1 \geq \cdots \geq \nu_m$ are called the left Wiener-Hopf factorization indices of $L_2(\lambda)$. They are uniquely determined by $L_2(\lambda)$. We call $L_1(\lambda)$ and $L_2(\lambda)$ *right Wiener-Hopf equivalent* with respect to Γ if $L_1(\lambda)$ and $L_2(\lambda)$ are invertible for each $\lambda \in \Gamma$ and

$$L_2(\lambda) = E_+(\lambda)L_1(\lambda)E_-(\lambda), \quad (\lambda \in \Gamma),$$

where $E_+(\lambda)$ and $E_-(\lambda)$ are as above. One defines *right Wiener-Hopf factorization* in the same way as it is done for the "left" case.

The next theorem is the first main result of this section.

Theorem 3.1. *For $i = 1, 2$ let (A_i, B_i) be a left null pair with respect to Ω of the regular $m \times m$ matrix polynomial $L_i(\lambda)$. Let P be the projection of $\mathbb{C}^{n+m}$ along $\{0\} \oplus \mathbb{C}^m$ onto $\mathbb{C}^n \oplus \{0\}$, and put*

$$Z_i = \begin{pmatrix} A_i & B_i \end{pmatrix} : \mathbb{C}^{n+m} \to \mathbb{C}^n \oplus \{0\}.$$

Then $L_1(\lambda)$ and $L_2(\lambda)$ are left Wiener-Hopf equivalent with respect to Γ if and only if the blocks $(Z_1; P, I_{n+m})$ and $(Z_2; P, I_{n+m})$ are block similar. More precisely, if

$$S = \begin{pmatrix} N & 0 \\ F & M \end{pmatrix} \tag{3.2}$$

is the block similarity of the blocks $(Z_1; P, I_{n+m})$ and $(Z_2; P, I_{n+m})$, then

$$E_-(\lambda) = M + F(\lambda I - A)^{-1}B_1, \quad E_+(\lambda) = L_1(\lambda)^{-1}E_-(\lambda)^{-1}L_2(\lambda)$$

establish the Wiener-Hopf equivalence

$$L_2(\lambda) = E_-(\lambda)L_1(\lambda)E_+(\lambda). \tag{3.3}$$

Conversely, if the Wiener-Hopf equivalence (3.3) is given, then S, given by (3.2) with

$$N = \frac{1}{2\pi i}\int_\Gamma (\mu I - A_2)^{-1}B_2L_2(\mu)E_+(\mu)^{-1}C_1(\mu I - A_1)^{-1}d\mu,$$

$$F = \frac{1}{2\pi i}\int_\Gamma E_-(\mu)L_1(\mu)C_1(\mu I - A_1)^{-1}d\mu,$$

and $M = E_-(\infty)$, is a block similarity of $(Z_1; P, I_{n+m})$ and $(Z_2; P, I_{n+m})$.

The proof of this theorem is the same for the infinite dimensional case. Therefore we postpone the proof to Section XIII.2, where it will appear as part of the proof of Theorem XIII.2.1.

The next result relates the Wiener-Hopf factorization indices of a regular matrix polynomial to the indices of the first kind of a related operator block.

Theorem 3.2. *Let (A, B) be left null pair with respect to Ω of the regular $m \times m$ matrix polynomial $L(\lambda)$. Assume that $L(\lambda)$ is invertible on Γ. Then the left Wiener-Hopf factorization indices of $L(\lambda)$ with respect to Γ are the indices of the first kind of full width block*

$$\left(\begin{pmatrix} A & B \end{pmatrix}; \begin{pmatrix} I_n & 0 \\ 0 & 0 \end{pmatrix}, I_{n+m} \right). \tag{3.4}$$

PROOF. Let $\kappa_1 \geq \cdots \geq \kappa_m$ be the indices of the first kind of the operator block (3.4). Then the block is block similar to the direct sum of shifts of the first kind $\bigoplus_{j=1}^m S_j$. Here

$$S_j = \left(\begin{pmatrix} I_{\kappa_j} & 0 \end{pmatrix}; \begin{pmatrix} 0 & 0 \\ 0 & I_{\kappa_j} \end{pmatrix}, I_{\kappa_j+1} \right).$$

For all j such that $\kappa_j > 0$, write $\begin{pmatrix} I_{\kappa_j} & 0 \end{pmatrix} = \begin{pmatrix} B_j & N_j \end{pmatrix}$, with

$$B_j = \begin{pmatrix} 1 \\ 0 \\ \vdots \\ 0 \\ 0 \end{pmatrix}, \quad N_j = \begin{pmatrix} 0 & 0 & \cdots & \cdots & 0 \\ 1 & 0 & \ddots & & \vdots \\ & \ddots & \ddots & \ddots & \vdots \\ & & \ddots & 0 & 0 \\ & & & 1 & 0 \end{pmatrix}.$$

Assume that $\kappa_r > 0 = \kappa_{r+1}$. Define the $n \times m$ matrix B_0 by $B_0 = \left(\bigoplus_{j=1}^r B_j \right) \oplus 0$, the $n \times n$ matrix A_0 by $A_0 = \bigoplus_{j=1}^m N_j$ and the $m \times n$ matrix C_0 by $C_0 = \left(\bigoplus_{j=1}^r C_j \right) \oplus 0$ where the $1 \times \kappa_j$ matrix C_j is defined by $C_j = \begin{pmatrix} 0 & \cdots & 0 & 1 \end{pmatrix}$. Put $D(\lambda) = \operatorname{diag}\left(\lambda^{\kappa_j}\right)_{j=1}^m$. We get that

$$D(\lambda)^{-1} - C_0(\lambda I - A_0)^{-1} B_0 = \operatorname{diag}\left(\delta_{0\kappa_j}\right)_{j=1}^m.$$

Since (A_0, B_0) is full range, (C_0, A_0) is zero kernel, and $D(\lambda)^{-1} - C_0(\lambda I - A_0)^{-1} B_0$ is analytic on $\mathbb{C}$, we conclude that (A_0, B_0) is a null pair of $D(\lambda)$ with respect to Ω. Since the blocks (3.4) and $\bigoplus_{j=1}^m S_j$ are block similar, we know from Theorem 3.1 that there exists a Wiener-Hopf factorization

$$L(\lambda) = E_-(\lambda) D(\lambda) E_+(\lambda), \quad \lambda \in \Gamma.$$

This proves that $\kappa_1, \ldots, \kappa_m$ are the factorization indices of the polynomial $L(\lambda)$. □

We recall that in Theorem V.1.2 there are formulas to compute the indices of the first kind of a full width block, and hence one may use these formulas to compute the Wiener-Hopf factorization indices.

Notes

The spectral theory of null pairs and pole pairs summarized in the first section originates in Gohberg-Lancaster-Rodman [1]; an independent exposition of this material with detailed proofs is presented in the appendix. The second section is taken from Gohberg-Kaashoek-Van Schagen [2]. Theorems 3.1 and 3.2 are finite dimensional versions of results in Gohberg-Kaashoek-Van Schagen [2]; we shall return to their infinite dimensional counter parts in Sections XIII.2 and XIII.3.

Chapter XI
Applications to Rational Matrix Functions

In this chapter we continue with applications of the general theory of operator blocks. We deal with two kind of homogeneous interpolation problems and with factorization indices for regular rational matrix functions. The first section contains preliminary material on the zero and pole structure of rational matrix functions.

XI.1 Preliminaries on pole pairs and null pairs

In this section we extend some of the results on the spectral theory of matrix polynomials to regular rational matrix functions on a domain Ω. Thus in this section we have to deal not only with poles of the inverse of the function, the zeros of the function, but also with poles of the function itself. It might even occur that a rational matrix function has a zero and a pole in the same point, as one sees in the example

$$W(\lambda) = \begin{pmatrix} 1 & \lambda^{-1} \\ 0 & 1 \end{pmatrix}.$$

As a starting point for our analysis we choose Appendix, Theorem 5.3.

Throughout this chapter γ is a rectifiable Jordan curve and Ω_+ is its inner domain. Let W be rational $m \times m$ matrix function which has no poles on γ. A zero kernel pair (C_π, A_π), with $C_\pi : \mathcal{X} \to \mathbb{C}^m$ and $A_\pi : \mathcal{X} \to \mathcal{X}$ linear operators and $\mathcal{X}$ finite dimensional, is called a *right pole pair* (*with main space* $\mathcal{X}$) of W with respect to Ω_+ if A_π has all its eigenvalues inside Ω_+ and there exists a linear operator $B_\pi : \mathbb{C}^m \to \mathcal{X}$ such that (A_π, B_π) is a full range pair and $W(\lambda) - C_\pi(\lambda I - A_\pi)^{-1}B_\pi$ has an analytic continuation to Ω_+. So $C_\pi(\lambda I - A_\pi)^{-1}B_\pi$ has the same Laurent principal parts in Ω_+ as W has. According to Lemma X.2.1 this implies that the eigenvalues of A_π are the poles of W in Ω_+. The full range pair (A_π, B_π), with $A_\pi : \mathcal{X} \to \mathcal{X}$ and $B_\pi : \mathbb{C}^m \to \mathcal{X}$ linear operators and $\mathcal{X}$ finite dimensional, is called a *left pole pair* (*with main space* $\mathcal{X}$) of W with respect to Ω_+ if there exists a linear operator $C_\pi : \mathcal{X} \to \mathbb{C}^m$ such that (C_π, A_π) is a zero kernel pair and $W(\lambda) - C_\pi(\lambda I - A_\pi)^{-1}B_\pi$ has an analytic continuation to Ω_+. In the case when $\Omega_+ = \mathbb{C}$ we call a left (right) pole pair with respect to Ω_+ a *global left (right) pole pair.*

We will show that left and right pole pairs do exist. Write W_γ for the sum of the Laurent principal parts of W corresponding to the poles inside γ. Since W_γ is a rational function it has (cf., Theorems 1.1 and 4.3 in Chapter IX) a minimal realization, $(A_\pi, B_\pi, C_\pi, 0)$ say. So $W(\lambda) - C_\pi(\lambda I - A_\pi)^{-1}B_\pi$ has an analytic continuation to Ω_+. By Theorem IX.4.1 it follows from the minimality of the realization $(A_\pi, B_\pi, C_\pi, 0)$ that the pair (C_π, A_π) is a zero kernel pair and the pair (A_π, B_π) is a full range pair. Therefore (C_π, A_π) is a right pole pair and (A_π, B_π) is a left pole pair of W with respect to Ω_+.

We call the pairs (C_1, A_1) and (C_2, A_2) similar if there exists an invertible linear operator S such that $C_2S = C_1$ and $A_2S = SA_1$.

Lemma 1.1. *Two right pole pairs of W with respect to Ω_+ are similar and the similarity is uniquely determined by the pairs.*

PROOF. Let $(C_\pi^{(1)}, A_\pi^{(1)})$ and $(C_\pi^{(2)}, A_\pi^{(2)})$ be right pole pairs of W with respect to Ω_+. Then there exist operators $B_\pi^{(1)}$ and $B_\pi^{(2)}$ such that the rational functions $W(\lambda) - C_\pi^{(1)}(\lambda I - A_\pi^{(1)})^{-1}B_\pi^{(1)}$ and $W(\lambda) - C_\pi^{(2)}(\lambda I - A_\pi^{(2)})^{-1}B_\pi^{(2)}$ are analytic in Ω_+. So the function

$$H(\lambda) = C_\pi^{(1)}(\lambda I - A_\pi^{(1)})^{-1}B_\pi^{(1)} - C_\pi^{(2)}(\lambda I - A_\pi^{(2)})^{-1}B_\pi^{(2)}$$

is analytic on Ω_+. Since the operators $A_\pi^{(1)}$ and $A_\pi^{(2)}$ have all eigenvalues inside Ω_+, it follows that the function H is analytic on $\mathbb{C}$. Moreover, the value of H at infinity is 0, and therefore, by Liouville's theorem,

$$C_\pi^{(1)}(\lambda I - A_\pi^{(1)})^{-1}B_\pi^{(1)} = C_\pi^{(2)}(\lambda I - A_\pi^{(2)})^{-1}B_\pi^{(2)}.$$

So the minimal systems $(A_\pi^{(1)}, B_\pi^{(1)}, C_\pi^{(1)}, 0)$ and $(A_\pi^{(2)}, B_\pi^{(2)}, C_\pi^{(2)}, 0)$ have the same transfer function, and hence, by Proposition IX.4.6, these systems are similar. In particular, the pairs $(C_\pi^{(1)}, A_\pi^{(1)})$ and $(C_\pi^{(2)}, A_\pi^{(2)})$ are similar. The uniqueness of the similarity follows from the fact that the $(C_\pi^{(1)}, A_\pi^{(1)})$ and $(C_\pi^{(2)}, A_\pi^{(2)})$ are zero kernel pairs (use the same arguments as in the last paragraph of the proof of Proposition IX.4.6). □

Let W be an $m \times m$ rational matrix function, and assume that W is regular (i.e., $\det W(\lambda)$ does not vanish identically). By definition a *zero of* W is a pole of the rational function W^{-1}. A *left null pair* of W with respect to Ω_+ is defined to be a left pole pair of W^{-1} with respect to Ω_+, and a *right null pair* of W with respect to Ω_+ is defined to be a right pole pair of W^{-1} with respect to Ω_+. In the case when $\Omega_+ = \mathbb{C}$ a left (right) null pair with respect to Ω_+ is called a *global left (right) null pair*. The next lemma is proved in the same way as Lemma 1.1. We omit the details.

Lemma 1.2. *Two right null pairs of W with respect to Ω_+ are similar and the similarity is uniquely determined by the pairs.*

From now on we assume that W is *proper* (i.e., is analytic at infinity) and the value of W at infinity is I. Let $\Theta = (A, B, C, I)$ be a minimal realization of the function W. In particular,

$$W(\lambda) = I + C(\lambda I - A)^{-1}B. \tag{1.1}$$

Assume that (C_π, A_π) is a right pole pair of W with respect to Ω_+ with main space $\mathcal{X}_\pi$, and that (A_ζ, B_ζ), with main space $\mathcal{X}_\zeta$, is a left null pair of W with respect to Ω_+. We will show that these pairs are coupled in a natural way. From (1.1) one obtains that

$$W(\lambda)^{-1} = I - C\big(\lambda I - (A - BC)\big)^{-1}B.$$

Put $A^\times = A - BC$. Let Q_Θ ($Q_\Theta^\times$) be the Riesz projection corresponding to the eigenvalues of A ($A^\times$) inside Ω_+. Consider

$$W(\lambda) = I + C|_{\operatorname{Im} Q_\Theta}(\lambda I - A|_{\operatorname{Im} Q_\Theta})^{-1}Q_\Theta B + C|_{\operatorname{Ker} Q_\Theta}(\lambda I - A|_{\operatorname{Ker} Q_\Theta})^{-1}(I - Q_\Theta)B. \tag{1.2}$$

In (1.2) the second term on the right hand side is analytic in Ω_+. It follows that the pair $(C|_{\operatorname{Im} Q_\Theta}, A|_{\operatorname{Im} Q_\Theta})$ is a right pole pair of W with respect to Ω_+, because $(C|_{\operatorname{Im} Q_\Theta}, A|_{\operatorname{Im} Q_\Theta})$ is a zero kernel pair and $(A|_{\operatorname{Im} Q_\Theta}, Q_\Theta B)$ is a full range pair. The main space of this pair is $\operatorname{Im} Q_\Theta$. Similarly, we see that $(A^\times|_{\operatorname{Im} Q_\Theta^\times}, Q_\Theta^\times B)$ is a left null pair of W. So there exist unique invertible operators $S_\pi(\Theta) : \mathcal{X}_\pi \to \operatorname{Im} Q$ and $S_\zeta(\Theta) : \mathcal{X}_\zeta \to \operatorname{Im} Q_\Theta^\times$ such that

$$(C_\pi S_\pi(\Theta)^{-1}, S_\pi(\Theta)A_\pi S_\pi(\Theta)^{-1}) = (C|_{\operatorname{Im} Q_\Theta}, A|_{\operatorname{Im} Q_\Theta}), \tag{1.3}$$

$$(S_\zeta(\Theta)A_\zeta S_\zeta(\Theta)^{-1}, S_\zeta(\Theta)B_\zeta) = (A^\times|_{\operatorname{Im} Q_\Theta^\times}, Q_\Theta^\times B). \tag{1.4}$$

One defines the *null-pole coupling operator* $\Gamma : \mathcal{X}_\pi \to \mathcal{X}_\zeta$ of the pairs (C_π, A_π) and (A_ζ, B_ζ) to be the operator

$$\Gamma = S_\zeta(\Theta)^{-1}Q_\Theta^\times S_\pi(\Theta). \tag{1.5}$$

The definition of Γ does not depend on the choice of the triple (A, B, C). To see this, let $\Psi = (A_1, B_1, C_1, I)$ be another minimal realization of W. Then there exists an invertible linear operator H such that $A_1 = HAH^{-1}$, $B_1 = HB$ and $C_1 = CH^{-1}$. Using the uniqueness of the similarities in Lemmas 1.1 and 1.2 one easily checks that

$$S_\pi(\Psi) = HS_\pi(\Theta), \quad S_\zeta(\Psi) = HS_\zeta(\Theta).$$

From these identities and $Q_\Psi^\times = HQ_\Theta^\times H^{-1}$ it follows that the definition of Γ does not depend on the choice of the realization Θ.

In the sequel we call $\tau = \big((C_\pi, A_\pi), (A_\zeta, B_\zeta), \Gamma\big)$ an Ω_+-*null-pole triple* of W. From the definition it is clear that the triple

$$\tau' = \big((C_\pi E, E^{-1}A_\pi E), (F^{-1}A_\zeta F, F^{-1}B_\zeta), E^{-1}\Gamma F\big), \tag{1.6}$$

where E and F are invertible operators, is again an Ω_+-null-pole triple for W. In fact, from the Lemmas 1.1 and 1.2 it follows that any Ω_+-null-pole triple may be obtained from τ in the way described by (1.6). In particular, by the construction described above, a minimal realization $\Theta = (A, B, C, I)$ of W yields the following Ω_+-null-pole triple

$$\big((C|_{\operatorname{Im} Q_\Theta}, A|_{\operatorname{Im} Q_\Theta}), (A^\times|_{\operatorname{Im} Q_\Theta^\times}, Q_\Theta^\times B), Q_\Theta^\times|_{\operatorname{Im} Q_\Theta}\big). \tag{1.7}$$

At this point we like to put emphasis on the fact that the realization Θ need not to be minimal, provided that we know that the pairs $(C|_{\operatorname{Im} Q_\Theta}, A|_{\operatorname{Im} Q_\Theta})$ and $(C|_{\operatorname{Im} Q_\Theta^\times}, A^\times|_{\operatorname{Im} Q_\Theta^\times})$ are zero kernel pairs and the pairs $(A^\times|_{\operatorname{Im} Q_\Theta^\times}, Q_\Theta^\times B)$ and $(A|_{\operatorname{Im} Q_\Theta}, Q_\Theta B)$ are full range pairs.

The coupling operator in a null-pole triple has an interesting additional property. Let Γ be defined by (1.5). One computes that

$$\begin{aligned}\Gamma A_\pi - A_\zeta \Gamma &= S_\zeta(\Theta)^{-1}(Q_\Theta^\times A - A^\times Q_\Theta^\times) S_\pi(\Theta)\\ &= S_\zeta(\Theta)^{-1} Q_\Theta^\times (A - A^\times) S_\pi(\Theta) = S_\zeta(\Theta)^{-1} Q_\Theta^\times BC S_\pi(\Theta) = B_\zeta C_\pi.\end{aligned}$$

Thus the null-pole coupling operator has the property that $\Gamma A_\pi - A_\zeta\Gamma = B_\zeta C_\pi$.

If the set Ω_+ contains all poles and zeros of W, then an Ω_+-null-pole triple τ of W is called a *global Ω_+-null-pole triple.* Moreover, we call the triple τ a *global null-pole triple* if for some set Ω_+ it is a global Ω_+-null-pole triple.

The main question in this chapter concerns the inverse problem, i.e., to construct a rational matrix function with prescribed zero and pole data. To be able to formulate one of the problems accurately, we call a triple $\big((C_\pi, A_\pi), (A_\zeta, B_\zeta), \Gamma\big)$ an *admissible Sylvester data set* if (C_π, A_π) is a zero kernel pair, (A_ζ, B_ζ) is a full range pair, and the operator Γ satisfies the Sylvester equation $\Gamma A_\pi - A_\zeta\Gamma = B_\zeta C_\pi$. If, moreover, A_π and A_ζ have all eigenvalues in the set Ω_+, then the triple is called a *Ω_+-admissible Sylvester data set.* Note that these defining properties implicitly put restrictions on the sizes of the matrices in an admissible Sylvester data set. From the remarks on Ω_+-null-pole triples made above it is clear that an Ω_+-null-pole triple is an Ω_+-admissible Sylvester data set. In Section 3 we consider the problem to find a rational matrix function for which a given Ω_+-admissible Sylvester data set is an Ω_+-null-pole triple. There the following result will be needed.

Theorem 1.3. *Let W_1 and W_2 be rational functions which are proper and have the value I_m at infinity. Then W_1 and W_2 have a common Ω_+-null-pole triple if and only $W_2^{-1}W_1$ has no poles and zeros in Ω_+.*

PROOF. Let $\Theta_1 = (A_1, B_1, C_1, I)$, with state space $\mathcal{X}_1$, be a minimal realization of W_1, and let $\Theta_2 = (A_2, B_2, C_2, I)$, with state space $\mathcal{X}_2$, be a minimal realization of W_2. For $i = 1, 2$ we consider the triple

$$\tau_i = \big((C_i|_{\operatorname{Im} Q_i}, A_i|_{\operatorname{Im} Q_i}), (A_i^\times|_{\operatorname{Im} Q_i^\times}, Q_i^\times B_i), Q_i^\times|_{\operatorname{Im} Q_i}\big)$$

where Q_i and $Q_i^\times$ are the Riesz projections of A_i and $A_i^\times = A_i - B_iC_i$, respectively, corresponding to the eigenvalues in Ω_+. Assume that W_1 and W_2 have a common Ω_+-null-pole triple. Then there exist invertible operators

$$E : \operatorname{Im} Q_1 \to \operatorname{Im} Q_2, \quad F : \operatorname{Im} Q_1^\times \to \operatorname{Im} Q_2^\times$$

such that the following identities hold true

$$(A_2|_{\operatorname{Im} Q_2})E = E(A_1|_{\operatorname{Im} Q_1}), \quad (C_2|_{\operatorname{Im} Q_2})E = C_1|_{\operatorname{Im} Q_1},$$

$$(A_2^\times|_{\operatorname{Im} Q_2^\times})F = F(A_1^\times|_{\operatorname{Im} Q_1^\times}), \quad Q_2^\times B_2 = FQ_1^\times B_1,$$

$$(Q_2^\times|_{\operatorname{Im} Q_2^\times})E = F(Q_1^\times|_{\operatorname{Im} Q_1^\times}).$$

Using these identities and the fact that $B_iC_i = A_i - A_i^\times$ for $i = 1, 2$, one gets by a straightforward calculation that

$$\begin{aligned} &W_2(\lambda)^{-1}W_1(\lambda) = \\ &\big(I - C_2(I - Q_2^\times)(\lambda I - A_2^\times)^{-1}B_2\big)\big(I + C_1(I - Q_1)(\lambda I - A_1)^{-1}B_1\big) + \\ &+ C_2(\lambda I - A_2^\times)^{-1}(I - Q_2^\times)EQ_1B_1 - C_2FQ_1^\times(I - Q_1)(\lambda I - A_1)^{-1}B_1 \end{aligned}$$

and

$$\begin{aligned} &W_1(\lambda)^{-1}W_2(\lambda) = \\ &\big(I - C_1(I - Q_1^\times)(\lambda I - A_1^\times)^{-1}B_1\big)\big(I + C_2(I - Q_2)(\lambda I - A_2)^{-1}B_2\big) + \\ &+ C_1(\lambda I - A_1^\times)^{-1}(I - Q_1^\times)EQ_2B_2 - C_1FQ_2^\times(I - Q_2)(\lambda I - A_2)^{-1}B_2. \end{aligned}$$

It follows that $W_2^{-1}W_1$ and $W_1^{-1}W_2$ are both analytic in Ω_+. Thus the function $W_2^{-1}W_1$ has no poles and zeros in Ω_+.

To prove the converse, assume that $W(\lambda) = W_2(\lambda)^{-1}W_1(\lambda)$ has no poles and zeros in Ω_+. Let $\Theta = (A, B, C, I)$ be a minimal realization of W. Then A and $A^\times = A - BC$ do not have eigenvalues in Ω_+ (cf., Lemma X.2.1). Consider the operators

$$\hat{A} = \begin{pmatrix} A_2 & B_2C \\ 0 & A \end{pmatrix}, \quad \hat{B} = \begin{pmatrix} B_2 \\ B \end{pmatrix}, \quad \hat{C} = \begin{pmatrix} C_2 & C \end{pmatrix}.$$

Then $\hat{\Theta} = (\hat{A}, \hat{B}, \hat{C}, I)$ is a realization of W_1. Let $\hat{Q}$ and $\hat{Q}^\times$ be the Riesz projections of $\hat{A}$ and $\hat{A}^\times$, respectively, corresponding to the eigenvalues inside Ω_+. Here $\hat{A}^\times = \hat{A} - \hat{B}\hat{C}$. We claim that

$$\hat{\tau} = \big((\hat{C}|_{\operatorname{Im}\hat{Q}}, \hat{A}|_{\operatorname{Im}\hat{Q}}), (\hat{A}^\times|_{\operatorname{Im}\hat{Q}^\times}, \hat{Q}^\times\hat{B}), \hat{Q}^\times|_{\operatorname{Im}\hat{Q}}\big)$$

is an Ω_+-null-pole triple of W_1. To prove this it suffices to show that the pairs $(\hat{C}|_{\operatorname{Im}\hat{Q}}, \hat{A}|_{\operatorname{Im}\hat{Q}})$ and $(\hat{C}|_{\operatorname{Im}\hat{Q}^\times}, \hat{A}^\times|_{\operatorname{Im}\hat{Q}^\times})$ are a zero kernel pairs, and the pairs

$(\hat{A}^\times|_{\operatorname{Im}\hat{Q}^\times}, \hat{Q}^\times\hat{B})$ and $(\hat{A}|_{\operatorname{Im}\hat{Q}}, \hat{Q}B)$ are a full range pairs. Let us check that $(\hat{C}|_{\operatorname{Im}\hat{Q}}, \hat{A}|_{\operatorname{Im}\hat{Q}})$ is a zero kernel pair. First note that A and $A^\times$ do not have eigenvalues in Ω_+. So the projections $\hat{Q}$ and $\hat{Q}^\times$ have the following representation:

$$\hat{Q} = \begin{pmatrix} Q_2 & R \\ 0 & 0 \end{pmatrix}, \quad \hat{Q}^\times = \begin{pmatrix} Q_2^\times & 0 \\ R^\times & 0 \end{pmatrix}.$$

Let $x \in \operatorname{Im}\hat{Q}$ be such that $\hat{A}x = \lambda x$ and $\hat{C}x = 0$. Then $x \in \mathcal{X}_2$, and therefore $A_2x = \lambda x$ and $C_2x = 0$. So $x = 0$. This proves that $(\hat{C}|_{\operatorname{Im}\hat{Q}}, \hat{A}|_{\operatorname{Im}\hat{Q}})$ is a zero kernel pair. Next, let us prove that the pair $(\hat{C}|_{\operatorname{Im}\hat{Q}^\times}, \hat{A}^\times|_{\operatorname{Im}\hat{Q}^\times})$ is a zero kernel pair. Let $y \in \operatorname{Im}\hat{Q}^\times$ be such that $\hat{A}^\times y = \lambda y$ and $\hat{C}y = 0$. If $\lambda \notin \Omega_+$, then it follows from $y \in \operatorname{Im}\hat{Q}^\times$ and $\hat{A}^\times y = \lambda y$ that $y = 0$. So we may assume that $\lambda \in \Omega_+$. Write

$$y = \begin{pmatrix} x_2 \\ x \end{pmatrix}, \quad \hat{C}y = C_2x_2 + Cx,$$

$$\hat{A}^\times y = \begin{pmatrix} A_2 - B_2C_2 & 0 \\ -BC_2 & A - BC \end{pmatrix}\begin{pmatrix} x_2 \\ x \end{pmatrix} = \begin{pmatrix} A_2^\times x_2 \\ -BC_2x_2 + A^\times x \end{pmatrix} = \begin{pmatrix} \lambda x_2 \\ \lambda x \end{pmatrix}.$$

From these equalities one computes that $Ax = A^\times x + BCx = A^\times x - BC_2x = \lambda x$. Since $\lambda \in \Omega_+$ and the operator A has no eigenvalues in Ω_+, this implies $x = 0$. Hence we have that $A_2^\times x_2 = \lambda x_2$ and $C_2x_2 = -Cx = 0$. Now use that the pair $(C_2, A_2^\times)$ is a zero kernel pair to conclude that $x_2 = 0$. We have proved that $y = 0$, and hence $(\hat{C}|_{\operatorname{Im}\hat{Q}^\times}, \hat{A}^\times|_{\operatorname{Im}\hat{Q}^\times})$ is a zero kernel pair. In much the same way one gets that $(\hat{A}^\times|_{\operatorname{Im}\hat{Q}^\times}, \hat{Q}^\times\hat{B})$ and $(\hat{A}|_{\operatorname{Im}\hat{Q}}, \hat{Q}B)$ are full range pairs. Therefore, we may conclude that the triple $\hat{\tau}$ is an Ω_+-null-pole triple of W_1.

Next, we show that $\hat{\tau}$ is also an Ω_+-null-pole triple of W_2. Note that $\operatorname{Im}\hat{Q} = \operatorname{Im}Q_2$. It follows that

$$\hat{A}|_{\operatorname{Im}\hat{Q}} = A_2|_{\operatorname{Im}Q_2}, \quad \hat{C}|_{\operatorname{Im}\hat{Q}} = C_2|_{\operatorname{Im}Q_2}. \tag{1.8}$$

Thus $(\hat{C}|_{\operatorname{Im}\hat{Q}}, \hat{A}|_{\operatorname{Im}\hat{Q}})$ is a pole pair of W_2 with respect to Ω_+. Since $\hat{Q}^\times$ is a projection, $R^\times Q_2^\times = R^\times$. Hence the operator

$$F : \operatorname{Im}Q_2^\times \to \operatorname{Im}\hat{Q}^\times, \quad Fx = \begin{pmatrix} x \\ R^\times x \end{pmatrix}$$

is well defined and invertible. One easily checks (use $\hat{A}^\times\hat{Q}^\times = \hat{Q}^\times\hat{A}^\times$) that the following identities hold true:

$$(\hat{A}^\times|_{\operatorname{Im}\hat{Q}^\times})F = F(\hat{A}^\times|_{\operatorname{Im}\hat{Q}^\times}), \quad \hat{Q}^\times\hat{B} = FQ_2^\times B_2, \tag{1.9}$$

$$Q^\times|_{\operatorname{Im}\hat{Q}} = F(Q_2^\times|_{\operatorname{Im}Q_2}). \tag{1.10}$$

We know that $\tau_2 = \big((C_2|_{\operatorname{Im}Q_2}, A_2|_{\operatorname{Im}Q_2}), (A_2^\times|_{\operatorname{Im}Q_2^\times}, Q_2^\times B_2), Q_2^\times|_{\operatorname{Im}Q_2}\big)$ is an Ω_+-null-pole triple of W_2. Therefore the identities (1.8), (1.9) and (1.10) show that the same is true for $\hat{\tau}$. Hence $\hat{\tau}$ is a common Ω_+-null-pole triple of W_1 and W_2. □

XI.2 The one sided homogeneous interpolation problem

Let (C_π, A_π) and (C_ζ, A_ζ) be zero kernel pairs of matrices. In this section we consider the question under what conditions does there exist a proper rational $m \times m$ matrix function W, invertible at infinity, such that (C_π, A_π) is a global right pole pair of W and (C_ζ, A_ζ) is a global right null pair of W. Remark that this clearly requires that the matrices C_π and C_ζ have m rows. The first theorem gives necessary conditions.

Theorem 2.1. *Let (C_π, A_π) be a right pole pair and (C_ζ, A_ζ) be a right null pair, with respect to $\mathbb{C}$, of the rational $m \times m$ matrix function W, which is analytic and invertible at infinity. Then*

(i) *A_π and A_ζ have the same order, n say,*

(ii) *the blocks $(Z_\pi; I, Q)$ and $(Z_\zeta; I, Q)$ are block similar. Here Q the projection of $\mathbb{C}^n \oplus \mathbb{C}^m$ along $\mathbb{C}^m$ onto $\mathbb{C}^n$, and*

$$Z_\pi = \begin{pmatrix} A_\pi \\ C_\pi \end{pmatrix} : \operatorname{Im} Q \to \mathbb{C}^n \oplus \mathbb{C}^m, \quad Z_\zeta = \begin{pmatrix} A_\zeta \\ C_\zeta \end{pmatrix} : \operatorname{Im} Q \to \mathbb{C}^n \oplus \mathbb{C}^m. \tag{2.1}$$

The similarity in (ii) *is of the form*

$$S = \begin{pmatrix} N & B_\pi \\ 0 & D \end{pmatrix}, \tag{2.2}$$

where N and D are invertible matrices of sizes $n \times n$ and $m \times m$, respectively, which can be chosen in such a way that

$$\begin{aligned} W(\lambda) &= D + C_\pi(\lambda I - A_\pi)^{-1} B_\pi, \\ W(\lambda)^{-1} &= D^{-1} + C_\zeta(\lambda I - A_\zeta)^{-1}(-N^{-1}B_\pi D^{-1}), \end{aligned} \tag{2.3}$$

(A_π, B_π) is a global left pole pair of W, and $(A_\zeta, -N^{-1}B_\pi D^{-1})$ is a global left null pair of W.

PROOF. There exists a matrix B_π such that $W(\lambda) - C_\pi(\lambda I - A_\pi)^{-1}B_\pi$ has an analytic continuation to $\mathbb{C}$ and the pair (A_π, B_π) is a left pole pair of W. It follows from Liouville's theorem that the function $W(\lambda) - C_\pi(\lambda I - A_\pi)^{-1}B_\pi$ is constant and therefore

$$W(\lambda) = D + C_\pi(\lambda I - A_\pi)^{-1} B_\pi, \tag{2.4}$$

where D is the value at infinity of W. According to our hypothesis, D is invertible, and it follows from (2.4) that

$$W(\lambda)^{-1} = D^{-1} - D^{-1}C_\pi\bigl(\lambda I - (A_\pi - B_\pi D^{-1} C_\pi)\bigr)^{-1} B_\pi D^{-1}.$$

Check that

$$\operatorname{Ker}\begin{pmatrix}\lambda I-(A_\pi-B_\pi D^{-1}C_\pi)\\ D^{-1}C_\pi\end{pmatrix}=\operatorname{Ker}\begin{pmatrix}\lambda I-A_\pi\\ C_\pi\end{pmatrix}=\{0\}$$

and that

$$\operatorname{Im}\big(\,[\lambda I-(A_\pi-B_\pi D^{-1}C_\pi)]\quad B_\pi\,\big)=\operatorname{Im}\big(\,[\lambda I-A_\pi]\quad B_\pi\,\big)=\mathbb{C}^n$$

for each value of λ. This proves that the pair $(D^{-1}C_\pi, A_\pi - B_\pi D^{-1}C_\pi)$ is a right pole pair of W^{-1}. From Lemma 1.1 we know that this pair is similar to the pair (C_ζ, A_ζ). So there exists a unique invertible operator N such that

$$C_\zeta = D^{-1}C_\pi N,\quad A_\zeta = N^{-1}(A_\pi - B_\pi D^{-1}C_\pi)N. \tag{2.5}$$

Then clearly (i) must be fulfilled. The formulas (2.5) can be rewritten as

$$\begin{pmatrix}A_\zeta\\ C_\zeta\end{pmatrix}N^{-1}=\begin{pmatrix}N^{-1} & -N^{-1}B_\pi D^{-1}\\ 0 & D^{-1}\end{pmatrix}\begin{pmatrix}A_\pi\\ C_\pi\end{pmatrix}$$

and are therefore equivalent to

$$\begin{pmatrix}A_\pi\\ C_\pi\end{pmatrix}N=\begin{pmatrix}N & B_\pi\\ 0 & D\end{pmatrix}\begin{pmatrix}A_\zeta\\ C_\zeta\end{pmatrix}. \tag{2.6}$$

We see that S given by (2.2) is a block similarity of the blocks $(Z_\pi; I, Q)$ and $(Z_\zeta; I, Q)$ and that (2.3) holds true.

Finally note that if S is any block similarity of $(Z_\pi; I, Q)$ and $(Z_\zeta; I, Q)$, then $S[\operatorname{Im} Q] = \operatorname{Im} Q$, and hence S has the form (2.2). □

The conditions (i) and (ii) in Theorem 2.1 are in general not sufficient to guarantee that the zero kernel pairs (C_π, A_π) and (C_ζ, A_ζ) are a global right pole pair and a global right null pair, respectively, of a proper rational matrix function that is invertible at infinity. To show this we give an example. Let

$$A_\pi = A_\zeta = \begin{pmatrix}0&1&0\\1&0&0\\0&0&0\end{pmatrix},\quad C_\pi = C_\zeta = \begin{pmatrix}0&1&0\\0&0&1\end{pmatrix}.$$

Then the pairs (C_π, A_π) and (C_ζ, A_ζ) are zero kernel pairs. Furthermore, the conditions (i) and (ii) are fulfilled. So assume that (C_π, A_π) and (C_ζ, A_ζ) are a global right pole pair and a global right null pair, respectively, of a proper rational matrix function W which is invertible at infinity. Then, according to Theorem 2.1, the function W can be written as $W(\lambda) = D + C_\pi(\lambda I - A_\pi)^{-1}B_\pi$, where D and B_π are such that (2.6) holds true with D and N invertible, while, in addition, (A_π, B_π) is a left pole pair of W. From (2.6) we derive that the matrices D and B_π have to be such that

$$\begin{pmatrix} N & B_\pi \\ 0 & D \end{pmatrix} = \begin{pmatrix} a & 0 & c & 0 & b \\ 0 & a & b & 0 & c \\ 0 & 0 & d & 0 & 0 \\ 0 & 0 & 0 & a & b \\ 0 & 0 & 0 & 0 & d \end{pmatrix}, \quad ad \neq 0. \tag{2.7}$$

However, for none of these possible values of B_π the pair (A_π, B_π) is a full range pair. Therefore, the pair (A_π, B_π) is not a left pole pair of W. This contradicts our assumptions. So there is no proper W, invertible at infinity, which has (C_π, A_π) and (C_ζ, A_ζ) as a global right pole pair and a global right null pair, respectively. From (2.7) we can also compute all rational matrix functions W given by (2.3). We obtain

$$W(\lambda) = \begin{pmatrix} a & b + (c\lambda + d)(\lambda^2 - 1)^{-1} \\ 0 & d \end{pmatrix}, \quad ad \neq 0.$$

None of these rational functions W has a pole at 0. Since 0 is an eigenvalue of A_π, this also shows that (C_π, A_π) is not a global pole pair of W.

According to Theorem 2.1 any proper rational matrix function, invertible at infinity, with (C_π, A_π) as global right pole pair and (C_ζ, A_ζ) as global right null pair can be expressed as in (2.3) in terms of a similarity between the blocks $(Z_\pi; I, Q)$ and $(Z_\zeta; I, Q)$ given by (2.1). But, in general, even if there exists a rational matrix function W with (C_π, A_π) as global right pole pair and (C_ζ, A_ζ) as global right null pair, it is not true that, conversely, each block similarity produces a rational matrix function with (C_π, A_π) as right pole pair and (C_ζ, A_ζ) as right null pair. This follows from the next example. Write

$$W_0(\lambda) = \begin{pmatrix} 1 & \lambda^{-2} \\ 0 & 1 \end{pmatrix}.$$

Put

$$A_\pi = \begin{pmatrix} 0 & 1 \\ 0 & 0 \end{pmatrix}, \quad C_\pi = \begin{pmatrix} 1 & 0 \\ 0 & 0 \end{pmatrix}, \quad B_\pi = \begin{pmatrix} 0 & 0 \\ 0 & 1 \end{pmatrix},$$

and

$$A_\zeta = \begin{pmatrix} 0 & 1 \\ 0 & 0 \end{pmatrix}, \quad C_\zeta = \begin{pmatrix} 1 & 0 \\ 0 & 0 \end{pmatrix}, \quad B_\zeta = \begin{pmatrix} 0 & 0 \\ 0 & -1 \end{pmatrix}.$$

Then $W_0(\lambda) = I + C_\pi(\lambda I - A_\pi)^{-1} B_\pi$, the pair (C_π, A_π) is a global right pole pair, and (C_ζ, A_ζ) is a global right null pair of W_0. In this case, the similarities between the associated blocks $(Z_\pi; I, Q)$ and $(Z_\zeta; I, Q)$ are given by

$$S = \begin{pmatrix} N & B \\ 0 & D \end{pmatrix} = \begin{pmatrix} a & 0 & 0 & c \\ 0 & a & 0 & d \\ 0 & 0 & a & b \\ 0 & 0 & 0 & e \end{pmatrix}, \quad ae \neq 0,$$

and a, b, c and d are complex numbers. Compute $W(\lambda) = I + C_\pi(\lambda I - A_\pi)^{-1}B$, with

$$B = \begin{pmatrix} 0 & c \\ 0 & d \end{pmatrix}.$$

This yields

$$W(\lambda) = \begin{pmatrix} a & b + c\lambda^{-1} + d\lambda^{-2} \\ 0 & e \end{pmatrix}, \qquad ae \neq 0.$$

If $d = 0$, then the pair (A_π, B) is not a full range pair, and therefore in this case the pair (C_π, A_π) is not a pole pair of W. Similarities with $d = 0$ do not lead to a rational matrix function with (C_π, A_π) as right pole pair. On the other hand, for $d \neq 0$ the pair (A_π, B) is a full range pair, and therefore in this case the pair (C_π, A_π) is a right pole pair of W. Note that

$$-N^{-1}BD^{-1} = \begin{pmatrix} 0 & \frac{c}{ae} \\ 0 & \frac{d}{ae} \end{pmatrix}.$$

So, if $d = 0$, then the pair $(A_\zeta, -N^{-1}BD^{-1})$ is not a full range pair. Hence in this case (C_ζ, A_ζ) is not a null pair of W.

The next theorem shows that the conditions (i) and (ii) in Theorem 2.1 are also sufficient if the matrices A_π and A_ζ do not have common eigenvalues.

Theorem 2.2. *Let (C_π, A_π) and (C_ζ, A_ζ) be zero kernel pairs, such that condition (i) of Theorem 2.1 holds. Suppose that A_π and A_ζ do not have common eigenvalues. In order that the pair (C_π, A_π) is a global right pole pair and (C_ζ, A_ζ) is a global right null pair of a proper rational $m \times m$ matrix function W, which is invertible at infinity, it is necessary and sufficient that the condition (ii) of Theorem 2.1 is fulfilled. In that case the general form of W is*

$$W(\lambda) = D + C_\pi(\lambda I - A_\pi)^{-1}B_\pi, \tag{2.8}$$

where

$$S = \begin{pmatrix} N & B_\pi \\ 0 & D \end{pmatrix} \tag{2.9}$$

is an arbitrary block similarity of the blocks $(Z_\pi; I, Q)$ and $(Z_\zeta; I, Q)$ given by (2.1).

PROOF. From Theorem 2.1 we know that the conditions (i) and (ii) are necessary and that W must have the form (2.8). It remains to show that the conditions (i) and (ii) are sufficient. So we choose the block similarity S given by (2.9) and define W by (2.8). We have to show that (C_π, A_π) is a right pole pair and (C_ζ, A_ζ) is a right null pair of W. Since S is a block similarity, the equation (2.6) is fulfilled. Therefore also (2.5) holds true. So we compute that W^{-1} is given by the second equality in (2.3). It only remains to show that the pairs (C_π, A_π)

and (C_ζ, A_ζ) are zero kernel pairs and the pairs (A_π, B_π) and $(A_\zeta, -N^{-1}B_\pi D^{-1})$ are full range pairs. To do this we use that, by our hypotheses, the matrices A_π and A_ζ have no common eigenvalues.

Put $\mathcal{V} = \bigcap_{k=0}^{\infty} C_\pi A_\pi^k$. Then $\mathcal{V}$ is invariant for A_π. Let $v \in \mathcal{V}$ be an eigenvector of A_π. Then $(A_\pi - B_\pi D^{-1}C_\pi)v = A_\pi v = \lambda v$, where λ is the eigenvalue of A_π corresponding to v. Thus λ is also an eigenvalue of $A_\pi - B_\pi D^{-1}C_\pi$. By the second identity in (2.5) the operator $A_\pi - B_\pi D^{-1}C_\pi$ is similar to A_ζ. So we see that λ is also an eigenvalue of A_ζ. Because A_π and A_ζ have no common eigenvalues, we reached a contradiction, and it follows that $\mathcal{V} = \{0\}$. Therefore (C_π, A_π) is a zero kernel pair. Using duality and a similar argument one shows that the pair (A_π, B_π) is a full range pair. This shows that (C_π, A_π) is a right pole pair of W. In a similar way one shows that (C_ζ, A_ζ) is a right zero pair of W. □

For the case when there are common eigenvalues of A_π and A_ζ extra conditions are needed for the existence of a rational matrix function having (C_π, A_π) is a right pole pair and $C_\zeta, A_\zeta)$ is a right null pair. Such a condition is described in Gohberg-Kaashoek-Van Schagen [3], Theorem 5.1.

XI.3 Homogeneous two sided interpolation

In this section the symbol W denotes a proper rational $m \times m$ matrix function which has value I at infinity. The first result in this section characterizes global null-pole triples of such a rational matrix function.

Theorem 3.1. *Let $\tau = \big((C_\pi, A_\pi), (A_\zeta, B_\zeta), \Gamma\big)$ be an Ω_+-admissible Sylvester data set. Then τ is a global Ω_+-null-pole triple for some proper rational matrix function W with value I at infinity if and only if Γ is invertible. In that case W is uniquely determined by τ, and $W(\lambda)$ and $W(\lambda)^{-1}$ are given by the following minimal realizations:*

$$W(\lambda) = I + C_\pi(\lambda I - A_\pi)^{-1}\Gamma^{-1}B_\zeta, \tag{3.1}$$

$$W(\lambda)^{-1} = I - C_\pi\Gamma^{-1}(\lambda I - A_\zeta)^{-1}B_\zeta. \tag{3.2}$$

PROOF. Assume that τ is a global Ω_+-null-pole triple for the proper rational matrix function W, which has the value I at infinity. We need to recall the construction of the null-pole coupling operator Γ. So let $\Theta = (A, B, C, I)$ be a minimal realization of W. Then the null-pole coupling operator $\Gamma : \mathcal{X}_\pi \to \mathcal{X}_\zeta$ is given by (1.5). To show that Γ is invertible, it is sufficient to show that the projection $Q_\Theta^\times$ is invertible. The minimality of the realization Θ gives that the eigenvalues of $A^\times$ are poles of W^{-1}. Since τ is a global Ω_+-null-pole triple, all the poles of W^{-1} are inside Ω_+. So all the eigenvalues of $A^\times$ are inside Ω_+, and thus the Riesz projection $Q_\Theta^\times = I$. We proved that Γ is invertible.

Next, let us show that in this case the function W is given by (3.1). We already obtained that $Q_\Theta^\times = I$. Since Θ is minimal, each eigenvalue of A is a pole of W. Now all poles of W are inside Ω_+, and hence all eigenvalues of A are inside Ω_+. It follows that $Q_\Theta = I$. From the formulas (1.2) and (1.3) we read off that

$$(C_\pi S_\pi(\Theta)^{-1}, S_\pi(\Theta)A_\pi S_\pi(\Theta)^{-1}) = (C, A) ,$$

$$(S_\zeta(\Theta)A_\zeta S_\zeta(\Theta)^{-1}, S_\zeta(\Theta)B_\zeta) = (A^\times, B) .$$

It follows that $S_\pi(\Theta)^{-1}B = \Gamma^{-1}B_\zeta$ and $CS_\zeta(\Theta) = C_\pi\Gamma^{-1}$. Inserting these equalities in the realization $W(\lambda) = I + C(\lambda I - A)^{-1}B$ yields the equality (3.1). Since $W(\lambda)^{-1} = I - C(\lambda I - A^\times)^{-1}B$, we also obtain (3.2). So W is uniquely determined by τ.

Next, assume that the coupling operator Γ of an Ω_+-admissible Sylvester data set τ is invertible. We have to prove that τ is a global Ω_+-null-pole triple for a rational matrix function W. Define W by (3.1). Then W is a proper rational $m \times m$ matrix function which has the value I at infinity. Since $\Gamma A_\pi - A_\zeta\Gamma = B_\zeta C_\pi$, it follows that W^{-1} is given by (3.2). The realizations (3.1) and (3.2) are minimal because (C_π, A_π) is a zero kernel pair and the pair (A_ζ, B_ζ) is full range. So the poles of W coincide with the eigenvalues of A_π and the zeros of W are the eigenvalues of A_ζ. Therefore the poles and zeros of W are inside the set Ω_+. Also we see that (C_π, A_π) is a right pole pair on Ω_+ and (A_ζ, B_ζ) is a left null pair of W on Ω_+. It remains to see that Γ is the null-pole coupling operator. This follows from the equality $(A_\pi - \Gamma^{-1}B_\zeta C_\pi, \Gamma^{-1}B_\zeta) = (\Gamma^{-1}A_\zeta\Gamma, \Gamma^{-1}B_\zeta)$. □

Recall that for a proper rational matrix function W the *McMillan degree* is defined to be the dimension of the state space of a minimal realization of W. It follows from Theorems IX.1.1, IX.4.3 and IX.4.4 that each proper rational matrix function has a minimal realization. The next theorem states that each Sylvester data set appears as the null-pole triple of some proper rational matrix function with the value I at infinity.

Theorem 3.2. *Let $\tau = \{(C_\pi, A_\pi), (A_\zeta, B_\zeta), \Gamma\}$ be an Ω_+-admissible Sylvester data set. Then τ is an Ω_+-null-pole triple for some proper rational matrix function with the value I at infinity. The minimal possible McMillan degree of such a function is* $\operatorname{order} A_\pi + \operatorname{order} A_\zeta - \operatorname{rank}\Gamma$.

The proof of this theorem is split into four lemmas.

Lemma 3.3. *Let $\tau = \{(C_\pi, A_\pi), (A_\zeta, B_\zeta), \Gamma\}$ be an Ω_+-admissible Sylvester data set. Decompose the main space $\mathcal{X}_\zeta$ of (A_ζ, B_ζ) as $\mathcal{X}_\zeta = \operatorname{Im}\Gamma \oplus \mathcal{K}$, and let $\rho_\zeta : \mathcal{X}_\zeta \to \mathcal{K}$ be the projection of $\mathcal{X}_\zeta$ along $\operatorname{Im}\Gamma$ onto $\mathcal{K}$ and let $\eta_\zeta : \mathcal{K} \to \mathcal{X}_\zeta$ be the canonical embedding. Let $\rho_\pi : \mathcal{X}_\pi \to \operatorname{Ker}\Gamma$ be a projection of the main space $\mathcal{X}_\pi$ of (C_π, A_π) onto $\operatorname{Ker}\Gamma$, and let $\eta_\pi : \operatorname{Ker}\Gamma \to \mathcal{X}_\pi$ be the canonical embedding. Then the pair $(C_\pi\eta_\pi, \rho_\pi A_\pi\eta_\pi)$, with main space $\operatorname{Ker}\Gamma$, is a zero kernel pair, and the pair $(\rho_\zeta A_\zeta\eta_\zeta, \rho_\zeta B_\zeta)$, with main space $\mathcal{K}$, is a full range pair.*

PROOF. We recall that the pair (C_π, A_π) is a zero kernel pair and that $B_\zeta C_\pi = \Gamma A_\pi - A_\zeta\Gamma$. Let $x \in \operatorname{Ker}\Gamma$ be such that $C_\pi\eta_\pi x = 0$ and $\lambda x - \rho_\pi A_\pi\eta_\pi x = 0$.

Using $\eta_\pi x \in \operatorname{Ker}\Gamma$ and $C_\pi\eta_\pi x = 0$, we get $\Gamma A_\pi \eta_\pi x = 0$. So $A_\pi\eta_\pi x \in \operatorname{Ker}\Gamma$. This means that $\eta_\pi\rho_\pi A_\pi\eta_\pi x = A_\pi\eta_\pi x$, and we conclude that $\lambda\eta_\pi x - A_\pi\eta_\pi x = 0$. Since (C_π, A_π) is a zero kernel pair, this proves that $\eta_\pi x = 0$, and thus also $x = 0$. So $(C_\pi\eta_\pi, \rho_\pi A_\pi \eta_\pi)$ is a zero kernel pair.

Next, we will prove that $(\rho_\varsigma A_\varsigma\eta_\varsigma, \rho_\varsigma B_\varsigma)$ is a full range pair. We write $A_\mathcal{K} = \rho_\varsigma A_\varsigma\eta_\varsigma$ and $B_\mathcal{K} = \rho_\varsigma B_\varsigma$. Since the pair $(A_\varsigma, B_\varsigma)$ is full range, it suffices to show that

$$\rho_\varsigma \operatorname{Im}\begin{pmatrix} A_\varsigma^j B_\varsigma & \cdots & A_\varsigma B_\varsigma & B_\varsigma \end{pmatrix} = \operatorname{Im}\begin{pmatrix} A_\mathcal{K}^j B_\mathcal{K} & \cdots & A_\mathcal{K} B_\mathcal{K} & B_\mathcal{K} \end{pmatrix} \tag{3.3}$$

for each $j \geq 0$. We prove by induction that

$$\rho_\varsigma A_\varsigma^j B_\varsigma x \in \operatorname{Im}\begin{pmatrix} A_\mathcal{K}^j B_\mathcal{K} & \cdots & A_\mathcal{K} B_\mathcal{K} & B_\mathcal{K} \end{pmatrix},$$

which implies (3.3). Obviously,

$$\rho_\varsigma A_\varsigma^j B_\varsigma x = \rho_\varsigma A_\varsigma (I - \eta_\varsigma\rho_\varsigma) A_\varsigma^{j-1} B_\varsigma x + \rho_\varsigma A_\varsigma \eta_\varsigma \rho_\varsigma A_\varsigma^{j-1} B_\varsigma x. \tag{3.4}$$

The second term in the right hand side is, according to the induction hypothesis, in the subspace

$$A_\mathcal{K}[\operatorname{Im}\begin{pmatrix} A_\mathcal{K}^{j-1} B_\mathcal{K} & \cdots & A_\mathcal{K} B_\mathcal{K} & B_\mathcal{K} \end{pmatrix}].$$

Remark that $(I - \eta_\varsigma\rho_\varsigma)A_\varsigma^{j-1}B_\varsigma x \in \operatorname{Im}\Gamma$. So the first term in the right hand side of (3.4) is of the form $\rho_\varsigma A_\varsigma \Gamma y$. Now use that $\Gamma A_\pi - A_\varsigma\Gamma = B_\varsigma C_\pi$ to get that

$$\rho_\varsigma A_\varsigma (I - \eta_\varsigma\rho_\varsigma) A_\varsigma^{j-1} B_\varsigma x = \rho_\varsigma(\Gamma A_\pi - B_\varsigma C_\pi) y = \rho_\varsigma B_\varsigma C_\pi y$$

This proves that the first term in the right hand side of (3.4) is in $\operatorname{Im} B_\mathcal{K}$. Thus we proved that the left hand side of (3.4) is in $\operatorname{Im}\begin{pmatrix} A_\mathcal{K}^j B_\mathcal{K} & \cdots & A_\mathcal{K} B_\mathcal{K} & B_\mathcal{K} \end{pmatrix}$. □

Recall that Ω_+ is the inner domain of the curve γ. We denote the domain outside γ by Ω_-.

Lemma 3.4. *Fix $\epsilon_- \in \Omega_-$. Let $C : \mathcal{M} \to \mathbb{C}^m$ and $A : \mathcal{M} \to \mathcal{M}$ be given operators such that the eigenvalues of A are in Ω_+, and assume that the full length block*

$$\left(\begin{pmatrix} A & 0 \\ C & 0 \end{pmatrix}; \begin{pmatrix} I & 0 \\ 0 & I \end{pmatrix}, \begin{pmatrix} I & 0 \\ 0 & 0 \end{pmatrix} \right)$$

has indices of the third kind $\alpha_1 \geq \cdots \geq \alpha_k$ and no invariant polynomials. Then there exists a pair (S, G), with $G : \mathbb{C}^m \to \mathcal{M}$ and $S = A - GC : \mathcal{M} \to \mathcal{M}$, such that S has ϵ_- as its only eigenvalue and Jordan blocks of sizes $\alpha_1 \geq \cdots \geq \alpha_k$.

Let $A : \mathcal{K} \to \mathcal{K}$ and $B : \mathbb{C}^m \to \mathcal{K}$ be given operators such that the eigenvalues of A are in Ω_+, and assume that the full width block

$$\left(\begin{pmatrix} A & B \\ 0 & 0 \end{pmatrix} \begin{pmatrix} I & 0 \\ 0 & 0 \end{pmatrix}, \begin{pmatrix} I & 0 \\ 0 & I \end{pmatrix} \right)$$

has indices of the first kind $\beta_1 \geq \cdots \geq \beta_\ell$ and no invariant polynomials. Then there exists a pair (F, T), with $F : \mathcal{K} \to \mathbb{C}^m$ and $T = A - BF : \mathcal{K} \to \mathcal{K}$ such that T has ϵ_- as its only eigenvalue and Jordan blocks of sizes $\beta_1 \geq \cdots \geq \beta_\ell$.

PROOF. Let Q be a projection of $\mathcal{M}$ along $\operatorname{Ker} C$. Then, according to Theorem III.1.5, the full length block $(A|_{\operatorname{Im} Q}; I, Q)$ which has $\mathcal{M}$ as its underlying space, has indices of the third kind equal to $\alpha_1 - 1 \geq \cdots \geq \alpha_k - 1$. For each G the operator $A + GC$ is completion of the block, and, conversely, each completion has this form. So, in order to prove the existence of the desired G, we only need to check that it is possible to complete the block such that the completion has eigenvalue ϵ_- with Jordan blocks of sizes $\alpha_1 \geq \cdots \geq \alpha_k$. To see that indeed this is possible we apply Theorem IV.1.1. Since the block has no non-constant invariant polynomials, the first condition in Theorem IV.1.1 is void. The second condition simplifies to

$$\sum_{i=1}^{j} \deg q_i \geq \sum_{i=1}^{j} \alpha_i,$$

where $q_i(\lambda) = (\lambda - \epsilon_-)^{\alpha_i}$ is the desired i-th invariant polynomial of the completion. Since the degree of this polynomial is α_i, the second condition in Theorem IV.1.1 is fulfilled.

The existence of F and T is proved in the same way as the existence of (S, G). Here we use Theorem V.5.1 in place of Theorem IV.1.1. Notice that in this case the i-th invariant polynomial of T is required to be $(\lambda - \epsilon_-)^{\beta_i}$. □

We apply Lemma 3.4 to the pairs $(C_\pi \eta_\pi, \rho_\pi A_\pi \eta_\pi)$ and $(\rho_\zeta A_\zeta \eta_\zeta, \rho_\zeta B_\zeta)$ appearing in Lemma 3.3. To see that this is possible first note that $(C_\pi \eta_\pi, \rho_\pi A_\pi \eta_\pi)$ is a zero kernel pair, and hence the corresponding block has no invariant polynomials. Also remark that $(\rho_\zeta A_\zeta \eta_\zeta, \rho_\zeta B_\zeta)$ is a full range pair, which implies that the corresponding block has no invariant polynomials. We construct for these pairs a pair (S, G) and a pair (F, T) with the properties described in Lemma 3.4. We will refer to such a pair (S, G) as an *ϵ_--pole correction pair* and to such a pair (F, T) as an *ϵ_--null correction pair* of the admissible Sylvester data set τ.

To state the third lemma we need an additional definition. If

$$\tau_1 = \{(C_\pi^{(1)}, A_\pi^{(1)}), (A_\zeta^{(1)}, B_\zeta^{(1)}), \Gamma_1\}$$

is an Ω_1-admissible Sylvester data set and

$$\tau_2 = \{(C_\pi^{(2)}, A_\pi^{(2)}), (A_\zeta^{(2)}, B_\zeta^{(2)}), \Gamma_2\}$$

is an Ω_2-admissible Sylvester data set, with $\Omega_1 \cap \Omega_2 = \emptyset$, then the *direct sum* $\tau_1 \oplus \tau_2$ is by definition the triple

$$\tau_1 \oplus \tau_2 = \big((C_\pi, A_\pi), (A_\zeta, B_\zeta), \Gamma\big),$$

with

$$(C_\pi, A_\pi) = \left(\begin{pmatrix} C_\pi^{(1)} & C_\pi^{(2)} \end{pmatrix}, \begin{pmatrix} A_\pi^{(1)} & 0 \\ 0 & A_\pi^{(2)} \end{pmatrix} \right),$$

$$(A_\zeta, B_\zeta) = \left(\begin{pmatrix} A_\zeta^{(1)} & 0 \\ 0 & A_\zeta^{(2)} \end{pmatrix}, \begin{pmatrix} B_\zeta^{(1)} \\ B_\zeta^{(2)} \end{pmatrix} \right),$$

and

$$\Gamma = \begin{pmatrix} \Gamma_1 & \Gamma_{12} \\ \Gamma_{21} & \Gamma_2, \end{pmatrix}$$

where Γ_{12} and Γ_{21} are the unique solutions of the Lyapunov equations

$$\Gamma_{12} A_\pi^{(2)} - A_\zeta^{(1)} \Gamma_{12} = B_\zeta^{(1)} C_\pi^{(2)}, \tag{3.5}$$

$$\Gamma_{21} A_\pi^{(1)} - A_\zeta^{(2)} \Gamma_{21} = B_\zeta^{(2)} C_\pi^{(1)}. \tag{3.6}$$

Note that the assumption $\Omega_1 \cap \Omega_2 = \emptyset$ implies that the matrices $A_\pi^{(2)}$ and $A_\zeta^{(1)}$ have no common eigenvalues, and hence (3.5) is indeed uniquely solvable. With a similar argument one sees that (3.6) has a unique solution. An admissible Sylvester data set τ_0 is called a *complement* of τ if the coupling operator of $\tau \oplus \tau_0$ is invertible. A complement τ_0 is called a *minimal complement* if the rank of the coupling operator of $\tau \oplus \tau_0$ is minimal among the ranks of the coupling operators of all possible complements of τ.

Lemma 3.5. *If τ is an Ω_+-null-pole triple of W, then there exists an admissible Sylvester data set τ' such that $\tau \oplus \tau'$ is a global null-pole triple of W.*

PROOF. We choose a minimal realization (A, B, C, D) of W. Let Q $(Q^\times)$ be the spectral projection of A $(A^\times)$ with respect to Ω_+. Let Ω_- be a bounded set that contains all the eigenvalues of A and $A^\times$ outside Ω_+, has an empty intersection with Ω_+, and has a boundary which is a rectifiable curve. We write P $(P^\times)$ for the spectral projection of A $(A^\times)$ with respect to Ω_-. Put

$$\begin{aligned} \tau_1 &= \big((C|_{\operatorname{Im} Q}, A|_{\operatorname{Im} Q}), (A^\times|_{\operatorname{Im} Q^\times}, Q^\times B), Q^\times|_{\operatorname{Im} Q} \big), \\ \tau_2 &= \big((C|_{\operatorname{Im} P}, A|_{\operatorname{Im} P}), (A^\times|_{\operatorname{Im} P^\times}, P^\times B), P^\times|_{\operatorname{Im} P} \big). \end{aligned}$$

Then $\tau_1 \oplus \tau_2$ is a global null-pole triple of W with coupling operator

$$\Gamma = \begin{pmatrix} Q^\times|_{\operatorname{Im} Q} & Q^\times|_{\operatorname{Im} P} \\ P^\times|_{\operatorname{Im} Q} & P^\times|_{\operatorname{Im} P} \end{pmatrix}.$$

Note that Γ is the identity operator on the state space of the minimal realization (A, B, C, D). Since τ and τ_1 are Ω_+-null-pole triples of W, there exists invertible operators S_π and S_ζ such that

$$(C_\pi S_\pi^{-1}, S_\pi A_\pi S_\pi^{-1}) = (C|_{\operatorname{Im} Q}, A|_{\operatorname{Im} Q}), \quad (S_\zeta A_\zeta S_\zeta^{-1}, S_\zeta B_\zeta) = (A^\times|_{\operatorname{Im} Q^\times}, Q^\times B).$$

and $\Gamma = S_\zeta^{-1} Q^\times S_\pi$. We will show that the triple $\tau \oplus \tau_2$ is a global null-pole triple of W. The coupling operator of $\tau \oplus \tau_2$ is

$$\Gamma_1 = \begin{pmatrix} S_\zeta^{-1} & 0 \\ 0 & I \end{pmatrix} \begin{pmatrix} Q^\times|_{\operatorname{Im} Q} & Q^\times|_{\operatorname{Im} P} \\ P^\times|_{\operatorname{Im} Q} & P^\times|_{\operatorname{Im} P} \end{pmatrix} \begin{pmatrix} S_\pi & 0 \\ 0 & I \end{pmatrix}.$$

which is invertible. It is now straight forward to check that also

$$W(\lambda) = I + \begin{pmatrix} C_\pi & C|_{\operatorname{Im} P} \end{pmatrix} \left(\lambda I - \begin{pmatrix} A_\pi & 0 \\ 0 & A|_{\operatorname{Im} P} \end{pmatrix} \right)^{-1} \Gamma_1^{-1} \begin{pmatrix} B_\zeta \\ P^\times B \end{pmatrix}.$$

This shows that indeed $\tau \oplus \tau_2$ is a global null-pole triple of W. □

With the operators η_ζ, ρ_ζ, η_π and ρ_π appearing in Lemma 3.3 we associate a generalized inverse $\Gamma^+ : \mathcal{X}_\zeta \to \mathcal{X}_\pi$ of Γ such that

$$\Gamma\Gamma^+ = I - \eta_\zeta \rho_\zeta, \quad \Gamma^+\Gamma = I - \eta_\pi \rho_\pi. \tag{3.7}$$

This generalized inverse is used in the final lemma of this section.

Lemma 3.6. *Let $\tau = \big((C_\pi, A_\pi), (A_\zeta, B_\zeta), \Gamma\big)$ be an Ω_+-admissible Sylvester data set, and fix ϵ_- in Ω_-. Let the pair (S, G) be an ϵ_--pole correction pair and (F, T) an ϵ_--null correction pair of the set τ. Then*

$$\tau_0 = \big((-C_\pi X - F, T), (S, -YB_\zeta + G), \Gamma_0\big)$$

is a minimal complement for τ. Here $X : \mathcal{K} \to \mathcal{X}_\pi$ and $Y : \mathcal{X}_\zeta \to \operatorname{Ker}\Gamma$ are the unique solutions of the Lyapunov equations

$$A_\pi X - XT = A_{12}, \quad YA_\zeta - SY = A_{21}, \tag{3.8}$$

with $A_{12} : \mathcal{K} \to \mathcal{X}_\pi$ and $A_{21} : \mathcal{X}_\zeta \to \operatorname{Ker}\Gamma$ being defined by

$$A_{21} = GF\rho_\zeta + (\rho_\pi A_\pi - S\rho_\pi - GC_\pi)\Gamma^+, \ A_{12} = \Gamma^+(A_\zeta\eta_\zeta - \eta_\zeta T - B_\zeta F). \tag{3.9}$$

The coupling operator Γ_0 is given by $\Gamma_0 = Y\Gamma X - Y\eta_\zeta - \rho_\pi X$. Furthermore, the direct sum $\tau \oplus \tau_0$ is a global null-pole triple of

$$\tilde{W}(\lambda) = I_m + C_\pi(\lambda I - A_\pi)^{-1}[(\Gamma_{11}^\times - \eta_\pi Y)B_\zeta + \eta_\pi G] + (-C_\pi X - F)(\lambda I - T)^{-1}\rho_\zeta B_\zeta, \tag{3.10}$$

of which the inverse is given by

$$\tilde{W}(\lambda)^{-1} = I_m + [-C_\pi(\Gamma_{11}^\times - X\rho_\zeta) + F\rho_\zeta)](\lambda I - A_\zeta)^{-1}B_\zeta + C_\pi\eta_\pi(\lambda I - S)^{-1}(YB_\zeta - G). \tag{3.11}$$

Here $\Gamma_{11}^\times = \eta_\pi Y + \Gamma^+ + X\rho_\zeta : \mathcal{X}_\zeta \to \mathcal{X}_\pi$.

PROOF. Remark that A_π and T do not have a common eigenvalue. Therefore the first Lyapunov equation of (3.8) is uniquely solvable. The second equation of (3.8) is uniquely solvable because also S and A_ζ do not have a common eigenvalue.

To show that τ_0 is an admissible Sylvester data set we first prove that the pair $(C_\pi X + F, T)$ is a zero kernel pair. So assume that $\lambda \in \mathbb{C}$ and $x \in \mathcal{K}$ are such that $(C_\pi X + F)x = 0$ and $Tx = \lambda x$. It follows that $B_\zeta(C_\pi X + F)x = 0$. We will rewrite $B_\zeta C_\pi X + B_\zeta F$. The first step is to use the coupling relation and (3.8) to get

$$B_\zeta C_\pi X = \Gamma A_\pi X - A_\zeta \Gamma X = \Gamma X T + \Gamma A_{12} - A_\zeta \Gamma X.$$

Next we use the definition of A_{12} in (3.9) and obtain

$$B_\zeta C_\pi X = \Gamma X T + \Gamma\Gamma^+(A_\zeta \eta_\zeta - \eta_\zeta T - B_\zeta F) - A_\zeta \Gamma X.$$

Since $\Gamma\Gamma^+ = I - \eta_\zeta \rho_\zeta$ and $\eta_\zeta \rho_\zeta \eta_\zeta = \eta_\zeta$, we get

$$B_\zeta C_\pi X = \Gamma X T + A_\zeta \eta_\zeta - B_\zeta F - \eta_\zeta \rho_\zeta A_\zeta \eta_\zeta + \eta_\zeta \rho_\zeta B_\zeta F - A_\zeta \Gamma X.$$

Now, using the definition of T, i.e., $T = \rho_\zeta A_\zeta \eta_\zeta - \rho_\zeta B_\zeta F$, we obtain

$$B_\zeta C_\pi X + B_\zeta F = \Gamma X T + A_\zeta \eta_\zeta - \eta_\zeta T - A_\zeta \Gamma X \tag{3.12}$$

But $(B_\zeta C_\pi X + B_\zeta F)x = 0$ and $Tx = \lambda x$, and so we get

$$(A_\zeta - \lambda I)(\eta_\zeta x - \Gamma X x) = 0.$$

If $\lambda \neq \epsilon_-$, then $x = 0$, and we are done. If $\lambda = \epsilon_-$, then λ is not an eigenvalue of A_ζ and thus $\eta_\zeta x = \Gamma X x$. Now $\operatorname{Im} \eta_\zeta \cap \operatorname{Im} \Gamma = \{0\}$, and thus $\eta_\zeta x = 0$, which proves that $x = 0$. So $(C_\pi X + F, T)$ is a zero kernel pair.

We also prove that the pair $(S, -YB_\zeta + G)$ is a full range pair. So let us assume that x^* is a right eigenvector of S, i.e., $x^*\lambda = x^*S$, such that $x^*(-YB_\zeta + G) = 0$. We intend to prove that $x^* = 0$. If $\lambda \neq \epsilon_-$, then it is clear that $x^* = 0$. So we may assume that $\lambda = \epsilon_-$. With a computation symmetric to the one leading to (3.12) we get

$$-YB_\zeta C_\pi + GC_\pi = SY\Gamma - Y\Gamma A_\pi + \rho_\pi A_\pi - S\rho_\pi. \tag{3.13}$$

Therefore

$$x^*(\rho_\pi - Y\Gamma)(A_\pi - \lambda I) = 0.$$

Since λ is not an eigenvalue of A_π, we obtain $x^*((\rho_\pi - Y\Gamma) = 0$. For any $y \in \operatorname{Ker} \Gamma$ we have that

$$x^* y = x^* \rho_\pi \eta_\pi = x^* Y \Gamma \eta_\pi y = x^* Y 0 = 0.$$

So we proved that $x^* = 0$.

Our next step is to show that

$$\Gamma_0 T - S\Gamma = (-YB_\zeta + G)(-C_\pi X - F). \tag{3.14}$$

In principle this is straightforward. Nevertheless we will perform the computation and derive some useful formulas. Remark that from the definitions of A_{12} and T it follows that

$$\begin{aligned}\Gamma A_{12} &= (I - \eta_\zeta \rho_\zeta)(A_\zeta \eta_\zeta - \eta_\zeta - TB_\zeta F) \\ &= A_\zeta \eta_\zeta - \eta_\zeta T - B_\zeta F + \eta_\zeta(\rho_\zeta A_\zeta \eta_\zeta - T - \rho_\zeta B_\zeta F)\end{aligned}$$

and thus

$$\Gamma A_{12} = A_\zeta \eta_\zeta - \eta_\zeta T - B_\zeta F. \tag{3.15}$$

From the definitions of S and A_{21} we get

$$\begin{aligned}\Gamma A_{21} &= GF\rho_\zeta \Gamma + (\rho_\pi A_\pi - S\rho_\pi - GC_\pi)(I - \eta_\pi \rho_\pi) \\ &= \rho_\pi A_\pi - S\rho_\pi - GC_\pi - (\rho_\pi A_\pi \eta_\pi - S - GC_\pi \eta_\pi)\rho_\pi\end{aligned}$$

and

$$\Gamma A_{21} = \rho_\pi A_\pi - S\rho_\pi - GC_\pi. \tag{3.16}$$

Furthermore, since $\Gamma^+ \eta_\zeta = 0$, $\rho_\pi \Gamma^+ = 0$ and $\rho_\zeta \eta_\zeta = I$, we have

$$A_{21}\eta_\zeta - \rho_\pi A_{12} = -GF \tag{3.17}$$

Use the definition of Γ_0 and (3.8) to see that

$$\Gamma_0 T = (Y\Gamma - \rho_\pi)(A_\pi X - A_{12}) - Y\eta_\zeta T,$$

$$S\Gamma_0 = (YA_\zeta - A_{21})(\Gamma X - \eta_\zeta) - S\rho_\pi X.$$

From these formulas, the formulas (3.15), (3.16) and (3.17), and the coupling relation $\Gamma A_\pi - A_\zeta \Gamma = B_\zeta C_\pi$ we obtain (3.14). We have proved that τ_0 is an admissible Sylvester data set.

The next step is to show that the coupling operator of the triple $\tau \oplus \tau_0$ is given by

$$\Gamma_{\tau\oplus\tau_0} = \begin{pmatrix} \Gamma & -\Gamma X + \eta_\zeta \\ -Y\Gamma + \rho_\pi & \Gamma_0 \end{pmatrix}.$$

We have to show that (3.5) and (3.6) hold true with τ substituted for τ_1 and τ_0 substituted for τ_2. In this case (3.5) and (3.6) coincide with (3.12) and (3.13), which finishes this step.

Now we are ready to show that $\Gamma_{\tau\oplus\tau_0}$ is invertible. It is sufficient to prove that the matrix

$$\begin{pmatrix} \eta_\pi Y + \Gamma^+ + X\rho_\zeta & \eta_\pi \\ \rho_\zeta & 0 \end{pmatrix} \tag{3.18}$$

is the inverse of $\Gamma_{\tau\oplus\tau_0}$. Let us check the four identities involved:

$$\Gamma(\eta_\pi Y + \Gamma^+ + X\rho_\zeta) - (-\Gamma X + \eta_\zeta)\rho_\zeta = \Gamma\Gamma^+ + \eta_\zeta\rho_\zeta = I,$$

$$\Gamma\eta_\pi = 0,$$

$$(-Y\Gamma+\rho_\pi)(\eta_\pi Y+\Gamma^+ + X\rho_\zeta)+\Gamma_0\rho_\zeta = -Y\Gamma\Gamma^+ + \rho_\pi\eta_\pi Y - Y\eta_\zeta\rho_\zeta = -Y+Y=0,$$

and

$$(-Y\Gamma+\rho_\pi)\eta_\pi = \rho_\pi\eta_\pi = I_{\mathrm{Ker}\,\Gamma}.$$

Remark that

$$\dim \mathcal{X}_\pi + \dim \mathcal{K} = \dim \mathcal{X}_\zeta + \dim \mathrm{Ker}\,\Gamma = \dim \mathcal{X}_\pi + \dim \mathcal{X}_\zeta - \mathrm{rank}\,\Gamma.$$

If τ_1 is a complement of τ, then it is easy to see from the invertibility of the coupling operator of $\tau \oplus \tau_1$ that the the rank of the coupling operator is at least $\dim \mathcal{X}_\zeta + \dim \mathrm{Ker}\,\Gamma$. Thus the complement τ_0 is indeed a minimal complement. It remains to check the formulas for $\tilde{W}$ and $\tilde{W}^{-1}$. These follow by specifying the formulas (3.1) and (3.2) in Theorem 3.1 for the triple $\tau \oplus \tau_0$. □

PROOF OF THEOREM 3.2. Apply Lemma 3.3 and Lemma 3.4 to construct an ϵ_--pole correction pair and an ϵ_--null correction pair of τ. Next apply Lemma 3.6 to τ and these pairs to construct a minimal complement τ_0 of τ and a corresponding function $\tilde{W}$. The representations (3.10) and (3.11) and the fact that both S and T have no eigenvalue in Ω_+ give that indeed τ is an Ω_+-null-pole triple of $\tilde{W}$. Note that the value of $\tilde{W}$ at infinity is I. Furthermore the McMillan degree of $\tilde{W}$ is equal to $\dim \mathcal{X}_\pi + \dim \mathcal{X}_\zeta - \mathrm{rank}\,\Gamma$. Lemma 3.5 states that any W that has τ as an Ω_+-null-pole triple has a global null-pole triple of the form $\tau \oplus \tau'$. Since τ_0 is a minimal complement of τ, the McMillan degree of any W that has τ as an Ω_+-null-pole triple is at least $\dim \mathcal{X}_\pi + \dim \mathcal{X}_\zeta - \mathrm{rank}\,\Gamma$. □

So far we did not use the information about the Jordan blocks for the main operators in the pole correction pair and the null correction pair. This additional information will turn out to be very useful in Section 5.

XI.4 An auxiliary result on block similarity

If a full length block $(B; I, Q)$ on $\mathcal{X}$ has no invariant polynomials and has indices of the third kind $\mu_1 \geq \cdots \geq \mu_k > 0 = \mu_{k+1} = \cdots = \mu_m$, then, by Theorem III.1.1, the block $(B; I, Q)$ is block similar to a block $(B'; I, Q')$ which is a direct sum of shifts of the third kind. Furthermore, in this case $Q'B' : \mathrm{Im}\, Q' \to \mathrm{Im}\, Q'$ is a nilpotent operator which has Jordan blocks of sizes $\mu_1, \ldots, \mu_k$. The next result shows that if we know in advance that $QB : \mathrm{Im}\, Q \to \mathrm{Im}\, Q$ is a nilpotent operator with Jordan blocks of sizes $\mu_1, \ldots, \mu_k$, then there already exists a similarity of the first kind transforming $(B; I, Q)$ into a direct sum of shifts of the third kind. In other words, in this case there exists an invertible operator R such that with

$$Q' = RQR^{-1}, \quad B' = RBR^{-1}|_{\mathrm{Im}\, Q'},$$

the block $(B'; I, Q')$ is a direct sum of shifts of the third kind.

Proposition 4.1. *Let $(B;I,Q)$ be a full length block on the space $\mathcal{X}$. Assume that the indices of the third kind of $(B;I,Q)$ are $\mu_1 \geq \cdots \geq \mu_k > 0 = \mu_{k+1} = \cdots = \mu_m$ and that $\sum_{i=1}^m \mu_i = \dim \operatorname{Im} Q$. If $QB : \operatorname{Im} Q \to \operatorname{Im} Q$ is nilpotent with Jordan blocks of sizes $\mu_1, \ldots, \mu_k$, then $(B;I,Q)$ is block similar, with a similarity of the first kind, to a direct sum of shifts of the third kind with indices $\mu_1, \ldots, \mu_k$.*

PROOF. First remark that the block $(B;I,Q)$ has no invariant polynomials. Choose a basis $\{g_{ij}\}_{i=1,\ j=1}^{\mu_j\ k}$ such that the matrix of $QB : \operatorname{Im} Q \to \operatorname{Im} Q$ with respect to this basis is in lower triangular Jordan normal form. In other words $QBg_{ij} = g_{i+1\ j}$ for $i = 1, \ldots, \mu_j - 1$, and $QBg_{\mu_j j} = 0$ for $j = 1, \ldots, k$. Define $g_{\mu_j+1\ j} = Bg_{\mu_j j}$ for $j = 1, \ldots, k$. Note that $B : \operatorname{Im} Q \to \mathcal{X}$ has full rank. Indeed, if $Bx = 0$, then $x \in \mathcal{D}_\infty$, the residual space of the block $(B;I,Q)$. But we know that $(B;I,Q)$ has no invariant polynomials, and hence we can conclude that $\mathcal{D}_\infty = \{0\}$. We see that $x = 0$. So $\operatorname{Ker} B = \{0\}$, which gives that $g_{\mu_1+1\ 1}, \ldots, g_{\mu_k+1\ k}$ is an independent set of vectors in $\mathcal{X}$. Since $Qg_{\mu_j+1\ j} = QBg_{\mu_j j} = 0$, we have that $g_{\mu_j+1\ j} \in \operatorname{Ker} Q$. Choose $g_{\mu_{k+1}+1\ k+1}, \ldots, g_{\mu_m+1\ m}$, $(\mu_{j+1} = 0$ if $j \geq k)$ such that $g_{\mu_1+1\ 1}, \ldots, g_{\mu_m+1\ m}$ is a basis of $\operatorname{Ker} Q$. With respect to the basis $\{g_{ij}\}_{i=1,\ j=1}^{\mu_j,\ k}$ of $\operatorname{Im} Q$ and $\{g_{\mu_j+1\ j}\}_{j=1}^m$ of $\operatorname{Ker} Q$ the matrix of B is given by

$$B = \begin{pmatrix} S \\ C \end{pmatrix}, \quad S = \operatorname{diag}(S_j)_{j=1}^k, \quad C = \begin{pmatrix} C_1 & \cdots & C_k \end{pmatrix},$$

with S_ℓ a lower triangular nilpotent $\mu_\ell \times \mu_\ell$ Jordan block and C_ℓ a $m \times \mu_\ell$ matrix of which the last column is the ℓ-th unit vector in $\mathbb{C}^m$. Let $C_{0\ell}$ be an $m \times \mu_\ell$ matrix of which the last column is the ℓ-th unit vector in $\mathbb{C}^m$ and the other columns are zero and let $C_0 = \begin{pmatrix} C_{01} & \cdots & C_{0k} \end{pmatrix}$. We have to prove that there exist an invertible $n \times n$ matrix N such that

$$\begin{pmatrix} N & 0 \\ 0 & I \end{pmatrix} \begin{pmatrix} S \\ C \end{pmatrix} N^{-1} = \begin{pmatrix} S \\ C_0 \end{pmatrix}. \tag{4.1}$$

Write $C_\ell = \left(c_{ij}^\ell\right)_{i,j=1}^{m,\mu_\ell}$. First, we will prove that $c_{ij}^\ell = 0$ if $j > \mu_i$. Let $\mathcal{D}_j$ be the j-th definition space of the block $(B;I,Q)$, i.e.,

$$\mathcal{D}_0 = \mathcal{X}, \quad \mathcal{D}_1 = \operatorname{Im} Q, \quad \mathcal{D}_i = \{x \in \mathcal{D}_{i-1} \mid Bx \in \mathcal{D}_{i-1}\}, \quad i = 2, 3, \ldots \quad ,$$

and let $q_k = \dim(\mathcal{D}_k/\mathcal{D}_{k+1})$. Since $\mathcal{D}_p = \operatorname{Ker} \operatorname{col}\left(CS^j\right)_{j=0}^{p-2}$ for $p \geq 2$ and $\mathcal{D}_1 = \operatorname{Im} Q$, we have that $q_1 = m - \dim \operatorname{Ker} C = \operatorname{rank} C$ and

$$q_p = \operatorname{rank} \operatorname{col}\left(CS^j\right)_{j=0}^{p-1} - \operatorname{rank} \operatorname{col}\left(CS^j\right)_{j=0}^{p-2}, \quad p \geq 2.$$

It follows that $\operatorname{rank} \operatorname{col}\left(CS^j\right)_{j=0}^{p-1} = \sum_{j=1}^p q_p$. Write $C_\ell = \operatorname{row}\left(c_j^\ell\right)_{j=1}^{\mu_\ell}$ where c_j^ℓ is the j-th column of C_ℓ. Then $c_{\mu_\ell}^\ell = e_\ell$, the ℓ-th unit vector in $\mathbb{C}^m$. Note that for $j < \mu_\ell$ we have

$$C_\ell S_\ell^j = \begin{pmatrix} c_{j+1}^\ell & \cdots & c_{\mu_\ell-1}^\ell & e_\ell & 0 & \cdots & 0 \end{pmatrix}$$

with j zero columns. Consider the matrix $\operatorname{col}(C_\ell S_\ell^j)_{j=0}^{p-1}$. The last $\min\{\mu_\ell, p\}$ columns of this matrix have the form

$$\operatorname{col}(c^\ell_{\mu_\ell - k}, \dots, c^\ell_{\mu_\ell - 1}, e_\ell, 0, \dots, 0), \quad 0 \le k \le \mu_\ell - 1, \quad 0 \le k \le p - 1. \tag{4.2}$$

Therefore the vectors of the form (4.2), with length mp, are columns of the matrix $\operatorname{col}(CS^j)_{j=0}^{p-1}$. Moreover, it is easy to see that these vectors are independent. Now $0 \le k \le \mu_\ell - 1$ if and only if $1 \le \ell \le q_{k+1}$. Thus there are precisely $\sum_{j=0}^{p-1} q_{j+1}$ of these vectors and this number is equal to $\operatorname{rank}\operatorname{col}(CS^j)_{j=0}^{p-1}$. So these vectors form a basis of the column space of the matrix $\operatorname{col}(CS^j)_{j=0}^{p-1}$. In particular, this means that the column space of CS^{p-1} is spanned by $e_1, \dots, e_{q_p}$. Therefore the rows $q_p + 1, \dots, m$ in CS^{p-1} are all zero rows. We proved that row i in $C_\ell S_\ell^{p-1}$ is zero if $i > q_p$. Recall that $i > q_p$ and $\mu_i < p$ are equivalent. Thus in $C_\ell S_\ell^{\mu_i}$ row i is zero. Since c^ℓ_{ij} is the $(j - \mu_i)$-th entry in row i of $C_\ell S_\ell^{\mu_i}$, this means that $c^\ell_{ij} = 0$ if $j > \mu_i$.

We now specify the matrix N required in (4.1). Put $N = (N_{\alpha\beta})_{\alpha,\beta=1}^k$ with $N_{\alpha\beta}$ the $\mu_\alpha \times \mu_\beta$ Toeplitz matrix given by $N_{\alpha\beta} = (n_{ij}^{(\alpha\beta)})_{i=1,j=1}^{\mu_\alpha,\mu_\beta}$ with

$$n_{ij}^{(\alpha\beta)} = \begin{cases} c^\beta_{\alpha\ \mu_\alpha - i + j} & \text{if } \max\{0, \mu_\alpha - \mu_\beta\} \le i - j \le \mu_\alpha - 1, \\ 0 & \text{otherwise.} \end{cases}$$

Note that in N the row $\mu_1 + \cdots + \mu_j$ is the j-th row of C. Moreover, the row with number $(\mu_1 + \cdots + \mu_j - p)$ in N is the j-th row in CS^p. Since $\operatorname{col}(CS^j)_{j=1}^n$ has full rank, it follows that the rows of N are independent, and hence N is invertible. It is an easy exercise in matrix multiplication to check that the Toeplitz character of $N_{\alpha\beta}$ yields $S_\alpha N_{\alpha\beta} = N_{\alpha\beta} S_\beta$, and hence $SN = NS$. Finally, multiplying N on the left by C_0 selects precisely those rows of N that form C, and thus $C_0 N = C$. □

Although we will not use this, it is worthwhile to note that equation (4.1) fully determines the matrix N. Hence the similarity of the first kind is unique, once the direct sum of shifts is fixed.

Corollary 4.2. *Let $(B; I, Q)$ be a full length block on the space $\mathcal{X}$. Assume that the indices of the third kind of $(B; I, Q)$ are $\mu_1 \ge \cdots \ge \mu_k > 0 = \mu_{k+1} = \cdots = \mu_m$ and that $\sum_{i=1}^m \mu_i = \dim \operatorname{Im} Q$. If $QB : \operatorname{Im} Q \to \operatorname{Im} Q$ is nilpotent with Jordan blocks of sizes $\mu_1, \dots, \mu_k$, then there exists a basis $\{f_{ij}\}_{i=1,\ j=1}^{\mu_j+1\ m}$ of $\mathcal{X}$ such that*

$$\operatorname{Im} Q = \operatorname{span}\{f_{ij}\}_{i=1,\ j=1}^{\mu_j\ k}, \quad \operatorname{Ker} Q = \operatorname{span}\{f_{\mu_{j+1}\ j}\}_{j=1}^k$$

and

$$B f_{ij} = f_{i+1\ j}, \quad i = 1, \dots, \mu_j, \quad j = 1, \dots, k.$$

PROOF. Consider the basis $\{g_{ij}\}_{i=1,\ j=1}^{\mu_j+1\ m}$ and the matrix

$$S = \begin{pmatrix} N & 0 \\ 0 & I \end{pmatrix}$$

constructed in the proof of Proposition 4.1. Define f_{ij} to be the vectors in $\mathcal{X}$ having the columns of S as coordinates with respect to the basis $\{g_{ij}\}_{i=1,\ j=1}^{\mu_j+1\ m}$. These vectors f_{ij} have the required properties. □

XI.5 Factorization indices for rational matrix functions

First, we recall the definitions of Wiener-Hopf factorization and Wiener-Hopf indices for rational matrix functions. Let W be a proper rational $m \times m$ matrix function with $W(\infty) = I$ and such that W has no zeros and no poles on the simple closed contour γ. Denote the interior domain of γ by Ω_+ and the region outside γ, which contains ∞, by Ω_-. Fix two points $\epsilon_+ \in \Omega_+$ and $\epsilon_- \in \Omega_-$. A *(left) Wiener-Hopf factorization* of W with respect to γ is defined as

$$W(\lambda) = W_-(\lambda)D(\lambda)W_+(\lambda), \quad \lambda \in \gamma, \tag{5.1}$$

where W_+ is a rational matrix function with no zeros and no poles in $\gamma \cup \Omega_+$, the factor W_- is a rational matrix function with no zeros and no poles in $\gamma \cup \Omega_-$ including ∞, and

$$D(\lambda) = \operatorname{diag}\big(\big(\tfrac{\lambda-\epsilon_+}{\lambda-\epsilon_-}\big)^{\kappa_j} \big)_{j=1}^{m}. \tag{5.2}$$

Here $\kappa_1 \geq \kappa_2 \geq \cdots \geq \kappa_m$ are integers, which are called the *(left) factorization indices*. They are uniquely determined by W. In case all $\kappa_j = 0$, the factorization is said to be *canonical*. Since $W(\infty) = I_m$, we may assume in what follows that also the functions W_+ and W_- are proper and normalized to I_m at infinity.

The main theorem of this section describes the factorization indices of W with respect to γ in terms of an Ω_+-null-pole triple.

Theorem 5.1. *Let W be a proper rational $m \times m$ matrix function with $W(\infty) = I$ and such that W has no zeros and no poles on the simple closed positively oriented contour γ, and Ω_+ be the interior domain of γ. Let*

$$\tau_+ = \big((C_\pi, A_\pi), (A_\zeta, B_\zeta), \Gamma\big)$$

be an Ω_+-null-pole triple of W. Let Γ^+ be a generalized inverse to Γ, and consider the projections

$$P_\zeta = \begin{pmatrix} I - \Gamma\Gamma^+ & 0 \\ 0 & 0 \end{pmatrix}, \quad Q_\zeta = \begin{pmatrix} I - \Gamma\Gamma^+ & 0 \\ 0 & I \end{pmatrix}$$

on $\mathcal{X}_\zeta \oplus \mathbb{C}^m$ and the projections

$$P_\pi = \begin{pmatrix} I - \Gamma^+\Gamma & 0 \\ 0 & I \end{pmatrix}, \quad Q_\pi = \begin{pmatrix} I - \Gamma^+\Gamma & 0 \\ 0 & 0 \end{pmatrix}$$

on $\mathcal{X}_\pi \oplus \mathbb{C}^m$. Here $\mathcal{X}_\pi$ is the main space of the pair (C_π, A_π) and $\mathcal{X}_\zeta$ is the main space of the pair (A_ζ, B_ζ). Then the positive factorization indices of W are the positive indices of the first kind of the block

$$\Theta_\zeta = \left(P_\zeta \begin{pmatrix} A_\zeta & B_\zeta \\ 0 & 0 \end{pmatrix} Q_\zeta; P_\zeta, Q_\zeta \right)$$

and the negative factorization indices of W are the negative of the positive indices of the third kind of the block

$$\Theta_\pi = \left(P_\pi \begin{pmatrix} A_\pi & 0 \\ C_\pi & 0 \end{pmatrix} Q_\pi; P_\pi, Q_\pi \right).$$

PROOF. Put

$$\begin{aligned}
\rho_\zeta &= I - \Gamma\Gamma^+ : \mathcal{X}_\zeta \to \operatorname{Im}(I - \Gamma\Gamma^+),\\
\eta_\zeta &= (I - \Gamma\Gamma^+)|_{\operatorname{Im}(I-\Gamma\Gamma^+)} : \operatorname{Im}(I - \Gamma\Gamma^+) \to \mathcal{X}_\zeta,\\
\rho_\pi &= I - \Gamma^+\Gamma : \mathcal{X}_\pi \to \operatorname{Im}(I - \Gamma^+\Gamma),\\
\eta_\pi &= (I - \Gamma^+\Gamma)|_{\operatorname{Im}(I-\Gamma^+\Gamma)} : \operatorname{Im}(I - \Gamma^+\Gamma) \to \mathcal{X}_\pi.
\end{aligned}$$

By applying Lemma 3.3 to τ_+, ρ_ζ, η_ζ, ρ_π and η_π we see that $(C_\pi\eta_\pi, \rho_\pi A_\pi \eta_\pi)$, with main space $\operatorname{Ker}\Gamma = \operatorname{Im}(I - \Gamma^+\Gamma)$, is a zero kernel pair and that $(\rho_\zeta A_\zeta \eta_\zeta, \rho_\zeta B_\zeta)$, with main space $\mathcal{K} = \operatorname{Im}(I - \Gamma\Gamma^+)$, is a full range pair.

Next fix $\epsilon_- \in \Omega_-$, the outer domain of γ, and apply Lemma 3.4 to construct an ϵ_--pole correction pair (S, G) and an ϵ_--null correction pair (T, F) of τ_+. Given τ_+ and these pairs we apply Lemma 3.6 to construct the function $\tilde{W}$ in (3.10) We know that τ_+ is an Ω_+-null-pole triple of $\tilde{W}$. We claim that $\tilde{W}$ has an additional interesting property. We shall show that $\tilde{W}$ has a Wiener-Hopf factorization of a special type, namely

$$\tilde{W}(\lambda) = W_-(\lambda)D(\lambda)E^{-1},$$

where E^{-1} is constant invertible, $D(\lambda)$ is given by (5.2), the positive factorization indices of $\tilde{W}$ are the positive indices of the first kind of Θ_ζ and the negative factorization indices of $\tilde{W}$ are minus the positive indices of the first kind of the block Θ_π. Once this factorization is established, the proof is almost finished.

From the minimal realizations (3.10) and (3.11) one reads off a global null-pole triple of $\tilde{W}^{-1}$. The coupling operator of this null-pole triple is the inverse to the coupling operator of $\tau_+ \oplus \tau_0$, and thus is given by (3.18). Also remark that ϵ_- is the only eigenvalue inside Ω_- of both $A_\pi \oplus T$ and $A_\zeta \oplus S$. With this knowledge it is clear that

$$\tilde{\tau}_0 = \big((C_\pi\eta_\pi, S), (T, \rho_\zeta B_\zeta), 0\big)$$

is an Ω_--null-pole triple of $\tilde{W}^{-1}$. We are going to construct a special minimal complement $\tau_0^\dagger$ of τ_0.

We denote the indices of the first kind of Θ_ζ by $\beta_1 \geq \cdots \geq \beta_\ell > 0 = \beta_{\ell+1} = \cdots = \beta_m$ and the indices of the third kind of Θ_π by $\alpha_1 \geq \cdots \geq \alpha_k > 0 = \alpha_{k+1} = \cdots = \alpha_m$. Note that $\Theta_\zeta = \Theta_\zeta^0 \oplus (0;0,0)$, where

$$\Theta_\zeta^0 = ((\,\rho_\zeta A_\zeta \eta_\zeta \quad \rho_\zeta B_\zeta\,) ; P_1, I)$$

is a full width block on $\mathcal{K} \oplus \mathbb{C}^m$ with P_1 being the projection of $\mathcal{K} \oplus \mathbb{C}^m$ along $\mathbb{C}^m$ onto $\mathcal{K}$. The indices of the first kind of Θ_ζ and Θ_ζ^0 are the same, since the block $(0;0,0)$ has only indices of the second kind (see Theorem VII.2.1). Similarly we have that $\Theta_\pi = \Theta_\pi^0 \oplus (0;0,0)$, where

$$\Theta_\pi^0 = (\begin{pmatrix} \rho_\pi A_\pi \eta_\pi \\ C_\pi \eta_\pi \end{pmatrix} ; I, Q_1)$$

is a full length block on $\operatorname{Ker}\Gamma \oplus \mathbb{C}^m$ with Q_1 being the projection of $\operatorname{Ker}\Gamma \oplus \mathbb{C}^m$ along $\mathbb{C}^m$ onto $\operatorname{Ker}\Gamma$. The indices of the third kind of Θ_π and Θ_π^0 are the same, because the block $(0;0,0)$ has no indices of the third kind.

Since the pair (S, G) is an ϵ_--pole correction pair, one has

$$\begin{pmatrix} I & -G \\ 0 & I \end{pmatrix} \begin{pmatrix} \rho_\pi A_\pi \eta_\pi - \epsilon_- I \\ C_\pi \eta_\pi \end{pmatrix} = \begin{pmatrix} S - \epsilon_- I \\ C_\pi \eta_\pi \end{pmatrix}.$$

Thus the blocks

$$(\begin{pmatrix} \rho_\pi A_\pi \eta_\pi - \epsilon_- I \\ C_\pi \eta_\pi \end{pmatrix} ; I, Q_1), \quad \text{and } (\begin{pmatrix} S - \epsilon_- I \\ C_\pi \eta_\pi \end{pmatrix} ; I, Q_1)$$

are block similar and have the same indices of the third kind. These indices are equal to the indices of Θ_π^0 (see Theorem III.1.4), and hence they are equal to $\alpha_1, \ldots, \alpha_m$. Moreover the matrix $S - \epsilon_- I$ is nilpotent and has Jordan blocks of sizes $\alpha_1, \ldots, \alpha_k$. Thus, according to Corollary 4.2, there exists a basis $\{e_{ij}\}_{i=1,j=1}^{\alpha_j+1,m}$ of $\operatorname{Ker}\Gamma \oplus \mathbb{C}^m$, such that

$$\operatorname{Ker}\Gamma = \operatorname{span}\{e_{ij}\}_{i=1,j=1}^{\alpha_j,k}, \quad \mathbb{C}^m = \operatorname{span}\{e_{\alpha_j+1,j}\}_{j=1}^m,$$

$$(S - \epsilon_- I)e_{ij} = e_{i+1\ j}, \quad (j = i, \ldots, \alpha_j - 1), \quad (S - \epsilon_- I)e_{\alpha_j j} = 0, \quad (j = 1, \ldots, k),$$

and

$$C_\pi \eta_\pi e_{ij} = 0 \quad (i = 1, \ldots, \alpha_j - 1), \quad C_\pi \eta_\pi e_{\alpha_j j} = e_{\alpha_j+1\ j} \quad (j = 1, \ldots, k).$$

Put $z_j = e_{\alpha_j+1\ j}$ for $j = 1, \ldots, k$. Clearly, the vectors $z_1, \ldots, z_k$ are linearly independent. Define

$$d_{ij} = \sum_{\nu=0}^{i-1} \binom{i-1}{\nu} (\epsilon_- - \epsilon_+)^\nu e_{i-\nu\ j},$$

and define the operator $\tilde{S}$ on $\operatorname{Ker}\Gamma$ by

$$(\tilde{S} - \epsilon_+ I)d_{ij} = d_{i+1\ j}, \quad j = 1, \ldots, \alpha_j - 1, \quad (\tilde{S} - \epsilon_+ I)d_{\alpha_j j} = 0.$$

So, in particular, we have that $\tilde{S}$ has only one eigenvalue, namely ϵ_+ with partial multiplicities $\alpha_1 \geq \cdots \geq \alpha_k$.

The pair (F, T) is an ϵ_--null correction pair, and thus

$$(\,\rho_\varsigma A_\varsigma \eta_\varsigma - \epsilon_- I \quad \rho_\varsigma B_\varsigma\,) \begin{pmatrix} I & 0 \\ -F & I \end{pmatrix} = (\,T - \epsilon_- I \quad \rho_\varsigma B_\varsigma\,).$$

Hence the blocks $((\,\rho_\varsigma A_\varsigma \eta_\varsigma - \epsilon_- I \quad \rho_\varsigma B_\varsigma\,); P_1, I)$ and $((\,T - \epsilon_- I \quad \rho_\varsigma B_\varsigma\,); P_1, I)$ are block similar, and thus they have the same indices of the first kind. These indices are equal to the indices of Θ_ς^0 (see Theorem V.1.2), and hence equal to $\beta_1, \ldots, \beta_m$. The matrix $T - \epsilon_- I$ is nilpotent with Jordan blocks of sizes $\beta_1, \ldots, \beta_m$. We use duality, Proposition V.4.1, and Corollary 4.2 to see that there exists a basis $\{g_{ij}\}_{i=1,j=1}^{\beta_j+1,m}$ of $\mathcal{K} \oplus \mathbb{C}^m$ such that

$$\mathcal{K} = \operatorname{span}\{g_{ij}\}_{i=2,j=1}^{\beta_j+1,m}, \quad \mathbb{C}^m = \operatorname{span}\{g_{1j}\}_{j=1}^m,$$

$$(T - \epsilon_- I)g_{ij} = g_{i+1\ j} \quad (i = 2, \ldots, \beta_j), \quad (T - \epsilon_- I)g_{\beta_j+1\ j} = 0 \quad (j = 1, \ldots, \ell),$$

and

$$\rho_\varsigma B_\varsigma g_{1j} = g_{2j} \quad (j = 1, \ldots, \ell), \quad \rho_\varsigma B_\varsigma g_{1j} = 0 \quad (j = \ell + 1, \ldots, m).$$

We define a basis $\{f_{ij}\}_{i=1,j=1}^{\beta_j,\ell}$ for $\mathcal{K}$ by

$$f_{ij} = \sum_{\nu=0}^{i-1} \binom{\nu + \beta_j - i}{\nu} (\epsilon_+ - \epsilon_-)^\nu g_{i-\nu+1\ j},$$

and an operator $\tilde{T}$ on $\mathcal{K}$ by setting

$$(\tilde{T} - \epsilon_+ I)f_{ij} = f_{i+1\ j}, \quad j = 1, \ldots, \beta_j - 1, \quad (\tilde{T} - \epsilon_+ I)f_{\beta_j j} = 0.$$

Remark that $\tilde{T}$ has only one eigenvalue, namely ϵ_+ with partial multiplicities $\beta_1 \geq \cdots \geq \beta_\ell$.

Put $y_j = g_{1j}$ for $j = 1, \ldots, \ell$. Then $y_1, \ldots, y_\ell$ is an independent set in $\mathbb{C}^m$. We will prove that $\{z_1, \ldots, z_k\} \cup \{y_1, \ldots, y_\ell\}$ is an independent set in $\mathbb{C}^m$. Assume that $x \in \operatorname{span}\{z_1, \ldots, z_k\} \cap \operatorname{span}\{y_1, \ldots, y_\ell\}$. It is sufficient to prove that $x = 0$. Remark that $\rho_\varsigma B_\varsigma C_\pi \eta_\pi = \rho_\varsigma(\Gamma A_\pi - A_\varsigma \Gamma)\eta_\pi = 0$. Furthermore, there exists a vector $v \in \operatorname{Ker}\Gamma$ such that $x = C_\pi \eta_\pi v$. It follows that $\rho_\varsigma B_\varsigma x = \rho_\varsigma B_\varsigma C_\pi \eta_\pi v = 0$. Since $\operatorname{span}\{y_1, \ldots, y_\ell\} \cap \operatorname{Ker} \rho_\varsigma B_\varsigma = \{0\}$, this proves that $x = 0$. Let $\mathcal{Y}_0$ be a direct complement in $\mathbb{C}^m$ of span$\{z_1,\dots,z_k,y_1,\dots,y_\ell\}$. We define operators $\tilde{F}:\mathcal{K}\to\mathbb{C}^m$ and $\tilde{G}:\mathbb{C}^m\to\operatorname{Ker}\Gamma$ by putting

$$\tilde{F}f_{ij}=\binom{\beta_j}{i}(\epsilon_+-\epsilon_-)^i y_j, \tag{5.3}$$

$$\begin{cases}\tilde{G}y=0 & \text{for } y\in\operatorname{span}\{y_1,\dots,y_\ell\}\oplus\mathcal{Y}_0,\\ \tilde{G}z_j=(\tilde{S}-S)d_{\alpha_j\,j} & \text{for } j=1,\dots,k\end{cases}$$

Consider $\tau_0^\dagger=\big((\tilde{F},\tilde{T}),(\tilde{S},\tilde{G}),0\big)$. The triple $\tau_0^\dagger$ is an Ω_+-admissible Sylvester data set, because $\epsilon_+\in\Omega_+$ and $\tilde{G}\tilde{F}=0=0\tilde{T}-\tilde{S}0$. We claim that $\tau_0^\dagger$ is a complement of $\tilde{\tau}_0$. To see this we show that the coupling operator of $\tau_0^\dagger\oplus\tilde{\tau}_0$ is given by

$$\begin{pmatrix}0 & I_{\mathcal{K}}\\ -I_{\operatorname{Ker}\Gamma} & 0\end{pmatrix}.$$

This requires to show that $\tilde{T}-T=\rho_\varsigma B_\varsigma\tilde{F}$ and $\tilde{S}-S=\tilde{G}C_\pi\eta_\pi$. We compute for $1\le i<\alpha_j$ that

$$(S-\tilde{S})d_{ij}=(\epsilon_--\epsilon_+)d_{ij}-d_{i+1\,j}+\sum_{\nu=0}^{i-1}\binom{i-1}{\nu}(\epsilon_--\epsilon_+)^\nu e_{i-\nu+1\,j}.$$

In this equality we insert the definitions of d_{ij} and $d_{i+1\,j}$, and we conclude that $(S-\tilde{S})d_{ij}=0$ if $1\le i<\alpha_j$. Furthermore, by the definition of $\tilde{G}$, we have that $(\tilde{S}-S)d_{\alpha_j j}=\tilde{G}z_j=\tilde{G}C_\pi\eta_\pi d_{\alpha_j j}$, because $C_\pi\eta_\pi d_{\alpha_j j}=C_\pi\eta_\pi e_{\alpha_j j}$. We have proved that $\tilde{S}-S=\tilde{G}C_\pi\eta_\pi$. Next, for $1\le i\le\beta_j$ we have that

$$(T-\tilde{T})f_{ij}=(\epsilon_--\epsilon_+)f_{ij}+f_{i+1\,j}-\sum_{\nu=0}^{i-1}\binom{\beta_j-i+\nu}{\nu}(\epsilon_--\epsilon_+)^\nu g_{i-\nu+2\,j}.$$

Here $g_{\beta_j+2\,j}=0$. Using the definition of f_{ij} and $f_{i+1\,j}$, one gets

$$(T-\tilde{T})f_{ij}=\binom{\beta_j}{i}(\epsilon_--\epsilon_+)^i g_{2j}.$$

This proves that $\tilde{T}-T=\rho_\varsigma B_\varsigma\tilde{F}$. So $\tau_0^\dagger$ is a complement of $\tilde{\tau}_0$, which is obviously a minimal complement.

From Theorem 3.1 we know now that the triple $\tilde{\tau}_0\oplus\tau_0^\dagger$ is a global null pole triple of a rational matrix function, $\tilde{D}(\lambda)^{-1}$ say. Our next step is to compute $\tilde{D}(\lambda)$ using the formulas in Theorem 3.1. From the form of $\tilde{\tau}_0\oplus\tau_0^\dagger$ we see that

$$\tilde{D}(\lambda)^{-1}=I-C_\pi\eta_\pi(\lambda I-S)^{-1}\tilde{G}+\tilde{F}(\lambda I-\tilde{T})^{-1}\rho_\varsigma B_\varsigma.$$

To compute $\tilde{D}$ we extend $z_1, \dots, z_k, y_1, \dots, y_\ell$ to a basis of $\mathbb{C}^n$ by adding a basis $x_1, \dots, x_{m-k-\ell}$ of $\mathcal{Y}_0$. Put

$$E = \begin{pmatrix} y_1 & \cdots & y_\ell & x_1 & \cdots & x_{m-k-\ell} & z_k & \cdots & z_1 \end{pmatrix} \tag{5.4}$$

Remark that $\rho_\varsigma B_\varsigma x_i = 0$ and $\tilde{G}x_i = 0$. So $\tilde{D}(\lambda)^{-1}x_i = x_i$ for $i = 1, \dots, m-k-\ell$. Next we compute $\tilde{D}(\lambda)^{-1}z_j$. Note that

$$\tilde{D}(\lambda)^{-1}z_j = z_j - C_\pi\eta_\pi(\lambda I - S)^{-1}\tilde{G}z_j = z_j + C_\pi\eta_\pi(\lambda I - S)^{-1}(S - \tilde{S})d_{\alpha_j j}. \tag{5.5}$$

We remark that

$$(S - \tilde{S})d_{\alpha_j j} = (\epsilon_- - \epsilon_+)d_{\alpha_j j} + \sum_{\nu=1}^{\alpha_j - 1} \binom{\alpha_j - 1}{\nu} (\epsilon_- - \epsilon_+)^\nu e_{\alpha_j - \nu + 1\ j}.$$

In this equality we replace $d_{\alpha_j j}$ by its definition. This yields

$$(S - \tilde{S})d_{\alpha_j j} = \sum_{\nu=0}^{\alpha_j - 1} \binom{\alpha_j}{\nu + 1} (\epsilon_- - \epsilon_+)^{\nu+1} e_{\alpha_j - \nu\ j}. \tag{5.6}$$

From the Jordan canonical form of S with respect to the basis $\{e_{ij}\}$, and from the equalities $C_\pi\eta_\pi e_{ij} = 0$ if $i < \alpha_j$, and $C_\pi\eta_\pi e_{\alpha_j j} = z_j$, we see that

$$C_\pi\eta_\pi(\lambda I - S)^{-1}e_{\alpha_j - \nu\ j} = (\lambda - \epsilon_-)^{-(\nu+1)}z_j. \tag{5.7}$$

Insert the equalities (5.6) and (5.7) into (5.5). We obtain

$$\begin{aligned} \tilde{D}(\lambda)^{-1}z_j &= \left(1 + \sum_{\nu=0}^{\alpha_j - 1} \binom{\alpha_j}{\nu + 1} (\epsilon_- - \epsilon_+)^{\nu+1}(\lambda - \epsilon_-)^{-(\nu+1)}\right) z_j \\ &= (\lambda - \epsilon_-)^{-\alpha_j} \left(\sum_{\nu=0}^{\alpha_j} \binom{\alpha_j}{\nu} (\epsilon_- - \epsilon_+)^\nu (\lambda - \epsilon_-)^{\alpha_j - \nu}\right) z_j = \left(\frac{\lambda - \epsilon_+}{\lambda - \epsilon_-}\right)^{\alpha_j} z_j. \end{aligned}$$

Next, one computes that

$$\begin{aligned} \tilde{D}(\lambda)^{-1}y_j &= y_j + \tilde{F}(\lambda I - \tilde{T})^{-1}\rho_\varsigma B_\varsigma y_j = y_j + \tilde{F}(\lambda I - \tilde{T})^{-1}f_{1j} \\ &= y_j + \sum_{\nu=1}^{\beta_j} (\lambda - \epsilon_+)^{-\nu}\tilde{F}f_{\nu j} = \sum_{\nu=0}^{\beta_j} \binom{\beta_j}{\nu} (\lambda - \epsilon_+)^{-\nu}(\epsilon_+ - \epsilon_-)^\nu y_j \\ &= \left(\frac{\lambda - \epsilon_-}{\lambda - \epsilon_+}\right)^{\beta_j} y_j. \end{aligned}$$

We have proved that

$$E^{-1}\tilde{D}(\lambda)E = D(\lambda)$$

with

$$D(\lambda) = \operatorname{diag}\Big(\Big(\tfrac{\lambda-\epsilon_+}{\lambda-\epsilon_-}\Big)^{\beta_1}, \ldots, \Big(\tfrac{\lambda-\epsilon_+}{\lambda-\epsilon_-}\Big)^{\beta_\ell}, 1, \ldots, 1, \Big(\tfrac{\lambda-\epsilon_+}{\lambda-\epsilon_-}\Big)^{-\alpha_k}, \ldots, \Big(\tfrac{\lambda-\epsilon_+}{\lambda-\epsilon_-}\Big)^{-\alpha_1} \Big). \tag{5.8}$$

The triple $\tilde{\tau}_0$ is an Ω_--null-pole triple for both $\tilde{W}^{-1}$ and $\tilde{D}^{-1}$. According to Theorem 1.3 this means that $\tilde{W}\tilde{D}^{-1}$ is analytic and invertible in Ω_-. We put $W_- = \tilde{W}\tilde{D}^{-1}E$. Then $\tilde{W} = W_-DE^{-1}$.

Since the rational matrix functions W and $\tilde{W}$ have the same Ω_+-null-pole triple, we know from Theorem 1.3 that $\tilde{W}^{-1}W$ is analytic and invertible on Ω_+. Put $W_+ = E^{-1}\tilde{W}^{-1}W$, where E is given by (5.4). Then W_+ is analytic and invertible in each point of Ω_+. With D given by (5.8), we get $W = W_-DW_+$, a Wiener-Hopf factorization of W. This proves that the Wiener-Hopf indices are as described in the theorem. □

Notes

The preliminary material on the pole and zero structure of a rational matrix function in Section 1 can be found in a more complete form in Ball-Gohberg-Rodman [1] or in Gohberg-Kaashoek [1]. Theorems 2.1 and 2.2 are due to Gohberg-Kaashoek-Van Schagen [3]. Theorem 3.1 is taken from Gohberg-Kaashoek-Lerer-Rodman [1] and Theorem 3.2 originates from Gohberg-Kaashoek-Ran [1] (see also Gohberg-Kaashoek [1] for an earlier version of this result). Proposition 4.1 is the operator block version of Proposition 4.1 in Ball-Kaashoek-Groenewald-Kim [1]. The main results of Section 5 are basically due to Gohberg-Kaashoek-Ran [1].

Chapter XII
Infinite Dimensional Operator Blocks

In this chapter we develop the theory of operator blocks and block similarity on infinite dimensional spaces. We do not treat the general case, but only consider the analogues of full width and full length blocks. As an application a solution of the Hilbert space analogue of the eigenvalue completion problem is obtained for full length blocks.

XII.1 Preliminaries

We consider a Banach space $\mathcal{X}$, and generalize the notions developed in the first two chapters for blocks on finite dimensional linear spaces to blocks on the space $\mathcal{X}$. In this chapter an operator will be linear and bounded unless explicitly stated otherwise. In analogy to Section I.3 we define an *operator block* to be a triple $(B; P, Q)$ with P and Q bounded projections of $\mathcal{X}$ and $B : \operatorname{Im} Q \to \operatorname{Im} P$ an operator. In this case we refer to $\mathcal{X}$ as the *underlying space*, and we speak of a *(P, Q)-block* on $\mathcal{X}$. For an operator $T : \mathcal{X} \to \mathcal{X}$ the block $(B; P, Q)$ is called the *(P, Q)-block of T* if $B = PTQ : \operatorname{Im} Q \to \operatorname{Im} P$. If $(B_1; P_1, Q_1)$ is another operator block on the Banach space $\mathcal{X}_1$, then $(B; P, Q)$ and $(B_1; P_1, Q_1)$ are called *block similar* if there exists an invertible bounded operator $S : \mathcal{X}_1 \to \mathcal{X}$ such that

(i) $S[\operatorname{Ker} P_1] = \operatorname{Ker} P$ and $S[\operatorname{Im} Q_1] = \operatorname{Im} Q$,

(ii) $(SB_1 - BS)x \in \operatorname{Ker} P$ for all $x \in \operatorname{Im} Q_1$.

We call the operator S a *block similarity*. Let us recall that the condition (i) is equivalent to

(i') $PS = PSP_1$, $P_1S^{-1} = P_1S^{-1}P$, $SQ_1 = QSQ_1$ and $S^{-1}Q = Q_1S^{-1}Q$.

If (i) is satisfied, then (ii) is equivalent to

(ii') $(P_1SP)B = B_1(Q_1SQ)$.

We define an operator block $(B; P, Q)$ on $\mathcal{X}$ to be *regularly decomposable* if there are nonzero closed subspaces $\mathcal{M}_1$ and $\mathcal{M}_2$ of $\mathcal{X}$ such that

(1) $\mathcal{X} = \mathcal{M}_1 \oplus \mathcal{M}_2$,

(2) $P[\mathcal{M}_i] \subset \mathcal{M}_i$, $Q[\mathcal{M}_i] \subset \mathcal{M}_i$, for $i = 1, 2$,

(3) $B[\mathcal{M}_i \cap \operatorname{Im} Q] \subset \mathcal{M}_i$, for $i = 1, 2$.

In this case we define blocks $(B_i; P_i, Q_i)$ by

$$Q_i = Q|_{\mathcal{M}_i} : \mathcal{M}_i \to \mathcal{M}_i, \quad P_i = P|_{\mathcal{M}_i} : \mathcal{M}_i \to \mathcal{M}_i,$$
$$B_i = B_{\mathcal{M}_i \cap \operatorname{Im} Q_i} : \mathcal{M}_i \cap \operatorname{Im} Q_i \to \mathcal{M}_i \cap \operatorname{Im} P_i$$

for $i = 1,2$. We write $(B;P,Q) = (B_1;P_1,Q_1) \oplus (B_2;P_2,Q_2)$, and we refer to $(B;P,Q)$ as the *direct sum* of the blocks $(B_1;P_1,Q_1)$ and $(B_2;P_2,Q_2)$.

Let us give two examples of blocks. Take $\mathcal{X} = \mathcal{Y}^n$, where $\mathcal{Y}$ is a Banach space and $\mathcal{Y}^n$ denotes the Banach direct sum of n copies of $\mathcal{Y}$. In the first example we define the operator block $(V;P,I)$ on $\mathcal{X}$ by putting

$$P(y_1, y_2, \dots, y_n) = (0, y_2, \dots, y_n),$$
$$V(y_1, y_2, \dots, y_n) = (0, y_1, \dots, y_{n-1}).$$

This block is called a *block shift of the first kind* with *base space* $\mathcal{Y}$ and *index* $n-1$. The second example concerns the operator block $(V;I,Q)$ on $\mathcal{X}$ defined by putting

$$Q(y_1, \dots, y_{n-1}, y_n) = (y_1, \dots, y_{n-1}, 0),$$
$$V(y_1, \dots, y_{n-1}, 0) = (0, y_1, \dots, y_{n-1}).$$

This block is called a *block shift of the third kind* with *base space* $\mathcal{Y}$ and *index* $n-1$.

We will not study operator blocks in general. One could generalize the theory presented in Chapter VII to the infinite dimensional setting, by supposing that the subspaces that occur are closed. We prefer however to restrict ourselves to blocks of the types $(B;P,I)$ and $(B;I,Q)$, which are generalizations of the blocks studied in Chapters V–VII. Therefore we do not introduce the block shifts of the second kind, which would be the natural generalization of the notion of a shift of the second kind in the finite dimensional theory.

XII.2 Main theorems about (P,I)-blocks

In this section we develop a theory which parallels the finite dimensional theory of Chapter V. We will need the notion of a generalized inverse, which is defined as follows. The operator $T^+ : \mathcal{Y} \to \mathcal{X}$ between the Banach space $\mathcal{X}$ and $\mathcal{Y}$ will be called *a generalized inverse of the operator* $T : \mathcal{X} \to \mathcal{Y}$ if $T = TT^+T$ and $T^+ = T^+TT^+$. Remark that this implies that TT^+ is a projection of $\mathcal{X}$ onto $\operatorname{Im} T$ along the $\operatorname{Ker} T^+$, and T^+T is a projection of $\mathcal{X}$ onto $\operatorname{Im} T^+$ and along $\operatorname{Ker} T$. It is well-known (see, e.g., Theorem XI.6.1 in Gohberg-Goldberg-Kaashoek [1]) that T has a generalized inverse if and only if $\operatorname{Ker} T$ and $\operatorname{Im} T$ are complemented in $\mathcal{X}$ and $\mathcal{Y}$, respectively.

The first result in this section is essentially a generalization of the properties (ii) and (iv) of Proposition V.1.3; it will be the key to the proof of the main decomposition theorem for (P,I)-blocks. In what follows the symbol $\mathcal{X}^j$ stands for the Banach space direct sum of j copies of the Banach space $\mathcal{X}$.

Lemma 2.1. *Let $(B;P,I)$ be an operator block on the Banach space $\mathcal{X}$. Put*

$$\mathcal{F}_0 = \{0\},\ \mathcal{F}_1 = \operatorname{Ker} P,\ \mathcal{F}_j = \mathcal{F}_1 \oplus B[\mathcal{F}_{j-1}], \tag{2.1}$$

for $j = 1, 2, \ldots$. If for $j = 1, 2, \ldots, \ell$ the operator

$$\Delta_j = \big((I-P) \quad B(I-P) \quad \cdots \quad B^{j-1}(I-P) \big) : \mathcal{X}^j \to \mathcal{X} \tag{2.2}$$

has a generalized inverse, then there exists closed subspaces U_{ij}, where $1 \leq i \leq j \leq \ell$, in $\mathcal{X}$ such that

(i) $\mathcal{F}_i = \mathcal{F}_{i-1} \oplus U_{ii} \oplus \cdots \oplus U_{il}$ for $i = 1, \ldots, \ell$;

(ii) $B[U_{ii}] = \{0\}$ for $i = 1, \ldots, \ell$;

(iii) $B[U_{ij}] = P[U_{i+1\ j}]$ for $1 \leq i \leq j-1 \leq \ell - 1$;

(iv) the operators $B|_{U_{ij}}$ and $P|_{U_{i+1\ j}}$ are injective and have closed range for $1 \leq i \leq j-1 \leq \ell-1$.

PROOF. We split the main body of the proof into four parts. The main conclusion of the first three parts will be that for $j = 1, \ldots, \ell - 1$ the operator

$$B_j = B|_{\mathcal{F}_j} : \mathcal{F}_j \to \mathcal{F}_{j+1} \tag{2.3}$$

has a generalized inverse.

In the proof we shall frequently use that $\operatorname{Im} \Delta_j = \mathcal{F}_j$ for $j = 1, \ldots, \ell$. For $j = 1$ this is obvious, and for the other values of j it follows by induction from $\operatorname{Im} \Delta_j = \operatorname{Ker} P \oplus \operatorname{Im} \Delta_{j-1}$ and the definition of $\mathcal{F}_j$. From $\operatorname{Im} \Delta_j = \mathcal{F}_j$ and the fact that Δ_j has a generalized inverse one deduces that $\mathcal{F}_j$ is a closed subspace of $\mathcal{X}$ Hence the spaces appearing in (2.3) are Banach spaces in their own right.

PART (A). Consider the operators

$$S_j = (I - \Delta_{j-1}\Delta_{j-1}^+)B^{j-1}(I-P) : \operatorname{Ker} P \to \mathcal{X}, \quad j = 1, \ldots, \ell,$$

where Δ_{j-1}^+ is a generalized inverse of Δ_{j-1}. So S_1 is the canonical embedding of $\operatorname{Ker} P$ into $\mathcal{X}$, and S_j is the operator that consists of the last component of Δ_j followed by a projection along the image of Δ_{j-1}.

In this part we shall show that $\operatorname{Ker} S_j$ is complemented in $\operatorname{Ker} P$. For $j = 1$ this is obvious. Therefore, take $j \geq 2$, and define

$$S_j^+ = (I-P)\pi_j \Delta_j^+ : \mathcal{X} \to \operatorname{Ker} P.$$

Here $\pi_j : \mathcal{X}^j \to \mathcal{X}$ is the canonical projection of $\mathcal{X}^j$ onto its last coordinate space. It suffices to show that $S_j S_j^+ S_j = S_j$. To prove this, take $x \in \operatorname{Ker} P$ and consider $z = \Delta_j^+ S_j x$. Write

$$z = \begin{pmatrix} y \\ y_0 \end{pmatrix}$$

where $y_0 = \pi_j z$ and $y \in \mathcal{X}^{j-1}$. Since $S_j x \in \mathcal{F}_j$, we have that $\Delta_j \Delta_j^+ S_j x = S_j x$ and thus

$$\begin{aligned} S_j x = \Delta_j \Delta_j^+ S_j x &= \begin{pmatrix} \Delta_{j-1} & B^{j-1}(I-P) \end{pmatrix} \begin{pmatrix} y \\ y_0 \end{pmatrix} \\ &= \Delta_{j-1} y + B^{j-1}(I-P) y_0. \end{aligned} \tag{2.4}$$

Now use the definition of S_j and apply $I - \Delta_{j-1}\Delta_{j-1}^+$ to both hand sides of (2.4) to see that

$$S_j x = S_j (I-P) y_0 \tag{2.5}$$

Since $(I-P)y_0 = (I-P)\pi_j \Delta_j^+ S_j x = S_j^+ S_j x$, we have proved that $S_j = S_j S_j^+ S_j$.

PART (B). From the previous part we know that $\operatorname{Ker} S_j$ is complemented in $\operatorname{Ker} P$. So there exist closed subspaces $\mathcal{M}_1, \dots, \mathcal{M}_\ell$ of $\operatorname{Ker} P$ such that

$$\operatorname{Ker} P = \operatorname{Ker} S_\ell \oplus \mathcal{M}_\ell, \quad \operatorname{Ker} S_{j+1} = \operatorname{Ker} S_j \oplus \mathcal{M}_j,$$

for $j = 1, \dots, \ell - 1$. Write

$$\mathcal{Z}_j = \left\{ \begin{pmatrix} x_1 \\ \vdots \\ x_j \end{pmatrix} \in \mathcal{X}^j \mid x_i \in \mathcal{M}_i \oplus \cdots \oplus \mathcal{M}_\ell \right\},$$

for $j = 1, \dots, \ell$. In this part we shall prove that Δ_j maps $\mathcal{Z}_j$ in a one to one manner onto $\mathcal{F}_j$. Since $\mathcal{Z}_1 = \operatorname{Ker} P$, the statement is trivial for $j = 1$. We proceed by induction. Assume that Δ_{j-1} is bijective from $\mathcal{Z}_{j-1}$ to $\mathcal{F}_{j-1}$. Suppose that $B^{j-1}(I-P)(m_j + \cdots + m_\ell) \in \Delta_{j-1}[\mathcal{Z}_{j-1}]$ with $m_i \in \mathcal{M}_i$ for $i = j, \dots, \ell$. Then

$$S_j(m_j + \cdots + m_\ell) = (I - \Delta_{j-1}\Delta_{j-1}^+) B^{j-1}(I-P)(m_j + \cdots + m_\ell) = 0.$$

The definition of $\mathcal{M}_j$ now gives that $m_j + \cdots + m_\ell = 0$. Using our induction hypothesis, we conclude that Δ_j is one-one on $\mathcal{Z}_j$. Next, we prove that Δ_j maps $\mathcal{Z}_j$ onto $\mathcal{F}_j$. Note that $\mathcal{F}_j = \operatorname{Im} \Delta_j = \operatorname{Im} \Delta_{j-1} + \operatorname{Im} B^{j-1}(I-P)$ and hence it follows that $\operatorname{Im} S_j = (I - \Delta_{j-1}\Delta_{j-1}^+)[\mathcal{F}_j]$. So

$$\mathcal{F}_j = \operatorname{Im} \Delta_{j-1}\Delta_{j-1}^+ \oplus \operatorname{Ker}(\Delta_{j-1}\Delta_{j-1}^+|_{\mathcal{F}_j}) = \mathcal{F}_{j-1} \oplus \operatorname{Im} S_j. \tag{2.6}$$

This proves that $\operatorname{Im} S_j = S_j[\mathcal{M}_j \oplus \cdots \oplus \mathcal{M}_\ell]$ is a direct complement to $\Delta_j[\mathcal{Z}_{j-1}]$. Since $S_j = (I - \Delta_{j-1}\Delta_{j-1}^+) B^{j-1}(I-P)$, we see that

$$\Delta_j \mathcal{Z}_j = \Delta_{j-1}(\mathcal{Z}_{j-1}) + B^{j-1}(I-P)(\mathcal{M}_j \oplus \cdots \oplus \mathcal{M}_\ell) = \mathcal{F}_{j-1} \oplus \operatorname{Im} S_j = \mathcal{F}_j.$$

PART (C). In this part we prove that for $j = 1, \dots, \ell - 1$ the operator B_j in (2.3) has a generalized inverse. For $j = 1$ this is obvious. We proceed by induction. Therefore, in what follows, we have $2 \le j \le \ell$, and we assume that B_{j-1} has a generalized inverse B^+_{j-1}. Put $\mathcal{K}_j = B^j(\mathcal{M}_{j+1} \oplus \cdots \oplus \mathcal{M}_\ell)$. Since Δ_j maps $\mathcal{Z}_j$ in a one to one manner onto $\mathcal{F}_j$, we have

$$\mathcal{F}_{j+1} = \mathcal{F}_j \oplus \mathcal{K}_j, \quad \mathcal{F}_j = \mathcal{F}_{j-1} \oplus \mathcal{K}_{j-1}.$$

Partition B_j as follows

$$B_j = \begin{pmatrix} B_{j-1} & C_{j-1} \\ 0 & D_{j-1} \end{pmatrix} : \mathcal{F}_{j-1} \oplus \mathcal{K}_{j-1} \to \mathcal{F}_j \oplus \mathcal{K}_j.$$

Here C_{j-1} and C_{j-1} are the operators defined by

$$\begin{aligned} C_{j-1}\Big(B^{j-1}(m_j + \cdots + m_\ell)\Big) &= B^j m_j, \\ D_{j-1}\Big(B^{j-1}(m_j + \cdots + m_\ell)\Big) &= B^j(m_{j+1} + \cdots + m_\ell), \end{aligned}$$

where $m_i \in \mathcal{M}_i$ for $i = j, \ldots, \ell$. Clearly, D_{j-1} is right invertible. In fact, a right inverse $D^+_{j-1} : \mathcal{K}_j \to \mathcal{K}_{j-1}$ of D_{j-1} is obtained by taking

$$D^+_{j-1}(B^j(m_{j+1} + \cdots + m_\ell)) = B^{j-1}(m_{j+1} + \cdots + m_\ell),$$

where $m_i \in \mathcal{M}_i$ for $i = j+1, \ldots, \ell$. Hence $C_{j-1}D^+_{j-1} = 0$.

Next we show that $\operatorname{Im} C_{j-1} \subset \operatorname{Im} B_{j-1}$. Notice that $B^j\mathcal{M}_j \subset \mathcal{F}_j$. Recall that $\Delta_j : \mathcal{Z}_j \to \mathcal{F}_j$ is bijective. Let $\Omega_j : \mathcal{F}_j \to \mathcal{Z}_j$ be the inverse of this operator. Thus for $1 \le i \le j$ we may define $E_{ij} : \mathcal{M}_j \to \mathcal{M}_i \oplus \cdots \oplus \mathcal{M}_\ell$ by setting

$$E_{ij} = \pi'_i \Omega_j B^j (I - P),$$

where π'_i is the natural projection of $\mathcal{Z}_j$ to its i-th coordinate vector. Then, for $x \in \mathcal{M}_j$, we have

$$(I-P)E_{1j}x + B(I-P)E_{2j}x + \cdots + B^{j-1}(I-P)E_{jj}x = \Delta_j\Omega_j B^j x = B^j x. \quad (2.7)$$

Since $\operatorname{Im} B \subset \operatorname{Im} P$, it follows that $E_{1j} = 0$. Thus

$$B^j x = B((I-P)E_{2j}x + \cdots + B^{j-2}(I-P)E_{jj}x) \in B\mathcal{F}_{j-1} = \operatorname{Im} B_{j-1}, \quad x \in \mathcal{M}_j,$$

which proves that $\operatorname{Im} C_{j-1} \subset \operatorname{Im} B_{j-1}$.

It is now easy to see that

$$\begin{pmatrix} B^+_{j-1} & 0 \\ 0 & D^+_{j-1} \end{pmatrix}$$

is a generalized inverse of B_j. Indeed

$$\begin{pmatrix} B_{j-1} & C_{j-1} \\ 0 & D_{j-1} \end{pmatrix} \begin{pmatrix} B_{j-1}^{+} & 0 \\ 0 & D_{j-1}^{+} \end{pmatrix} \begin{pmatrix} B_{j-1} & C_{j-1} \\ 0 & D_{j-1} \end{pmatrix}$$
$$= \begin{pmatrix} B_{j-1}B_{j-1}^{+} & 0 \\ 0 & I \end{pmatrix} \begin{pmatrix} B_{j-1} & C_{j-1} \\ 0 & D_{j-1} \end{pmatrix} = \begin{pmatrix} B_{j-1} & C_{j-1} \\ 0 & D_{j-1} \end{pmatrix},$$

and

$$\begin{pmatrix} B_{j-1}^{+} & 0 \\ 0 & D_{j-1}^{+} \end{pmatrix} \begin{pmatrix} B_{j-1} & C_{j-1} \\ 0 & D_{j-1} \end{pmatrix} \begin{pmatrix} B_{j-1}^{+} & 0 \\ 0 & D_{j-1}^{+} \end{pmatrix} = \begin{pmatrix} B_{j-1} & 0 \\ 0 & D_{j-1} \end{pmatrix}.$$

PART (D). We start now with the construction of the spaces $\mathcal{U}_{ij}$. Choose

$$\mathcal{U}_{\ell\ell} = \operatorname{Ker}(\Delta_{\ell-1}\Delta_{\ell-1}^{+}) \cap \mathcal{F}_{\ell}.$$

Since $\mathcal{F}_{\ell-1} = \operatorname{Im}\Delta_{\ell-1}$, it is clear that $\mathcal{F}_{\ell} = \mathcal{F}_{\ell-1} \oplus \mathcal{U}_{\ell\ell}$. As $\operatorname{Ker} P \subset \mathcal{F}_{\ell-1}$, the operator $P|_{\mathcal{U}_{\ell\ell}}$ is injective and has closed range. Now assume that we have constructed closed subspaces $\mathcal{U}_{ij}$, for $k \leq i \leq j \leq \ell$, of $\mathcal{X}$ such that

(i') $\mathcal{F}_i = \mathcal{F}_{i-1} \oplus U_{ii} \oplus \cdots \oplus U_{i\ell}$ for $i = k+1, \ldots, \ell$;

(ii') $B[U_{ii}] = \{0\}$ for $i = k+1, \ldots, \ell$;

(iii') $B[U_{ij}] = P[U_{i+1\ j}]$ for $k+1 \leq i \leq j-1 \leq \ell-1$;

(iv') the operators $B|_{U_{ij}}$ and $P|_{U_{i+1\ j}}$ are injective and have close range for $k+1 \leq i \leq j-1 \leq \ell-1$.

From $\mathcal{F}_{k+1} = \mathcal{F}_k \oplus \mathcal{U}_{k+1\ k+1} \oplus \cdots \oplus \mathcal{U}_{k+1\ \ell}$ and $\operatorname{Ker} P \subset \mathcal{F}_k$ it is clear that for $j = k+1, \ldots, \ell$ the operators $P|_{\mathcal{U}_{k+1\ j}}$ are injective and have closed range. Put $\mathcal{V}_{k+1\ j} = P[\mathcal{U}_{k+1\ j}]$ for $j = k+1, \ldots, \ell$, and write $\mathcal{V}_{k+1} = \mathcal{V}_{k+1\ k+1} \oplus \cdots \oplus \mathcal{V}_{k+1\ \ell}$. Observe that $P[\mathcal{F}_{k+1}] = P[\mathcal{F}_k] \oplus \mathcal{V}_{k+1}$. Also, $P[\mathcal{F}_{k+1}] = P(\operatorname{Ker} P \oplus B[\mathcal{F}_k]) = PB[\mathcal{F}_k] = B[\mathcal{F}_k]$, and similarly $P[\mathcal{F}_k] = B[\mathcal{F}_{k-1}]$. It follows that

$$B[\mathcal{F}_k] = P[\mathcal{F}_{k+1}] = P[\mathcal{F}_k] \oplus \mathcal{V}_{k+1} = B[\mathcal{F}_{k-1}] \oplus \mathcal{V}_{k+1}. \tag{2.8}$$

Put $\mathcal{U}_{kk} = \Delta_k T_k[\mathcal{M}_k]$. Then $B\mathcal{U}_{kk} = \{0\}$ and

$$(\operatorname{Ker} B) \cap \mathcal{F}_k = (\operatorname{Ker} B) \cap \mathcal{F}_{k-1} \oplus \mathcal{U}_{kk}.$$

Let B_k^{+} be the generalized inverse of $B|_{\mathcal{F}_k} : \mathcal{F}_k \to \mathcal{F}_{k+1}$. Define $\mathcal{U}_{kj} = B_k^{+}[\mathcal{V}_{k+1\ j}]$ for $j = k+1, \ldots, \ell$. Then $\mathcal{U}_{kj}$ is closed, and $B|_{\mathcal{U}_{kj}}$ is injective and has closed range for $j = k+1, \ldots, \ell$. Use (2.8) to see that $\mathcal{V}_k \subset \operatorname{Im} B|_{\mathcal{F}_k}$. So $BB_k^{+}[\mathcal{V}_{k+1\ j}] = \mathcal{V}_{k+1\ j}$, and thus $B[\mathcal{U}_{kj}] = \mathcal{V}_{k+1\ j} = P[\mathcal{U}_{k+1\ j}]$ for $j = k+1, \ldots, \ell$.

It remains to show that

$$\mathcal{F}_k = \mathcal{F}_{k-1} \oplus \mathcal{U}_{kk} \oplus \mathcal{U}_{kk+1} \oplus \cdots \oplus \mathcal{U}_{k\ell}.$$

Let $x \in \mathcal{F}_k$. Then $Bx = Bf_{k-1} + v_{k+1} + \cdots + v_\ell$ with $f_{k-1} \in \mathcal{F}_{k-1}$ and $v_j \in \mathcal{V}_{k+1\ j}$ for $j = k+1, \ldots, \ell$. Let $u_j = B_k^+ v_j \in \mathcal{U}_{kj}$. Then $Bu_j = v_j$ since BB_k^+ is a projection onto $\operatorname{Im} B|_{\mathcal{F}_k}$ and $\mathcal{V}_{k+1\ j} \subset \operatorname{Im} B|_{\mathcal{F}_k}$. On the other hand,

$$x - (f_{k-1} + u_{k+1} + \cdots + u_\ell) \in (\operatorname{Ker} B) \cap \mathcal{F}_k,$$

and $(\operatorname{Ker} B) \cap \mathcal{F}_k \subset \mathcal{F}_{k-1} + \mathcal{U}_{kk}$. So $x - (f_{k-1} + u_{k+1} + \cdots + u_\ell) = f'_{k-1} + u_k$ for some $f'_{k-1} \in \mathcal{F}_{k-1}$ and $u_k \in \mathcal{U}_{kk}$. Since also $U_{kj} \subset \mathcal{F}_k$ and $\mathcal{F}_{k-1} \subset \mathcal{F}_k$, this proves that

$$\mathcal{F}_k = \mathcal{F}_{k-1} + \mathcal{U}_{kk} + \mathcal{U}_{kk+1} + \cdots + \mathcal{U}_{k\ell}.$$

Next assume that

$$f_{k-1} + u_k + \cdots + u_\ell = 0, \tag{2.9}$$

with $f_{k-1} \in \mathcal{F}_{k-1}$ and $u_j \in \mathcal{U}_{kj}$. Apply B to the equality to get

$$Bf_{k-1} + Bu_{k+1} + \cdots + Bu_\ell = 0 \in \mathcal{F}_k \oplus \mathcal{V}_{k+1\ k+1} \oplus \cdots \oplus \mathcal{V}_{k+1\ \ell}.$$

This proves that $Bf_{k-1} = 0$ and $Bu_j = 0$ for $j = k+1, \ldots, \ell$. So $u_j = 0$ for $j = k+1, \ldots, \ell$, and $f_{k-1} \in (\operatorname{Ker} B) \cap \mathcal{F}_{k-1}$. We see that $f_{k-1} + u_k = 0$, and therefore $u_k \in (\operatorname{Ker} B) \cap \mathcal{F}_{k-1} \cap \mathcal{U}_{kk} = \{0\}$. So all terms in the sum (2.9) are equal to zero, which proves that

$$\mathcal{F}_k = \mathcal{F}_{k-1} \oplus \mathcal{U}_{kk} \oplus \mathcal{U}_{kk+1} \oplus \cdots \oplus \mathcal{U}_{k\ell}. \qquad \square$$

The next theorem will be a generalization of Theorem V.1.1 for the special case when there are no invariant polynomials; it is the main decomposition theorem for (P, I)-blocks.

Theorem 2.2. *Let $(B; P, I)$ be an operator block on the Banach space $\mathcal{X}$. In order that $(B; P, I)$ is block similar to a direct sum of block shifts of the first kind it is necessary and sufficient that there exists a natural number $\ell \geq 1$ such that the transformation Δ_j given by (2.2) has a generalized inverse for $j = 1, \ldots, \ell$ and is surjective for $j = \ell$.*

Proof. Let $(B_1; P_1, I_1)$ be a block shift of the first kind with base space $\mathcal{Y}_1$ and index $\nu - 1$. Then it is straightforward to check that the operator

$$\Delta_{1j} = \big((I_1 - P_1) \quad B_1(I_1 - P_1) \quad \cdots \quad B_1^{j-1}(I_1 - P_1) \big) : \mathcal{Y}_1^j \to \mathcal{Y}_1$$

has a generalized inverse for $j = 1, \ldots, \nu$ and is surjective for $j = \nu$. Now suppose that the operator block $(B_0; P_0, I_0)$ on the space $\mathcal{X}_0$ is a direct sum of block shifts of the first kind. It follows that there exists a number ℓ such that

$$\Delta_{0j} = \big((I_0 - P_0) \quad B_0(I_0 - P_0) \quad \cdots \quad B_0^{j-1}(I_0 - P_0) \big) : \mathcal{X}_0^j \to \mathcal{X}_0 \tag{2.10}$$

has a generalized inverse for $j = 1, \dots, \ell$ and is surjective for $j = \ell$. Assume that $(B; P, I)$ is block similar to $(B_0; P_0, I_0)$. Let $S : \mathcal{X} \to \mathcal{X}_0$ be the block similarity. So S is an invertible and bounded linear operator such that

$$S[\operatorname{Ker} P] = \operatorname{Ker} P_0, \quad P_0(SA - A_0 S) = 0.$$

Let Ω_j be the block Toeplitz matrix

$$\Omega_j = \begin{pmatrix} (I-P_0)S(I-P) & (I-P_0)SA(I-P) & \cdots & (I-P_0)SA^{j-1}(I-P) \\ 0 & (I-P_0)S(I-P) & \cdots & (I-P_0)SA^{j-2}(I-P) \\ \vdots & & \ddots & \vdots \\ 0 & 0 & & (I-P_0)S(I-P) \end{pmatrix}. \tag{2.11}$$

Obviously, Ω_j is a bijection from $(\operatorname{Ker} P)^j$ to $(\operatorname{Ker} P_0)^j$. By induction one proves that $\Delta_{0j}\Omega_j = S\Delta_j$. Write Λ_{0j} for the restriction of Δ_{0j} to $(\operatorname{Ker} P_0)^j$ as an operator from $(\operatorname{Ker} P_0)^j$ to $\mathcal{X}_0$, and write Λ_j for the restriction of Δ_j to $(\operatorname{Ker} P)^j$ as an operator from $(\operatorname{Ker} P)^j$ to $\mathcal{X}$. Since $\operatorname{Im} P_0 \subset \operatorname{Ker} \Delta_{0j}$, the operator $\Lambda_{0j} : (\operatorname{Ker} P_0)^j \to \mathcal{X}_0$ has a generalized inverse for $j = 1, \dots, \ell$ and is surjective for $j = \ell$. So we conclude that the same is true for Λ_j. Since also $(\operatorname{Im} P)^j \subset \operatorname{Ker} \Delta_j$, it follows that the operator Δ_j has the desired properties.

Now conversely, suppose that Δ_j has a generalized inverse for $j = 1, \dots, \ell$ and is surjective for $j = \ell$. With $\mathcal{F}_\ell$ given by (2.1) and Δ_ℓ by (2.2) we have that $\mathcal{X} = \operatorname{Im} \Delta_\ell = \mathcal{F}_\ell$. Therefore, by Lemma 2.1, we may write

$$\mathcal{X} = \bigoplus_{i=1}^{\ell} \bigoplus_{j=i}^{\ell} \mathcal{U}_{ij} = \bigoplus_{j=1}^{\ell} \bigoplus_{i=1}^{j} \mathcal{U}_{ij}, \quad \operatorname{Ker} P = \bigoplus_{j=1}^{\ell} \mathcal{U}_{1j}.$$

Define a projection P_1 by putting $\operatorname{Ker} P_1 = \operatorname{Ker} P$ and $\operatorname{Im} P_1 = \bigoplus_{j=2}^{\ell} \bigoplus_{i=1}^{j} \mathcal{U}_{ij}$. Consider the operator block $(P_1 B; P_1, I)$. This operator block is block similar to the block $(B; P, I)$, the block similarity being given by the identity transformation. For $1 \le i < j \le \ell$ we compute that

$$P_1 B[\mathcal{U}_{ij}] = P_1 P[\mathcal{U}_{i+1\ j}] = \mathcal{U}_{i+1\ j},$$

and $P_1 B|_{\mathcal{U}_{ij}} : \mathcal{U}_{ij} \to \mathcal{U}_{i+1\ j}$ is a bounded bijective operator. Put $\mathcal{U}_j = \mathcal{U}_{1j}$ and define the operator

$$S : \bigoplus_{j=1}^{\ell} \mathcal{U}_j^j \to \mathcal{X} = \bigoplus_{j=1}^{\ell} \bigoplus_{i=1}^{j} \mathcal{U}_{ij}$$

by

$$S|_{\mathcal{U}_j^j} = \begin{pmatrix} I & & & & \\ & P_1 B & & & \\ & & \ddots & & \\ & & & P_1 B^{j-2} & \\ & & & & P_1 B^{j-1} \end{pmatrix} \Big|_{\mathcal{U}_j^j} : \mathcal{U}_j^j \to \bigoplus_{i=1}^{j} \mathcal{U}_{ij}.$$

Then S is invertible and

$$S^{-1}(P_1 B)S|_{\mathcal{U}_j^j} = \begin{pmatrix} 0 & 0 & \cdots & 0 & 0 \\ I & 0 & \cdots & 0 & 0 \\ & \ddots & \ddots & \vdots & \vdots \\ 0 & 0 & \ddots & 0 & 0 \\ 0 & 0 & & I & 0 \end{pmatrix} : \mathcal{U}_j^j \to \mathcal{U}_j^j.$$

It follows that S induces a block similarity of $(P_1 B; P_1, I)$ and a direct sum of block shifts of the first kind. □

A block $(B; P, I)$ is said to be of *finite type* if there exists a positive number ℓ such that $\Delta_j = \operatorname{row}\big(B^{i-1}(I-P)\big)_{i=1}^{j}$ has a generalized inverse for $j = 1, \ldots, \ell$ and is surjective for $j = \ell$. According to the previous theorem a block $(B; P, I)$ of finite type is block similar to a direct sum $(B_1; P_1, I_1) \oplus \cdots \oplus (B_r; P_r, I_r)$ of block shifts of the first kind. Let $\mathcal{Y}_j$ be the base space of the block shift $(B_j; P_j, I_j)$, and let ν_j be its index. The proof of Theorem 2.2 shows us that we may assume that $\operatorname{Ker} P = \mathcal{Y}_1 \oplus \cdots \oplus \mathcal{Y}_r$. Without loss of generality we may also assume that the spaces $\mathcal{Y}_j$ are non-trivial, that is $\mathcal{Y}_j \neq \{0\}$. If the indices ν_p and ν_q are equal, then $(B_p; P_p, I_p) \oplus (B_q; P_q, I_q)$ is again a block shift with index $\nu = \nu_p = \nu_q$, but now with base space $\mathcal{Y}_p \oplus \mathcal{Y}_q$. So, without loss of generality $\nu_1 > \cdots > \nu_r \geq 0$. Recall that $\mathcal{Y}_j \neq \{0\}$. The set of pairs $\{(\mathcal{Y}_1, \nu_1), \ldots, (\mathcal{Y}_r, \nu_r)\}$, where $\mathcal{Y}_i \neq \{0\}$ and $\nu_1 > \cdots > \nu_r \geq 0$, is called the *characteristics* of the block $(B; P, I)$. Two sets of characteristics $\{(\mathcal{Y}_1, \nu_1), \ldots, (\mathcal{Y}_r, \nu_r)\}$ and $\{(\mathcal{Y}_1', \nu_1'), \ldots, (\mathcal{Y}_s', \nu_s')\}$ will be called *equal* if $s = r$, the Banach spaces $\mathcal{Y}_i$ and $\mathcal{Y}_i'$ are isomorphic and $\nu_i = \nu_i'$ for $i = 1, \ldots, s$. This terminology is justified by the following result.

Theorem 2.3. *The operator blocks $(B; P, I)$ and $(B'; P', I')$ of finite type are block similar of and only if they have the same characteristics.*

PROOF. Assume $(B; P, I)$ and $(B'; P', I')$ have the same characteristics. Then, as we have seen above, there is a direct sum of block shifts of the first kind which is similar to both $(B; P, I)$ and $(B'; P', I')$. Since block similarity is a transitive relation it follows that $(B; P, I)$ and $(B'; P', I')$ are block similar.

To prove the converse, let $\{(\mathcal{Y}_1, \nu_1), \ldots, (\mathcal{Y}_r, \nu_r)\}$ be a set off characteristics for the block $(B; P, I)$. We have to show that this set is uniquely determined by the block $(B; P, I)$. Let $(B_j; P_j, I_j)$ be the block shift of the first kind with base space $\mathcal{Y}_j$ and index ν_j. Put

$$(B_0; P_0, I_0) = (B_1; P_1, I_1) \oplus \cdots \oplus (B_r; P_r, I_r).$$

Then the blocks $(B_0; P_0, I_0)$ and $(B; P, I)$ are block similar with similarity $S : \mathcal{X} \to \mathcal{X}_0$, say. Let Δ_j be given by formula (2.2) and Δ_{0j} by formula (2.10). With Ω_j as in (2.11) we have that $\Delta_{0j}\Omega_j = S\Delta_j$. Since Ω_j is a bijection from $(\operatorname{Ker} P)^j$ onto $(\operatorname{Ker} P_0)^j$, we see that $S[\operatorname{Im} \Delta_j] = \operatorname{Im} \Delta_{0j}$. If $Bx = 0$, then $B_0 Sx = P_0 B_0 Sx =$

$P_0SBx = 0$. So $S[\operatorname{Ker} B] \subset \operatorname{Ker} B_0$. Using the symmetry in the similarity relation we conclude that $S[\operatorname{Ker} B] = \operatorname{Ker} B_0$. So

$$S[\operatorname{Ker} B \cap \operatorname{Im} \Delta_j] = \operatorname{Ker} B_0 \cap \operatorname{Im} \Delta_{0j}. \tag{2.12}$$

One checks that

$$\operatorname{Ker} B_0 \cap \operatorname{Im} \Delta_{0j} = \operatorname{Ker} B_0 \cap \operatorname{Im} \Delta_{0\ j-1} \oplus \begin{cases} \mathcal{Y}_i & \text{if } i = \nu_j + 1, \\ \{0\} & \text{otherwise.} \end{cases}$$

Next use (2.12) to see that this implies that the numbers $\nu_1, \ldots, \nu_r$ are uniquely determined by the block $(B; P, I)$. Also the spaces $\mathcal{Y}_1, \ldots, \mathcal{Y}_r$ are determined up to an isomorphism. Indeed, $\mathcal{Y}_j$ is isomorphic to the quotient space

$$(\operatorname{Ker} B \cap \operatorname{Im} \Delta_j)/(\operatorname{Ker} B \cap \operatorname{Im} \Delta_{j-1}).$$

□

The last part of the proof of the previous theorem gives us also the following result.

Corollary 2.4. *Let $(B; P, I)$ be a block of finite type. Put $\Delta_0 = 0$ and $\Delta_j = \operatorname{row}\big(B^{i-1}(I-P)\big)_{i=1}^{j}$ for $j = 1, 2, \ldots$. Let $\omega_1 > \cdots > \omega_r$ be the natural numbers j such that the quotient space*

$$\mathcal{Y}_j = \frac{\operatorname{Ker} B \cap \operatorname{Im} \Delta_j}{\operatorname{Ker} B \cap \operatorname{Im} \Delta_{j-1}} \neq \{0\}.$$

Then $(\mathcal{Y}_{\omega_1}, \omega_1 - 1), \ldots, (\mathcal{Y}_{\omega_r}, \omega_r - 1)$ is the set of characteristics of the block $(B; P, I)$.

Let $(\mathcal{Y}_{\nu_1}, \nu_1), \ldots, (\mathcal{Y}_{\nu_r}, \nu_r)$ be the set of characteristics of the block $(B; P, I)$. Then $(B; P, I)$ is similar to the (P_0, I)-block $(P_0A_0; P_0, I)$ of the operator A_0, where A_0 and P_0 are given by

$$A_0 = \begin{pmatrix} A_1 & 0 & \cdots & 0 \\ 0 & A_2 & & 0 \\ \vdots & & \ddots & \\ 0 & 0 & & A_r \end{pmatrix}, \quad P_0 = \begin{pmatrix} P_1 & 0 & \cdots & 0 \\ 0 & P_2 & & 0 \\ \vdots & & \ddots & \\ 0 & 0 & & P_r \end{pmatrix},$$

where

$$A_i = \begin{pmatrix} 0 & 0 & \cdots & 0 & 0 \\ I_i & 0 & \cdots & 0 & 0 \\ & \ddots & \ddots & \vdots & \vdots \\ 0 & 0 & \ddots & 0 & 0 \\ 0 & 0 & & I_i & 0 \end{pmatrix}, \quad P_i = \begin{pmatrix} 0 & 0 & \cdots & 0 & 0 \\ 0 & I_i & & 0 & 0 \\ \vdots & & \ddots & & \vdots \\ 0 & 0 & & I_i & 0 \\ 0 & 0 & \cdots & 0 & I_i \end{pmatrix},$$

with I_i the identity on $\mathcal{Y}_i$.

The following theorem represents a special case of Theorem 2.2.

Theorem 2.5. *Let $\mathcal{X}$ and $\mathcal{U}$ be Banach spaces, and let $F:\mathcal{X}\to\mathcal{X}$ and $G:\mathcal{U}\to\mathcal{X}$ be bounded linear operators. Assume that*

$$\begin{pmatrix} G & FG & \cdots & F^{j-1}G \end{pmatrix} : \mathcal{U}^j \to \mathcal{X}$$

has a generalized inverse for $j=1,\dots,\ell$ and is surjective for $j=\ell$. Then there exist closed subspaces $\mathcal{U}_0,\dots,\mathcal{U}_\ell$ of $\mathcal{U}$ with $\mathcal{U}=\mathcal{U}_0\oplus\cdots\oplus\mathcal{U}_\ell$, and there exist bounded linear operators

$$N:\mathcal{U}_1^1\oplus\mathcal{U}_2^2\oplus\cdots\oplus\mathcal{U}_\ell^\ell\to\mathcal{X},\quad K:\mathcal{X}\to\mathcal{U},$$

such that N is invertible and

(i) $N^{-1}G=E_0\oplus\cdots\oplus E_\ell:\mathcal{U}_0\oplus\mathcal{U}_2\oplus\cdots\oplus\mathcal{U}_\ell\to\mathcal{U}_1^1\oplus\mathcal{U}_2^2\oplus\cdots\oplus\mathcal{U}_\ell^\ell$ *with* $E_0=0$ *and*

$$E_j=\begin{pmatrix} I_j\\ 0\\ \vdots\\ 0\end{pmatrix}:\mathcal{U}_j\to\mathcal{U}_j^j,\quad j=1,\dots,\ell\,, \tag{2.13}$$

(ii) $N^{-1}(F-GK)N=J_1\oplus\cdots\oplus J_\ell:\mathcal{U}_1^1\oplus\mathcal{U}_2^2\oplus\cdots\oplus\mathcal{U}_\ell^\ell\to\mathcal{U}_1^1\oplus\mathcal{U}_2^2\oplus\cdots\oplus\mathcal{U}_\ell^\ell$, *with* $J_1=0$ *and*

$$J_j=\begin{pmatrix} 0 & 0 & \cdots & 0 & 0\\ I_j & 0 & \cdots & 0 & 0\\ & \ddots & \ddots & \vdots & \vdots\\ 0 & 0 & \ddots & 0 & 0\\ 0 & 0 & & I_j & 0\end{pmatrix}:\mathcal{U}_j^j\to\mathcal{U}_j^j, \tag{2.14}$$

for $j=2,\dots,\ell$.

In (i) and (ii) the symbol I_j denotes the identity operator on $\mathcal{U}_j$. Note that it may happen that $\mathcal{U}_j=\{0\}$ for some values of j.

PROOF. Consider the operator block $(B;P,I)$ on the space $\mathcal{X}\oplus\mathcal{U}$ given by

$$P=\begin{pmatrix} I & 0\\ 0 & 0\end{pmatrix}:\mathcal{X}\oplus\mathcal{U}\to\mathcal{X}\oplus\mathcal{U},$$
$$B=\begin{pmatrix} F & G\end{pmatrix}:\mathcal{X}\oplus\mathcal{U}\to\operatorname{Im}P\subset\mathcal{X}\oplus\mathcal{U}.$$

We compute that

$$\Delta_j=\begin{pmatrix}(I-P) & B(I-P) & \cdots & B^{j-1}(I-P)\end{pmatrix}:(\mathcal{X}\oplus\mathcal{U})^j\to\mathcal{X}\oplus\mathcal{U}$$

can be written as

$$\Delta_j=\left(\begin{pmatrix} 0 & 0\\ 0 & I\end{pmatrix}\ \begin{pmatrix} 0 & G\\ 0 & 0\end{pmatrix}\ \begin{pmatrix} 0 & FG\\ 0 & 0\end{pmatrix}\ \cdots\ \begin{pmatrix} 0 & F^{j-2}G\\ 0 & 0\end{pmatrix}\right).$$

After a reordering of the components in $(\mathcal{X}\oplus\mathcal{U})^j$, this operator can be represented as

$$\Delta_j = \begin{pmatrix} 0 & \Lambda_{j-1} & 0 \\ 0 & 0 & I \end{pmatrix} : \mathcal{X}^j \oplus \mathcal{U}^{j-1} \oplus \mathcal{U} \to \mathcal{X} \oplus \mathcal{U},$$

with

$$\Lambda_{j-1} = \begin{pmatrix} G & FG & \cdots & F^{j-2}G \end{pmatrix} : \mathcal{U}^{j-1} \to \mathcal{X}.$$

According to our hypothesis Λ_j has a generalized inverse for $j = 1, \dots, \ell$ and is surjective for $j = \ell$. It follows that Δ_j has a generalized inverse for $j = 2, \dots, \ell+1$ and is surjective for $j = \ell + 2$. Since Δ_1 clearly has a generalized inverse, the conditions of Theorem 2.2 are satisfied. So the block $(B; P, I)$ is block similar to a direct sum of block shifts of the first kind with indices $1, \dots, \ell+1$ and base spaces $\mathcal{U}_0, \dots, \mathcal{U}_\ell$. The proof of Theorem 2.2 shows us that we may assume without loss of generality that $\mathcal{U} = \operatorname{Ker} P = \mathcal{U}_0 \oplus \cdots \oplus \mathcal{U}_\ell$. Let $(B_i; P_i, I_i)$ be the block shift of the kind with index $i + 1$ and base space $\mathcal{U}_i$. The matrix representation of B_i can be written as $B_i = \begin{pmatrix} J_i & E_i \end{pmatrix} : \mathcal{U}_i^i \oplus \mathcal{U}_i \to \mathcal{U}_i^i$. Put $\overline{P} = P_0 \oplus \cdots \oplus P_\ell$ and $\overline{B} = B_0 \oplus \cdots \oplus B_\ell$. The operator $\overline{B}$ can be partitioned as

$$\overline{B} = \begin{pmatrix} F_0 & G_0 \end{pmatrix} : \left(\bigoplus_{i=1}^{\ell} \mathcal{U}_i^i \right) \oplus \mathcal{U} \to \bigoplus_{i=1}^{\ell} \mathcal{U}_i^i,$$

where $G_0 = E_0 \oplus \cdots \oplus E_\ell$ and $F_0 = J_1 \oplus \cdots \oplus J_\ell$, with E_i as in (2.13) and J_i as in (2.14). It remains to prove that there exist operators N and K such that $F_0 = N^{-1}(F - GK)N$ and $G_0 = N^{-1}G$.

The blocks $(B; P, I)$ on $\mathcal{X} \oplus \mathcal{U}$ and $(\overline{B}; \overline{P}, \overline{I})$ on $(\bigoplus_{i=1}^{\ell} \mathcal{U}_i^i) \oplus \mathcal{U}$ are block similar. Since $\operatorname{Ker} P = \mathcal{U}$ and $\operatorname{Ker} \overline{P} = \mathcal{U}$, this block similarity has the form

$$S = \begin{pmatrix} N & 0 \\ -KN & I \end{pmatrix}.$$

We see that $N^{-1}\begin{pmatrix} F & G \end{pmatrix} S = \begin{pmatrix} F_0 & G_0 \end{pmatrix}$, and thus $F_0 = N^{-1}(F - GK)N$ and $G_0 = N^{-1}G$. □

XII.3 Main theorems for (I, Q)-blocks

In this section we treat the case dual to the one of the previous section. So we consider (I, Q)-blocks and we will present the main representation theorems for such blocks. The finite dimensional case is treated in Chapter III. The ideas will be essentially the same although we have to account here for the extra problems arising from the fact that the underlying spaces are infinite dimensional. The first result will be basic for the main theorems. Before we state the result we remark that given the block $(B; Q, I)$ the map BQ can be considered to be an operator defined on the full space $\mathcal{X}$.

Lemma 3.1. *Let $(B;I,Q)$ be an operator block on the space $\mathcal{X}$. Put*

$$\mathcal{D}_0=\mathcal{X},\quad \mathcal{D}_1=\operatorname{Im}Q,\quad \mathcal{D}_j=\{x\in\mathcal{D}_{j-1}\mid Bx\in\mathcal{D}_{j-1}\}\quad (j=2,\dots,\ell). \tag{3.1}$$

If for $j=1,\dots,\ell$ the operator

$$\Omega_j=\operatorname{col}\big((I-Q)(BQ)^{i-1}\big)_{i=1}^{j}:\mathcal{X}\to\mathcal{X}^j \tag{3.2}$$

has a generalized inverse, then there exist closed subspaces $\mathcal{U}_{ij}$ of $\mathcal{X}$ for $1\le i\le j\le\ell$, such that

(i) *$\mathcal{D}_{i-1}=\mathcal{D}_i\oplus\mathcal{U}_{ii}\oplus\cdots\oplus\mathcal{U}_{i\ell}$ for $i=1,\dots,\ell$;*

(ii) *$B[\mathcal{U}_{ij}]=U_{i-1\ j}$ and the restriction $B|_{U_{ij}}$ is injective for $2\le i\le j\le\ell$.*

PROOF. First we check that $\mathcal{D}_j=\operatorname{Ker}\Omega_j$ for $j=1,\dots,\ell$. Clearly $\mathcal{D}_1=\operatorname{Ker}\Omega_1$. Assume that $\mathcal{D}_{j-1}=\operatorname{Ker}\Omega_{j-1}$. Then

$$\operatorname{Ker}\Omega_j=\operatorname{Ker}\Omega_{j-1}\cap\operatorname{Ker}\big(\Omega_{j-1}(BQ)\big)=\{x\in\mathcal{D}_{j-1}\mid Bx\in\mathcal{D}_{j-1}\}.$$

The operators Ω_j have generalized inverses by assumption, and therefore the subspaces $\mathcal{D}_j$ are closed and complemented in $\mathcal{X}$. For a moment fix a number j with $1\le j\le\ell$. We proved that there exists a subspace $\mathcal{W}$ such that

$$\mathcal{D}_{j-1}=\mathcal{D}_j\oplus\mathcal{W}. \tag{3.3}$$

Define $S_j:\mathcal{D}_{j-1}\to\operatorname{Ker}Q$ by $S_jx=(I-Q)(BQ)^{j-1}x$. Since

$$\mathcal{D}_j=\operatorname{Ker}\Omega_j=\operatorname{Ker}\Omega_{j-1}\cap\operatorname{Ker}\big((I-Q)(BQ)^{j-1}\big)$$

and $\mathcal{D}_{j-1}=\operatorname{Ker}\Omega_{j-1}$, it follows that $\operatorname{Ker}S_j=\mathcal{D}_j$. We conclude that $\operatorname{Ker}S_j$ has a closed complement.

Next we prove that S_j has a generalized inverse. Define $\Omega_j^0:\mathcal{X}\to(\operatorname{Ker}Q)^j$ by $\Omega_j^0x=\Omega_jx$. Because Ω_j has a generalized inverse for $j=1,\dots,\ell$, the same is true for Ω_j^0. Now let $\mathcal{V}$ be a closed complement to $\mathcal{D}_{j-1}$ in $\mathcal{X}$. Then $\mathcal{X}=\mathcal{D}_j\oplus\mathcal{W}\oplus\mathcal{V}$. Applying Ω_j^0 to this equality yields $\Omega_j^0[\mathcal{X}]=\Omega_j^0[\mathcal{V}]\oplus\Omega_j^0[\mathcal{W}]$ and all spaces involved are closed. We see that $\Omega_j^0[\mathcal{W}]$ is complemented in $\Omega_j^0[\mathcal{X}]$. Since $\Omega_j^0[\mathcal{X}]$ is complemented in $(\operatorname{Ker}Q)^j$, this gives that $\Omega_j^0[\mathcal{W}]$ is complemented in $(\operatorname{Ker}Q)^j$. Observe that

$$\Omega_j^0x=\begin{pmatrix}\Omega_{j-1}^0x\\ S_jx\end{pmatrix}=\begin{pmatrix}0\\ S_jx\end{pmatrix},\quad x\in\mathcal{W}\subset\mathcal{D}_{j-1}.$$

Next use that $\Omega_j^0[\mathcal{W}]$ is complemented in $(\operatorname{Ker}Q)^j$ to conclude that $\operatorname{Im}S_j$ is complemented in $\operatorname{Ker}Q$. Hence S_j has a generalized inverse.

The construction of the spaces $\mathcal{U}_{ij}$ is carried out by induction. We take $\mathcal{U}_{\ell\ell}$ such that $\mathcal{D}_{\ell-1} = \mathcal{D}_\ell \oplus \mathcal{U}_{\ell\ell}$ (see (3.3)). For the induction step assume that we have $\mathcal{U}_{ij}$, for $k \le i \le j \le \ell$, such that

(i') $\mathcal{D}_{i-1} = \mathcal{D}_i \oplus \mathcal{U}_{ii} \oplus \cdots \oplus \mathcal{U}_{i\ell}$ for $i = k, \ldots, \ell$;

(ii') $B[\mathcal{U}_{ij}] = U_{i-1\ j}$ and the restriction $B|_{U_{ij}}$ is injective for $k+1 \le i \le j \le \ell$.

We shall construct the spaces $\mathcal{U}_{k-1\ k-1}, \ldots, \mathcal{U}_{k-1\ \ell}$.

Put

$$\mathcal{V} = \mathcal{U}_{kk} \oplus \cdots \oplus \mathcal{U}_{k\ell}.$$

This subspace $\mathcal{V}$ is closed in $\mathcal{X}$ and $\mathcal{D}_{k-1} = \mathcal{D}_k \oplus \mathcal{V}$. Since $\operatorname{Ker} S_k = \mathcal{D}_k$, one has $\operatorname{Im} S_k = S_k[\mathcal{V}]$. Next we show that $B|_\mathcal{V} : \mathcal{V} \to \mathcal{D}_{k-2}$ is one-one and has closed range. To prove this we choose a sequence $(x_n)_{n=1}^\infty$ in $\mathcal{V}$ such that $\|x_n\| = 1$ for $n = 1, 2, \ldots$, and $\lim_{n\to\infty} Bx_n = 0$. Let S_k^+ be a generalized inverse to S_k such that $\operatorname{Im} S_k^+ S_k = \mathcal{V}$. Then

$$x_n = S_k^+ S_k x_n = S_k^+ S_{k-1} B x_n \to 0, \quad (n \to \infty),$$

which contradicts $\|x_n\| = 1$ for all n. So $B|_\mathcal{V}$ is one-one and has closed range.

Since the operator S_k has a generalized inverse and $\operatorname{Ker} S_k = \mathcal{D}_k$, we see that the subspace $S_k[\mathcal{V}]$ is complemented in $\operatorname{Ker} Q$. On the other hand, for $x \in \mathcal{D}_{k-1}$, by definition, $Bx \in \mathcal{D}_{k-2}$ and $S_k x = S_{k-1} B x$. So $S_{k-1}B[\mathcal{V}] = S_k[\mathcal{V}]$ is a subspace of $\operatorname{Im} S_{k-1}$ and is complemented in $\operatorname{Ker} Q$. The operator S_{k-1} has a generalized inverse. Thus $\operatorname{Im} S_{k-1}$ is complemented in $\operatorname{Ker} Q$, and hence there exist closed subspaces $\mathcal{Z}$ and $\mathcal{U}$ of $\operatorname{Ker} Q$ such that

$$\operatorname{Ker} Q = \operatorname{Im} S_{k-1} \oplus \mathcal{Z} = S_{k-1}B[\mathcal{V}] \oplus \mathcal{U} \oplus \mathcal{Z}.$$

Let S_{k-1}^+ be a generalized inverse to S_{k-1} such that $\operatorname{Ker} S_{k-1} S_{k-1}^+ = \mathcal{Z}$. Let Q_{k-1} be the projection along $S_{k-1}B[\mathcal{V}] \oplus \mathcal{Z}$ onto $\mathcal{U}$. Then $Q_{k-1}(I - S_{k-1}S_{k-1}^+) = 0$, and with this equality one easily checks that $S_{k-1}^+ Q_{k-1}$ is a generalized inverse of $Q_{k-1}S_{k-1}$. So $\operatorname{Ker}(Q_{k-1}S_{k-1})$ is a complemented subspace of $\mathcal{D}_{k-2}$. If $Q_{k-1}S_{k-1}x = 0$, then $S_{k-1}x \in \operatorname{Ker} Q_{k-1}$, which means that $x \in B[\mathcal{V}] + \operatorname{Ker} S_{k-1}$. We see that $\operatorname{Ker}(Q_{k-1}S_{k-1}) = B[\mathcal{V}] + \mathcal{D}_{k-1}$. This shows that there exists a closed subspace $\mathcal{U}_{k-1\ k-1}$ such that

$$\mathcal{D}_{k-2} = (\mathcal{D}_{k-1} + B[\mathcal{V}]) \oplus \mathcal{U}_{k-1\ k-1}.$$

We will prove that $\mathcal{D}_{k-1} + B[\mathcal{V}] = \mathcal{D}_{k-1} \oplus B[\mathcal{V}]$. Since $B[\mathcal{V}]$ is closed, it suffices to prove that $\mathcal{D}_{k-1} \cap B[\mathcal{V}] = \{0\}$. Indeed, if $x \in \mathcal{V}$ and $Bx \in \mathcal{D}_{k-1}$, then $S_k x = S_{k-1}Bx = 0$. So $x \in (\operatorname{Ker} S_k \cap \mathcal{V}) = \mathcal{D}_k \cap \mathcal{V} = \{0\}$.

Now choose, again using that $B|_\mathcal{V}$ is one-one and has closed range, $\mathcal{U}_{k-1\ j} = B[\mathcal{U}_{kj}]$ for $k \le j \le \ell$. Together with the already chosen $\mathcal{U}_{k-1\ k-1}$ these spaces satisfy the conditions (i') and (ii'), and the proof may be completed by induction. □

The next result is the main structure theorem for (I,Q)-blocks.

Theorem 3.2. *Let $(B;I,Q)$ be an operator block on the Banach space $\mathcal{X}$. Then $(B;I,Q)$ is similar to a direct sum of shifts of the third kind if and only if there exists a number ℓ such that the operator Ω_j defined by (3.2) has a generalized inverse for $j=1,\dots,\ell$ and is injective for $j=\ell$.*

PROOF. Suppose that the operator block $(B_0;I_0,Q_0)$ on the space $\mathcal{X}_0=\mathcal{Y}_0^\nu$ is a block shift of the third kind of index $\nu-1$. Then it is clear that the operator

$$\Omega_{0j}=\operatorname{col}\big((I-Q_0)(B_0Q_0)^{i-1}\big)_{i=1}^{j}:\mathcal{X}_0\to\mathcal{X}_0^j$$

has a generalized inverse for $j=1,\dots,\nu$ and is surjective for $j=\nu$. It follows that also for a direct sum $(B';I',Q')$ of block shifts of the third kind on a space $\mathcal{X}'$ there exists a number ℓ such that

$$\Omega_j'=\operatorname{col}\big((I'-Q')(B'Q')^{i-1}\big)_{i=1}^{j}:\mathcal{X}'\to(\mathcal{X}')^j \tag{3.4}$$

has a generalized inverse for $j=1,\dots,\ell$ and is injective for $j=\ell$. Now let S be a block similarity of $(B;I,Q)$ and $(B';I',Q')$. This means that $S:\mathcal{X}\to\mathcal{X}'$ is an invertible operator satisfying

$$S[\operatorname{Im}Q]=\operatorname{Im}Q',\quad (SB-B'S)Q=0.$$

One computes that $T_j\Omega_j=\Omega_j'S$ with

$$T_j=\begin{pmatrix}(I'-Q')S & 0 & \cdots & 0 & 0\\ (I'-Q')B'Q'S & (I'-Q')S & & 0 & 0\\ \vdots & \ddots & \ddots & & \\ \vdots & & & \ddots & \\ \vdots & & \ddots & & \\ \vdots & & & \ddots & \ddots \\ (I'-Q')(B'Q')^{j-1}S & \cdots & \cdots & (I'-Q')B'Q'S & (I'-Q')S\end{pmatrix}.$$

We conclude that $\Omega_j^0:\mathcal{X}\to(\operatorname{Ker}Q)^j$, defined by $\Omega_j^0x=\Omega_jx$, has a generalized inverse for $j=1,\dots,\ell$ and is injective for $j=\ell$. So the same holds for Ω_j.

Now, conversely, assume that Ω_j has a generalized inverse for $j=1,\dots,\ell$ and is injective for $j=\ell$. Let $\mathcal{D}_j$ be defined by (3.1). Since $\operatorname{Ker}\Omega_\ell=\{0\}$, one sees that $\mathcal{D}_\ell=\{0\}$. Using Lemma 3.1, we have

$$\mathcal{X}=\bigoplus_{i=1}^{\ell}\bigoplus_{j=i}^{\ell}\mathcal{U}_{ij}=\bigoplus_{j=1}^{\ell}\bigoplus_{i=1}^{j}\mathcal{U}_{ij}\qquad\text{and}\qquad\operatorname{Im}Q=\bigoplus_{i=2}^{\ell}\bigoplus_{j=i}^{\ell}\mathcal{U}_{ij}=\bigoplus_{j=2}^{\ell}\bigoplus_{i=2}^{j}\mathcal{U}_{ij}.$$

Put $\mathcal{U}_j = \mathcal{U}_{jj}$ for $j = 1, \dots, \ell$. Recall that $B|_{\mathcal{U}_{ij}} : \mathcal{U}_{ij} \to \mathcal{U}_{i-1\ j}$ is a invertible operator. Therefore $B^{i-1}|_{\mathcal{U}_j} : \mathcal{U}_j \to \mathcal{U}_{i-1\ j}$ is invertible. Define the operator

$$S : \bigoplus_{j=1}^{\ell} \mathcal{U}_j^j \to \mathcal{X} = \bigoplus_{j=1}^{\ell} \bigoplus_{i=1}^{j} \mathcal{U}_{j-i+1\ j}$$

by

$$S|_{\mathcal{U}_{j'}^j} = \begin{pmatrix} I_j & 0 & \cdots & 0 \\ 0 & B|_{\mathcal{U}_j} & & 0 \\ \vdots & & \ddots & \\ 0 & 0 & & B^{j-1}|_{\mathcal{U}_j} \end{pmatrix}.$$

We showed that $S|_{\mathcal{U}_j^j}$ is an invertible operator, and thus S is also invertible. Let Q' be the projection of $\mathcal{X}$ with $\operatorname{Im} Q' = \operatorname{Im} Q$ and $\operatorname{Ker} Q' = \bigoplus_{j=1}^{\ell} \mathcal{U}_{1j}$. One computes that

$$S^{-1}(BQ')S|_{\mathcal{U}_j^j} = \begin{pmatrix} 0 & 0 & \cdots & 0 & 0 \\ I_j & 0 & \cdots & 0 & 0 \\ & \ddots & \ddots & \vdots & \vdots \\ 0 & 0 & \ddots & 0 & 0 \\ 0 & 0 & & I_j & 0 \end{pmatrix},$$

where I_j is the identity on $\mathcal{U}_j$. We see that the block $(BQ'; I, Q')$ is block similar to a direct sum of block shifts of the third kind, with S acting as the block similarity. To finish the proof we just have to remark that the blocks $(B; I, Q)$ and $(BQ'; I, Q')$ are block similar, the block similarity being given by the identity $I_{\mathcal{X}}$. □

The block $(B; I, Q)$ is said to be of *finite type* if there exists a positive integer ℓ such that the operator $\Omega_j = \operatorname{col}\big((I - Q)(BQ)^{i-1}\big)_{i=1}^{j}$ has a generalized inverse for $j = 1, \dots, \ell$ and is injective for $j = \ell$. So a block of finite type is always block similar to a direct sum $(B_1; I_1, Q_1) \oplus \cdots \oplus (B_s; I_s, Q_s)$, where $(B_j; I_j, Q_j)$ is a block shift of the third kind with base space $\mathcal{Z}_j$ and index κ_j. Without loss of generality we may assume that $\mathcal{Z}_j \neq \{0\}$ and $\kappa_1 > \cdots > \kappa_r \geq 0$. Then the set $\{(\mathcal{Z}_1, \kappa_1), \dots, (\mathcal{Z}_r, \kappa_r)\}$ is called the *characteristics* of the block $(B; I, Q)$. Two sets of characteristics $\{(\mathcal{Z}_1, \kappa_1), \dots, (\mathcal{Z}_r, \kappa_r)\}$ and $\{(\mathcal{Z}_1', \kappa_1'), \dots, (\mathcal{Z}_s', \kappa_s')\}$ will be called *equal* if $s = r$, the Banach spaces $\mathcal{Z}_i$ and $\mathcal{Z}_i'$ are isomorphic and $\kappa_i = \kappa_i'$ for $i = 1, \dots, s$.

The next result is the analogue of Theorem 2.3 in the previous section.

Theorem 3.3. *The operator blocks $(B; I, Q)$ and $(B'; I', Q')$ of finite type are block similar if and only if they have the same characteristics.*

PROOF. The proof of the "if part" is the same as the corresponding part in the proof of Theorem 2.3. To prove the converse, let $\{(\mathcal{Z}_1, \kappa_1), \dots, (\mathcal{Z}_r, \kappa_r)\}$ be a set of characteristics of the block $(B; I, Q)$. We have to show that this set is uniquely determined by $(B;I,Q)$. Let $(B_j;I_j,Q_j)$ be the block shift of the third kind with index κ_j and base space $\mathcal{Z}_j$. Put

$$(B';I',Q') = (B_1;I_1,Q_1) \oplus \cdots \oplus (B_r;I_r,Q_r).$$

and let $S : \mathcal{X} \to \mathcal{X}'$ be the block similarity of the blocks $(B';I',Q')$ and $(B;I,Q)$. So S is an invertible operator, $S[\operatorname{Ker} Q] = \operatorname{Ker} Q'$ and $(SB - B'S)Q = 0$. Let Ω_j be given by formula (3.2), and Ω'_j by formula (3.4). Write $\mathcal{D}_j = \operatorname{Ker}\Omega_j$ and $\mathcal{D}'_j = \operatorname{Ker}\Omega'_j$. From the special form of $\mathcal{D}'_j$ it follows that, for $j = 2, \ldots,$

$$\mathcal{D}'_{j-1} = (\mathcal{D}'_j + B[\mathcal{D}'_j]) \oplus \begin{cases} \mathcal{Z}_i & \text{if } j = \kappa_i + 1, \\ \{0\} & \text{otherwise.} \end{cases}$$

Remark that $S\Omega_j = \Omega'_j S$. It follows that $S[\mathcal{D}_j] = \mathcal{D}'_j$ and $SB[\mathcal{D}_j] = B'[\mathcal{D}'_j]$, for $j = 1, 2, \ldots$. We see that, for $j = 2, \ldots,$

$$S[\mathcal{D}_{j-1}] = (S[\mathcal{D}_j] + SB[\mathcal{D}'_j]) \oplus \begin{cases} \mathcal{Z}_i & \text{if } j = \kappa_i + 1, \\ \{0\} & \text{otherwise.} \end{cases} \tag{3.5}$$

This shows that the spaces $\mathcal{Z}_i$ are determined up to similarity by the block $(B;I,Q)$ and the numbers $\kappa_1, \ldots, \kappa_r$ are defined uniquely by $(B;I,Q)$. □

Formula (3.5) provides a formula for the characteristics of the block $(B;I,Q)$. This is our next result.

Corollary 3.4. *Let $(B;I,Q)$ be an operator block of finite type on the space $\mathcal{X}$, put $\Omega_0 = I_{\mathcal{X}}$, and let Ω_j be given by (3.2). Let $\kappa_1 > \cdots > \kappa_r$ be the natural numbers j such that for $j = \kappa_i$ the quotient space*

$$\mathcal{Z}_j = \frac{\operatorname{Ker}\Omega_j}{\operatorname{Ker}\Omega_{j+1} + B[\operatorname{Ker}\Omega_{j+1}]}$$

is nontrivial. Then $(\mathcal{Z}_{\kappa_1}, \kappa_1), \ldots, (\mathcal{Z}_{\kappa_r}, \kappa_r)$ is the set of characteristics of the block (B, I, Q).

As in the previous section we apply the results in a particular situation.

Theorem 3.5. *Let $\mathcal{X}$ and $\mathcal{Y}$ be Banach spaces, and let $F : \mathcal{X} \to \mathcal{X}$ and $H : \mathcal{X} \to \mathcal{Y}$ be bounded linear operators. Assume that*

$$\begin{pmatrix} H \\ HF \\ \vdots \\ HF^{j-1} \end{pmatrix} : \mathcal{X} \to \mathcal{Y}^j$$

has a generalized inverse for $j = 1, \dots, \ell$ and is injective for $j = \ell$. Then there exist Banach spaces $\mathcal{Z}_0, \dots, \mathcal{Z}_\ell$ and bounded linear operators

$$M : \mathcal{Z}_0 \oplus \cdots \oplus \mathcal{Z}_\ell \to \mathcal{Y}, \quad N : \mathcal{Z}_1^1 \oplus \mathcal{Z}_2^2 \oplus \cdots \oplus \mathcal{Z}_\ell^\ell \to \mathcal{X}, \quad L : \mathcal{Y} \to \mathcal{X},$$

such that M and N are invertible and

(i) $M^{-1}HN = E_0 \oplus \cdots \oplus E_\ell : \mathcal{Z}_1^1 \oplus \mathcal{Z}_2^2 \oplus \cdots \oplus \mathcal{Z}_\ell^\ell \to \mathcal{Z}_0 \oplus \mathcal{Z}_2 \oplus \cdots \oplus \mathcal{Z}_\ell$ *with* $E_0 = 0$ *and*

$$E_j = \begin{pmatrix} 0 & 0 & \cdots & I_j \end{pmatrix} : \mathcal{Z}_j^j \to \mathcal{Z}_j, \quad j = 1, \dots, \ell\,, \tag{3.6}$$

(ii) $N^{-1}(F - LH)N = J_1 \oplus \cdots \oplus J_\ell : \mathcal{Z}_1^1 \oplus \mathcal{Z}_2^2 \oplus \cdots \oplus \mathcal{Z}_\ell^\ell \to \mathcal{Z}_1^1 \oplus \mathcal{Z}_2^2 \oplus \cdots \oplus \mathcal{Z}_\ell^\ell$, *with* $J_1 = 0$ *and*

$$J_j = \begin{pmatrix} 0 & 0 & \cdots & 0 & 0 \\ I_j & 0 & \cdots & 0 & 0 \\ & \ddots & \ddots & \vdots & \vdots \\ 0 & 0 & \ddots & 0 & 0 \\ 0 & 0 & & I_j & 0 \end{pmatrix} : \mathcal{Z}_j^j \to \mathcal{Z}_j^j, \tag{3.7}$$

for $j = 2, \dots, \ell$.

In (i) and (ii) the symbol I_j denotes the identity operator on $\mathcal{Z}_j$. Note that it may happen that $\mathcal{Z}_j = \{0\}$ for some values of j.

PROOF. Consider the operator block $(B; I, Q)$ on the space $\mathcal{X} \oplus \mathcal{Y}$ given by

$$Q = \begin{pmatrix} I & 0 \\ 0 & 0 \end{pmatrix} : \mathcal{X} \oplus \mathcal{Y} \to \mathcal{X} \oplus \mathcal{Y}, \quad B = \begin{pmatrix} F \\ H \end{pmatrix} : \operatorname{Im} Q \to \mathcal{X} \oplus \mathcal{Y}.$$

We compute that, for $j = 1, \dots,$ the operator

$$\Omega_j = \operatorname{col}\big((I - Q)(BQ)^{i-1}\big)_{i=1}^j : \mathcal{X} \oplus \mathcal{Y} \to (\mathcal{X} \oplus \mathcal{Y})^j$$

can be written as

$$\Omega_j = \begin{pmatrix} \begin{pmatrix} 0 & 0 \\ 0 & I \end{pmatrix} \\ \begin{pmatrix} 0 & 0 \\ H & 0 \end{pmatrix} \\ \begin{pmatrix} 0 & 0 \\ HF & 0 \end{pmatrix} \\ \vdots \\ \begin{pmatrix} 0 & 0 \\ HF^{j-2} & 0 \end{pmatrix} \end{pmatrix} : \mathcal{X} \oplus \mathcal{Y} \to (\mathcal{X} \oplus \mathcal{Y})^j.$$

After a reordering of the components in $(\mathcal{X}\oplus\mathcal{Y})^j$, this operator can be represented as

$$\Omega_j = \begin{pmatrix} 0 & 0 \\ \Lambda_{j-1} & 0 \\ 0 & I \end{pmatrix} : \mathcal{X}\oplus\mathcal{Y} \to \mathcal{X}^j \oplus \mathcal{Y}^{j-1} \oplus \mathcal{Y},$$

with

$$\Lambda_{j-1} = \operatorname{col}\big(HF^{i-1}\big)_{i=1}^{j-1} : \mathcal{X} \to \mathcal{Y}^{j-1}.$$

It is given that Λ_j has a generalized inverse for $j = 1,\ldots,\ell$ and is injective for $j = \ell$. It follows that Ω_j has a generalized inverse for $j = 2,\ldots,\ell+1$ and is injective for $j = \ell+2$. Since Ω_1 clearly has a generalized inverse, the block $(B; I, Q)$ satisfies the conditions of Theorem 3.2. So the block $(B; I, Q)$ is block similar to a direct sum of block shifts of the third kind with indices $1,\ldots,\ell+1$ and base spaces $\mathcal{Z}_0,\ldots,\mathcal{Z}_\ell$. Let $(B_i; I_i, Q_i)$ be the block shift of the third kind with index $i+1$ and base space $\mathcal{Z}_i$. The operator B_i can be partitioned as

$$B_i = \begin{pmatrix} J_i \\ E_i \end{pmatrix} : \mathcal{Z}_i^i \to \mathcal{Z}_i^i \oplus \mathcal{Z}_i.$$

We write $\overline{Q} = Q_0 \oplus \cdots \oplus Q_\ell$ and $\overline{B} = B_0 \oplus \cdots \oplus B_\ell$. The operator $\overline{B}$ has the following block matrix representation:

$$\overline{B} = \begin{pmatrix} F_0 \\ H_0 \end{pmatrix} : \bigoplus_{i=1}^{\ell} \mathcal{Z}_i^i \to \Big(\bigoplus_{i=1}^{\ell} \mathcal{Z}_i^i\Big) \oplus \Big(\bigoplus_{i=1}^{\ell} \mathcal{Z}_i\Big).$$

where $H_0 = E_0 \oplus \cdots \oplus E_\ell$ and $F_0 = J_1 \oplus \cdots \oplus J_\ell$, with E_i as in (3.6) and J_i as in (3.7).

Next we prove that there exists operators N, M and L such that $F_0 = N^{-1}(F - LH)N$ and $H_0 = M^{-1}HN$. The blocks $(B; I, Q)$ on $\mathcal{X}\oplus\mathcal{Y}$ and $(\overline{B}; \overline{I}, \overline{Q})$ on $(\bigoplus_{i=1}^{\ell} \mathcal{Z}_i^i)\oplus(\bigoplus_{i=1}^{\ell} \mathcal{Z}_i)$ are block similar. Since $\operatorname{Im} Q = \mathcal{X}$ and $\operatorname{Im}\overline{Q} = \bigoplus_{i=1}^{\ell} \mathcal{Z}_i^i$, this block similarity has the form

$$S = \begin{pmatrix} N & LM \\ 0 & M \end{pmatrix}.$$

But then it follows from

$$\begin{pmatrix} F_0 \\ H_0 \end{pmatrix} = S^{-1} \begin{pmatrix} F \\ H \end{pmatrix} N$$

that $F_0 = N^{-1}(F - LH)N$ and $H_0 = M^{-1}HN$. □

Remark that without loss of generality we may assume that the spaces $\mathcal{Z}_i$, for $i = 0,\ldots,\ell$, occuring in Theorem 3.5, are closed subspaces of $\mathcal{Y}$ and that $\mathcal{Y} = \mathcal{Z}_0 \oplus \cdots \oplus \mathcal{Z}_\ell$. If we choose $\mathcal{Z}_0,\ldots,\mathcal{Z}_\ell$ in this way, then the invertible linear operator linear M in Theorem 3.5 will be the identity on $\mathcal{Y}$.

XII.4 Operator blocks on a separable Hilbert space

First we introduce some notions that will play an important role in this section. Let $(B;P,Q)$ and $(B';P',Q')$ be operator blocks on the Banach space $\mathcal{X}$. The block $(B';P',Q')$ is called an *extension* of the block $(B;P,Q)$ if $\operatorname{Im} Q' \supset \operatorname{Im} Q$, $\operatorname{Ker} P' \subset \operatorname{Ker} P$ and $B = PB'|_{\operatorname{Im} Q}$. In this case we also refer to $(B;P,Q)$ as a *restriction* of $(B';P',Q')$. The block $(B;P,I)$ on the Banach space $\mathcal{X}$ is called a *full range block* if

$$\operatorname{Ker} P + B[\operatorname{Ker} P] + \cdots + B^{n-1}[\operatorname{Ker} P] = \mathcal{X}$$

for some natural number n. The block $(B;I,Q)$ is called a *zero kernel block* if

$$\bigcap_{j=0}^{n-1} \operatorname{Ker}(I-Q)(BQ)^j = \{0\}$$

for some natural number n. In the previous sections we introduced blocks of finite type. It is clear that if a block $(B;P,I)$ is of finite type, then it is a full range block, and if $(B;I,Q)$ is of finite type, then $(B;I,Q)$ is a zero kernel block.

In this section we will show that in the case when the underlying space $\mathcal{X}$ is a separable Hilbert space a full range block has an extension to a block of finite type and that the same is true for a zero kernel block.

Theorem 4.1. *Let $\mathcal{H}$ be a separable Hilbert space. A full range block $(B;P,I)$ on $\mathcal{H}$ has an extension of finite type.*

PROOF. Since the statement is trivial in the finite dimensional case, we assume that $\mathcal{H}$ is infinite dimensional. According to our hypothesis there is a number n such that

$$\operatorname{Ker} P + B[\operatorname{Ker} P] + \cdots + B^{n-1}[\operatorname{Ker} P] = \mathcal{H}.$$

We will prove that there exists a block $(B';P',I)$ such that:

(i) $\operatorname{Ker} P' \subset \operatorname{Ker} P$;

(ii) $B = PB'$

(iii) $\operatorname{Ker} P' + B'[\operatorname{Ker} P'] + \cdots + (B')^k[\operatorname{Ker} P']$ is closed in $\mathcal{H}$ for each k;

(iv) $\operatorname{Ker} P' + B'[\operatorname{Ker} P'] + \cdots + (B')^N[\operatorname{Ker} P'] = \mathcal{H}$ for some natural number N.

The proof will be by induction on the number n. If $n = 1$, then we choose $B' = B$ and $P' = P$. Assume that the theorem is proved for $n-1$. The rest of the prove will consist of several steps.

STEP 1. In this step we construct a closed subspace $\mathcal{U}_0$ of $\operatorname{Ker} P$ such that $B[\mathcal{U}_0]$ is closed in $\operatorname{Im} P$ and $B[\mathcal{U}_0] + \cdots + B^{n-1}[\mathcal{U}_0] = \operatorname{Im} P$. If $B[\operatorname{Ker} P]$ is closed, we take $\mathcal{U}_0 = \operatorname{Ker} P$, and we are ready. Therefore assume that $B[\operatorname{Ker} P]$ is not closed in $\operatorname{Im} P$. Since $B[\operatorname{Ker} P] + \cdots + B^{n-1}[\operatorname{Ker} P] = \operatorname{Im} P$, there exists a real $\epsilon > 0$ such that for each operator $B_1 : \mathcal{H} \to \operatorname{Im} P$ the inequality $\|B_1 - B\| < 2\epsilon$ implies that $B_1[\operatorname{Ker} P] + \cdots + B_1^{n-1}[\operatorname{Ker} P] = \operatorname{Im} P$. Write

$$B = \begin{pmatrix} B_0 & A \end{pmatrix} : \operatorname{Ker} P \oplus \operatorname{Im} P \to \operatorname{Im} P.$$

Let $\{E_t\}_{t \geq 0}$ be the spectral measure of $|B_0| = (B_0^* B_0)^{1/2}$. So

$$|B_0| = \int_0^\infty t dE_t.$$

Put $\mathcal{U}_0 = (I - E_\epsilon)[\operatorname{Ker} P]$ and $B_{01} = B_0(I - E_\epsilon)$. Then for $u \in \mathcal{U}_0$ one computes

$$\|B_0 u\| = \|B_{01} u\| = \| \, |B_0| u\| \geq \frac{\epsilon}{2}\|u\|.$$

This proves that $B[\mathcal{U}_0] = B_0[\mathcal{U}_0]$ is closed. On the other hand, with B_1 defined as $B_1 = \begin{pmatrix} B_{01} & A \end{pmatrix}$, we get that $\|B_1 - B\| = \|B_0 - B_{01}\| \leq \epsilon$, and thus

$$B_1[\operatorname{Ker} P] + \cdots + B_1^{n-1}[\operatorname{Ker} P] = \operatorname{Im} P.$$

Since $B_1[\operatorname{Ker} P] = B[\mathcal{U}_0]$, we have $B_1^k[\mathcal{U}_0] = B^k[\mathcal{U}_0]$, and the first step is finished.

STEP 2. In this step we construct a block $(B_4; P_4, I_4)$ of finite type on $\mathcal{U}_0 \oplus \operatorname{Im} P$ which fits almost our requirements. Put $\mathcal{H}_2 = \operatorname{Im} P$ and let P_2 be the orthogonal projection of $\mathcal{H}_2$ along $B[\mathcal{U}_0]$. Put $B_2 = P_2 B|_{\mathcal{H}_2}$. Since

$$\operatorname{Ker} P_2 + B_2[\operatorname{Ker} P_2] + \cdots + B_2^{n-2}[\operatorname{Ker} P_2] = \mathcal{H}_2,$$

we may apply the induction hypothesis in order to get a block $(B_3; P_3, I_2)$ and a number N such that:

$\operatorname{Ker} P_3 \subset \operatorname{Ker} P_2$;

$B_2 = P_2 B_3$

$\operatorname{Ker} P_3 + B_3[\operatorname{Ker} P_3] + \cdots + (B_3)^k[\operatorname{Ker} P_3]$ is closed in $\mathcal{H}_2$ for each k;

$\operatorname{Ker} P_3 + B_3[\operatorname{Ker} P_3] + \cdots + (B_3)^{N-3}[\operatorname{Ker} P_3] = \mathcal{H}_2$ for some N.

Put $\mathcal{U}_3 = \operatorname{Ker} P_3$. Remark that

$$\operatorname{Im}(B_3 - B_2) = \operatorname{Im}(I - P_2)B_3 \subset (I - P_2)[\operatorname{Im} P_3] \subset \operatorname{Ker} P_2 = B[\mathcal{U}_0].$$

Thus

$$\operatorname{Im}(B_3 - P_2 B|_{\mathcal{H}_2}) \subset B[\mathcal{U}_0]$$

Put $F = B_3 - B|_{\operatorname{Im} P} : \operatorname{Im} P \to \operatorname{Im} P$. Then

$$F = B_3 - P_2 B|_{\operatorname{Im} P} - (I - P_2) B|_{\operatorname{Im} P},$$

and thus $\operatorname{Im} F \subset \operatorname{Ker} P_2 = B[\mathcal{U}_0]$. Extend F to an operator on $\mathcal{H}$ by putting $F|_{\operatorname{Ker} P} = 0$. Then $B+F$ is an extension of B_3 to an operator defined on the whole of $\mathcal{H}$. We denote this operator again by B_3. Choose $\mathcal{U}_4$ to be the inverse image of $\mathcal{U}_3$ under $B_3|_{\mathcal{U}_0}$, i.e., $\mathcal{U}_4 = (B_3|_{\mathcal{U}_0})^{-1}[\mathcal{U}_3]$. One observes that

$$\mathcal{U}_4 + B_3[\mathcal{U}_4] + \cdots + B_3^k[\mathcal{U}_4] \subset \mathcal{H}$$

is closed for $1 \leq k \leq N-2$ and

$$\mathcal{U}_4 + B_3[\mathcal{U}_4] + \cdots + B_3^{N-2}[\mathcal{U}_4] = \mathcal{U}_4 \oplus \operatorname{Im} P.$$

Consider the block $(B_3; P_3, I)$ on $\mathcal{U}_4 \oplus \operatorname{Im} P$, with P_3 the projection along $\mathcal{U}_4$ onto $\operatorname{Im} P$. This block is of finite type. Apply Lemma 2.1 to this block. So we obtain closed subspaces $\mathcal{U}_{ij}$ of $\mathcal{U}_4 \oplus \operatorname{Im} P$ for $1 \leq i \leq j \leq N-2$, such that $B_3[\mathcal{U}_{ii}] = \{0\}$ for $i = 1, \ldots, N-2$. One of the subspaces $\mathcal{U}_{ii}$, say $\mathcal{U}_{i_0 i_0}$, is infinite dimensional. Define the operator G by putting $G[\mathcal{U}_{i_0 i_0}] = \mathcal{U}_0 \ominus \mathcal{U}_4$, $G[\mathcal{U}_{ij}] = 0$ if $(i, j \neq (i_0, i_0)$ and $G[\mathcal{U}_0 \ominus \mathcal{U}_4] = \{0\}$. Then $B_4 = B_3 + G$ has the property that

$$\mathcal{U}_4 + B_4[\mathcal{U}_4] + \cdots + B_4^k[\mathcal{U}_4] \subset \mathcal{H}$$

is closed for $1 \leq k \leq N-1$ and

$$\mathcal{U}_4 + B_4[\mathcal{U}_4] + \cdots + B_4^{N-1}[\mathcal{U}_4] = \mathcal{U}_0 \oplus \operatorname{Im} P.$$

Now we have on the space $\mathcal{U}_0 \oplus \operatorname{Im} P$ the equality $B_4 = B + F + G$. We would like to have that $PB_4 = B$. This however is not the case.

STEP 3. Next we construct an invertible transformation S on $\mathcal{U}_0 \oplus \operatorname{Im} P$ such that $PSB_4 = BS$ and $S[\mathcal{U}_0] = \mathcal{U}_0$. Remark that with respect to the decomposition $\mathcal{U}_0 \oplus \operatorname{Ker} P_2 \oplus \operatorname{Im} P_2$ of the space $\mathcal{U}_0 \oplus \operatorname{Im} P$ the operator matrix representations of B, F, G and $P|_{\mathcal{U}_0 \oplus \operatorname{Im} P}$ are

$$B = \begin{pmatrix} 0 & 0 & 0 \\ B_{21} & B_{22} & B_{23} \\ 0 & B_{32} & B_{33} \end{pmatrix}, \quad F = \begin{pmatrix} 0 & 0 & 0 \\ 0 & F_{22} & F_{23} \\ 0 & 0 & 0 \end{pmatrix},$$

$$G = \begin{pmatrix} G_{11} & G_{12} & G_{13} \\ 0 & 0 & 0 \\ 0 & 0 & 0 \end{pmatrix}, \quad P = \begin{pmatrix} 0 & 0 & 0 \\ 0 & I & 0 \\ 0 & 0 & I \end{pmatrix}.$$

Since $B[\mathcal{U}_0] = \operatorname{Ker} P_2$, the operator B_{21} has a right inverse, which we denote by B_{21}^{+}. Put

$$S = \begin{pmatrix} I & B_{21}^{+} F_{22} & B_{21}^{+} F_{23} \\ 0 & I & 0 \\ 0 & 0 & I \end{pmatrix}.$$

It requires an easy computation to verify that S has the desired properties. Now put $\mathcal{U}_5 = S[\mathcal{U}_4]$, take P_5 to be a projection of $\mathcal{U}_0 \oplus \operatorname{Im} P$ along $\mathcal{U}_5$ and choose $B_5 = P_5 S B_4 S^{-1}$. Then

$$\mathcal{U}_5 + B_5[\mathcal{U}_5] + \cdots + B_5^k[\mathcal{U}_5] \subset \mathcal{H}$$

is closed for $1 \leq k \leq N-1$ and

$$\mathcal{U}_5 + B_5[\mathcal{U}_5] + \cdots + B_5^{N-1}[\mathcal{U}_5] = \mathcal{U}_0 \oplus \operatorname{Im} P.$$

Moreover $PB_5 = PP_5SB_4S^{-1} = PSB_4S^{-1} = B$. So $(B_5; P_5, I)$ on $\mathcal{U}_0 \oplus \operatorname{Im} P$ is of finite type. To extend this block to $\mathcal{H}$ one proceeds like we did in extending $(B_3; P_3, I)$ to (B_4, P_4, I). So we obtain the block $(B'; P', I')$ with the properties (i), (ii), (iii) and (iv). □

The next result is the dual result to Theorem 4.1 for full length blocks.

Theorem 4.2. *Let $\mathcal{H}$ be a separable Hilbert space. A zero kernel block $(B; I, Q)$ on $\mathcal{H}$ has an extension of finite type.*

PROOF. We consider the dual block $\big((BQ)^*; Q^*, I\big)$ of the block $(B; I, Q)$. Here we treat BQ as a linear operator defined for each $x \in \mathcal{H}$. Since $(B; I, Q)$ is a zero kernel block, we have that for some number n

$$\operatorname{Ker} \begin{pmatrix} I-Q \\ (I-Q)BQ \\ \vdots \\ (I-Q)(BQ)^{n-1} \end{pmatrix} = \{0\}.$$

So for the dual block we get

$$\operatorname{Im}\big(I - Q^* \quad (BQ)^*(I-Q^*) \quad \cdots \quad ((BQ)^*)^{n-1}(I-Q^*)\big) = \mathcal{H}$$

and thus

$$\operatorname{Ker} Q^* + (BQ)^*[\operatorname{Ker} Q^*] + \cdots + ((BQ)^*)^{n-1}[\operatorname{Ker} Q^*] = \mathcal{H}.$$

Thus the block $\big((BQ)^*; Q^*, I\big)$ is a full range block. So there exists an extension $(B_1; P_1, I)$ of finite type of $\big((BQ)^*; Q^*, I\big)$. Consider its dual $(B_1^*; I, P_1^*)$. This is an extension of $(B; I, Q)$. Moreover, for each j the subspace

$$\operatorname{Ker} \begin{pmatrix} I-P_1^* \\ (I-P_1^*)B_1^* \\ \vdots \\ (I-P_1^*)(B_1^*)^j \end{pmatrix}$$

has as a closed complement the space

$$\operatorname{Ker} P_1 + B_1[\operatorname{Ker} P_1] + \cdots + B_1^j[\operatorname{Ker} P_1].$$

So $(B_1^*; I, P_1^*)$ is also of finite type. □

XII.5 Spectral completion and assignment problems

In this section we consider an analogue of the eigenvalue completion problem for full length blocks on infinite dimensional separable Hilbert spaces. Assume we have a block $(B; P, I)$ on the infinite dimensional Hilbert space $\mathcal{H}$ and a nonempty compact set K in the complex plane. The question that we treat in this section is whether or not there exists an operator $A : \mathcal{H} \to \mathcal{H}$ such that the spectrum of A is K and $B = PA$. The answer to this question is provided by the next result. As in the finite dimensional case, we call an operator A such that $B = PA$ a *completion of the block* $(B; P, I)$.

Theorem 5.1. *The block $(B; P, I)$ on the infinite dimensional separable Hilbert space $\mathcal{H}$ has an extension of finite type if and only if for each nonempty compact set K in the complex plane there exists a completion $A : \mathcal{H} \to \mathcal{H}$ of $(B; P, I)$ such that the spectrum of A is K.*

PROOF. Let $(B'; P', I)$ be an extension of $(B; P, I)$. Thus $P'P = P$ and $PB' = B$. It follows that any completion A of $(B'; P', I)$ is also a completion of $(B; P, I)$. Indeed if $B' = P'A$, then

$$PA = PP'A = PB' = B.$$

Thus to prove the only if part of the theorem we may as well assume that $(B; P, I)$ is of finite type.

Assume that $(B; P, I)$ is a block of finite type, and let K be an arbitrary nonempty compact set. We want to construct a completion A with $\sigma(A) = K$. According to our assumption $(B; P, I)$ is block similar to a direct sum of shifts of the first kind. Let us assume that this direct sum is the block $(B'; P', I')$ on the space $\mathcal{H}'$. So there exists an invertible linear operator $S : \mathcal{H} \to \mathcal{H}'$ such that $P'S = P'SP$ and $P'(SB - B'S) = 0$. Assume that we have an operator A', which is a completion of $(B'; P', I')$ with $\sigma(A') = K$. So $P'A' = B'$, and thus

$$PS^{-1}P'A'S = PS^{-1}P'B'S = PS^{-1}P'SB,$$

and

$$PS^{-1}A'S = PS^{-1}SB = PB = B.$$

We see that $A = S^{-1}A'S$ is a completion of $(B; P, I)$ with $\sigma(A) = K$. We conclude that it is sufficient to prove that there exists an operator A' such that A' is a completion of $(B'; P', I')$ and $\sigma(A) = K$. So let

$$(B'; P', I') = \bigoplus_{i=1}^{k} (S_i; P_i, I_i),$$

where $(S_i; P_i, I_i)$ is a shift of the first kind with index $i - 1$ and base space $\mathcal{U}_i$ for $i = 1, \ldots, k$. So, in particular, $\mathcal{H}_i$ is the direct sum of the spaces $\mathcal{U}_i^i$. At least one of the spaces $\mathcal{U}_i$ is infinite dimensional. Assume that $\mathcal{U}_\ell$ is infinite dimensional. Let $\phi_1, \phi_2, \ldots$ be an orthogonal basis for $\mathcal{U}_\ell$, and let $\lambda_1, \lambda_2, \ldots$ be a countable dense set in K. Define the bounded linear operator A_ℓ on $\mathcal{U}_\ell^\ell$ by

$$A_\ell S_\ell^{j-1}\phi_\nu = S_\ell^j \phi_\nu - \binom{\ell}{j}(-\lambda_\nu)^j \phi_\nu, \quad j = 1, \ldots, \ell - 1, \quad A_\ell S_\ell^{\ell-1}\phi_\nu = (-\lambda_\nu)^\ell \phi_\nu.$$

The restriction of A_ℓ to the subspace $\operatorname{span}\{\phi_\nu, S_\ell\phi_\nu, \ldots, S_\ell^{\ell-1}\phi_\nu\}$ has λ_ν as its only eigenvalue. Moreover, A_ℓ is a completion of the block $(S_\ell; P_\ell, I_\ell)$. Notice that $\sigma(A) = K$. In a similar way we define A_i on $\mathcal{U}_i^i$ such that $\sigma(A_i) \subset K$. (In fact, we can even construct A_i in such a way that $\sigma(A_i)$ consists of a single point from K). Then the direct sum of the operators $A_1, \ldots, A_k$ is a completion A' of $(B'; P', I')$ with $\sigma(A') = K$.

Conversely, assume that $(B; P, I)$ is such that for each nonempty compact set K in the complex plane there exists a completion $A : \mathcal{H} \to \mathcal{H}$ of $(B; P, I)$ with the spectrum of A equal to K. Define $A_0 : \mathcal{H} \to \mathcal{H}$ by putting $A_0 x = Bx$ for each $x \in \mathcal{H}$. Choose $r \in \mathbb{R}$ such that $\|A_0\| \le r - \epsilon$ with $\epsilon > 0$, and choose a completion $A : \mathcal{H} \to \mathcal{H}$ of $(B; P, I)$ such that $\sigma(A) = \{r\}$. Then A is invertible, and $\lim_{n\to\infty} \|A^{-n}A_0^n\| = 0$, because $\lim_{n\to\infty} \|A^{-n}\|^{\frac{1}{n}} = \frac{1}{r}$ and $\|A_0\|^n < (r - \epsilon)^n$. Hence there exists a natural number N such that $I - A^{-N}A_0^N$ is invertible. Since A is invertible and $A^N - A_0^N = A^N(I - A^{-N}A_0^N)$, it is clear that $A_0^N - A^N$ is invertible. Use

$$A_0^N - A^N = \sum_{k=0}^{N-1} A_0^k (A_0 - A) A^{N-k-1},$$

and $P(A - A_0) = 0$ to obtain that

$$I = \sum_{k=0}^{N-1} A_0^k (I - P) X_k, \tag{5.1}$$

with $X_k = (A - A_0)A^{N-k-1}(A^N - A_0^N)^{-1}$. From (5.1) it is clear that

$$\mathcal{H} = \operatorname{Ker} P + A_0[\operatorname{Ker} P] + \cdots + A_0^{N-1}[\operatorname{Ker} P],$$

and thus

$$\mathcal{H} = \operatorname{Ker} P + B[\operatorname{Ker} P] + \cdots + B^{N-1}[\operatorname{Ker} P].$$

So we conclude from Theorem 4.1 that there exists an extension of finite type of $(B; P, I)$. □

The next result is dual to Theorem 5.1.

Theorem 5.2. *The block $(B; I, Q)$ on the infinite dimensional separable Hilbert space $\mathcal{H}$ has an extension of finite type if and only if for each nonempty compact set K in the complex plane there exists a completion $A : \mathcal{H} \to \mathcal{H}$ of $(B; I, Q)$ such that the spectrum of A is K.*

One can prove this theorem by either using duality and applying Theorem 5.1 or adapt the proof of Theorem 5.1 to the present situation. We omit further details.

Notes

The material of Sections 1, 2 and 3 is taken from Gohberg-Kaashoek-Van Schagen [2]. Section 4, which serves as a preparation for the solution of the spectral completion and assignment problems in Section 5, is based on Eckstein [1]. Theorem 5.1 is due to Eckstein [1], and in a somewhat weaker version to Kaashoek-Van der Mee-Rodman [1] (see also Takahashi [1], Rodman [1], Section 7.2, and Rodman [2]).

Chapter XIII
Factorization of Operator Polynomials

As applications of the results in the previous chapter we treat here Wiener-Hopf equivalence and Wiener-Hopf factorization for operator polynomials with coefficients acting on a Banach space, and strict equivalence for infinite dimensional operator pencils. These applications also serve as a motivation for the concepts introduced in the previous chapter. The starting point is the infinite dimensional analogue of a null pair for an operator polynomial, which is introduced and analyzed in the first section.

XIII.1 Preliminaries on null pairs and spectral triples

Throughout this chapter Γ is a closed rectifiable Jordan curve in the complex plane with a bounded inner domain Ω, and $\mathcal{X}$ and $\mathcal{Y}$ are complex Banach spaces. All operators are assumed to linear and bounded. An important tool will be the notion of Γ-null pair for a matrix polynomial. This notion generalizes that of a null pair for a matrix polynomial (see Section 5 of the Appendix).

In this section we introduce the notion of Γ-null pair and derive some of its properties. We use Theorems 5.1 and 5.3 of the Appendix as a starting point for our definition. A pair (A, B) with $A : \mathcal{X} \to \mathcal{X}$ and $B : \mathcal{Y} \to \mathcal{X}$ is called a *left Γ-null pair* for the operator polynomial $L(\lambda) = \sum_{i=0}^{\ell} \lambda^i L_i$ on $\mathcal{Y}$ if

(a) $\sigma(A) \subset \Omega$;

(b) $(\lambda I_{\mathcal{X}} - A)^{-1} B L(\lambda)$ has an analytic continuation on Ω;

(c) $\begin{pmatrix} B & AB & \cdots & A^{\ell-1}B \end{pmatrix}$ is right invertible;

(d) there exists a bounded linear operator $C : \mathcal{X} \to \mathcal{Y}$ such that the function $L(\lambda)^{-1} - C(\lambda I_{\mathcal{X}} - A)^{-1} B$ has an analytic continuation to Ω.

The Banach space $\mathcal{X}$ is called the *base space* of the pair (A, B). In general a pair of bounded linear operators (A, B) is called a *left admissible pair* if $A : \mathcal{X} \to \mathcal{X}$ and $B : \mathcal{Y} \to \mathcal{X}$. Note that if $L(\lambda)$ is a matrix polynomial, and the pair (A, B) is a left null pair with respect to Ω in the sense of Section X.1, then the conditions (a), (b) and (c) are fulfilled by definition, and condition (d) follows from (a), (b) and (c). Thus (A, B) is a left Γ-null pair. Conversely, if (A, B) is a left Γ-null pair of the matrix polynomial $L(\lambda)$, then (A, B) is a left null pair with respect to Ω in the sense of Section X.1, because the conditions (i), (ii) and (iii) of the definition of left null pair with respect to Ω are implied by (a), (b) and (c) above.

Our first concern is to prove that left Γ-null pairs do exist.

Theorem 1.1. *Let $L(\lambda)$ be an operator polynomial of degree ℓ on $\mathcal{Y}$ such that $L(\lambda)$ is invertible on Γ. Let α be a point in $\mathbb{C}$ outside Γ and such that $L(\alpha)$ is invertible. Put*

$$Q_{\ell-n} = \frac{1}{n!}L^{(n)}(\alpha)L(\alpha)^{-1}, \quad n = 1, \dots, \ell$$

$$T = \begin{pmatrix} 0 & I & & & \\ 0 & 0 & \ddots & & \\ \vdots & \vdots & & \ddots & \\ 0 & 0 & & & I \\ -Q_0 & -Q_1 & \cdots & \cdots & -Q_{\ell-1} \end{pmatrix} : \mathcal{Y}^\ell \to \mathcal{Y}^\ell, \quad Y = \begin{pmatrix} 0 \\ 0 \\ \vdots \\ 0 \\ I \end{pmatrix} : \mathcal{Y} \to \mathcal{Y}^\ell, \tag{1.1}$$

$$\Gamma_1 = \{\mu \in \mathbb{C} \mid \alpha + \mu^{-1} \in \Gamma\},$$

$$P = \frac{1}{2\pi i}\int_{\Gamma_1} (\lambda I - T)^{-1} d\lambda, \quad A_P = PT|_{\operatorname{Im} P} : \operatorname{Im} P \to \operatorname{Im} P,$$

$$B = PY : \mathcal{Y} \to \operatorname{Im} P, \quad A = \alpha I + A_P^{-1}.$$

Then (A, B) is a Γ-null pair of $L(\lambda)$.

PROOF. The proof consists of several steps.

STEP 1. In this step we consider the monic polynomial

$$Q(\mu) = \mu^\ell I + \sum_{i=0}^{\ell-1} \mu^i Q_i.$$

It is easy to verify that $Q(\mu) = \mu^\ell L(\alpha + \mu^{-1})L(\alpha)^{-1}$. Define the operator polynomials $E(\mu)$ and $F(\mu)$ on $\mathcal{Y}^\ell$ as follows:

$$F(\mu) = \begin{pmatrix} I & 0 & \cdots & 0 & 0 \\ -\mu I & I & & 0 & 0 \\ & \ddots & \ddots & & \vdots \\ 0 & 0 & \ddots & I & 0 \\ 0 & 0 & & -\mu I & I \end{pmatrix},$$

$$E(\mu) = \begin{pmatrix} B_{\ell-1}(\mu) & B_{\ell-2}(\mu) & \cdots & B_1(\mu) & B_0(\mu) \\ -I & 0 & \cdots & 0 & 0 \\ 0 & -I & & 0 & 0 \\ \vdots & & \ddots & & \vdots \\ 0 & 0 & & -I & 0 \end{pmatrix},$$

where $B_0(\mu) = I$ and $B_{r+1}(\mu) = \mu B_r(\mu) + Q_{\ell-r-1}$ for $r = 0, 1, \dots, \ell-2$. It is immediate that $E(\mu)$ and $F(\mu)$ are invertible for each value of μ. Direct multiplication on both sides shows that

$$E(\mu)(\mu I - T) = \begin{pmatrix} Q(\mu) & 0 \\ 0 & I \end{pmatrix} F(\mu). \tag{1.2}$$

Put

$$X = \begin{pmatrix} I & 0 & \cdots & 0 \end{pmatrix} : \mathcal{Y}^\ell \to \mathcal{Y}.$$

Formula (1.2) shows that $\mu I - T$is invertible whenever $Q(\mu)$ is invertible. From (1.2) we compute that $Q(\mu)^{-1} = XF(\mu)(\mu I - T)^{-1}E(\mu)^{-1}X^T$. So we get that

$$Q(\mu)^{-1} = X(\mu I - T)^{-1}Y. \tag{1.3}$$

STEP 2. In this step we show that the pair (A_P, B) is a Γ_1-null pair of $Q(\mu)$. Write $\lambda = \alpha + (1/\mu)$. Since $L(\lambda)$ is invertible on Γ, the definition of Γ_1 implies that $Q(\mu)$ is invertible on Γ_1. Thus $\mu I - T$ is invertible on Γ_1, which shows that the Riesz projection P is well defined. Moreover, since α is outside Γ the point zero is outside Γ_1, and the point λ is inside Γ if and only if μ is inside Γ_1. From (1.3) we get that

$$\begin{aligned} Q(\mu)^{-1} =& X|_{\operatorname{Im} P}(\mu I_{\operatorname{Im} P} - PT|_{\operatorname{Im} P})^{-1}PY + \\ & X|_{\operatorname{Ker} P}(\mu I_{\operatorname{Ker} P} - (I-P)T|_{\operatorname{Ker} P})^{-1}(I-P)Y. \end{aligned} \tag{1.4}$$

We now check that the pair (A_P, B) is a Γ_1-null pair of $Q(\mu)$. Obviously, $\sigma(A_P) \subset \Omega_1$, where Ω_1 is the inner domain of Γ_1. Put $C_P = X_{\operatorname{Im} P}$. Since the second term on the right hand side of (1.4) is analytic in Ω_1, formula (1.4) gives that $Q(\mu)^{-1} - C_P(\mu I_{\operatorname{Im} P} - A_P)^{-1}B$ has an analytic continuation to Ω_1. Next remark that from

$$(\mu I - T)^{-1}E(\mu)^{-1}\begin{pmatrix} Q(\mu) & 0 \\ 0 & I \end{pmatrix} = F(\mu)^{-1}.$$

By taking the first column on the left and the right one gets that $(\mu I - T)^{-1}YQ(\mu)$ has an analytic continuation on $\mathbb{C}$. It follows that $(\mu I_{\operatorname{Im} P} - A_P)^{-1}BQ(\mu)$ has an analytic continuation to Ω_1. An easy computation shows that

$$\begin{pmatrix} Y & TY & \cdots & T^{\ell-1}Y \end{pmatrix} = I_{\mathcal{Y}^\ell}.$$

So we get that $\begin{pmatrix} B & A_PB & \cdots & A_P^{\ell-1}B \end{pmatrix}$ is right invertible.

STEP 3. Put $\Gamma_\alpha = \{\mu \in \mathbb{C} \mid \mu + \alpha \in \Gamma\}$. Denote the inner domain of this curve by Ω_α. Remark that this means that $\mu \in \Gamma_1$ if and only if $\mu^{-1} \in \Gamma_\alpha$. The third step is to show that the pair (A_P^{-1}, B) is a Γ_α-null pair for the polynomial $R(\mu) = \mu^\ell Q(\mu^{-1})L(\alpha)$. First remark that A_P is invertible since 0 is not in the inner domain of Γ_1. So we may put $A_R = A_P^{-1}$, and the pair (A_R, B) is well defined indeed. Put $C_R = -L(\alpha)^{-1}C_PA_R^{-\ell+2}$. Let us check the properties of a Γ_α-null pair for the pair (A_R, B). First we use $\mu I - A_R = -\mu A_P^{-1}(\mu^{-1}I - A_P)$ to compute

$$\begin{aligned} & R(\mu)^{-1} + L(\alpha)^{-1}C_PA_R^{-\ell+2}(\mu I - A_R)^{-1}B \\ & = L(\alpha)^{-1}\mu^{-\ell}Q(\mu^{-1})^{-1} - L(\alpha)^{-1}C_P\mu^{-\ell}\left(\mu^{-1}I - A_P\right)^{-1}B + \\ & + L(\alpha)^{-1}C_P\mu^{-1}(\mu^{-\ell+1}I - A_P^{\ell-1})\left(\mu^{-1}I - A_P\right)^{-1}B. \end{aligned}$$

The sum of the first two terms on the right hand side has an analytic continuation for μ in Ω_α, because $Q(\mu)^{-1} - C_P(\mu I - A_P)^{-1}B$ has an analytic continuation to Ω_1 and $\mu^{-\ell}$ is analytic in Ω_α. The third term has an analytic continuation to Ω_α if $\ell \geq 2$ and it is zero if $\ell = 1$. So we proved condition (d) of the definition of a Γ_α-null pair for the polynomial $R(\mu)$ and the pair (A_R, B). Next remark that $\sigma(A_R) \subset \Omega_\alpha$, because $\sigma(A_P) \subset \Omega_1$. The operator $\begin{pmatrix} B & A_P B & \cdots & A_P^{\ell-1}B \end{pmatrix}$ has a right inverse, and, since A_R is invertible, $\begin{pmatrix} B & A_R B & \cdots & A_R^{\ell-1}B \end{pmatrix}$ also has a right inverse. To complete Step 3 we compute that

$$(\mu I - A_R)^{-1} B R(\mu) = \mu^{\ell-1} A_R (A_R^{-1} - \mu^{-1})^{-1} B Q(\mu^{-1}) L(\alpha),$$

and use the analyticity of the right hand side for μ^{-1} in Ω_1 to conclude that the left hand side is analytic for μ in Ω_α.

STEP 4. For the fourth and final step we recall that $R(\mu) = L(\alpha + \mu)$. We will prove that (A, B), with $A = \alpha I + A_R$, is a Γ-null pair of $L(\lambda)$. Put $C = C_R$. Then

$$L(\lambda)^{-1} - C(\lambda I - A)^{-1}B = R(\lambda - \alpha)^{-1} - C\big((\lambda - \alpha)I - (-\alpha I + A)\big)^{-1}B. \quad (1.5)$$

Put $\lambda = \mu + \alpha$, and notice that the right hand side of (1.5) is

$$R(\mu)^{-1} - C_R(\mu I - A_R)^{-1}B,$$

which is analytic for $\mu \in \Omega_\alpha$. Therefore the left hand side of (1.5) is analytic for $\lambda \in \Omega$. To prove that $\begin{pmatrix} B & AB & \cdots & A^{\ell-1}B \end{pmatrix}$ is right invertible we use that

$$\begin{pmatrix} B & AB & \cdots & A^{\ell-1}B \end{pmatrix} = \begin{pmatrix} B & A_R B & \cdots & A_R^{\ell-1}B \end{pmatrix} (X_{ij})_{i,j=1}^{\ell},$$

where the invertible matrix $(X_{ij})_{i,j=1}^{\ell}$ is given by

$$X_{ij} = \begin{cases} \binom{j}{i}\alpha^{j-i} I & \text{if } j \geq i, \\ 0 & \text{if } j < i. \end{cases}$$

It is clear that $\sigma(A) \subset \Omega$, and finally

$$(\lambda I - A)^{-1} B L(\lambda) = (\mu I - A_R)^{-1} B R(\mu)$$

which is analytic for $\mu \in \Omega_\alpha$ or equivalently for $\lambda \in \Omega$. □

We introduce right pairs in the same way as introduced left pairs. So a pair of operators (C, A) is called a *right admissible pair* if $A : \mathcal{X} \to \mathcal{X}$ and $C : \mathcal{X} \to \mathcal{Y}$ are bounded linear operators acting between Banach spaces. The Banach space $\mathcal{X}$ is called the *base space* of the pair (C, A). A right admissible pair is called a *right* Γ*-null pair* for the operator polynomial $L(\lambda) = \sum_{i=0}^{\ell} \lambda^i L_i$ on $\mathcal{Y}$ if

(a) $\sigma(A) \subset \Omega$;

(b') $L(\lambda)C(\lambda I_{\mathcal{X}} - A)^{-1}$ has an analytic continuation on Ω;

(c') the operator $\operatorname{col}\big(CA^{j-1}\big)_{j=1}^{\ell} : \mathcal{X} \to \mathcal{Y}^{\ell}$ is left invertible;

(d') there exists a bounded linear operator $B : \mathcal{Y} \to \mathcal{X}$ such that $L(\lambda)^{-1} - C(\lambda I_{\mathcal{X}} - A)^{-1}B$ has an analytic continuation to Ω.

The next result states that right Γ-null pairs exist and shows how one may construct such a pair.

Theorem 1.2. *Let $L(\lambda)$ be an operator polynomial of degree ℓ on $\mathcal{Y}$ such that $L(\lambda)$ is invertible on Γ. Let α be a point in $\mathbb{C}$ outside Γ and such that $L(\alpha)$ is invertible. Put*

$$Q_{\ell-n} = \frac{1}{n!} L^{(n)}(\alpha) L(\alpha)^{-1}, \quad n = 1, \ldots, \ell,$$

$$T = \begin{pmatrix} 0 & I & & & \\ 0 & 0 & \ddots & & \\ \vdots & \vdots & & \ddots & \\ 0 & 0 & & & I \\ -Q_0 & -Q_1 & \cdots & \cdots & -Q_{\ell-1} \end{pmatrix} : \mathcal{Y}^{\ell} \to \mathcal{Y}^{\ell},$$

$$X = \begin{pmatrix} I & 0 & \cdots & 0 \end{pmatrix} : \mathcal{Y}^{\ell} \to \mathcal{Y},$$

$$\Gamma_1 = \{\mu \in \mathbb{C} \mid \alpha + \mu^{-1} \in \Gamma\},$$

$$P = \frac{1}{2\pi i} \int_{\Gamma_1} (\lambda I - T)^{-1} d\lambda, \quad A_P = PT|_{\operatorname{Im} P} : \operatorname{Im} P \to \operatorname{Im} P,$$

$$A = \alpha I + A_P^{-1}, \quad C = -L(\alpha)^{-1} X|_{\operatorname{Im} P} A_P^{\ell-2}.$$

Then (C, A) is a right Γ-null pair of $L(\lambda)$.

PROOF. Define Y by (1.1) and B by $B = PY : \mathcal{Y} \to \operatorname{Im} P$. Let us check the properties (a), (b'), (c') and (d') for the pair (C, A). Since A, B and C are defined in the same way here as they were defined in (the proof) of Theorem 1.1, we can rely on the results derived in the proof of Theorem 1.1. Remark that (A, B) is a left Γ-null pair of $L(\lambda)$. So the properties (a) and (d') require no further argument. Let us check property (b'). From (d') it follows that $L(\lambda)C(\lambda I - A)^{-1}B$ has an analytic continuation to Ω. Compute

$$L(\lambda)C(\lambda I - A)^{-1}A^k B = L(\lambda)C(\lambda I - A)^{-1}(A^k - \lambda^k I)B + L(\lambda)C(\lambda I - A)^{-1}\lambda^k B.$$

Since both terms on the right hand side have an analytic continuation to Ω, so has $L(\lambda)C(\lambda I - A)^{-1}A^kB$. Now, use the fact that $\begin{pmatrix} B & AB & \cdots & A^{\ell-1}B \end{pmatrix}$ is right invertible to see that $L(\lambda)C(\lambda I - A)^{-1}$ has an analytic continuation to Ω. We check (c'). Remark that $\operatorname{col}\bigl(XT^{j-1}\bigr)_{j=1}^{\ell}$ is invertible. Therefore the operator $\operatorname{col}\bigl(X|_{\operatorname{Im} P}(PT|_{\operatorname{Im} P})^{j-1}\bigr)_{j=1}^{\ell}$ is right invertible. Since A_P and $-L(\alpha)^{-1}$ are invertible and

$$\begin{pmatrix} CA_P^{\ell-1} \\ \vdots \\ C \end{pmatrix} = \begin{pmatrix} -L(\alpha)^{-1} & & \\ & \ddots & \\ & & -L(\alpha)^{-1} \end{pmatrix} \begin{pmatrix} X|_{\operatorname{Im} P} \\ \vdots \\ X|_{\operatorname{Im} P}(PT|_{\operatorname{Im} P})^{\ell-1} \end{pmatrix} A_P^{-1},$$

the operator $\operatorname{col}\bigl(C(A_P^{-1})^{j-1}\bigr)_{j=1}^{\ell}$ is left invertible. Finally remark that

$$\operatorname{col}\bigl(C(A_P^{-1})^{j-1}\bigr)_{j=1}^{\ell} = \operatorname{col}\bigl(CA^{j-1}\bigr)_{j=1}^{\ell}\bigl(X_{ij}\bigr)_{i,j=1}^{\ell},$$

with the invertible matrix $\bigl(X_{ij}\bigr)_{i,j=1}^{\ell}$ given by

$$X_{ij} = \begin{cases} \binom{i}{j}\alpha^{i-j}I & \text{if } i \geq j, \\ 0 & \text{if } i < j. \end{cases}$$

We conclude that (c') holds true. □

The next result shows that a right Γ-null pair is unique up to a similarity transformation.

Theorem 1.3. *Assume that (C_1, A_1) with base space $\mathcal{X}_1$ and (C_2, A_2) with base space $\mathcal{X}_2$ are right Γ-null pairs for the operator polynomial $L(\lambda)$ on Ω. Then there exists a unique invertible operator $S : \mathcal{X}_1 \to \mathcal{X}_2$ with the property that*

$$C_2S = C_1, \quad A_2S = SA_1.$$

Furthermore, if B_1 and B_2 are such that the functions $L(\lambda) - C(\lambda I - A_1)^{-1}B_1$ and $L(\lambda) - C_2(\lambda I - A_2)^{-1}B_2$ are analytic in Ω, then also $B_2 = B_1S$.

Since the proof is the same for the case when $L(\lambda)$ is an analytic operator function, we postpone the proof to Chapter XIV, where Theorem 1.3 will appear as a corollary of Theorem XIV.1.3.

The following result relates left Γ-null pairs to right Γ-null pairs.

Corollary 1.4. *If (A, B) is a left Γ-null pair of $L(\lambda)$ and C is such that $L(\lambda) - C(\lambda I - A)^{-1}B$ has an analytic continuation to Ω, then (C, A) is a right Γ-null pair. Conversely, if (C, A) is a right Γ-null pair and B is such that $L(\lambda) - C(\lambda I - A)^{-1}B$ has an analytic continuation to Ω, then (A, B) is a left Γ-null pair.*

PROOF. Let (A, B) be a left Γ-null pair, and assume that $L(\lambda)-C(\lambda I-A)^{-1}B$ has an analytic continuation to Ω. Let (A_1, B_1) be the left Γ-null pair given in Theorem 1.1, and let (C_1, A_1) be the right Γ-null pair given in Theorem 1.2. Then Theorem 1.3 gives that there exists an invertible operator S such that $AS = SA_1$, $CS = C_1$ and $B = B_1 S$. Since (C_1, A_1) is a right Γ-null pair, it follows that also (C, A) is a right Γ-null pair.

The second part of the corollary is proved in the same way. □

We call (C, A, B) a *Γ-spectral triple* of $L(\lambda)$ if (A, B) is a left Γ-null pair of $L(\lambda)$ and $L(\lambda) - C(\lambda I - A)^{-1}B$ has an analytic continuation to Ω. Corollary 1.4 shows that if (C, A) is a right Γ-null pair and $L(\lambda)-C(\lambda I-A)^{-1}B$ has an analytic continuation to Ω, then (C, A, B) is a Γ-spectral triple.

The next result is the 'left' version of Theorem 1.3.

Corollary 1.5. *Assume that (A_1, B_1) with base space $\mathcal{X}_1$ and (A_2, B_2) with base space $\mathcal{X}_2$ are right Γ-null pairs for the operator polynomial $L(\lambda)$ on Ω. Then there exists a unique invertible operator $S : \mathcal{X}_1 \to \mathcal{X}_2$ with the property that*

$$A_2 S = S A_1, \quad B_2 = S B_1.$$

PROOF. Choose C_1 and C_2 such that (C_1, A_1, B_1) and (C_2, A_2, B_2) are Γ-spectral triples. Then (C_1, A_1) and (C_2, A_2) are right Γ-null pairs for the matrix polynomial $L(\lambda)$ on Ω. The result now follows from Theorem 1.3. □

In the sequel a maximality property of Γ-spectral triples will play an important role. The next result states this property.

Lemma 1.6. *Let (C, A, B) be a Γ-spectral triple for $L(\lambda)$. Then*

$$I_{\mathcal{X}} = \frac{1}{2\pi i}\int_\Gamma (\lambda I_{\mathcal{X}} - A)^{-1} B L(\lambda) C (\lambda I_{\mathcal{X}} - A)^{-1} d\lambda. \tag{1.6}$$

PROOF. We compute that

$$C(\lambda - A)^{-1} A^n B = -\sum_{j=1}^{n} \binom{n}{j} \lambda^{n-j} C(\lambda - A)^{j-1} B + \lambda^n C(\lambda - A)^{-1} B.$$

Then

$$\frac{1}{2\pi i}\int_\Gamma (\lambda I_{\mathcal{X}} - A)^{-1} B L(\lambda) C(\lambda I_{\mathcal{X}} - A)^{-1} d\lambda (A^n B)$$
$$= \frac{1}{2\pi i}\int_\Gamma (\lambda I_{\mathcal{X}} - A)^{-1} B L(\lambda)\Big(-\sum_{j=1}^{n}\binom{n}{j}\lambda^{n-j} C(\lambda - A)^{j-1} B\Big) d\lambda +$$
$$\frac{1}{2\pi i}\int_\Gamma \lambda^n (\lambda I_{\mathcal{X}} - A)^{-1} B L(\lambda) C(\lambda - A)^{-1} B d\lambda.$$

The first term on the right hand side of this equality is equal to zero since both $(\lambda I_{\mathcal{X}} - A)^{-1}BL(\lambda)$ and $C(\lambda - A)^{j-1}B$ are analytic inside the curve Γ. So we compute the second term. Note $L(\lambda)^{-1} - C(\lambda - A)^{-1}B = -A(\lambda)$ is analytic inside Γ. Thus $L(\lambda)C(\lambda - A)^{-1}B = I + L(\lambda)A(\lambda)$. We get

$$\begin{aligned}
&\frac{1}{2\pi i}\int_\Gamma \lambda^n(\lambda I_{\mathcal{X}} - A)^{-1}BL(\lambda)C(\lambda - A)^{-1}B d\lambda \\
&= \frac{1}{2\pi i}\int_\Gamma \lambda^n(\lambda I_{\mathcal{X}} - A)^{-1}B d\lambda + \frac{1}{2\pi i}\int_\Gamma \lambda^n(\lambda I_{\mathcal{X}} - A)^{-1}BL(\lambda)A(\lambda) d\lambda.
\end{aligned}$$

The second term on the right hand side gives 0 since $(\lambda I_{\mathcal{X}} - A)^{-1}BL(\lambda)$ and $A(\lambda)$ are analytic inside Γ. The first term on the right hand side gives A^nB. We showed that for each natural number n

$$\frac{1}{2\pi i}\int_\Gamma (\lambda I_{\mathcal{X}} - A)^{-1}BL(\lambda)C(\lambda I_{\mathcal{X}} - A)^{-1} d\lambda (A^nB) = A^nB.$$

Then it follows from property (c) that (1.6) holds. □

XIII.2 Wiener-Hopf equivalence

We begin with the definition of Wiener-Hopf equivalence. Recall that Γ is a closed rectifiable Jordan curve in the complex plane with a bounded inner domain Ω. For simplicity we shall assume that $0 \in \Omega$. In this chapter $\mathcal{Y}$ will denote a complex Banach space. Let $L_1(\lambda)$ and $L_2(\lambda)$ be operator polynomials on $\mathcal{Y}$, i.e., polynomials whose coefficients are bounded linear operators on $\mathcal{Y}$. We call $L_1(\lambda)$ and $L_2(\lambda)$ *left Wiener-Hopf equivalent* with respect to Γ if $L_1(\lambda)$ and $L_2(\lambda)$ are invertible for each $\lambda \in \Gamma$ and

$$L_2(\lambda) = E_-(\lambda)L_1(\lambda)E_+(\lambda), \quad (\lambda \in \Gamma), \tag{2.1}$$

where $E_+(\lambda)$ is holomorphic on Ω and $E_-(\lambda)$ is holomorphic on $\mathbb{C}_\infty \setminus \overline{\Omega}$, both are continuous up to the boundary Γ, for each $\lambda \in \overline{\Omega}$ the operator $E_+(\lambda)$ is invertible, and for each $\lambda \in \mathbb{C}_\infty \setminus \Omega$ the operator $E_-(\lambda)$ is invertible. Here $\mathbb{C}_\infty = \mathbb{C} \cup \{\infty\}$. The operator polynomials $L_1(\lambda)$ and $L_2(\lambda)$ are called *right Wiener-Hopf equivalent* with respect to Γ if $L_1(\lambda)$ and $L_2(\lambda)$ are invertible for each $\lambda \in \Gamma$ and

$$L_2(\lambda) = E_+(\lambda)L_1(\lambda)E_-(\lambda), \quad (\lambda \in \Gamma),$$

where $E_+(\lambda)$ and $E_-(\lambda)$ are as above.

In this section we present the infinite dimensional generalizations of results on matrix polynomials obtained earlier in Sections V.2, IX.5, and X.3. To state these results we need to extend some definitions to the Banach space setting.

Let $F_1 : \mathcal{X}_1 \to \tilde{\mathcal{X}}_1$, $G_1 : \mathcal{X}_1 \to \tilde{\mathcal{X}}_1$, $F_2 : \mathcal{X}_2 \to \tilde{\mathcal{X}}_2$ and $G_2 : \mathcal{X}_2 \to \tilde{\mathcal{X}}_2$, be bounded linear operators acting between Banach spaces. The linear operator pencils $\lambda F_1 + G_1$ and $\lambda F_2 + G_2$ are said to be *strictly equivalent* (in the sense of

Kronecker) if there exists invertible bounded linear operators $S : \mathcal{X}_1 \to \mathcal{X}_2$ and $\tilde{S} : \tilde{\mathcal{X}}_1 \to \tilde{\mathcal{X}}_2$ such that $S(\lambda F_1 + G_1) = (\lambda F_2 + G_2)\tilde{S}$.

Let $A_1 : \mathcal{X}_1 \to \mathcal{X}_1$, $B_1 : \mathcal{U}_1 \to \mathcal{X}_1$, $A_2 : \mathcal{X}_2 \to \mathcal{X}_2$ and $B_2 : \mathcal{U}_2 \to \mathcal{X}_2$ be bounded linear operators acting between Banach spaces. The pairs (A_1, B_1) and (A_2, B_2) are called *feedback equivalent* if there exist an operator $F_1 : \mathcal{X}_1 \to \mathcal{U}_1$ and invertible operators $S : \mathcal{X}_1 \to \mathcal{X}_2$ and $T : \mathcal{U}_1 \to \mathcal{U}_2$ such that

$$A_2 = S(A_1 + B_1 F_1)S^{-1}, \quad B_2 T = S B_1.$$

The next theorem provides the relation between block similarity and Wiener-Hopf equivalence.

Theorem 2.1. *For $i = 1, 2$ let (A_i, B_i) be a left Γ-null pair with base space $\mathcal{X}_i$ of the operator polynomial $L_i(\lambda)$ on the space $\mathcal{Y}$. Let P_i be the projection of $\mathcal{X}_i \oplus \mathcal{Y}$ along $\{0\} \oplus \mathcal{Y}$ onto $\mathcal{X}_i \oplus \{0\}$, and put*

$$Z_i = \begin{pmatrix} A_i & B_i \end{pmatrix} : \mathcal{X}_i \oplus \mathcal{Y} \to \mathcal{X}_i \oplus \{0\}.$$

Then the following statements are equivalent:

(1) the polynomials $L_1(\lambda)$ and $L_2(\lambda)$ are left Wiener-Hopf equivalent with respect to Γ;

(2) the blocks $(Z_1; P_1, I_{\mathcal{X}_1 \oplus \mathcal{Y}})$ and $(Z_2; P_2, I_{\mathcal{X}_2 \oplus \mathcal{Y}})$ are block similar;

(3) the linear pencils $\lambda\begin{pmatrix} I_{\mathcal{X}_1} & 0 \end{pmatrix} + \begin{pmatrix} A_1 & B_1 \end{pmatrix}$ and $\lambda\begin{pmatrix} I_{\mathcal{X}_2} & 0 \end{pmatrix} + \begin{pmatrix} A_2 & B_2 \end{pmatrix}$ are strictly equivalent;

(4) the pairs (A_1, B_1) and (A_2, B_2) are feedback equivalent.

PROOF. First we prove that (2) implies (3). Let

$$\begin{pmatrix} S_{11} & S_{12} \\ S_{21} & S_{22} \end{pmatrix} : \mathcal{X}_1 \oplus \mathcal{Y} \to \mathcal{X}_2 \oplus \mathcal{Y}$$

be the block similarity of $(Z_1; P_1, I_{\mathcal{X}_1})$ and $(Z_1; P_1, I_{\mathcal{X}_1})$. Since $S[\operatorname{Ker} P_1] = \operatorname{Ker} P_2$, the entry S_{12} must be 0 and S_{22} must be invertible. Therefore S_{11} is invertible and

$$S_{11}\Big(\lambda\begin{pmatrix} I_{\mathcal{X}_1} & 0 \end{pmatrix} + \begin{pmatrix} A_1 & B_1 \end{pmatrix}\Big) = \Big(\lambda\begin{pmatrix} I_{\mathcal{X}_2} & 0 \end{pmatrix} + \begin{pmatrix} A_2 & B_2 \end{pmatrix}\Big) S$$

gives the strict equivalence of the pencils.

We prove that (3) implies (4). So let

$$E\Big(\lambda\begin{pmatrix} I_{\mathcal{X}_1} & 0 \end{pmatrix} + \begin{pmatrix} A_1 & B_1 \end{pmatrix}\Big) = \Big(\lambda\begin{pmatrix} I_{\mathcal{X}_2} & 0 \end{pmatrix} + \begin{pmatrix} A_2 & B_2 \end{pmatrix}\Big) F,$$

where

$$E : \mathcal{X}_1 \to \mathcal{X}_2, \quad F = \begin{pmatrix} F_{11} & F_{12} \\ F_{21} & F_{22} \end{pmatrix} : \mathcal{X}_1 \oplus \mathcal{Y} \to \mathcal{X}_2 \oplus \mathcal{Y}$$

are invertible operators. By comparing the coefficients of λ one sees that $E = F_{11}$ and $F_{12} = 0$, and hence F_{22} is invertible. So we get

$$EA_1 = (A_2 + B_2F_{21}E^{-1})E, \quad EB_1 = B_2F_{22}.$$

This proves that (A_1, B_1) and (A_2, B_2) are feedback equivalent.

The third step in the proof is to show that (4) implies (2). So assume that

$$A_2 = S(A_1 + B_1F_1)S^{-1}, \quad B_2T = SB_1.$$

Then choose

$$R = \begin{pmatrix} S & 0 \\ -F_1S^{-1} & T \end{pmatrix}.$$

It is trivial to check that $R[\operatorname{Ker} P_1] = \operatorname{Ker} P_2$ and that $RZ_1 = Z_2R$. So R is a block similarity of the blocks $(Z_1; P_1, I_{\mathcal{X}_1\oplus\mathcal{Y}})$ and $(Z_2; P_2, I_{\mathcal{X}_2\oplus\mathcal{Y}})$.

The fourth step in the proof is to show that (2) implies (1). So assume that S is a block similarity of the blocks $(Z_1; P_1, I_{\mathcal{X}_1\oplus\mathcal{Y}})$ and $(Z_2; P_2, I_{\mathcal{X}_2\oplus\mathcal{Y}})$. According to the first paragraph of the proof this means that

$$S = \begin{pmatrix} S_{11} & 0 \\ S_{21} & S_{22} \end{pmatrix} : \mathcal{X}_1 \oplus \mathcal{Y} \to \mathcal{X}_2 \oplus \mathcal{Y}$$

with S_{11} and S_{22} invertible, and that $S_{11}A_1 = A_2S_{11} + B_2S_{21}$ and $S_{11}B_1 = B_2S_{22}$. We will define functions $E_-(\lambda)$ and $E_+(\lambda)$ which will establish the Wiener-Hopf equivalence of the polynomials $L_1(\lambda)$ and $L_2(\lambda)$ with respect to the curve Γ. Put

$$E_-(\lambda) = S_{22} + S_{21}(\lambda I - A_1)^{-1}B_1. \tag{2.2}$$

Then

$$\begin{aligned} E_-(\lambda)^{-1} &= S_{22}^{-1} - \big(\lambda I - (A_1 - B_1S_{22}^{-1}S_{21})\big)^{-1}B_1S_{22}^{-1} \\ &= S_{22}^{-1} - S_{22}^{-1}S_{21}S_{11}^{-1}(\lambda I - A_2)^{-1}B_2. \end{aligned}$$

Since both $\sigma(A_1)$ and $\sigma(A_2)$ are subsets of Ω, the functions $E_-(\lambda)$ and $E_-(\lambda)^{-1}$ are analytic outside the curve Γ and continuous towards Γ. Next we put

$$E_+(\lambda) = L_1(\lambda)^{-1}E_-(\lambda)^{-1}L_2(\lambda). \tag{2.3}$$

Let us show that $E_+(\lambda)$ has an analytic continuation on Ω. Choose C_1 such that (C_1, A_1, B_1) is a Γ-spectral triple. Write

$$\begin{aligned} E_+(\lambda) =& \big(L_1(\lambda)^{-1} - C_1(\lambda I - A_1)^{-1}B_1\big)E_-(\lambda)^{-1}L_2(\lambda) \\ &+ C_1(\lambda I - A_1)^{-1}B_1E_-(\lambda)^{-1}L_2(\lambda). \end{aligned} \tag{2.4}$$

In the first term on the right hand side the factor $L_1(\lambda)^{-1} - C_1(\lambda I - A_1)^{-1}B_1$ has an analytic continuation to Ω, because (C_1, A_1, B_1) is a Γ-spectral triple for $L_1(\lambda)$. The second factor of the first term is

$$\big(S_{22}^{-1} - S_{22}^{-1}S_{21}S_{11}^{-1}(\lambda I - A_2)^{-1}B_2\big)L_2(\lambda).$$

The function $(\lambda I - A_2)^{-1}B_2L_2(\lambda)$ has an analytic continuation to Ω, and hence the first term in the right hand side of (2.4) has an analytic continuation on Ω. For the second term we compute

$$\begin{aligned}
&C_1(\lambda I - A_1)^{-1}B_1E_-(\lambda)^{-1}\\
&=C_1(\lambda I - A_1)^{-1}B_1S_{22}^{-1} - C_1(\lambda I - A_1)^{-1}B_1S_{22}^{-1}S_{21}S_{11}^{-1}(\lambda I - A_2)^{-1}B_2\\
&=C_1(\lambda I - A_1)^{-1}B_1S_{22}^{-1}+\\
&\quad - C_1(\lambda I - A_1)^{-1}\big(S_{11}^{-1}(\lambda I - A_2) - (\lambda I - A_1)S_{11}^{-1}\big)(\lambda I - A_2)^{-1}B_2\\
&=C_1S_{11}^{-1}(\lambda I - A_2)^{-1}B_2
\end{aligned}$$

Thus the second term in the right hand side of (2.4) is

$$C_1(\lambda I - A_1)^{-1}B_1E_-(\lambda)^{-1}L_2(\lambda) = C_1S_{11}^{-1}(\lambda I - A_2)^{-1}B_2L_2(\lambda)$$

which has an analytic continuation to Ω, because (A_2, B_2) is a Γ-null pair of $L_2(\lambda)$. Similarly we treat

$$E_+(\lambda)^{-1} = L_2(\lambda)^{-1}E_-(\lambda)L_1(\lambda).$$

So we write

$$\begin{aligned}
E_+(\lambda)^{-1} =&\big(L_2(\lambda)^{-1} - C_2(\lambda I - A_2)^{-1}B_2\big)E_-(\lambda)L_1(\lambda)+\\
&+ C_2(\lambda I - A_2)^{-1}B_2E_-(\lambda)L_1(\lambda),
\end{aligned}$$

where C_2 is such that (C_2, A_2, B_2) is a Γ-spectral triple of $L_2(\lambda)$. Again we see that the first term has an analytic continuation on Ω. The second term we rewrite with

$$\begin{aligned}
&C_2(\lambda I - A_2)^{-1}B_2E_-(\lambda)\\
&=C_2(\lambda I - A_2)^{-1}B_2S_{22} + C_2(\lambda I - A_2)^{-1})B_2S_{21}(\lambda I - A_1)^{-1}B_1\\
&=C_2(\lambda I - A_2)^{-1}B_2S_{22}+\\
&\quad + C_2(\lambda I - A_2)^{-1}\big((\lambda I - A_2)S_{11} - S_{11}(\lambda I - A_1)\big)(\lambda I - A_1)^{-1}B_1\\
&=C_2S_{11}(\lambda I - A_1)^{-1}B_1
\end{aligned}$$

So

$$C_2(\lambda I - A_2)^{-1}B_2E_-(\lambda)L_1(\lambda) = C_2S_{11}(\lambda I - A_1)^{-1}B_1L_1(\lambda),$$

which is analytic on Ω, because the pair (A_1, B_1) is a Γ-null pair of $L_1(\lambda)$. We proved that $E_+(\lambda)$ and $E_+(\lambda)^{-1}$ are analytic on the domain Ω and continuous towards the boundary Γ.

Finally we prove that (1) implies (2). So we assume that

$$L_2(\lambda) = E_+(\lambda)L_1(\lambda)E_-(\lambda), \quad \lambda \in \Gamma.$$

Define operators $N : \mathcal{X}_1 \to \mathcal{X}_2$, $M : \mathcal{Y} \to \mathcal{Y}$ and $F : \mathcal{X}_1 \to \mathcal{Y}$ by

$$N = \frac{1}{2\pi i}\int_\Gamma (\mu I - A_2)^{-1}B_2L_2(\mu)E_+(\mu)^{-1}C_1(\mu I - A_1)^{-1}d\mu, \tag{2.5}$$

$$F = \frac{1}{2\pi i}\int_\Gamma E_-(\mu)L_1(\mu)C_1(\mu I - A_1)^{-1}d\mu, \tag{2.6}$$

and

$$M = E_-(\infty). \tag{2.7}$$

We shall show that

$$S = \begin{pmatrix} N & 0 \\ F & M \end{pmatrix} : \mathcal{X}_1 \oplus \mathcal{Y} \to \mathcal{X}_2 \oplus \mathcal{Y} \tag{2.8}$$

is a block similarity of the blocks $(Z_1; P_1, I_{\mathcal{X}_1\oplus\mathcal{Y}})$ and $(Z_2; P_2, I_{\mathcal{X}_2\oplus\mathcal{Y}})$. So we have to prove that

$$NA_1 = A_2N + B_2F, \quad NB_1 = B_2M,$$

and that N and M are invertible. Let Γ_1 be a closed rectifiable curve in Ω such that $L_1(\lambda)$ and $L_2(\lambda)$ are analytic on Γ_1 and $\sigma(A_1)$ and $\sigma(A_2)$ are contained in the inner domain of Γ_1. Put

$$N_1 = \frac{1}{2\pi i}\int_{\Gamma_1} (\lambda I - A_1)^{-1}B_1L_1(\lambda)E_+(\lambda)C_2(\lambda I - A_2)^{-1}d\lambda.$$

We will prove that $N_1 = N^{-1}$. Use the resolvent equation

$$(\mu I - A_1)^{-1}(\lambda I - A_1)^{-1} = \frac{1}{\lambda - \mu}((\mu I - A_1)^{-1} - (\lambda I - A_1)^{-1})$$

to see that

$$\begin{aligned} NN_1 = \left(\frac{1}{2\pi i}\right)^2 \int_\Gamma\int_{\Gamma_1} &(\mu I - A_2)^{-1}B_2L_2(\mu)E_+(\mu)^{-1}C_1\frac{1}{\lambda - \mu} \\ &((\mu I - A_1)^{-1} - (\lambda I - A_1)^{-1})B_1L_1(\lambda)E_+(\lambda)C_2(\lambda I - A_2)^{-1}d\lambda d\mu. \end{aligned} \tag{2.9}$$

The functions

$$F_1(\mu) = (\mu I - A_2)^{-1}B_2L_2(\mu)E_+(\mu)^{-1}, \quad F_2(\lambda) = L_1(\lambda)E_+(\lambda)C_2(\lambda I - A_2)^{-1}$$

are analytic on Ω. So if we add in (2.9) to the function

$$\frac{1}{\lambda-\mu}\big((\mu I - A_1)^{-1} - (\lambda I - A_1)^{-1}\big)$$

an analytic function in both λ and μ, then the value of the integral still will be NN_1. Now use that $H_1(\lambda) = -C_1(\lambda I - A_1)^{-1}B_1 + L_1(\lambda)^{-1}$ is analytic in Ω. Thus also the function

$$\frac{H_1(\lambda) - H_1(\mu)}{\lambda-\mu}$$

is analytic on Ω in both λ and μ. We get

$$NN_1 = \left(\frac{1}{2\pi i}\right)^2 \int_\Gamma\int_{\Gamma_1} F_1(\mu)\frac{1}{\lambda-\mu}\big(L_1(\mu)^{-1} - L_1(\lambda)^{-1}\big)F_2(\lambda)d\lambda d\mu.$$

Now rewrite $F_1(\mu)$ as $F_1(\mu) = (\mu I - A_2)^{-1}B_2E_-(\mu)L_1(\mu)$. We obtain for NN_1 the expression

$$(2\pi i)^2 NN_1$$
$$= \int_\Gamma\int_{\Gamma_1} (\mu I - A_2)^{-1}B_2E_-(\mu)\frac{1}{\lambda-\mu}\big(L_1(\lambda) - L_1(\mu)\big)E_+(\lambda)C_2(\lambda I - A_2)^{-1}d\lambda d\mu. \tag{2.10}$$

Observe that

$$\int_\Gamma \frac{1}{\lambda-\mu}(\mu I - A_2)^{-1}B_2E_-(\mu)d\mu = 0,$$

because the integrand is analytic outside Γ and has a second order zero at infinity. Use that $E_-(\mu)L_1(\mu) = L_2(\mu)E_+(\mu)^{-1}$ and that $(\mu I - A_2)^{-1}B_2L_2(\mu)E_+(\mu)^{-1}$ is analytic inside Γ, to see that

$$\frac{1}{2\pi i}\int_\Gamma \frac{1}{\lambda-\mu}(\mu I - A_2)^{-1}B_2E_-(\mu)L_1(\mu)d\mu = (\lambda I - A_2)^{-1}B_2L_2(\lambda)E_+(\lambda)^{-1}.$$

We change the order of integration in (2.9) and find

$$NN_1 = \frac{1}{2\pi i}\int_{\Gamma_1} (\lambda I - A_2)^{-1}B_2L_2(\lambda)C_2(\lambda I - A_2)^{-1}d\lambda = I_{\mathcal{X}_2}.$$

From the symmetry in the definitions of N and N_1 it is clear that in a similar way one proves that $N_1N = I_{\mathcal{X}_1}$.

Next, we prove $NB_1 = B_2M$. From the definition of N we get

$$NB_1 = \frac{1}{2\pi i}\int_\Gamma (\mu I - A_2)^{-1}B_2L_2(\mu)E_+(\mu)^{-1}C_1(\mu I - A_1)^{-1}B_1d\mu.$$

Recall that $C_1(\mu I - A_1)^{-1}B_1 = L_1(\mu)^{-1} - H_1(\mu)$ with $H_1(\mu)$ analytic on Ω. So

$$NB_1 = \frac{1}{2\pi i}\int_\Gamma (\mu I - A_2)^{-1}B_2L_2(\mu)E_+(\mu)^{-1}\big(L_1(\mu)^{-1} - H_1(\mu)\big)d\mu.$$

Use that $(\mu I - A_2)^{-1}B_2L_2(\mu)E_+(\mu)^{-1}$ and $H_1(\mu)$ are analytic in Ω and that $E_-(\mu) = L_2(\mu)E_+(\mu)^{-1}L_1(\mu)^{-1}$, to see that

$$NB_1 = \frac{1}{2\pi i}\int_\Gamma (\mu I - A_2)^{-1}B_2E_-(\mu)d\mu = B_2E_-(\infty) = B_2M.$$

Finally, we check $A_2N - NA_1 = -B_2F$. Recall that

$$N = \frac{1}{2\pi i}\int_\Gamma F_1(\mu)C_1(\mu I - A_1)^{-1}d\mu.$$

So $A_2N - NA_1$ is equal to

$$\frac{1}{2\pi i}\int_\Gamma (A_2 - \mu I)F_1(\mu)C_1(\mu I - A_1)^{-1} + F_1(\mu)C_1(\mu I - A_1)^{-1}(\mu I - A_1)d\mu$$

$$\begin{aligned}&= \frac{1}{2\pi i}\int_\Gamma -B_2L_2(\mu)E_+(\mu)^{-1}C_1(\mu I - A_1)^{-1}d\mu + \\ &\quad + \frac{1}{2\pi i}\int_\Gamma (\mu I - A_2)^{-1}B_2L_2(\mu)E_+(\mu)^{-1}C_1 d\mu.\end{aligned} \tag{2.11}$$

Since $L_2(\mu)E_+(\mu)^{-1} = E_-(\mu)L_1(\mu)$, the first integral in (2.11) is equal to $-B_2F$. The second integral in (2.10) is equal to zero, because $(\mu I - A_2)^{-1}B_2L_2(\mu)$ and $E_+(\mu)^{-1}$ both are analytic in Ω. So we get $A_2N - NA_1 = -B_2F$. □

Note that the proof of this theorem is constructive. Indeed, if the polynomials are left Wiener-Hopf equivalent with respect to the curve Γ, then the formula (2.8) (with (2.5), (2.6) and (2.7)) gives the block similarity of the blocks made from the left Γ-null pairs of the polynomials. Conversely, given the block similarity of these blocks the function needed for the Wiener-Hopf equivalence are provided by (2.2) and (2.3).

For the next theorem we need to introduce one more notion. Let

$$A_1 : \mathcal{X}_1 \to \mathcal{X}_1, \quad C_1 : \mathcal{X}_1 \to \mathcal{Y}_1, \quad A_2 : \mathcal{X}_2 \to \mathcal{X}_2, \quad C_2 : \mathcal{X}_2 \to \mathcal{Y}_2$$

be operators acting between Banach spaces. The pairs (C_1, A_1) and (C_2, A_2) are called *output injection equivalent* if there exist an operator $G_1 : \mathcal{Y}_1 \to \mathcal{X}_1$ and invertible operators $S : \mathcal{X}_1 \to \mathcal{X}_2$ and $T : \mathcal{Y}_1 \to \mathcal{Y}_2$ such that

$$A_2 = S(A_1 + G_1C_1)S^{-1}, \quad TC_1 = C_2S.$$

The next theorem is a transposed version of Theorem 2.1. We omit the proof, because it easily can be derived from the proof of the Theorem 2.1.

Theorem 2.2. *For $i = 1, 2$ let (C_i, A_i) be right Γ-null pair with base space $\mathcal{X}_i$ of the operator polynomial $L_i(\lambda)$ on the space $\mathcal{Y}$. Let Q_i be the projection of $\mathcal{X}_i \oplus \mathcal{Y}$ along $\{0\} \oplus \mathcal{Y}$ onto $\mathcal{X}_i \oplus \{0\}$, and put*

$$Z_i = \begin{pmatrix} A_i \\ C_i \end{pmatrix} : \operatorname{Im} Q_i \to \mathcal{X}_i \oplus \mathcal{Y}.$$

Then the following statements are equivalent:

(1) the polynomials $L_1(\lambda)$ and $L_2(\lambda)$ are right Wiener-Hopf equivalent with respect to Γ;

(2) the blocks $(Z_1; I_{\mathcal{X}_1 \oplus \mathcal{Y}}, Q_1)$ and $(Z_2; I_{\mathcal{X}_2 \oplus \mathcal{Y}}, Q_2)$ are block similar;

(3) the linear pencils

$$\lambda \begin{pmatrix} I_{\mathcal{X}_1} \\ 0 \end{pmatrix} + \begin{pmatrix} A_1 \\ C_1 \end{pmatrix} \quad \text{and} \quad \lambda \begin{pmatrix} I_{\mathcal{X}_2} \\ 0 \end{pmatrix} + \begin{pmatrix} A_2 \\ C_2 \end{pmatrix}$$

are strictly equivalent;

(4) the pairs (C_1, A_1) and (C_2, A_2) are output injection equivalent.

XIII.3 Wiener-Hopf factorization

We call (2.1) a *left Wiener-Hopf factorization* of $L_2(\lambda)$ if in the right side of (2.1) the middle factor $L_1(\lambda) = D(\lambda) = \sum_{i=1}^{r} \lambda^{\nu_i} P_i$, where $\nu_1 \geq \cdots \geq \nu_r \geq 0$ are integers and $P_1, \ldots, P_r$ are mutually disjoint projections of Y such that $\sum_{i=1}^{r} P_i = I_{\mathcal{Y}}$. If $\mathcal{Y}$ is finite dimensional, we prefer for obvious reasons to choose the projections P_i to be one dimensional, and in that case $\nu_1 \geq \cdots \geq \nu_r$ are called the left Wiener-Hopf factorization indices of $L_2(\lambda)$. In the infinite dimensional case we shall assume that $\nu_1 > \nu_2 > \cdots > \nu_r \geq 0$. One defines *right Wiener-Hopf factorization* in the same way as it is done for the "left" case.

In this section we provide necessary and sufficient conditions for the existence of Wiener-Hopf factorizations of operator polynomials with respect to Γ. To state the results we make use of some notions that we will introduce first. Let $A : \mathcal{X} \to \mathcal{X}$ and $B : \mathcal{Y} \to \mathcal{X}$ be a pair of bounded linear operators acting between Banach spaces. With the pair (A, B) we associate a block $(Z; P, I)$ on $\mathcal{X} \oplus \mathcal{Y}$ by putting

$$P = \begin{pmatrix} I & 0 \\ 0 & 0 \end{pmatrix} : \mathcal{X} \oplus \mathcal{Y} \to \mathcal{X} \oplus \mathcal{Y}, \quad Z = \begin{pmatrix} A & B \\ 0 & 0 \end{pmatrix} : \mathcal{X} \oplus \mathcal{Y} \to \operatorname{Im} P.$$

We call the pair (A, B) of *finite type* if the block is of finite type. The *characteristics of the pair* are defined to be the characteristics of the block. We will express the property of being of finite type directly in terms of the operators A and B. First recall that the block is of finite type if $\operatorname{row}\big(Z^i(I-P)\big)_{i=0}^{j-1}$ has a generalized inverse for each j and is surjective for $j=\ell$. Remark that

$$Z^i(I-P)=\begin{pmatrix} A^i & A^{i-1}B \\ 0 & 0\end{pmatrix}\begin{pmatrix} 0 & 0 \\ 0 & I\end{pmatrix}=\begin{pmatrix} 0 & A^{i-1}B \\ 0 & 0\end{pmatrix}.$$

Thus

$$\operatorname{row}\big(Z^i(I-P)\big)_{i=0}^{j-1}= \\ =\left(\begin{pmatrix} 0 & 0 \\ 0 & I\end{pmatrix}\ \begin{pmatrix} 0 & B \\ 0 & 0\end{pmatrix}\ \begin{pmatrix} 0 & AB \\ 0 & 0\end{pmatrix}\ \cdots\ \begin{pmatrix} 0 & A^{j-2}B \\ 0 & 0\end{pmatrix}\right).$$

So it is easy to see that $F_j=\operatorname{row}\big(Z^i(I-P)\big)_{i=0}^{j-1}$ has a generalized inverse if and only if

$$E_j=\begin{pmatrix} B & AB & \cdots & A^{j-2}B\end{pmatrix}$$

has a generalized inverse and F_ℓ is left invertible if and only if E_ℓ is left invertible.

We formulate the main result of this section.

Theorem 3.1. *Let $L(\lambda)$ be an operator polynomial, and let (A,B) be a left Γ-null pair for $L(\lambda)$. Then $L(\lambda)$ admits a left Wiener-Hopf factorization with respect to Γ,*

$$L(\lambda)=E_-(\lambda)\left(\sum_{i=1}^{r}\lambda^{\nu_i}P_i\right)E_+(\lambda), \tag{3.1}$$

if and only if the pair (A,B) is of finite type. Furthermore, if in (3.1) the projections $P_1,\dots,P_r$ are different from zero and $\nu_1>\nu_2>\cdots>\nu_r$, then the set

$$\{(\operatorname{Im}P_1,\nu_1),\dots,(\operatorname{Im}P_r,\nu_r)\}$$

is equal to the set of characteristics of the pair (A,B).

PROOF. Assume that the operator polynomial $L(\lambda)$ admits a Wiener-Hopf factorization (3.1). Without loss of generality we may assume that the projections $P_1,\dots,P_r$ are different from zero and that $\nu_1>\nu_2>\cdots>\nu_r$. Put $\mathcal{X}_0=\bigoplus_{i=1}^r(\operatorname{Im}P_i)^{\nu_i}$. We define the operators $A_0:\mathcal{X}_0\to\mathcal{X}_0$, $B_0:\mathcal{Y}\to\mathcal{X}_0$ and $C_0:\mathcal{X}_0\to\mathcal{Y}$ as follows:

$$A_0=\bigoplus_{i=1}^r S_i:\bigoplus_{i=1}^r(\operatorname{Im}P_i)^{\nu_i}\to\bigoplus_{i=1}^r(\operatorname{Im}P_i)^{\nu_i}, \tag{3.2}$$

where $S_i:(\operatorname{Im}P_i)^{\nu_i}\to(\operatorname{Im}P_i)^{\nu_i}$ is given by $S_i(x_1,\dots,x_r)=(0,x_1,\dots,x_{r-1})$;

$$B_0=\bigoplus_{i=1}^r T_i:\bigoplus_{i=1}^r\operatorname{Im}P_i\to\bigoplus_{i=1}^r(\operatorname{Im}P_i)^{\nu_i}, \tag{3.3}$$

with $T_i : \operatorname{Im} P_i \to (\operatorname{Im} P_i)^{\nu_i}$ given by $T_i(x) = (x, 0, \dots, 0)$;

$$C_0 = \bigoplus_{i=1}^{r} R_i : \bigoplus_{i=1}^{r} (\operatorname{Im} P_i)^{\nu_i} \to \bigoplus_{i=1}^{r} \operatorname{Im} P_i, \tag{3.4}$$

where $R_i : (\operatorname{Im} P_i)^{\nu_i} \to \operatorname{Im} P_i$ is given by $R_i(x_1, \dots, x_r) = x_r$. If $\nu_r = 0$, then one has to understand $(\operatorname{Im} P_r)^{\nu_r}$ as the space $\{0\}$, therefore also S_1, T_1 and R_1 as the corresponding zero operators. The triple (C_0, A_0, B_0) is a Γ-spectral triple for the operator polynomial $\sum_{i=1}^{r} \lambda^{\nu_i} P_i$. This polynomial is Wiener-Hopf equivalent to $L(\lambda)$. So from Theorem 2.1 we get that the blocks $(Z; P, I)$ and $(Z_0; P_0, I_0)$ given by

$$P = \begin{pmatrix} I & 0 \\ 0 & 0 \end{pmatrix} : \mathcal{X} \oplus \mathcal{Y} \to \mathcal{X} \oplus \mathcal{Y}, \quad Z = \begin{pmatrix} A & B \\ 0 & 0 \end{pmatrix} : \mathcal{X} \oplus \mathcal{Y} \to \operatorname{Im} P,$$

$$P_0 = \begin{pmatrix} I & 0 \\ 0 & 0 \end{pmatrix} : \mathcal{X}_0 \oplus \mathcal{Y} \to \mathcal{X}_0 \oplus \mathcal{Y}, \quad Z_0 = \begin{pmatrix} A_0 & B_0 \\ 0 & 0 \end{pmatrix} : \mathcal{X}_0 \oplus \mathcal{Y} \to \operatorname{Im} P_0,$$

are block similar. Since $(Z_0; P_0, I_0)$ is given as a direct sum of shifts of the first kind it follows from Theorem XII.2.2 that $(Z; P, I)$ is of finite type. Moreover, it is clear that the characteristics of the block $(Z_0; P_0, I_0)$ are $\{(\operatorname{Im} P_1, \nu_1), \dots, (\operatorname{Im} P_r, \nu_r)\}$, and thus Theorem XII.2.3 gives that the characteristics of $(Z; P, I)$ are also $\{(\operatorname{Im} P_1, \nu_1), \dots, (\operatorname{Im} P_r, \nu_r)\}$. So this set is the set of characteristics of the pair (A, B).

Conversely, assume that the pair (A, B) is of finite type. We apply Theorem XII.2.5 to (A, B). First we derive the projections $P_1, \dots, P_\ell$ from the decomposition $\mathcal{Y} = \bigoplus_{i=0}^{\ell} \mathcal{U}_i$ by putting P_i to be the projection of $\mathcal{Y}$ onto $\mathcal{U}_i$ along $\bigoplus_{j \neq i}^{\ell} \mathcal{U}_i$. Let N, K and M be as in Theorem XII.2.5, and write $A_0 = N^{-1}(A - BK)N$ and $B_0 = N^{-1}B$. Then (A_0, B_0) is a Γ-spectral pair of the polynomial $L_0(\lambda) = \sum_{i=0}^{\ell} \lambda^i P_i$. Since the pairs (A, B) and (A_0, B_0) are feedback equivalent, we get that $L(\lambda)$ and $L_0(\lambda)$ are left Wiener-Hopf equivalent with respect to Γ by Theorem 2.1. So $L(\lambda)$ admits a left Wiener-Hopf factorization with respect to Γ. □

The analogue of Theorem 3.1 for right Wiener-Hopf factorization and right null pairs is the next result. The proof is similar to the proof of Theorem 3.1 and is therefore omitted.

Theorem 3.2. *Let $L(\lambda)$ be an operator polynomial, and let (C, A) be a right Γ-null pair for $L(\lambda)$. Then $L(\lambda)$ admits a right Wiener-Hopf factorization with respect to Γ,*

$$L(\lambda) = E_+(\lambda) \left(\sum_{i=1}^{r} \lambda^{\kappa_i} P_i \right) E_-(\lambda), \tag{3.5}$$

if and only if the pair (C, A) is of finite type. Furthermore, if in (3.5) the projections $P_1, \dots, P_r$ are different from zero and $\kappa_1 > \kappa_2 > \dots > \kappa_r$, then the set

$$\{(\operatorname{Im} P_1, \kappa_1), \dots, (\operatorname{Im} P_r, \kappa_r)\}$$

is equal to the set of characteristics of the pair (C, A).

Let (A,B) be a left Γ-spectral pair of $L(\lambda)=\sum_{j=0}^{\ell}\lambda^j L_j$. From Theorem 3.1 we know that $L(\lambda)$ admits a left Wiener-Hopf factorization with respect to Γ if and only if the pair (A,B) is a pair of finite type. In what follows we show that this condition may be formulated in terms of the moments of $L(\lambda)^{-1}$ with respect to Γ. To do this, choose C such that (C,A,B) is a Γ-spectral triple for $L(\lambda)$, with respect to Γ. Since $L(\lambda)^{-1}-C(\lambda I-A)^{-1}B$ has an analytic continuation to the inner domain Ω of Γ, we can use condition (a) in the definition of a left Γ-null pair to show that

$$R_{-j}=\frac{1}{2\pi i}\int_\Gamma \lambda^{j-1}L(\lambda)^{-1}d\lambda = CA^{j-1}B,\quad j\geq 1. \tag{3.6}$$

It follows that

$$\begin{pmatrix} C\\ CA\\ \vdots\\ CA^{\ell-1}\end{pmatrix}\begin{pmatrix} B & AB & \cdots & A^{j-1}B\end{pmatrix}=\begin{pmatrix} R_{-1} & R_{-2} & \cdots & R_{-j}\\ R_{-2} & R_{-3} & \cdots & R_{-j-1}\\ \vdots & \vdots & & \vdots\\ R_{-\ell} & R_{-\ell-1} & \cdots & R_{-\ell-j+1}\end{pmatrix}. \tag{3.7}$$

Recall that $\operatorname{col}\big(CA^{i-1}\big)_{i=1}^{\ell}$ is left invertible. So $\Omega_j=\begin{pmatrix} B & AB & \cdots & A^{j-1}B\end{pmatrix}$ has a generalized inverse if and only if the operator matrix in the right hand side of (3.7) has a generalized inverse. Since Ω_ℓ is right invertible, as can be seen form condition (c) in the definition of Γ-null pair, we conclude that the pair (A,B) is of finite type if and only if the following operator matrices

$$\begin{pmatrix} R_{-1} & R_{-2} & \cdots & R_{-j}\\ R_{-2} & R_{-3} & \cdots & R_{-j-1}\\ \vdots & \vdots & & \vdots\\ R_{-\ell} & R_{-\ell-1} & \cdots & R_{-\ell-j+1}\end{pmatrix},\quad j=1,\ldots,\ell-1, \tag{3.8}$$

have generalized inverses. By combining this with Theorem 3.1, one obtains the next corollary.

Corollary 3.3. *Let $L(\lambda)$ be an operator polynomial, and let $R_{-1},R_{-2},\ldots$ be given by* (3.6). *Then $L(\lambda)$ admits a left Wiener-Hopf factorization if and only if the matrices* (3.8) *have generalized inverses.*

In a similar way one can prove that $L(\lambda)$ admits a right Wiener-Hopf factorization if and only if the matrices

$$\begin{pmatrix} R_{-1} & R_{-2} & \cdots & R_{-\ell}\\ R_{-2} & R_{-3} & \cdots & R_{-\ell-1}\\ \vdots & \vdots & & \vdots\\ R_{-j} & R_{-j-1} & \cdots & R_{-\ell-j+1}\end{pmatrix},\quad j=1,\ldots,\ell-1,$$

have generalized inverses.

XIII.4 Wiener-Hopf factorization and strict equivalence

The direct sum of operator pencils is defined in the usual way, that is, for pencils

$$\lambda E_1 + F_1 : \mathcal{X}_1 \to \mathcal{Y}_1, \quad \lambda E_2 + F_2 : \mathcal{X}_2 \to \mathcal{Y}_2$$

we define the direct sum $(\lambda E_1 + F_1) \oplus (\lambda E_2 + F_2)$ by

$$(\lambda E_1 + F_1) \oplus (\lambda E_2 + F_2) = \lambda(E_1 \oplus E_2) + (F_1 \oplus F_2) : \mathcal{X}_1 \oplus \mathcal{X}_2 \to \mathcal{Y}_1 \oplus \mathcal{Y}_2.$$

The next result gives a canonical form for pencils related to pairs of finite type.

Theorem 4.1. *Let $\lambda\begin{pmatrix} I & 0 \end{pmatrix} + \begin{pmatrix} A & B \end{pmatrix}$ be a pencil of bounded operators from $\mathcal{X} \oplus \mathcal{Y}$ to $\mathcal{X}$. Then the following two conditions are equivalent:*

(i) *the pair (A, B) is of finite type and $\{(\mathcal{Y}_1, \nu_1), \ldots, (\mathcal{Y}_r, \nu_r)\}$ is its set of characteristics;*

(ii) *the pencil $\lambda\begin{pmatrix} I & 0 \end{pmatrix} + \begin{pmatrix} A & B \end{pmatrix}$ is strictly equivalent to a direct sum $N_1(\lambda) \oplus \cdots \oplus N_r(\lambda)$, where for $i = 1, \ldots, r$*

$$N_i(\lambda) = \begin{pmatrix} \lambda I_i & I_i & 0 & \cdots & 0 & 0 \\ 0 & \lambda I_i & I_i & & 0 & 0 \\ \vdots & & \ddots & \ddots & & \vdots \\ 0 & 0 & & \ddots & I_i & 0 \\ 0 & 0 & & & \lambda I_i & I_i \end{pmatrix} : \mathcal{Y}^{\nu_i+1} \to \mathcal{Y}^{\nu_i}$$

and $\nu_1 > \cdots > \nu_r \geq 0$ with $\mathcal{Y}^0 = \{0\}$.

PROOF. The fact that $\{(\mathcal{Y}_1, \nu_1), \ldots, (\mathcal{Y}_r, \nu_r)\}$ is the set of characteristics of the finite type pair (A, B) means that the block

$$\left(\begin{pmatrix} A & B \\ 0 & 0 \end{pmatrix}; \begin{pmatrix} I & 0 \\ 0 & 0 \end{pmatrix}, I\right),$$

on $\mathcal{X} \oplus \mathcal{Y}$, is block similar to the direct sum $(A_0; P_0, I_0)$ of shifts of the first kind of index ν_i,

$$(A_0; P_0, I_0) = \left(\bigoplus_{i=1}^{r} V_i; \bigoplus_{i=1}^{r} P_i, \bigoplus_{i=1}^{r} I_i^{\nu_i+1}\right)$$

on $\bigoplus_{i=1}^{r} \mathcal{Y}^{\nu_i+1}$. According to Theorem 2.1, the latter is equivalent to the fact that the pencil $\lambda\begin{pmatrix} I & 0 \end{pmatrix} + \begin{pmatrix} A & B \end{pmatrix}$ is strictly equivalent to the pencil $\lambda\begin{pmatrix} I & 0 \end{pmatrix} + \begin{pmatrix} A_0|_{\operatorname{Im} P_0} & A_0|_{\operatorname{Ker} P_0} \end{pmatrix}$ on $\operatorname{Im} P_0 \oplus \operatorname{Ker} P_0$. This last pencil is, up to the ordering of the factors in the direct sum $\bigoplus_{i=1}^{r} \mathcal{Y}^{\nu_i+1}$, equal to the pencil $N_1(\lambda) \oplus \cdots \oplus N_r(\lambda)$. □

We define the *right Kronecker characteristics* of the pencil $\lambda\begin{pmatrix} I & 0 \end{pmatrix} + \begin{pmatrix} A & B \end{pmatrix}$ to be the characteristics of the pair (A, B). From the statements (2) and (3) in Theorem 2.1 and from Theorem XII.2.3 one obtains the following result.

Theorem 4.2. *Let (A_1, B_1) and (A_2, B_2) be pairs of finite type. The pencils $\lambda\begin{pmatrix} I_1 & 0 \end{pmatrix} + \begin{pmatrix} A_1 & B_1 \end{pmatrix}$ and $\lambda\begin{pmatrix} I_2 & 0 \end{pmatrix} + \begin{pmatrix} A_2 & B_2 \end{pmatrix}$ are strictly equivalent if and only if they have the same Kronecker characteristics.*

The next theorem follows from Theorem 3.1 and the definition of right Kronecker characteristics.

Theorem 4.3. *Let (A, B) be a left Γ-null pair of the operator polynomial $L(\lambda)$. Assume that (A, B) is of finite type. Let*

$$L(\lambda) = E_-(\lambda)\left(\sum_{i=1}^{r} \lambda^{\nu_i} P_i\right) E_+(\lambda), \quad \lambda \in \Gamma$$

be a left Wiener-Hopf factorization of $L(\lambda)$ with respect to Γ, where the projections P_i are nonzero and $\nu_1 > \cdots > \nu_r \geq 0$. Then the set of right Kronecker characteristics of the pencil $\lambda\begin{pmatrix} I & 0 \end{pmatrix} + \begin{pmatrix} A & B \end{pmatrix}$ is equal to the set of pairs $\{(\operatorname{Im} P_1, \nu_1), \ldots, (\operatorname{Im} P_r, \nu_r)\}$.

XIII.5 The Fredholm case

In this section we consider factorization problems for a rational operator polynomial $L(\lambda) = \sum_{i=-s}^{t} \lambda^i L_i$. Throughout this section we assume the coefficients $L_{-s}, \ldots, L_t$ to be bounded linear operators on the Banach space $\mathcal{Y}$. Let Γ be closed rectifiable curve with inner domain Ω. A *left Wiener-Hopf factorization* of $L(\lambda)$ with respect to Γ is a factorization

$$L(\lambda) = E_-(\lambda)\left(\sum_{i=-s}^{t} \lambda^i P_i\right) E_+(\lambda), \quad \lambda \in \Gamma, \tag{5.1}$$

where $E_+(\lambda)$ and $E_+(\lambda)^{-1}$ are holomorphic on Ω and continuous towards Γ, the functions $E_-(\lambda)$ and $E_-(\lambda)^{-1}$ are holomorphic on the complement of $\bar{\Omega}$ (including the point ∞) and continuous towards Γ, and $P_{-s}, \ldots, P_t$ are mutually disjoint projections that add up to the identity on $\mathcal{Y}$. Note that we choose the ranges of the exponents of λ in $L(\lambda)$ and of the middle factor in the right hand side of (5.1) to be equal. This is not a restriction, since we do not assume the coefficients L_i and P_i to be nonzero.

Theorem 5.1. *Assume that $L(\lambda) = \sum_{i=-s}^{t} \lambda^i L_i$ admits a left Wiener-Hopf factorization (5.1). Let (A, B) be a left Γ-null pair of $\lambda^s L(\lambda)$ with underlying space $\mathcal{X}$, and consider the operator*

$$\Delta_s = \begin{pmatrix} B & AB & \cdots & A^{s-1}B \end{pmatrix} : \mathcal{Y}^s \to \mathcal{X}.$$

Then for each $i \neq 0$ the projection P_i is of finite rank if and only if Δ_s is a Fredholm operator, i.e., $\dim \operatorname{Ker} \Delta_s$ *and* $\operatorname{codim} \operatorname{Im} \Delta_s = \dim(\mathcal{X}/\operatorname{Im} \Delta_s)$ *are both finite. More generally,*

(a) $\dim \operatorname{Ker} \Delta_s = \sum_{j=1}^{s} j(\operatorname{rank} P_{-j})$,

(b) $\operatorname{codim} \operatorname{Im} \Delta_s = \sum_{j=1}^{t} j(\operatorname{rank} P_j)$.

PROOF. Let (A_0, B_0) be a left Γ-null pair for $\sum_{i=0}^{s+t} \lambda^i P_{i-s}$. We may assume that A_0 and B_0 are given by formulas (3.2) and (3.3) (replace r by $s+t$, and ν_i by i). Put

$$\Delta_{0s} = \begin{pmatrix} B_0 & A_0B_0 & \cdots & A_0^{s-1}B_0 \end{pmatrix}.$$

Then it is easy to check (a) and (b) hold for this pair Δ_{0s} instead of Δ_s. Next we will relate the operators Δ_{0s} and Δ_s. Remark that $\lambda^s L(\lambda)$ and $\sum_{i=0}^{s+t} \lambda^i P_{i-s}$ are Wiener-Hopf equivalent. Apply Theorem 2.1 to see that the pairs (A, B) and (A_0, B_0) are feedback equivalent. Thus there exist invertible S and T such that $A_0S = S(A+BF)$ and $B_0T = SB$. Put $F_0 = -TF$ to get

$$B_0F_0 = SA - A_0S, \quad B_0T = SB.$$

Then one proves by induction that

$$\Delta_{0s} \begin{pmatrix} T & F_0B & F_0AB & \cdots & \cdots & \cdots & F_0A^{s-2}B \\ 0 & T & F_0B & \ddots & & & F_0A^{s-3}B \\ 0 & 0 & T & \ddots & \ddots & & F_0A^{s-4}B \\ \vdots & \vdots & & \ddots & \ddots & \ddots & \vdots \\ \vdots & \vdots & & & \ddots & \ddots & F_0AB \\ 0 & 0 & 0 & & & \ddots & F_0B \\ 0 & 0 & 0 & & & & T \end{pmatrix} = S\Delta_s.$$

From this formula it is clear that

$$\dim \operatorname{Ker} \Delta_s = \dim \operatorname{Ker} \Delta_{0s}, \quad \operatorname{codim} \operatorname{Im} \Delta_s = \operatorname{codim} \operatorname{Im} \Delta_{0s}.$$

This proves the theorem. □

The next theorem is the analogue of Theorem 5.1 for the right Wiener-Hopf factorization

$$L(\lambda) = E_+(\lambda) \left(\sum_{i=-s}^{t} \lambda^i Q_i \right) E_-(\lambda), \quad \lambda \in \Gamma, \tag{5.2}$$

where $E_+(\lambda)$ and $E_+(\lambda)^{-1}$ are holomorphic on Ω and continuous towards Γ, $E_-(\lambda)$ and $E_-(\lambda)^{-1}$ are holomorphic on the complement of $\bar{\Omega}$ and continuous towards Γ, and $Q_{-s}, \ldots, Q_t$ are mutually disjoint projections that add up to the identity on $\mathcal{Y}$. The proof of the next theorem is similar to the proof of Theorem 5.1, and therefore it is omitted.

Theorem 5.2. *Assume that $L(\lambda) = \sum_{i=-s}^{t} \lambda^i L_i$ admits a right Wiener-Hopf factorization (5.2). Let (C, A) be a right Γ-null pair of $\lambda^s L(\lambda)$ with underlying space $\mathcal{X}$, and consider the operator*

$$\Omega_s = \operatorname{col}\left(CA^{i-1}\right)_{i=1}^{s} : \mathcal{X} \to \mathcal{Y}^s.$$

Then for each $i \neq 0$ the projection Q_i is of finite rank if and only if Ω_s is a Fredholm operator. More generally,

(a) $\dim \operatorname{Ker} \Omega_s = \sum_{j=1}^{s} j(\operatorname{rank} Q_j)$,

(b) $\operatorname{codim} \operatorname{Im} \Omega_s = \sum_{j=1}^{s} j(\operatorname{rank} Q_{-j})$.

Theorem 5.3. *Assume that $L(\lambda) = \sum_{i=-s}^{s} \lambda^i L_i$ admits a left and a right Wiener-Hopf factorization with respect to the unit circle γ*

$$L(\lambda) = E_-(\lambda)\left(\sum_{i=-s}^{s} \lambda^i P_i\right) E_+(\lambda) = F_+(\lambda)\left(\sum_{i=-s}^{s} \lambda^i Q_i\right) F_-(\lambda), \quad \lambda \in \gamma. \tag{5.3}$$

Then for each $i \neq 0$ the projections P_i and Q_i are of finite rank if and only if the operator

$$H_s = \begin{pmatrix} R_{s-1} & \cdots & R_0 \\ \vdots & & \vdots \\ R_0 & \cdots & R_{1-s} \end{pmatrix} : \mathcal{Y}^s \to \mathcal{Y}^s \tag{5.4}$$

is Fredholm. Here

$$R_{-j} = \frac{1}{2\pi i} \int_\gamma \lambda^{j-1} L(\lambda)^{-1} d\lambda.$$

PROOF. Let (C_1, A_1, B_1) be a γ-spectral triple for the polynomial $\lambda^s L(\lambda)$, and let (C_2, A_2, B_2) be a γ-spectral triple for the polynomial $\lambda^s L(1/\lambda)$. Suppose that for each $i \neq 0$ the projections P_i and Q_i are of finite rank. From Theorems 5.1 and 5.2 we conclude that the operators

$$\Omega_s = \operatorname{col}\left(C_1 A_1^{i-1}\right)_{i=1}^{s}, \quad \Delta_s = \begin{pmatrix} B_1 & A_1 B_1 & \cdots & A_1^{s-1} B_1 \end{pmatrix}$$

are Fredholm. So $\Omega_s \Delta_s$ is a Fredholm operator. Furthermore, since $\lambda^{-s} L(\lambda)^{-1} - C_1(\lambda I - A_1)^{-1} B_1$ is analytic inside γ we get for $\alpha \geq 0$ that

$$C_1 A_1^\alpha B_1 = \frac{1}{2\pi i} \int_\gamma \lambda^\alpha C_1(\lambda I - A_1)^{-1} B_1 d\lambda = \frac{1}{2\pi i} \int_\gamma \lambda^\alpha \lambda^{-s} L(\lambda)^{-1} d\lambda = R_{s-\alpha-1},$$

and thus $H_s = \Omega_s \Delta_s$. So H_s is Fredholm.

Conversely, assume that H_s is Fredholm. Then $\operatorname{Ker}\Delta_s$ is finite dimensional and $\operatorname{Im}\Omega_s$ has finite codimension. Theorem 5.1 gives that $P_{-s}, \dots, P_{-1}$ are of finite rank and Theorem 5.2 gives that $Q_{-s}, \dots, Q_{-1}$ are of finite rank. Next note that

$$L(\frac{1}{\lambda}) = E_-(\frac{1}{\lambda})\left(\sum_{i=-s}^{s} \lambda^i P_{-i}\right)E_+(\frac{1}{\lambda}) = F_+(\frac{1}{\lambda})\left(\sum_{i=-s}^{s} \lambda^i Q_{-i}\right)F_-(\frac{1}{\lambda}), \qquad (5.5)$$

and these factorizations are, respectively, right and left Wiener-Hopf factorizations of $L(\frac{1}{\lambda})$ with respect to γ. Since $\lambda^{-s}L(1/\lambda)^{-1} - C_2((\lambda I - A_2)B_2$ is analytic inside γ, it follows that for $\alpha \geq 0$

$$\begin{aligned} C_2A_2^\alpha B_2 &= \frac{1}{2\pi i}\int_\gamma \lambda^\alpha C_2(\lambda I - A_2)B_2 d\lambda \\ &= \frac{1}{2\pi i}\int_\gamma \lambda^{\alpha-s}L(1/\lambda)^{-1}d\lambda \\ &= \frac{1}{2\pi i}\int_\gamma \lambda^{-\alpha+s-2}L(\lambda)^{-1}d\lambda = R_{-s+\alpha+1}. \end{aligned}$$

So we get that

$$H_s = \operatorname{col}\big(C_2A_2^{i-1}\big)_{i=1}^{s} \operatorname{row}\big(A_2^{i-1}B_2\big)_{i=1}^{s}.$$

As H_s is Fredholm, we see that $\operatorname{Ker}(\operatorname{row}\big(A_2^{i-1}B_2\big)_{i=1}^{s})$ is finite dimensional and that the space $\operatorname{Im}(\operatorname{col}\big(C_2A_2^{i-1}\big)_{i=1}^{s})$ has finite codimension. From (5.5) and Theorems 5.1 and 5.2 we may conclude that $P_1, \dots, P_s$ and $Q_1, \dots, Q_s$ are of finite rank. □

We conclude this section with a remark about the uniqueness of the diagonal term in (5.1). Suppose that $L(\lambda) = \sum_{i=-s}^{t} \lambda^i L_i$ admits a Wiener-Hopf factorization (5.1), and let us assume that for each $i \neq 0$ the projections P_i are of finite rank. Then the numbers $\operatorname{rank} P_{-s}, \dots, \operatorname{rank} P_{-1}$ and $\operatorname{rank} P_1, \dots, \operatorname{rank} P_t$ determine uniquely the left Wiener-Hopf equivalence class of $L(\lambda)$. Indeed, consider two diagonal factors

$$D(\lambda) = \sum_{i=-s}^{t} \lambda^i P_i, \quad D'(\lambda) = \sum_{i=-s}^{t} \lambda^i P_i',$$

and assume that $\operatorname{rank} P_i = \operatorname{rank} P_i'$ for $i \neq 0$. Then $\operatorname{rank}(I - P_O) = \operatorname{rank}(I - P_0')$. So for each value of i we can find an invertible operator S_i such that

$$S_iP_iS_i^{-1} = P_i', \quad i = -s, \dots, t.$$

Put $E = \sum_{j=-s}^{t} P_j'S_jP_j$. Then E is invertible and $ED(\lambda) = D'(\lambda)E$.

Notes

The material of Section 1 has its roots in Gohberg-Lerer-Rodman [1] and Rowley [1]. The results on Wiener-Hopf equivalence in Section 2 are taken from Gohberg-Kaashoek-Van Schagen [2]. The first theorems on Wiener-Hopf factorization of operator polynomials were given in Rowley [1]. The theorems and their proofs as presented here in Section 3 are taken from Gohberg-Kaashoek-Van Schagen [2]. Corollary 3.3 is due to Rowley [1]. An earlier version of Corollary 3.3, dealing with spectral factorization of matrix polynomials, is due to Lopatinskii [1]. The explicit description of the factorization indices as well as the results in Sections 4 and 5 are due to Gohberg-Kaashoek-Van Schagen [2]. The first results of this type (relating the indices to the spectral pair) were obtained by Rowley [1]. For a comprehensive treatment of operator polynomials we refer to Rodman [1].

Chapter XIV
Factorization of Analytic Operator Functions

This chapter carries out a similar program as in the previous chapter, but now for analytic operator-valued functions. Wiener-Hopf equivalence and Wiener-Hopf factorization are described in terms block similarity of certain operator blocks and their canonical forms.

XIV.1 Preliminaries on spectral triples

Throughout this chapter Γ will be a closed rectifiable Jordan curve in the complex plane with a bounded inner domain Ω. Also we assume that $0 \in \Omega$. Operators in this chapter are assumed to be linear and bounded. Throughout this chapter $\mathcal{Y}$ will denote a fixed Banach space. We consider $\mathcal{L}(\mathcal{Y})$-valued operator functions on the closure $\overline{\Omega}$ of the domain Ω, i.e., $W : \overline{\Omega} \to L(\mathcal{Y})$, with W analytic in Ω and continuous up to the boundary. Furthermore, we assume that $W(\lambda)$ is invertible for all λ on the boundary Γ.

A triple (C, A, B) of operators $A : \mathcal{X} \to \mathcal{X}, \quad B : \mathcal{Y} \to \mathcal{X}, \quad C : \mathcal{Y} \to \mathcal{X}$, is called a Γ-*spectral triple* for the function W if

(a) $\sigma(A) \subset \Omega$;

(b) $W(\lambda)C(\lambda I - A)^{-1}$ has an analytic continuation on the whole of Ω;

(c) $\bigcap_{j=0}^{\infty} \operatorname{Ker} CA^j = \{0\}$;

(d) $W(\lambda)^{-1} - C(\lambda I - A)^{-1}B$ has an analytic continuation on the whole of Ω.

The space $\mathcal{X}$ may depend on the function W and will be referred to as the *base space* of the triple.

If W is an operator polynomial, then the above definition coincides with the one given in the previous chapter. Indeed, if (a)–(c) hold for an operator polynomial W, then the pair (C, A) is a right Γ-null pair of W, and, by Corollary XIII.1.4, the additional condition (d) implies that (C, A, B) is a Γ-spectral triple as defined in Section XIII.1. Conversely, if (C, A, B) is a Γ-spectral triple of an operator polynomial W in the sense of Section XIII.1, then (a)–(d) above are fulfilled. We conclude that the definition of a Γ-spectral triple for an analytic operator function extends the definition of a Γ-spectral triple of an operator polynomial.

The first result shows that Γ-spectral triples exist and how one may construct such a triple.

Theorem 1.1. *Let $W : \Omega \cup \Gamma \to L(\mathcal{Y})$ be analytic on Ω, continuous up to the boundary Γ of Ω, and invertible on Γ. Let $C(\Gamma, \mathcal{Y})$ be the Banach space of all*

$\mathcal{Y}$-valued functions endowed with the supremum norm. Let $S : C(\Gamma,\mathcal{Y}) \to \mathcal{Y}$, $V : C(\Gamma,\mathcal{Y}) \to C(\Gamma,\mathcal{Y})$ and $R : \mathcal{Y} \to C(\Gamma,\mathcal{Y})$ be defined by

$$Sf = \frac{1}{2\pi i}\int_\Gamma (I - W(z))f(z)dz, \quad Vf(z) = zf(z), \quad (Ry)(z) = y. \tag{1.1}$$

Then there exists a contour γ in Ω around $\sigma(V-RS)\cap\Omega$ and with

$$\Pi = \frac{1}{2\pi i}\int_\gamma \big(\lambda I - (V - RS)\big)^{-1} d\lambda, \tag{1.2}$$

the triple $(-S\Pi, \Pi(V-RS)\Pi, \Pi R)$ with base space $\mathcal{X} = \operatorname{Im}\Pi$ is a Γ-spectral triple of W.

PROOF. According to Theorem 2.4 and Corollary 2.7 from Bart-Gohberg-Kaashoek [1] we have that for each $\lambda \in \Omega$,

$$W(\lambda) = I + S(\lambda I - V)^{-1}R, \tag{1.3}$$

where S, V and R are defined by (1.1). Moreover,

$$\sigma(V - RS)\cap\Omega = \{\lambda\in\Omega \mid W(\lambda) \text{ is not invertible}\}. \tag{1.4}$$

Since W is invertible on Γ and continuous on $\Omega\cup\Gamma$, it follows from (1.4) that $\sigma(V - RS)\cap\Omega$ is an open and closed subset of $\sigma(V-RS)$. Hence the Riesz projection Π in (1.2) is well defined. Put

$$C = -S\Pi, \quad A = \Pi(V - RS)\Pi, \quad B = \Pi R. \tag{1.5}$$

From the definition of A it is clear that $\sigma(A) = \sigma(V-RS)\cap\Omega$. In particular we have that $\sigma(A)\subset\Omega$. Write $T = V - RS$. For $\lambda\notin\sigma(A)$ we find that

$$W(\lambda)C(\lambda I - A)^{-1} = -W(\lambda)S\Pi(\lambda I - T)^{-1}\Pi.$$

Since $\operatorname{Im}\Pi$ is T-invariant,

$$W(\lambda)C(\lambda I - A)^{-1} = -W(\lambda)S(\lambda I - T)^{-1}\Pi$$

The representation (1.3) implies $W(\lambda)^{-1} = I - R\big(\lambda I - (V-RS)\big)^{-1}S$. So we have

$$W(\lambda)C(\lambda I - A)^{-1} = -S(\lambda I - T)^{-1}\Pi - S(\lambda I - V)^{-1}RS(\lambda I - T)^{-1}\Pi.$$

Substituting $RS = (\lambda I - T) - (\lambda I - V)$ in this equality results in

$$W(\lambda)C(\lambda I - A)^{-1} = -S(\lambda I - V)^{-1}\Pi.$$

Since $\sigma(V)\cap\Omega = \emptyset$, we see that $W(\lambda)C(\lambda I - A)^{-1}$ has an analytic extension on Ω.

Put $\mathcal{M} = \bigcap_{j=0}^{\infty} \operatorname{Ker} CA^j$. As $\mathcal{M} \subset \operatorname{Im}\Pi$, we have that $\mathcal{M}$ is T-invariant and $\sigma(T|_{\mathcal{M}}) \subset \Omega$. For $x \in \mathcal{M}$ one has that $-ST^jx = -ST^j\Pi x = CA^jx$. It follows that $\mathcal{M} \subset \bigcap_{j=0}^{\infty} \operatorname{Ker} ST^j$. So $Tx = (V - RS)x = Vx$ for each $x \in \mathcal{M}$, and thus also $Vx \in \mathcal{M}$ for each $x \in \mathcal{M}$. Furthermore, $\sigma(V|_{\mathcal{M}}) = \sigma(T|_{\mathcal{M}}) \subset \Omega$. Let γ' be a contour in Ω around $\sigma(V|_{\mathcal{M}})$. Then for each $x \in \mathcal{M}$

$$x = \frac{1}{2\pi i}\int_{\gamma'} (\lambda I - V|_{\mathcal{M}})^{-1} x d\lambda = \frac{1}{2\pi i}\int_{\gamma'} (\lambda I - V)^{-1} x d\lambda = 0.$$

In this last equality we use that $\sigma(V) \cap \Omega = \emptyset$. So $\mathcal{M} = \bigcap_{j=0}^{\infty} \operatorname{Ker} CA^j = \{0\}$.

Finally, for $\lambda \in \Omega$ one has

$$W(\lambda)^{-1} - C(\lambda I - A)^{-1}B = I - S(\lambda I - T)^{-1}\Pi R,$$

and hence this function has an analytic continuation on Ω. □

The conditions in the definition of a Γ-spectral triple look non-symmetric. The following result resolves some of this asymmetry.

Lemma 1.2. *Let (C, A, B) be a Γ-spectral triple of W. Then $(\lambda I - A)^{-1}BW(\lambda)$ has an analytic continuation on the whole of Ω.*

PROOF. Let γ be a contour in Ω containing $\sigma(A)$ in its interior domain. Let $z \in \Omega$ be a point outside γ. We have to show that for each $y \in \mathcal{Y}$ the function

$$\phi(z) = \frac{1}{2\pi i}\int_{\gamma} (z - \lambda)^{-1}(\lambda I - A)^{-1}BW(\lambda) y d\lambda = 0.$$

Use that $\lambda^n I - A^n = \left(\sum_{k=0}^{n-1} \lambda^k A^{n-1-k}\right)(\lambda I - A)$ for each $n \geq 0$ to see that

$$A^n\phi(z) = \frac{1}{2\pi i}\int_{\gamma} \frac{A^n(\lambda I - A)^{-1}BW(\lambda)}{z - \lambda} y d\lambda = \frac{1}{2\pi i}\int_{\gamma} \frac{\lambda^n(\lambda I - A)^{-1}BW(\lambda)}{z - \lambda} y d\lambda$$

Thus

$$CA^n\phi(z) = \frac{1}{2\pi i}\int_{\gamma} \frac{\lambda^n C(\lambda I - A)^{-1}BW(\lambda)}{z - \lambda} y d\lambda = \frac{1}{2\pi i}\int_{\gamma} \frac{\lambda^n}{z - \lambda} y d\lambda = 0,$$

since $C(\lambda I - A)^{-1}B - W(\lambda)^{-1}$ has an analytic continuation on Ω. So property (c) gives that $\phi(z) = 0$. □

Although spectral triples may be constructed in many different ways, they are all similar. This is the contents of the next theorem.

Theorem 1.3. *Assume that (C_1, A_1, B_1) with base space $\mathcal{X}_1$ and (C_2, A_2, B_2) with base space $\mathcal{X}_2$ are Γ-spectral triples for $W(\lambda)$ on Ω. Then there exists a unique invertible bounded linear operator $S : \mathcal{X}_1 \to \mathcal{X}_2$ with the property that*

$$C_2S = C_1, \quad A_2S = SA_1, \quad B_2 = SB_1.$$

PROOF. Let γ be a contour in Ω containing $\sigma(A_1)$ and $\sigma(A_2)$ in its interior. Put

$$S = \frac{1}{2\pi i}\int_\gamma (\lambda I - A_2)^{-1}B_2W(\lambda)C_1(\lambda I - A_1)^{-1}d\lambda, \tag{1.6}$$

$$T = \frac{1}{2\pi i}\int_\gamma (\lambda I - A_1)^{-1}B_1W(\lambda)C_2(\lambda I - A_2)^{-1}d\lambda.$$

We shall prove that S is invertible with $S^{-1} = T$, and that S provides the desired similarity. Note that the definition of the operators T and S does not depend on the particular choice of contour γ, provided it meets the conditions. So we choose a contour γ' with $\sigma(A_1)$, $\sigma(A_2)$ and γ in its interior. Then

$$\begin{aligned} TS &= \frac{1}{2\pi i}\int_{\gamma'} (\mu I - A_1)^{-1}B_1W(\mu)C_2(\mu I - A_2)^{-1}S d\mu \\ &= \left(\frac{1}{2\pi i}\right)^2 \int_{\gamma'}\Big(\int_\gamma (\mu I - A_1)^{-1}B_1W(\mu)C_2(\mu I - A_2)^{-1}(\lambda I - A_2)^{-1}B_2 \cdot \\ &\qquad\qquad \cdot W(\lambda)C_1(\lambda I - A_1)^{-1}d\lambda\Big)d\mu. \end{aligned}$$

Use the resolvent identity

$$(\mu I - A_2)^{-1}(\lambda I - A_2)^{-1} = \frac{1}{\lambda - \mu}\left((\mu I - A_2)^{-1} - (\lambda I - A_2)^{-1}\right)$$

to rewrite the integrand in the right hand side as

$$\begin{aligned} &\frac{1}{\lambda-\mu}(\mu I - A_1)^{-1}B_1W(\mu)C_2(\mu I - A_2)^{-1}B_2W(\lambda)C_1(\lambda I - A_1)^{-1} + \\ &- \frac{1}{\lambda-\mu}(\mu I - A_1)^{-1}B_1W(\mu)C_2(\lambda I - A_2)^{-1}B_2W(\lambda)C_1(\lambda I - A_1)^{-1}. \end{aligned}$$

For a fixed μ on the contour γ' the first term is analytic in the interior of γ. So the double integral of the first term is equal to zero. To integrate the second term we interchange the order of integration. By Lemma 1.2 the operator function $(\mu I - A_1)^{-1}B_1W(\mu)$ is analytic on the whole of Ω. So

$$TS = \frac{1}{2\pi i}\int_\gamma (\lambda I - A_1)^{-1}B_1W(\lambda)C_2\lambda I - A_2)^{-1}B_2W(\lambda)C_1(\lambda I - A_1)^{-1}d\lambda.$$

Since $W(\lambda)^{-1} - C_2\lambda I - A_2)^{-1}B_2$ is analytic in Ω, we obtain that

$$TS = \frac{1}{2\pi i}\int_\gamma (\lambda I - A_1)^{-1}B_1W(\lambda)C_1(\lambda I - A_1)^{-1}d\lambda.$$

For $n \geq 0$ we multiply this identity by the operator $C_1A_1^n$ from the left. Next apply the formula $A_1(\lambda I - A_1)^{-1} = \lambda(\lambda I - A_1)^{-1} - I$ and use n times condition (b) in the definition of Γ-spectral triple. This gives for $n \geq 0$

$$C_1A_1^nTS = \frac{1}{2\pi i}\int_\gamma \lambda^n C_1(\lambda I - A_1)^{-1}B_1W(\lambda)C_1(\lambda I - A_1)^{-1}d\lambda.$$

Use that $C_1(\lambda I - A_1)^{-1}B_1 - W(\lambda)^{-1}$ and $W(\lambda)C_1(\lambda I - A_1)^{-1}$ are analytic in Ω, to see that

$$C_1A_1^nTS = C_1\frac{1}{2\pi i}\int_\gamma \lambda^n(\lambda I - A_1)^{-1}d\lambda \quad (n \geq 0).$$

Since γ contains $\sigma(A_1)$ in its interior, we get $C_1A_1^nTS = C_1A_1^n$, $\quad (n \geq 0)$. This together with condition (c) gives $TS = I_{\mathcal{X}_1}$. In the same way one proves that $ST = I_{\mathcal{X}_2}$ Hence, S is invertible and $S^{-1} = T$.

Next we check that $C_2S = C_1$. We compute

$$\begin{aligned} C_2S &= \frac{1}{2\pi i}\int_\gamma C_2(\lambda I - A_2)^{-1}B_2W(\lambda)C_1(\lambda I - A_1)^{-1}d\lambda \\ &= \frac{1}{2\pi i}\int_\gamma C_1(\lambda I - A_1)^{-1}d\lambda + \\ &+ \frac{1}{2\pi i}\int_\gamma (C_2(\lambda I - A_2)^{-1}B_2 - W(\lambda)^{-1})W(\lambda)C_1(\lambda I - A_1)^{-1}d\lambda. \end{aligned}$$

The second integral on the right hand side of this equality is equal to zero because of conditions (b) and (d) in the definition of Γ-spectral triple, and the first integral on the right hand side gives C_1 since $\sigma(A)$ is in the interior domain of γ. So $C_2S = C_1$. In a similar way one proves that $B_2 = SB_1$. Furthermore

$$\begin{aligned} A_2S &= \frac{1}{2\pi i}\int_\gamma A_2(\lambda I - A_2)^{-1}B_2W(\lambda)C_1(\lambda I - A_1)^{-1}d\lambda \\ &= \frac{1}{2\pi i}\int_\gamma \lambda(\lambda I - A_2)^{-1}B_2W(\lambda)C_1(\lambda I - A_1)^{-1}d\lambda, \end{aligned}$$

because the difference between the integrands is $B_2W(\lambda)C_1(\lambda I - A_1)^{-1}$, which is analytic in Ω. A similar argument, based on Lemma 1.2, gives that

$$\begin{aligned} SA_1 &= \frac{1}{2\pi i}\int_\gamma (\lambda I - A_2)^{-1}B_2W(\lambda)C_1(\lambda I - A_1)^{-1}A_1d\lambda \\ &= \frac{1}{2\pi i}\int_\gamma (\lambda I - A_2)^{-1}B_2W(\lambda)C_1(\lambda I - A_1)^{-1}\lambda d\lambda, \end{aligned}$$

and hence $A_2S = SA_1$.

Finally, assume that also $C_2S' = C_1$ and $A_2S' = S'A_1$. Then $C_2A_2^nS = C_1A_1^n = C_2A_2^nS'$ for each $n \geq 0$. So $C_2A_2^n(S - S')x = 0$ for each $n \geq 0$ and each $x \in \mathcal{X}$. Condition (c) shows that $S = S'$, and thus the uniqueness of the invertible operator S is proven. □

From formula (1.6) for the similarity S and the uniqueness of S we get the following identity

$$\frac{1}{2\pi i}\int_\gamma (\lambda I - A)^{-1}BW(\lambda)C(\lambda I - A)^{-1}d\lambda = I.$$

XIV.2 Wiener-Hopf equivalence

Let $W_1 : \Omega \cup \Gamma \to L(\mathcal{Y})$ and $W_2 : \Omega \cup \Gamma \to L(\mathcal{Y})$ be analytic on Ω and continuous on $\Omega \cup \Gamma$, with invertible values on Γ. The functions W_1 and W_2 are called *left Wiener-Hopf equivalent with respect to* Γ if W_1 and W_2 are related in the following way:

$$W_2(\lambda) = E_-(\lambda) W_1(\lambda) E_+(\lambda), \quad \lambda \in \Gamma, \tag{2.1}$$

where E_- and E_+ are analytic on $\mathbb{C}_\infty \setminus (\Omega \cup \Gamma)$ and Ω, respectively, both E_- and E_+ are continuous up to the boundary Γ, the operator $E_-(\lambda)$ is invertible for each $\lambda \in \mathbb{C}_\infty \setminus \Omega$ and $E_+(\lambda)$ is invertible for each $\lambda \in \Omega \cup \Gamma$. We call (2.1) a *left Wiener-Hopf factorization* if

$$W_1(\lambda) = D(\lambda) = \sum_{j=1}^{r} \lambda^{\nu_j} P_j,$$

where $\nu_1 \geq \nu_2 \geq \cdots \geq \nu_r \geq 0$ are integers and $P_1, \ldots, P_r$ are mutually disjoint projections of $\mathcal{Y}$ such that $\sum_{j=1}^{r} P_j = I_{\mathcal{Y}}$. To define right Wiener-Hopf equivalence with respect to Γ and right Wiener-Hopf factorization we just have to reverse the order of the factors in the right hand side of (2.1). The following theorem is the analogue of Theorem XIII.2.1 and can be proved in exactly the same way as in the operator polynomial case.

Theorem 2.1. *For $i = 1, 2$, let (C_i, A_i, B_i) be a Γ-spectral triple for the operator function W_i, and let $\mathcal{X}_i$ be its base space. Let P_i be the projection of $\mathcal{X}_i \oplus \mathcal{Y}$ along $\{0\} \oplus \mathcal{Y}$ onto $\mathcal{X}_i \oplus \{0\}$, and put*

$$Z_i = \begin{pmatrix} A_i & B_i \end{pmatrix} : \mathcal{X}_i \oplus \mathcal{Y} \to \mathcal{X}_i \oplus \{0\}.$$

Then the following statements are equivalent:

1. *the operator functions W_1 and W_2 are left Wiener-Hopf equivalent with respect to Γ;*
2. *the blocks $(Z_1; P_1, I_{\mathcal{X}_1 \oplus \mathcal{Y}})$ and $(Z_2; P_2, I_{\mathcal{X}_2 \oplus \mathcal{Y}})$ are block similar;*
3. *the linear pencils $\lambda\begin{pmatrix} I_{\mathcal{X}_1} & 0 \end{pmatrix} + \begin{pmatrix} A_1 & B_1 \end{pmatrix}$ and $\lambda\begin{pmatrix} I_{\mathcal{X}_2} & 0 \end{pmatrix} + \begin{pmatrix} A_2 & B_2 \end{pmatrix}$ are strictly equivalent;*
4. *the pairs (A_1, B_1) and (A_2, B_2) are feedback equivalent.*

Assume that condition (1) in Theorem 2.1 is fulfilled, and let (2.1) be the left Wiener-Hopf equivalence. Put

$$N = \frac{1}{2\pi i} \int_\Gamma (\mu I - A_2)^{-1} B_2 W_2(\mu) E_+(\mu)^{-1} C_1 (\mu I - A_1)^{-1} d\mu,$$

$$F = \frac{1}{2\pi i} \int_\Gamma E_-(\mu) W_1(\mu) C_1 (\mu I - A_1)^{-1} d\mu,$$

and $M = E_-(\infty)$. Then we know from the proof of Theorem 2.1 in Section XIII.2 that the block similarity in (2) is given by

$$S = \begin{pmatrix} N & 0 \\ F & M \end{pmatrix} : \mathcal{X}_1 \oplus \mathcal{Y} \to \mathcal{X}_2 \oplus \mathcal{Y}.$$

Conversely, if condition (2) holds true, then we can express the function $E_-(\lambda)$ in terms of the block similarity and the Γ-spectral triple:

$$E_-(\lambda) = M + F(\lambda I - A_1)^{-1}B_1.$$

We omit the formulation of the analogous results for right Wiener-Hopf equivalence.

XIV.3 Wiener-Hopf factorization

The first result in this section is analogous to the Theorem XIII.3.1 for matrix polynomials. Also the proof is the same. Therefore we just state the result and omit the proof.

Theorem 3.1. *Let $W : \Omega \cup \Gamma \to L(\mathcal{Y})$ be analytic in Ω, continuous on $\Omega \cup \Gamma$, and assume that $W(\lambda)$ is invertible for each $\lambda \in \Gamma$. Let (C, A, B) be a Γ-spectral triple with base space $\mathcal{X}$ for W. Then W admits a left Wiener-Hopf factorization with respect to Γ*

$$W(\lambda) = E_-(\lambda)\left(\sum_{i=1}^{r} \lambda^{\nu_i} P_i\right) E_+(\lambda) \tag{3.1}$$

if and only if the pair (A, B) is of finite type. Furthermore, if the projections $P_1, \ldots, P_r$ are different from zero and $\nu_1 > \nu_2 > \cdots > \nu_r$, then the set

$$\{(\operatorname{Im} P_1, \nu_1), \ldots, (\operatorname{Im} P_r, \nu_r)\}$$

is equal to the characteristics of the pair (A, B).

There is also a similar result for right Wiener-Hopf factorization (cf., Theorem XIII.3.2).

For the remainder of this section we fix a realization of W, namely

$$W(\lambda) = I + R(\lambda I - V)^{-1}S, \quad \lambda \in \Omega,$$

where $\sigma(V) \cap \Omega = \emptyset$. If $W(\lambda)$ is invertible for each λ in a neighbourhood of Γ in Ω, then there exists a contour γ in Ω around $\sigma(V - SR) \cap \Omega$. To see this remark that

$$\begin{pmatrix} I & -R \\ 0 & \lambda I - V \end{pmatrix} \begin{pmatrix} W(\lambda) & 0 \\ (\lambda I - V)^{-1}S & I \end{pmatrix} = \begin{pmatrix} I & 0 \\ S & I \end{pmatrix} \begin{pmatrix} I & -R \\ 0 & \lambda I - (V - RS) \end{pmatrix}.$$

Since $\lambda I - V$ is invertible for each $\lambda \in \Omega$, the first factors in the left hand and right hand sides of this equality are invertible. So the second factor of the left hand side is invertible if and only if the second factor of the right hand is invertible. Thus we see that $\lambda I - (V - RS)$ is invertible for each λ in a neighbourhood of Γ in Ω, and hence there exists a contour γ in Ω around $\sigma(V-SR)\cap\Omega$. Our aim now is to give explicit formulas for the factors $E_-(\lambda)$ and $E_+(\lambda)$ in terms of the operators S, T and V.

Theorem 3.2. *Let $W:\Omega\cup\Gamma\to L(\mathcal{Y})$ be analytic in Ω , continuous on $\Omega\cup\Gamma$, and let $W(\lambda)$ be invertible for each $\lambda\in\Gamma$. Let $W(\lambda)=I+R(\lambda I-V)^{-1}S$ for all $\lambda\in\Omega$, where $V:\mathcal{Z}\to\mathcal{Z}$ is a bounded linear operator on the Banach space $\mathcal{Z}$ with no spectrum in Ω. Let γ be a contour in Ω around $\sigma(V-SR)\cap\Omega$, the part of the spectrum of $V-RS$ inside Ω. Put*

$$\Delta_j=\operatorname{row}\left(\frac{1}{2\pi i}\int_\gamma \lambda^{k-1}\big(\lambda I-(V-RS)\big)^{-1}Rd\lambda\right)_{k=1}^{j}:\mathcal{Y}^j\to\mathcal{Z}.$$

Then $W(\lambda)$ admits a left Wiener-Hopf factorization with respect to Γ if and only if for some integer $\ell\geq 1$ the operators $\Delta_1,\cdots,\Delta_\ell$ have generalized inverses and the following equality holds

$$\operatorname{Im}\Delta_\ell=\operatorname{Im}\left(\frac{1}{2\pi i}\int_\gamma\big(\lambda I-(V-RS)\big)^{-1}Rd\lambda\right). \tag{3.2}$$

Furthermore, in that case there exist mutually disjoint projections $P_0,P_1,\ldots,P_\ell$ of $\mathcal{Y}$ and bounded linear operators $F,G:\mathcal{Z}\to\mathcal{Y}$ such that with

$$E_-(\lambda)=I-F\big(\lambda I-(V-RS-RF)\big)^{-1}R, \tag{3.3}$$

$$E_+(\lambda)=P_0+(P_0S+P_0F+G)(\lambda I-V)^{-1}R \tag{3.4}$$

the function W has the following left Wiener-Hopf factorization with respect to Γ:

$$W(\lambda)=E_-(\lambda)\left(\sum_{i=1}^{r}\lambda^iP_i\right)E_+(\lambda).$$

PROOF. Put $T=V-RS$. Let, as before, Π denote the Riesz projection corresponding to $\sigma(T)\cap\Omega$,

$$\Pi=\frac{1}{2\pi i}\int_\gamma(\lambda I-(V-RS))^{-1}d\lambda.$$

Put $\mathcal{X}=\operatorname{Im}\Pi$. Then the triple (C,A,B) defined as

$$C=-S\Pi:\mathcal{X}\to\mathcal{Y},\quad A=\Pi T\Pi:\mathcal{X}\to\mathcal{X},\quad B=\Pi R:\mathcal{Y}\to\mathcal{X}$$

is a Γ-spectral triple of W. This one proves by the arguments that were used to derive Theorem 1.1 from Theorem 1.2. Let $\tau:\mathcal{X}\to\mathcal{Z}$ denote the canonical embedding. Then

$$\frac{1}{2\pi i}\int_\gamma T^k(\lambda I-T)^{-1}Rd\lambda=T^k\Pi R=(\Pi T\Pi)^k\Pi R=\tau A^kB.$$

So

$$\Delta_j = \tau \cdot \operatorname{row}\left(A^{k-1}B\right)_{k=1}^{j}.$$

Therefore Δ_j has a generalized inverse if and only if row $\left(A^{k-1}B\right)_{k=1}^{j}$ has a generalized inverse. Furthermore, $\operatorname{Im}\Delta_\ell = \operatorname{Im}\Pi$ if and only if row $\left(A^{k-1}B\right)_{k=1}^{\ell}$ is surjective. So (3.2) is equivalent to the surjectivity of row $\left(A^{k-1}B\right)_{k=1}^{\ell}$. We conclude that the first part of the theorem follows from Theorem 3.1.

Next assume that $W(\lambda)$ admits a left Wiener-Hopf factorization with respect to Γ. Thus we may assume that the pair (A, B) is of finite type. We apply Theorem XII.2.5, and get closed subspaces $\mathcal{U}_0, \ldots, \mathcal{U}_\ell$ of $\mathcal{Y}$ with $\mathcal{Y} = \mathcal{U}_0 \oplus \cdots \oplus \mathcal{U}_\ell$, and bounded linear operators

$$N : \mathcal{U}_1^1 \oplus \mathcal{U}_2^2 \oplus \cdots \oplus \mathcal{U}_\ell^\ell \to \mathcal{X}, \quad F_0 : \mathcal{X} \to \mathcal{Y},$$

(where $\mathcal{U}_j^j$ stands for the direct sum of j copies of $\mathcal{U}_j$) such that N is invertible and

(i) $N^{-1}B = E_0 \oplus \cdots \oplus E_\ell : \mathcal{U}_1^1 \oplus \mathcal{U}_2^2 \oplus \cdots \oplus \mathcal{U}_\ell^\ell \to \mathcal{U}_1^1 \oplus \mathcal{U}_2^2 \oplus \cdots \oplus \mathcal{U}_\ell^\ell$, with $E_0 = 0$ and

$$E_j = \begin{pmatrix} I_j \\ 0 \\ \vdots \\ 0 \end{pmatrix} : \mathcal{U}_j \to \mathcal{U}_j^j, \quad j = 1, \ldots, \ell$$

(ii) $N^{-1}(A - BF_0)N = J_1 \oplus \cdots \oplus J_\ell : \mathcal{U}_1^1 \oplus \mathcal{U}_2^2 \oplus \cdots \oplus \mathcal{U}_\ell^\ell \to \mathcal{U}_1^1 \oplus \mathcal{U}_2^2 \oplus \cdots \oplus \mathcal{U}_\ell^\ell$, with $J_1 = 0$ and

$$J_j = \begin{pmatrix} 0 & 0 & \cdots & 0 & 0 \\ I_j & 0 & \cdots & 0 & 0 \\ & \ddots & \ddots & \vdots & \vdots \\ 0 & 0 & \ddots & 0 & 0 \\ 0 & 0 & & I_j & 0 \end{pmatrix} : \mathcal{U}_j^j \to \mathcal{U}_j^j,$$

for $j = 2, \ldots, \ell$ and $\mathcal{U}_0 = \operatorname{Ker} B$.

Let P_j be the projection of $\mathcal{Y}$ onto $\mathcal{U}_j$ along the direct sum of the subspaces $\mathcal{U}_i$, $i \neq j$. Define $G_0 : \mathcal{X} \to \mathcal{Y}$ by

$$G_0 N = D_1 \oplus \cdots \oplus D_\ell : \mathcal{U}_1^1 \oplus \mathcal{U}_2^2 \oplus \cdots \oplus \mathcal{U}_\ell^\ell \to \mathcal{U}_0 \oplus \mathcal{U}_1 \oplus \cdots \oplus \mathcal{U}_\ell$$

with

$$D_j = \begin{pmatrix} 0 & 0 & \cdots & I_j \end{pmatrix} : \mathcal{U}_j^j \to \mathcal{U}_j.$$

Then

$$P_0 + G_0\left(\lambda I - (A - BF_0)\right)^{-1} B = \sum_{i=0}^{\ell} \lambda^{-i} P_i. \tag{3.5}$$

Write $D(\lambda) = \sum_{i=0}^{\ell} \lambda^i P_i$. Then the left hand side of (3.5) is equal to $D(\lambda)^{-1}$. Furthermore $(G_0, A - BF_0, B)$ is a Γ-spectral triple for $D(\lambda)$. Apply Theorem 2.1 (see also the remarks made after Theorem 2.1) and conclude that

$$W(\lambda) = E_-(\lambda)D(\lambda)E_+(\lambda), \quad \lambda \in \Gamma,$$

where $E_-(\lambda) = I - F_0\big(\lambda I - (A - BF_0)\big)^{-1}B$, and $E_-(\lambda)^{-1} = I + F_0(\lambda I - A)^{-1}B$. Next we want to replace the operators A and B by the operators S, T and R. Define $F = F_0\Pi : \mathcal{Z} \to \mathcal{Y}$. From $B = \Pi R$ and $A = \Pi T \Pi$ we get

$$E_-(\lambda)^{-1} = I + F(\lambda I - T)^{-1}R,$$

and

$$E_-(\lambda) = I - F\big(\lambda I - (T - RF)\big)^{-1}R.$$

We have proved formula (3.3).

Next we will prove (3.4). We first compute that

$$\begin{aligned} D(\lambda)^{-1}E_-(\lambda)^{-1} =& \big(P_0 + G_0\big(\lambda I - (A - BF_0)\big)^{-1}B\big)\big(I + F_0(\lambda I - A)^{-1}B\big) = \\ =& P_0 + P_0F_0(\lambda I - A)^{-1}B + G_0\big(\lambda I - (A - BF_0)\big)^{-1}B + \\ & + G_0\big(\lambda I - (A - BF_0)\big)^{-1}BF_0(\lambda I - A)^{-1}B = \\ =& P_0 + (P_0F_0 + G_0)(\lambda I - A)^{-1}B. \end{aligned}$$

Define $G : \mathcal{Z} :\to \mathcal{Y}$ by $G = G_0\Pi$, and recall that $\tau(\lambda I - A)^{-1}B = \Pi(\lambda I - T)^{-1}R$. It follows that

$$D(\lambda)^{-1}E_-(\lambda)^{-1} = P_0 + (P_0F + G)(\lambda I - T)^{-1}R.$$

Finally, we multiply this equality on the right by $W(\lambda)$ and obtain

$$\begin{aligned} E_+(\lambda) =& D(\lambda)^{-1}E_-(\lambda)^{-1}W(\lambda) \\ =& \big(P_0 + (P_0F + G)(\lambda I - T)^{-1}R\big)\big(I + S(\lambda I - V)^{-1}R\big) \\ =& P_0 + P_0S(\lambda I - V)^{-1}R + (P_0F + G)\big(\lambda I - (V - RS)\big)^{-1}R + \\ & + (P_0F + G)\big(\lambda I - (V - RS)\big)^{-1}RS(\lambda I - V)^{-1}R \\ =& P_0 + (P_0S + P_0F + G)(\lambda I - V)^{-1}R, \end{aligned}$$

which proves (3.4). □

Notes

Theorem 1.1 is based on Theorem 2.4 and Corollary 2.7 in Bart-Gohberg-Kaashoek [1]. Theorem 1.3 is Theorem 1.2 from Kaashoek-Van der Mee-Rodman [1]. The results in Sections 2 and 3 are taken from Chapter II in Gohberg-Kaashoek-Van Schagen [2].

Chapter XV
Eigenvalue Completion Problems for Triangular Matrices

In this chapter the theory developed in the previous chapters is extended to a class of partially specified matrices with a non-block pattern. It deals with matrices of which the upper triangular part is given and the elements in the strictly lower triangular part are considered as unspecified. For such partially specified matrices we consider problems that are similar in nature to the ones considered earlier for operator blocks. The class of admissible similarities consists here of the U-similarities, which are the natural analogies of block similarities in the context of this chapter. In principle, we carry out the same program as for operator blocks. For example, we describe the invariants and the canonical form for certain equivalence classes under U-similarity. The analogue of the eigenvalue completion problem appears in a natural way and is solved for the case when the multiplicities are not taken into account. As an application a full solution of the spectral radius completion problem is obtained in this triangular setting.

XV.1 U-similar and decomposed U-specified matrices

In this chapter we consider upper triangular partially specified matrices, i.e., partially specified matrices of the form

$$A_V = \begin{pmatrix} a_{11} & \cdots & \cdots & a_{1n} \\ ? & \ddots & & \vdots \\ \vdots & \ddots & \ddots & \vdots \\ ? & \cdots & ? & a_{nn} \end{pmatrix},$$

where the pattern $V = \{(i,j) \mid 1 \leq i \leq j \leq n\}$. Here, as before, the question marks denote the unspecified entries. We call such a matrix a *U-specified matrix.*

First we want to find the natural similarities for this class of partially specified matrices. These similarities should play for the present structure the role that block similarities have for operator blocks. Therefore we are looking for the invertible matrices S, which have the property that SA_VS^{-1} is well defined as a U-specified matrix. This means that each entry in the upper triangular part of SA_VS^{-1} can be computed from the entries in the upper triangular part of A_V and the entries of S. The next lemma makes clear which class of invertible matrices S is the natural choice for the similarities. Recall (cf., Section I.1) that a full matrix $B = (b_{ij})$ is a completion of the partially specified matrix A_V with entries a_{ij} for $(i,j) \in V$ if $b_{ij} = a_{ij}$ for each $(i,j) \in V$.

Lemma 1.1. *Let A_V be a U-specified matrix, and let S be an invertible $n \times n$ matrix. If for any completion A of A_V the elements in the upper triangular part of SAS^{-1} depend on the elements in the upper triangular part of A only, then S is a lower triangular matrix. Conversely, if S is a lower triangular matrix, then the elements in the upper triangular part of SAS^{-1}, where A is a completion of A_V, depend on the elements in the upper triangular part of A only.*

PROOF. Assume that for any completion A of A_V the elements in the upper triangular part of SAS^{-1} depend on the elements in the upper triangular part of A only. Let A_0 be the completion of A_V with $a_{ij} = 0$ if $(i,j) \notin V$, and let A_1 be an arbitrary completion of A_V. This means that $E = A_1 - A_0$ is an arbitrary strictly lower triangular matrix. Since SA_1S^{-1} and SA_0S^{-1} have the same upper triangular part, the matrix SES^{-1} is strictly lower triangular. Thus we have to show that SES^{-1} is strictly lower triangular for any strictly lower triangular matrix E implies that S is lower triangular.

To prove this write $S = \left(s_{ij}\right)_{i,j=1}^{n}$. Assume that all entries in the upper triangular part above row i are zero. In other words, if $k < i$ and $\ell > k$, then $s_{k\ell} = 0$. We show that $s_{ij} = 0$ if $i < j$. Let $E = \left(e_{k\ell}\right)_{k,\ell=1}^{n}$ be the matrix with $e_{k\ell} = 0$ if $(k,\ell) \neq (j,i)$ and $e_{ji} = 1$. Put $L_0 = SES^{-1}$. By assumption L_0 is strictly lower triangular and $L_0S = SE$. Now compute the entry in place (i,i) for the left hand and the right hand side of this equality separately. In the left hand side we multiply row i of L_0 with column i of S. Row i of L_0 has zero entries from place i on and column i of S has, according to the assumption on i, zero entries up to place $i-1$. Therefore the entry in place (i,i) in L_0S is equal to 0. From the right hand side we see that the entry in place (i,i) is s_{ij}. This proves that $s_{ij} = 0$. So indeed S is lower triangular.

Conversely, assume that S is an invertible lower triangular matrix. Therefore for each lower triangular matrix L_0, with entries zero on the main diagonal, we have that SL_0S^{-1} is lower triangular with entries zero on the main diagonal. Thus the upper triangular parts of SAS^{-1} and $S(A+L_0)S^{-1}$ coincide for each L_0. This means that the entries in the upper triangular part of SAS^{-1} do not depend on the entries below the diagonal of A. □

Let A_V and B_V be U-specified $n \times n$ matrices. We say that B_V *is U-similar to* A_V if there exists an invertible lower triangular matrix L such that $B_V - LA_VL^{-1}$ has only zero entries in its upper triangular part. In this case L is called an *U-similarity* of B_V and A_V. Note that if B_V is U-similar to A_V, then A_V is U-similar to B_V.

Let A be a U-specified $n \times n$ matrix. We call A *decomposed* (as a U-specified matrix) if there exists a block partition $A = \left(A_{pq}\right)_{p,q=1}^{r}$ with $r \geq 2$ such that

(i) A_{pp} is a U-specified $n_p \times n_p$ matrix;

(ii) $A_{pq} = 0$ if $p < q$.

Thus, if A is decomposed, then A has block entries equal to 0 above the main block diagonal, fully unspecified block entries below the main block diagonal, and U-specified matrices on the main block diagonal. Note that in a decomposed U-specified matrix one of the (scalar) entries $a_{i\ i+1}$ is zero. The next lemma shows that the notion of decomposed is invariant under U-similarity.

Lemma 1.2. *Let A and B be U-similar U-specified matrices. Then A is decomposed if and only if B is decomposed.*

PROOF. Let A be decomposed. Assume that B and SAS^{-1} have the same upper triangular parts, with S a U-similarity. Write $A = \left(A_{pq}\right)_{p,q=1}^{r}$ with $r > 1$, $A_{pq} = 0$ if $p < q$ and A_{pp} is a U-specified $n_p \times n_p$ matrix. Decompose S and S^{-1} accordingly. So $S = \left(S_{pq}\right)_{p,q=1}^{r}$ and $S^{-1} = T = \left(T_{pq}\right)_{p,q=1}^{r}$. Since S and T are lower triangular, we have that $S_{pq} = 0$ and $T_{pq} = 0$ if $p < q$. Thus SAS^{-1} is the product of three block lower triangular matrices and is therefore itself a block lower triangular matrix. Since the upper triangular parts of B and SAS^{-1} are the same, this means that B is decomposed.

The converse implication is proved in the same manner. □

The next lemma, which will play a role later on, provides in each class of U-similar not decomposed matrices a (non unique) representative of a special form.

Lemma 1.3. *The U-specified matrix A is not decomposed if and only if it is U-similar to a U-specified matrix B with entries b_{ij} such that*

(a) *$b_{i\ i+1} \neq 0$ for all $i = 1, \ldots, n-1$, and*

(b) *whenever $b_{ij} \neq 0$ for a pair of indices $1 \leq i < j \leq n$, then $b_{k\ell} \neq 0$ for all the pairs (k, ℓ) satisfying $i \leq k < \ell \leq j$.*

PROOF. First assume that A is U-similar to a B as in the lemma. If A would be decomposed, then B would be decomposed which, by definition, implies that for some k the entry $b_{k\ k+1}$ is zero. This contradicts (a), and hence A is not decomposed.

Next, assume that $A = \left(a_{ij}\right)_{i,j=1}^{n}$ is not decomposed. Define the set

$$\Omega = \{(i,j) \mid 1 \leq i,j \leq n,\ a_{ij} \neq 0 \text{ and } a_{k\ell} = 0 \text{ if } (k,l) \neq (i,j),\ k \leq i,\ j \leq \ell\}$$

So $(i,j) \in \Omega$ if and only if $a_{ij} \neq 0$ and

$$A = \begin{pmatrix} A_{11} & A_{12} \\ A_{21} & A_{22} \end{pmatrix}, \quad A_{12} = \begin{pmatrix} 0 & 0 & \cdots & 0 \\ \vdots & \vdots & & \vdots \\ 0 & 0 & \cdots & 0 \\ a_{ij} & 0 & \cdots & 0 \end{pmatrix}.$$

If (i,j) and (k,ℓ) are elements of Ω, then $(i,j) \neq (k,\ell)$ implies that $i \neq k$ and $j \neq \ell$. Moreover, in this case $i < k$ if and only if $j < \ell$. Therefore we can write

$\Omega = \{(i_1, j_1), \dots, (i_p, j_p)\}$ with $i_1 < \cdots < i_p$, and thus also $j_1 < \cdots < j_p$. Recall that A is not decomposed if and only if for $k = 1, \dots, n-1$ the right upper $k \times (n-k)$ submatrix of A is nonzero. This gives for $k = 1$ that $i_1 = 1$ and for $k = n-1$ that $j_p = n$. Furthermore, it follows that $i_{r+1} \leq j_r$. In order to see this, assume that $i_{r+1} > j_r$. Thus

$$i_r < j_r < i_{r+1} < j_{r+1}.$$

First we show that there exist k and ℓ such that $i_r < k < i_{r+1}$, $j_r < \ell < j_{r+1}$, and $a_{k\ell} \neq 0$.

Assume otherwise. Then $a_{k\ell} = 0$ if $i_r < k \leq j_r$ and $j_r + 1 \leq \ell < j_{r+1}$. Since $(i_r, j_r) \in \Omega$, we have that

$$a_{k\ell} = 0, \quad \text{if } k \leq i_r \quad \text{and} \quad j_r + 1 \leq \ell, \tag{1.1}$$

and since $(i_{r+1}, j_{r+1}) \in \Omega$ also

$$a_{k\ell} = 0, \quad \text{if } k < i_{r+1} \quad \text{and} \quad j_{r+1} \leq \ell. \tag{1.2}$$

We conclude that $a_{k\ell} = 0$ if $k \leq j_r$ and $\ell \geq j_r + 1$. This contradicts the assumption that A is not decomposed.

Now choose k_0 to be the minimal k and ℓ_0 to be the maximal ℓ such that $i_r < k < i_{r+1}$, $j_r < \ell < j_{r+1}$, and $a_{k\ell} \neq 0$. We will see that $(k_0, \ell_0) \in \Omega$. Indeed, $a_{k\ell} = 0$ if $i_r < k \leq k_0$, $\ell_0 \leq \ell < j_{r+1}$ and $(k, \ell) \neq (k_0, \ell_0)$. From (1.2) one sees that $a_{k\ell} = 0$ if $k \leq k_0$ and $\ell \geq j_{r+1}$, and from (1.1) that $a_{k\ell} = 0$ if $k \leq i_r$ and $\ell \geq \ell_0$. So $(k_0, \ell_0) \in \Omega$, which contradicts the ordering we have chosen in the set Ω. Indeed we have $i_{r+1} \leq j_r$. We conclude that

$$i_1 = 1, \quad j_p = n, \quad i_{r+1} \leq j_r \quad (r = 1, \dots, p-1).$$

(For completeness we note that on the other hand the arguments given above could also be used to prove that if either $i_1 > 1$ or $j_p < n$ or $i_{r+1} > j_r$ for some $1 \leq r \leq n$, then A is decomposed. This last fact will not be used in the sequel.)

Let $(k, \ell) \in \Omega$, and assume that $k + 1 < \ell$. We shall prove that then there exists a U-specified matrix B with entries b_{ij} for $1 \leq i \leq j \leq n$, U-similar to A, and such that $b_{k\,\ell-1} \neq 0$ and $b_{ij} \neq 0$ whenever $a_{ij} \neq 0$. In case $a_{k\,\ell-1} \neq 0$ we take $B = A$. So assume that $a_{k\,\ell-1} = 0$. Put $S_1(x) = I_n + xE_{\ell\,\ell-1}$, where $E_{\ell\,\ell-1}$ is the matrix with all entries equal to zero except for the entry on place $(\ell, \ell-1)$, which is taken to be equal to one. Then $S_1(x)$ is lower triangular and invertible, and hence $S_1(x)$ is a U-similarity for each value of x. Let the entries in the upper triangular part of the U-specified matrix $\left(S_1(x)\right)^{-1} A S_1(x)$ be $b_{ij}(x)$ for $1 \leq i \leq j \leq n$. Then

$$b_{ij}(x) = \begin{cases} a_{ij} & \text{for } i \neq \ell \text{ and } j \neq \ell - 1, \\ a_{\ell j} + x a_{\ell-1\,j} & \text{for } i = \ell \text{ and } j \neq \ell - 1, \\ a_{i\,\ell-1} - x a_{i\ell} & \text{for } i \neq \ell \text{ and } j = \ell - 1, \\ a_{\ell\,\ell-1} + x a_{\ell-1\,\ell-1} - x a_{\ell\ell} - x^2 a_{\ell-1\,\ell} & \text{for } i = \ell \text{ and } j = \ell - 1. \end{cases}$$

Since we assume that $a_{k\ \ell-1}=0$, it follows that $b_{k\ \ell-1}=-xa_{k\ell}\neq 0$ for any $x\neq 0$. So in order to get that also $b_{ij}\neq 0$ whenever $a_{ij}\neq 0$ we must satisfy the following conditions

(1) $a_{\ell j}+xa_{\ell-1\ j}\neq 0$ for $i=\ell$ and $j\neq \ell-1$ and $a_{\ell j}\neq 0$,

(2) $a_{i\ \ell-1}-xa_{i\ell}\neq 0$ for $j=\ell-1$ and $i\neq \ell$ and $a_{i\ \ell-1}\neq 0$,

(3) $a_{\ell\ \ell-1}+xa_{\ell-1\ \ell-1}-xa_{\ell\ell}-x^2a_{\ell-1\ \ell}\neq 0$ and $a_{\ell\ \ell-1}\neq 0$.

Clearly it is possible to choose x such these inequalities are satisfied.

In much the same way one proves the existence of a matrix C, U-similar to A, and such that $c_{k+1\ \ell}\neq 0$ and $c_{ij}\neq 0$ whenever $a_{ij}\neq 0$.

Now apply the constructions given in the two previous paragraphs step by step for each $(i_k,j_k)\in\Omega$. In this way we obtain a matrix $B=\left(b_{ij}\right)_{i,j=1}^{n}$, U-similar to A, and such that if $(k,\ell)\in\Omega$, then for each (i,j) with $k\leq i<j\leq \ell$ we have that $b_{ij}\neq 0$. By the choice of the set Ω this gives that B satisfies condition (b). Finally remark that for each i with $1\leq i<n$ there exists a number r such that $i_r\leq i<i+1\leq j_r$. So we see that B also satisfies (a). □

Recall that a decomposed U-specified matrix A admits a block partitioning $A=\left(A_{pq}\right)_{p,q=1}^{r}$ with $r>1$ such that $A_{pq}=0$ if $p<q$ and A_{pp} is an $n_p\times n_p$ U-specified matrix which is not decomposed. By applying Lemma 1.3 to each A_{pp} one may obtain a representative in the U-similarity class of A, which is a direct sum of matrices B of the type appearing in Lemma 1.3.

XV.2 Invariants for U-similarity

In this section we introduce some data of a U-specified matrix that remain unchanged if a U-similarity is applied to the matrix. So let A be a U-specified $n\times n$ matrix with entries a_{ij} in its upper triangular part. For every $k=1,\dots,[\frac{n}{2}]$ we define the matrices

$$M_k(A)=\begin{pmatrix} a_{1\ n-k+1} & \cdots & a_{1n}\\ a_{2\ n-k+1} & \cdots & a_{2n}\\ \vdots & & \vdots\\ a_{k\ n-k+1} & \cdots & a_{kn}\end{pmatrix}$$

(Here and elsewhere in this chapter $[\alpha]$ stands for the greatest integer less than or equal to α.) Since $k\leq[\frac{n}{2}]$, the matrix $M_k(A)$ is the fully specified $k\times k$ matrix appearing in the right upper corner of A. We set

$$\mu_k(A)=\det M_k(A),\quad k=1,\dots,\left[\frac{n}{2}\right],\qquad \mu_0(A)=1. \tag{2.1}$$

For any pair of integers (k,ℓ) with $1\leq k\leq[\frac{n-1}{2}]$ and $k<\ell<n-k+1$ we define

the matrices $N_{k\ell}(A)$ by

$$N_{k\ell}(A) = \begin{pmatrix} a_{1\ell} & a_{1\ n-k+1} & \cdots & a_{1n} \\ a_{2\ell} & a_{2\ n-k+1} & \cdots & a_{2n} \\ \vdots & \vdots & & \vdots \\ a_{k\ell} & a_{k\ n-k+1} & \cdots & a_{kn} \\ a_{\ell\ell} & a_{\ell\ n-k+1} & \cdots & a_{\ell n} \end{pmatrix}.$$

Again note that $N_{k\ell}(A)$ is a fully specified matrix appearing in the upper triangular part of A. Also notice that $M_k(A)$ is a submatrix of $N_{k\ell}(A)$. Now put

$$\nu_{k\ell}(A) = \begin{cases} \det N_{k\ell}(A) & 1 \le k \le [\frac{n-1}{2}], \quad k < \ell < n-k+1, \\ a_{\ell\ell} & 1 \le \ell \le n, \quad k = 0. \end{cases}$$

Furthermore, for each integer k with $0 \le k \le [\frac{n-1}{2}]$ and $\mu_k(A) \neq 0$ let us define the quantities

$$I_{k+1}(A) = \frac{1}{\mu_k(A)} \sum_{\ell=k+1}^{n-k} \nu_{k\ell}(A). \tag{2.2}$$

Note that in particular $I_1(A) = \operatorname{trace} A$. The next theorem shows that the quantities $I_k(A)$ are invariant under U-similarities.

Theorem 2.1. *Let A and B be U-similar U-specified $n \times n$ matrices. Then*

(a) *for every $k = 0, 1, \ldots, [\frac{n}{2}]$ one has $\mu_k(A) = 0$ if and only if $\mu_k(B) = 0$;*

(b) *$I_{k+1}(A) = I_{k+1}(B)$ for each $k = 0, 1, \ldots, [\frac{n-1}{2}]$ such that $\mu_k(A) \neq 0$.*

PROOF. The properties (a) and (b) are evident for $k = 0$.

Let k be an integer with $1 \le k \le [\frac{n}{2}]$. Let L be the U-similarity of B and A and partition L and A in the following way

$$L = \begin{pmatrix} L_1 & 0 \\ L_2 & L_3 \end{pmatrix} = \begin{pmatrix} L_1' & 0 \\ L_2' & L_3' \end{pmatrix}, \quad A = \begin{pmatrix} A_1 & A_2 \\ A_3 & A_4 \end{pmatrix},$$

where L_1, L_3' and A_2 are $k \times k$ matrices. Note that A_3 is fully unspecified, that A_1 and A_4 are U-specified matrices and A_2 is fully known, in fact $A_2 = M_k(A)$. The matrix $L^{-1}AL$ has the form

$$\begin{pmatrix} L_1^{-1} & 0 \\ -L_3^{-1}L_2L_1^{-1} & L_3^{-1} \end{pmatrix} \begin{pmatrix} A_1 & A_2 \\ A_3 & A_4 \end{pmatrix} \begin{pmatrix} L_1' & 0 \\ L_2' & L_3' \end{pmatrix} = \begin{pmatrix} * & L_1^{-1}A_2L_3' \\ * & * \end{pmatrix}.$$

Since $B - L^{-1}AL$ is strictly lower triangular, the matrix $L_1^{-1}A_2L_3' = M_k(B)$. Hence

$$\mu_k(B) = \det(L_1^{-1})\mu_k(A) \det L_3'.$$

Both $\det(L_1^{-1})$ and $\det L_3'$ are nonzero and therefore (a) is proved.

Now assume that k is such that $1 \leq k \leq [\frac{n-1}{2}]$ and $\mu_k(A) \neq 0$. Assume that the $n \times n$ matrix L gives a U-similarity of A and B. From the Gauss elimination process it follows that L is the product of elementary matrices of the the following two types:

$$F(m,\alpha) = \big(f(m,\alpha)_{ij}\big)_{i,j=1}^{n}, \quad 1 \leq m \leq n, \quad 0 \neq \alpha \in \mathbb{C}, \tag{2.3}$$

where

$$f(m,\alpha)_{ij} = \begin{cases} 0 & \text{for } i \neq j, \\ 1 & \text{for } i = j \neq m, \\ \alpha & \text{for } i = j = m, \end{cases}$$

and

$$G(q,p,\beta) = \big(g(q,p,\beta)_{ij}\big)_{i,j=1}^{n}, \quad 1 \leq p < q \leq n, \quad \beta \in \mathbb{C}, \tag{2.4}$$

where

$$g(q,p,\beta)_{ij} = \begin{cases} 1 & \text{for } i = j, \\ 0 & \text{for } i \neq j \text{ and } (i \neq q \text{ or } j \neq p), \\ \beta & \text{for } i = q \text{ and } j = p. \end{cases}$$

It is sufficient to show that (b) is true if the U-similarity of A and B is of the type (2.3) or of the type (2.4).

Assume that B and $F(s,\alpha)^{-1}AF(s,\alpha)$ have the same upper triangular part. If $1 \leq s \leq k$, then $M_k(B)$ is obtained from $M_k(A)$ by means of a multiplication by α^{-1} of its s-th row, and for every ℓ in the range $k < \ell \leq n-k+1$, the matrix $N_{k\ell}(B)$ is obtained in the same way from $N_{k\ell}(A)$. Hence in this case $\mu_k(B) = \alpha^{-1}\mu_k(A)$ and $\nu_{k\ell}(B) = \alpha^{-1}\nu_{k\ell}(A)$, and therefore $I_{k+1}(B) = I_{k+1}(A)$ if $1 \leq s \leq k$.

If $k < s < n-k+1$, then $M_k(B) = M_k(A)$. So again $\mu_k(B) = \mu_k(A)$. If $\ell \neq s$, then $N_{k\ell}(B) = N_{k\ell}(A)$. If $\ell = s$, then $N_{k\ell}(B)$ is obtained from $N_{k\ell}A)$ by multiplying the last row by α^{-1} and the first column by α. Thus, in general, we find for $k < s < n-k+1$ that $\nu_{k\ell}(B) = \nu_{k\ell}(A)$ if $k < \ell < n-k+1$. This proves that $I_{k+1}(B) = I_{k+1}(A)$ if $k < s < n-k+1$.

Next, we consider the case when $k \geq n-k+1$. In this case the matrix $M_k(B)$ is obtained from $M_k(A)$ by multiplying the column $k-(n-s)$ by α and $N_{k\ell}(B)$ is obtained from $N_{k\ell}(A)$ by multiplying the column $k+1-(n-s)$ by α. So $\mu(B) = \alpha\mu_k(A)$ and $\nu_{k\ell}(B) = \alpha\nu_{k\ell}A)$. This proves that $I_{k+1}(B) = I_{k+1}(A)$ if $k \geq n-k+1$. So in the case when the U-similarity is of the type $F(s,\alpha)$ we proved (b).

The next step is to assume that B and $G(q,p,\beta)^{-1}AG(q,p,\beta)$ have the same upper triangular part. If $1 \leq p < q < k$, then the matrices $M_k(B)$ and $N_{k\ell}(B)$ are obtained from the matrices $M_k(A)$ and $N_{k\ell}(A)$, respectively, by adding to their q-th row $(-\beta)$ times their p-th row. If $n-k+1 \leq p < q \leq n$, then the matrix $M_k(B)$ is obtained from the matrix $M_k(A)$ by adding to its $(p+k-n)$-th column β times its $(q+k-n)$-th column, and the matrix $N_{k\ell}(B)$ is obtained from the matrix $N_{k\ell}(A)$ by adding to its $(p+k-n+1)$-th column β times its $(q+k-n+1)$-th column. Hence in either of these cases $\mu_k(B) = \mu_k(A)$ and $\nu_{k\ell}(B) = \nu_{k\ell}(A)$. Here ℓ is in the range $k < \ell \leq n-k+1$. So if $1 \leq p < q < k$ or $n-k+1 \leq p < q \leq n$, then $I_{k+1}(B) = I_{k+1}(A)$.

If $k+1 \leq p < q \leq n-k$, then

$$\nu_{kp}(B) = \det \begin{pmatrix} a_{1p} + \beta a_{1q} & a_{1\ n-k+1} & \cdots & a_{1n} \\ \vdots & \vdots & & \vdots \\ a_{kp} + \beta a_{kq} & a_{k\ n-k+1} & \cdots & a_{kn} \\ a_{pp} + \beta a_{pq} & a_{p\ n-k+1} & \cdots & a_{pn} \end{pmatrix}$$

$$= \nu_{kp}(A) + \beta \ \det \begin{pmatrix} a_{1q} & a_{1\ n-k+1} & \cdots & a_{1n} \\ \vdots & \vdots & & \vdots \\ a_{kq} & a_{k\ n-k+1} & \cdots & a_{kn} \\ a_{pq} & a_{p\ n-k+1} & \cdots & a_{pn} \end{pmatrix},$$

$$\nu_{kq}(B) = \det \begin{pmatrix} a_{1q} & a_{1\ n-k+1} & \cdots & a_{1n} \\ \vdots & \vdots & & \vdots \\ a_{kq} & a_{k\ n-k+1} & \cdots & a_{kn} \\ a_{qq} - \beta a_{pq} & a_{q\ n-k+1} - \beta a_{p\ n-k+1} & \cdots & a_{qn} - \beta a_{pn} \end{pmatrix}$$

$$= \nu_{qk}(A) - \beta \ \det \begin{pmatrix} a_{1q} & a_{1\ n-k+1} & \cdots & a_{1n} \\ \vdots & \vdots & & \vdots \\ a_{kq} & a_{k\ n-k+1} & \cdots & a_{kn} \\ a_{pq} & a_{p\ n-k+1} & \cdots & a_{pn} \end{pmatrix},$$

and $N_{k\ell}(B) = N_{k\ell}(A)$ for $\ell \neq p$ or $\ell \neq q$, where ℓ is in the range $k < \ell \leq n-k+1$. Therefore

$$\sum_{\ell=k+1}^{n-k} \nu_{k\ell}(B) = \sum_{\ell=k+1}^{n-k} \nu_{k\ell}(A).$$

Since for any p, q for which $k+1 \leq p < q \leq n-k$ we have $M_k(B) = M_k(A)$, it follows that also in this case $I_{k+1}(B) = I_{k+1}(A)$.

If $1 \leq p \leq k < q \leq n$ or $1 \leq p \leq k < n-k+1 \leq q \leq n$ or $k+1 \leq p \leq n-k < q \leq n$ it is easy to see that $M_k(B) = M_k(A)$ and $N_{k\ell}(B) = N_{k\ell}(A)$ for ℓ in the range $k < \ell \leq n-k+1$. So we have proved that $I_{k+1}(B) = I_{k+1}(A)$ for $1 \leq k \leq [\frac{n-1}{2}]$.

Since each lower triangular matrix is a product of matrices of the form (2.3) and of the form (2.4), this shows that any B that is U-similar to A has the property (b). $\square$

Theorem 2.1 provides necessary conditions for the U-similarity of two U-specified matrices. However in general these conditions are not sufficient, as the next example shows. Write

$$A=\begin{pmatrix}0&1&0&0\\ ?&0&0&1\\ ?&?&0&0\\ ?&?&?&0\end{pmatrix},\quad B=\begin{pmatrix}0&0&1&0\\ ?&0&0&0\\ ?&?&0&1\\ ?&?&?&0\end{pmatrix}.$$

Let $L=\left(\ell_{ij}\right)_{i,j=1}^{n}$ be an invertible lower triangular matrix. In AL the entry in place (1,2) is ℓ_{22}. On the other hand the (1,2)-entry in of LB is zero. This shows that $AL-LB$ is not lower triangular. If A and B were U-similar, then $B-L^{-1}AL$ would be strictly lower triangular and thus also $AL-LB$ would be strictly lower triangular. Hence A and B are not U-similar. Nevertheless the conditions (a) and (b) in Theorem 2.1 are fulfilled. Indeed, since B is the mirror image of A with respect to the second main diagonal, it follows that $\mu_k(A)=\mu_k(B)$ for $k=0,1,2$ and trace $A=$ trace B. Since $\mu_1(A)=\mu_2(A)=0$, this shows that (a) and (b) are fulfilled.

In the above example A and B are not decomposed. To see this assume that A is decomposed. Let $A=\left(A_{pq}\right)_{p,q=1}^{r}$ be a block partition of A such that $A_{pq}=0$ if $1\le p<q\le r$ and A_{pp} is a U-specified $n_p\times n_p$ matrix for $1\le p\le r$. One sees that this implies that $n_1\ge 2$ and $n_r\ge 3$. Since however $n_1+n_r\le 4$, this means that $r=1$ which is a contradiction. Therefore A is not decomposed. Similarly one proves that B is not decomposed.

XV.3 Invariants and U-canonical form in the generic case

In this section we consider U-specified matrices A that have the property $\mu_k(A)\neq 0$. Here $\mu_k(A)$ denotes the number defined by (2.1). The class of these matrices is generic in the sense that the upper triangular matrices with this property form an open and dense subset in the vector space of all upper triangular matrices. For this subclass of partially specified matrices we present a full set of invariants and a canonical form under U-similarity. This is the contents of the next result.

Theorem 3.1. *Let A be an U-specified $n\times n$ matrix, and assume that $\mu_k(A)\neq 0$ for $k=1,\dots,\left[\frac{n}{2}\right]$. Then A is U-similar to a U-specified $n\times n$ matrix B with entries b_{ij}, $1\le i\le j\le n$, satisfying*

(a) $b_{ij}=0$ *if* $i<j$ *and* $i+j\neq n+1$;

(b) $b_{ij}=1$ *if* $i<j$ *and* $i+j=n+1$;

(c) $b_{ii}=0$ *if* $i>\left[\frac{n+1}{2}\right]$.

Moreover, in this case

$$b_{ii}=(-1)^{i+1}(I_i(A)+I_{i+1}(A)),\quad i=1,\dots,\left[\frac{n+1}{2}\right].$$

Here $I_\ell(A)$ is given by (2.2) if $0 \le \ell \le \left[\frac{n+1}{2}\right]$, and $I_\ell(A) = 0$ if $\ell = \left[\frac{n+1}{2}\right] + 1$. In particular, the upper triangular part of a U-specified matrix B, which has the properties (a), (b) and (c) above and which is U-similar to A, is uniquely determined by A.

For the case when $n = 2p - 1$ the U-specified matrix B in Theorem 3.1 has the form

$$B = \begin{pmatrix} b_{11} & 0 & & \cdots & & & 0 & 1 \\ ? & b_{22} & \ddots & & & \cdot^{\cdot^{\cdot}} & 1 & 0 \\ \vdots & & \ddots & \ddots & \cdot^{\cdot^{\cdot}} & \cdot^{\cdot^{\cdot}} & \cdot^{\cdot^{\cdot}} & \vdots \\ & & & \ddots & 0 \quad 1 & \cdot^{\cdot^{\cdot}} & & \\ & & & & b_{pp} \quad 0 & & & \\ & & & & \quad\; 0 & & & \\ \vdots & & & & & \ddots & & \vdots \\ & & & & & & 0 & 0 \\ ? & & & \cdots & & & ? & 0 \end{pmatrix}, \tag{3.1}$$

and in the case when $n = 2p$ the form of B is

$$B = \begin{pmatrix} b_{11} & 0 & & \cdots & & & 0 & 1 \\ ? & b_{22} & \ddots & & & \cdot^{\cdot^{\cdot}} & 1 & 0 \\ \vdots & & \ddots & 0 & 0 & \cdot^{\cdot^{\cdot}} & \cdot^{\cdot^{\cdot}} & \vdots \\ & & & b_{pp} & 1 & \cdot^{\cdot^{\cdot}} & & \\ & & & & 0 & & & \\ \vdots & & & & & \ddots & & \vdots \\ & & & & & & 0 & 0 \\ ? & & & \cdots & & & ? & 0 \end{pmatrix} \tag{3.2}$$

A matrix B, satisfying conditions (a), (b) and (c) in Theorem 3.1 is said to be in the *U-canonical form*. So, if the order of B is odd, then the U-canonical form is (3.1), and, if the order is even, then it is (3.2).

Proof of Theorem 3.1. The proof will be given by induction on the order n of the U-specified matrix A. Let a_{ij} $(1 \le i \le j \le n)$ be in the entries in the upper triangular part of A. For $n = 1$ there is nothing to prove because the conditions (a), (b) and (c) are void. For $n = 2$ one uses that $a_{12} = \mu_1(A)$ to compute

$$\begin{pmatrix} a_{12} & 0 \\ a_{22} & 1 \end{pmatrix}^{-1} \begin{pmatrix} a_{11} & a_{12} \\ ? & a_{22} \end{pmatrix} \begin{pmatrix} a_{12} & 0 \\ a_{22} & 1 \end{pmatrix} = \begin{pmatrix} a_{11} + a_{22} & 1 \\ ? & 0 \end{pmatrix}. \tag{3.3}$$

Let n be a natural number with $n \ge 3$. Assume that any $p \times p$ matrix of size $p \le n - 2$ is U-similar to a matrix satisfying the conditions (a), (b) and (c). Let A be a U-specified $n \times n$ matrix such that $\mu_k(A) \ne 0$ for $k = 1, \dots, \left[\frac{n}{2}\right]$. We apply to A a sequence of elementary similarities to get zero entries in the positions $(1,2),\ldots,(1,n-1)$ and $(2,n),\ldots,(n,n)$, and a unit entry in position $(1,n)$. Let $A(1)=F(1,a_{1n}^{-1})AF(1,a_{1n})$, with F given by (2.3). Then the first row of $A(1)$ is

$$\begin{pmatrix} a_{11} & a_{12}a_{1n}^{-1} & \cdots & a_{1\ n-1}a_{1n}^{-1} & 1 \end{pmatrix}, \tag{3.4}$$

and the other (specified) entries of $A(1)$ are equal to those of A. Next we change the last column of $A(1)$ by elementary row operations. We put

$$A(i)=G(i,1,-a_{in})A(i-1)G(i,1,a_{in}), \quad i=2,\ldots,n.$$

Here $G(i,1,a_{in})$ is given by (2.4). Then $A(i)$ and $A(i-1)$ differ only in the (1,1) entry and in the entries in row i. Moreover, the (i,n)-th entry in $A(i)$ is zero and in $A(i-1)$ it is equal to a_{in}. As a result the U-specified matrix $A(n)$ has a last column equal to $\begin{pmatrix} 1 & 0 & \cdots & 0 \end{pmatrix}^T$ and a first row given by (3.4). Next we compute

$$A(n+i)=G(n,n-i,a_{1\ n-i}a_{1n}^{-1})A(n+i-1)G(n,n-i,-a_{1\ n-i}a_{1n}^{-1})$$

for $i=1,\ldots,n-i$. Thus in each step we add $a_{1\ n-i}a_{1n}^{-1}$ times row $n-i$ to row n, which leaves every specified entry unchanged, and we subtract $a_{1\ n-i}a_{1n}^{-1}$ times column n from column $n-i$, which generates a zero entry in position $(1,n-i)$ and leaves every other entry unchanged. We end up with

$$A'=A(2n-1)=\begin{pmatrix} a'_{11} & 0 & 0 & \cdots & 0 & 1 \\ ? & a'_{22} & a'_{23} & \cdots & a'_{2\ n-1} & 0 \\ ? & ? & a'_{33} & \cdots & a_{3\ n-1} & 0 \\ \vdots & \vdots & \ddots & \ddots & \vdots & \vdots \\ ? & ? & ? & \ddots & a'_{n-1\ n-1} & 0 \\ ? & ? & ? & \cdots & ? & 0 \end{pmatrix}.$$

Let us consider the U-specified $(n-2)\times(n-2)$ matrix A'' with entries a''_{ij} in its upper triangular part, which is A' without its first and last row and without its first and last column. So $a''_{ij}=a'_{i+1\ j+1}$ for $1\le i,j\le n-2$. Since, according to Theorem 2.1, $\mu_{k+1}(A')\neq 0$ if and only if $\mu_{k+1}(A)\neq 0$, it follows that $\mu_k(A'')\neq 0$ for $k=1,\ldots,\left[\frac{n-3}{2}\right]$. By the induction hypothesis there exists a lower triangular invertible $(n-2)\times(n-2)$ matrix L' such that $(L')^{-1}A''L'=B'$ has entries b'_{ij} in its upper triangular part such that

(a') $b'_{ij}=0$ if $i<j$ and $i+j\neq n-1$;

(b') $b'_{ij}=1$ if $i<j$ and $i+j=n-1$;

(c') $b'_{ii}=0$ if $i>\left[\frac{n-1}{2}\right]$.

Define the invertible lower triangular $n\times n$ matrix L by

$$L=\begin{pmatrix} 1 & 0 & 0 \\ 0 & L' & 0 \\ 0 & 0 & 1 \end{pmatrix}.$$

Then the U-specified matrix $B=L^{-1}A'L$ satisfies the conditions (a), (b) and (c).

Now let $B'' = \left(b''_{ij}\right)_{i,j=1}^{n}$ be a U-specified matrix with entries b''_{ij} for $1 \leq i \leq j \leq n$. Assume that B'' is U-similar to A and satisfies the conditions (a), (b) and (c). From the very definition of the numbers $I_k(B'')$ one obtains

$$\begin{aligned} I_1(B'') &= \sum_{i=1}^{\left[\frac{n+1}{2}\right]} b''_{ii}, \\ I_2(B'') &= -\sum_{i=2}^{\left[\frac{n+1}{2}\right]} b''_{ii}, \\ &\vdots \\ I_{\left[\frac{n+1}{2}\right]}(B'') &= (-1)^{\left[\frac{n-1}{2}\right]} b''_{\left[\frac{n+1}{2}\right]\left[\frac{n+1}{2}\right]}. \end{aligned} \tag{3.5}$$

Theorem 2.1 gives that $I_k(B'') = I_k(A)$ for $1 \leq k \leq \left[\frac{n+1}{2}\right]$. Thus the system (3.5) uniquely determines the values of the diagonal elements of B'' as functions of $I_k(A)$:

$$b''_{kk} = (-1)^{k+1}\left(I_k(A) + I_{k+1}(A)\right) \quad \text{for} \quad 1 \leq i \leq \left[\frac{n-1}{2}\right],$$

$$b''_{\left[\frac{n+1}{2}\right]\left[\frac{n+1}{2}\right]} = (-1)^{\left[\frac{n+3}{2}\right]} I_{\left[\frac{n+1}{2}\right]}(A).$$

This finishes the proof of Theorem 3.1. □

Theorem 3.1 yields a criterion for U-similarity. Namely, if for a U-specified matrix A of order n the numbers $\mu_k(A) \neq 0$ for $k = 1, \ldots, \left[\frac{n}{2}\right]$, then the U-similarity class of A is fully determined by the numbers $I_k(A)$ for $k = 1, \ldots, \left[\frac{n+1}{2}\right]$.

Corollary 3.2. *Let A be a U-specified matrix of order n for which $\mu_k(A) \neq 0$ for $k = 1, \ldots, \left[\frac{n}{2}\right]$. A U-specified matrix B is U-similar to A if and only if the following two conditions are satisfied:*

(a) $\mu_k(B) \neq 0$ *for* $k = 1, \ldots, \left[\frac{n}{2}\right]$,

(b) $I_k(B) = I_k(A)$ *for* $k = 1, \ldots, \left[\frac{n+1}{2}\right]$.

PROOF. This result follows immediately from Theorems 2.1 and 3.1. □

XV.4 The diagonal of U-similar matrices

In this section we describe all possible diagonals of U-specified matrices which are U-similar to a given U-specified matrix. Let A be a given U-specified matrix with entries a_{ij} for $1 \leq i \leq j \leq n$, and let $\alpha_1, \alpha_2, \ldots, \alpha_n$ be given numbers (not necessarily all different). It is clear that if there exists a U-specified matrix B which is U-similar to A and has diagonal entries $\alpha_1, \alpha_2, \dots, \alpha_n$, then

$$\operatorname{trace} A = \sum_{i=1}^{n} \alpha_i. \tag{4.1}$$

For not decomposed matrices A (not decomposed matrices are defined in Section 1) the necessary condition (4.1) is also sufficient. This is the contents of the next theorem.

Theorem 4.1. *Let A be a not decomposed U-specified $n \times n$ matrix, and let $\alpha_1, \alpha_2, \dots, \alpha_n$ be complex numbers (not necessarily all different). Then A is U-similar to an U-specified matrix B with diagonal entries $b_{ii} = \alpha_i$ for $i = 1, \dots, n$ if and only if* trace $A = \sum_{i=1}^{n} \alpha_i$.

PROOF. We have already seen that condition (4.1) is necessary. Conversely, assume that (4.1) holds true. Denote the entries of A by a_{ij} for $1 \le i \le j \le n$. Since A is not decomposed we may apply Lemma 1.3. So without loss of generality we may assume that $a_{i\ i+1} \ne 0$ for all $1 \le i \le n-1$ and that for any pair (i,j) with $1 \le i < j \le n$ for which $a_{ij} \ne 0$ also the entries $a_{k\ell} \ne 0$ if $i \le k < \ell \le j$. We shall prove the sufficiency of condition (4.1) by induction on the order n of A. For $n = 1$ this is trivially true. Assume that $n > 1$, and let i be such that $a_{in} \ne 0$ and $a_{kn} = 0$ for all $k < i$. Since $a_{n-1\ n} \ne 0$, we must have that $1 \le i \le n-1$. Note that A is U-similar to

$$A' = G(n,i,x)^{-1} A G(n,i,x), \tag{4.2}$$

where $G(n,i,x)$ is defined by (2.4). Let the entries of A' be a'_{ij} for $1 \le i \le j \le n$. Put $x = (a_{nn} - \alpha_n) a_{in}^{-1}$. We find

$$\begin{aligned} a'_{nn} &= a_{nn} - x a_{in} = \alpha_n, \\ a'_{ii} &= a_{ii} + x a_{in} = a_{ii} + a_{nn} - \alpha_n, \\ a'_{kk} &= a_{kk} \quad (k \ne i, n), \\ a'_{k\ell} &= a_{k\ell} \quad (1 \le k < \ell \le n). \end{aligned} \tag{4.3}$$

Let A'' be the U-specified principal $n-1 \times n-1$ submatrix in the left upper corner of A'. The entries of A'' are a'_{ij} for $1 \le i \le j \le n-1$. It follows from (4.3) and the fact that A is not decomposed that A'' is not decomposed. Moreover (4.1) and (4.3) imply that trace $A'' = \sum_{k=1}^{n-1} \alpha_k$. Hence, A'' is U-similar, with a U-similarity L', to a matrix $B'' = (L')^{-1} A'' L'$ with diagonal entries $b'_{kk} = \alpha_k$ for $k = 1, \dots, n-1$. Let B be the U-specified $n \times n$ matrix defined by

$$B = \begin{pmatrix} L' & 0 \\ 0 & 1 \end{pmatrix}^{-1} A' \begin{pmatrix} L' & 0 \\ 0 & 1 \end{pmatrix}$$

Then the diagonal entries b_{kk} of B are given by $b_{kk} = b'_{kk} = \alpha_k$ for $k < n$ and $b_{nn} = \alpha_n$. □

For a general $n \times n$ matrix the following result holds.

Theorem 4.2. *Let A be a U-specified $n \times n$ matrix. Assume that A admits a block lower triangular partitioning as $A = \left(A_{pq}\right)_{p,q=1}^{r}$ with A_{pp} a not decomposed U-specified $n_p \times n_p$ matrix for $p = 1, \dots, r$ and $A_{pq} = 0$ for $1 \le p < q \le r$. Let $\alpha_1, \alpha_2, \dots, \alpha_n$ be given numbers (not necessarily all different). Then A is U-similar to a U-specified $n \times n$ matrix B with on the main diagonal entries $\alpha_1, \alpha_2, \dots, \alpha_n$ (in this order) if and only if*

$$\operatorname{trace} A_{pp} = \sum_{i=m_{p-1}+1}^{m_p} \alpha_i \quad ,(p = 1, \dots, r), \tag{4.4}$$

where $m_0 = 0$ and $m_p = m_{p-1} + n_p$ for $p = 1, \dots, r$.

PROOF. First we prove that condition (4.4) is necessary. Assume that A is U-similar to B. Consider the block partitioning of $B = \left(B_{pq}\right)_{p,q=1}^{r}$, with B_{pp} a not decomposed U-specified $n_p \times n_p$ matrix for $p = 1, \dots, r$. Then $B_{pq} = 0$ for $1 \le p < q \le r$ and A_{pp} is U-similar to B_{pp} for $p = 1, \dots, r$, because $A_{pq} = 0$ for $p < q$ and a U-similarity is a lower triangular matrix. Apply Theorem 4.1 to each of the not decomposed U-specified matrices A_{pp} and the sequence $\alpha_{m_{p-1}+1}, \dots, \alpha_{m_p}$.

Conversely, assume that (4.4) holds. Again apply Theorem 4.1 to each A_{pp}, and conclude that there exist lower triangular invertible matrices L_p such that $L_p^{-1} A_{pp} L_p$ has the numbers $\alpha_{m_{p-1}+1}, \dots, \alpha_{m_p}$ on its main diagonal. Let L be the $n \times n$ invertible block matrix having the matrices $L_1, \dots, L_r$ as its block diagonal entries. Then L is a U-similarity and $B = L^{-1} A L$ satisfies the conditions of the theorem. □

Notice that in Theorems 4.1 and 4.2 the lower triangular invertible matrix that achieves the U-similarity may without loss of generality assumed to have diagonal entries equal to one. This follows from the fact the for an invertible diagonal matrix D the diagonals of A and $D^{-1} A D$ are identical.

In the generic case, i.e., when the U-specified matrix A has the additional property that

$$\mu_k(A) = \det \begin{pmatrix} a_{1\ n-k+1} & \cdots & a_{1n} \\ a_{2\ n-k+1} & \cdots & a_{2n} \\ \vdots & & \vdots \\ a_{k\ n-k+1} & \cdots & a_{kn} \end{pmatrix} \neq 0, \quad k = 1, \dots, \left[\frac{n}{2}\right], \tag{4.5}$$

a more precise result can be obtained. Notice that (4.5) implies that A is not decomposed, and hence Theorem 4.1 is applicable in this case. The next result shows that the additional condition (4.5) allows one to give an explicit form of a matrix with a prescribed diagonal in the U-similarity class of the U-specified matrix A.

Theorem 4.3. *Let A be a U-specified $n \times n$ matrix with $\mu_k(A) \neq 0$ for $k = 1, \ldots, \left[\frac{n}{2}\right]$. Let $\alpha_1, \alpha_2, \ldots, \alpha_n$ be given numbers (not necessarily all different) such that* trace $A = \sum_{i=1}^{n} \alpha_i$. *Then A is U-similar to a unique U-specified matrix of the form*

$$T(A) = \begin{pmatrix} P & Q \\ ? & R \end{pmatrix},$$

where the $\left[\frac{n+1}{2}\right] \times \left[\frac{n+1}{2}\right]$ U-specified matrix P and the $\left[\frac{n}{2}\right] \times \left[\frac{n}{2}\right]$ U-specified matrix R have the form

$$P = \begin{pmatrix} \alpha_1 & \beta_2 & \cdots & \beta_{\left[\frac{n-1}{2}\right]} & \beta_{\left[\frac{n+1}{2}\right]} \\ ? & \alpha_2 & & 0 & 0 \\ \vdots & & \ddots & & \vdots \\ ? & ? & & \alpha_{\left[\frac{n-1}{2}\right]} & 0 \\ ? & ? & \cdots & ? & \alpha_{\left[\frac{n+1}{2}\right]} \end{pmatrix},$$

$$R = \begin{pmatrix} \alpha_{\left[\frac{n+3}{2}\right]} & 0 & \cdots & 0 & 0 \\ ? & \alpha_{\left[\frac{n+5}{2}\right]} & & 0 & 0 \\ \vdots & & \ddots & & \vdots \\ ? & ? & & \alpha_{n-1} & 0 \\ ? & ? & \cdots & ? & \alpha_n \end{pmatrix},$$

and the $\left[\frac{n+1}{2}\right] \times \left[\frac{n}{2}\right]$ matrix Q has the form

$$Q = \begin{pmatrix} 0 & \cdots & 0 & 1 \\ & & & 1 \\ & Z_{\frac{n-2}{2}} & & \vdots \\ & & & 1 \end{pmatrix} \text{ for } n \text{ even}, \quad Q = \begin{pmatrix} 0 & \cdots & 0 & 1 \\ & & & 1 \\ & Z_{\frac{n-3}{2}} & & \vdots \\ & & & 1 \\ 0 & \cdots & 0 & 1 \end{pmatrix} \text{ for } n \text{ odd}.$$

Here the $k \times k$ matrix Z_k is given by

$$Z_k = \begin{pmatrix} & & 1 \\ & \iddots & \\ 1 & & \end{pmatrix},$$

where the blank spots stand for zero entries. The quantities β_k in the matrix P are given by

$$\beta_k = \alpha_k + \alpha_{n-k+1} + (-1)^k \big(I_k(A) + I_{k+1}(A)\big), \quad k = 2, \ldots, \left[\frac{n-1}{2}\right], \tag{4.6}$$

$$\beta_k = \alpha_k + \alpha_{k+1} + (-1)^k I_k(A), \quad k = \left[\frac{n+1}{2}\right], \quad n \text{ even}, \tag{4.7}$$

$$\beta_k = \alpha_k + (-1)^k I_k(A), \quad k = \left[\frac{n+1}{2}\right], \quad n \text{ odd}. \tag{4.8}$$

Here the numbers $I_k(A)$, $0 \leq k \leq \left[\frac{n+1}{2}\right]$, are defined by (2.2).

PROOF. To prove this result we apply Corollary 3.2 and show that the conditions (a) and (b) are fulfilled for the matrices A and $B = T(A)$. To check (a) for $T(A)$ one easily computes that $\mu_k(T(A)) = (-1)^{\frac{k(k-1)}{2}}$ for $1 \leq k \leq \left[\frac{n}{2}\right]$. To check (b) one computes for $k+1 \leq \ell \leq \left[\frac{n+1}{2}\right]$ that

$$\nu_{k\ell}(T(A)) = \begin{pmatrix} \beta_\ell & 0 & \cdots & 0 & 1 \\ 0 & 0 & & 1 & 1 \\ \vdots & & \iddots & & \vdots \\ 0 & 1 & & 0 & 1 \\ \alpha_\ell & 0 & \cdots & 0 & 1 \end{pmatrix} = (-1)^{\frac{k(k+1)}{2}}(\alpha_\ell - \beta_\ell),$$

and for $\left[\frac{n+1}{2}\right] < \ell < n-k+1$ that

$$\nu_{k\ell}(T(A)) = \begin{pmatrix} 0 & 0 & \cdots & 0 & 1 \\ 0 & 0 & & 1 & 1 \\ \vdots & & \iddots & & \vdots \\ 0 & 1 & & 0 & 1 \\ \alpha_\ell & 0 & \cdots & 0 & 0 \end{pmatrix} = (-1)^{\frac{k(k+1)}{2}}\alpha_\ell.$$

It follows that

$$I_{k+1}(T(A)) = (-1)^{k+1}\left(\sum_{i=k+1}^{\left[\frac{n+1}{2}\right]}\beta_i - \sum_{i=k+1}^{n-k}\alpha_i\right),$$

for $1 \leq k \leq \left[\frac{n-1}{2}\right]$. Hence the relations $I_{k+1}(A) = I_{k+1}(T(A))$ are satisfied for $1 \leq k \leq \left[\frac{n-1}{2}\right]$ if and only if the quantities β_i $(2 \leq i \leq \left[\frac{n+1}{2}\right])$ satisfy the linear system of equations

$$\sum_{i=k+1}^{\left[\frac{n+1}{2}\right]}\beta_i = \sum_{i=k+1}^{n-k}\alpha_i + (-1)^{k+1}I_{k+1}(A), \quad k = 1,\dots,\left[\frac{n-1}{2}\right].$$

This system has a unique solution, which is given by (4.6)–(4.8). Therefore (b) is fulfilled and $T(A)$ is unique. □

XV.5 An eigenvalue completion problem

In this section we consider the following eigenvalue completion problem. Given a U-specified matrix A, find the sets of eigenvalues of all possible completions of A. In other words, given a U-specified matrix A and entries a_{ij} for $1 \leq i \leq j \leq n$, find the possible eigenvalues of the matrices $B = \left(b_{ij}\right)_{i,j=1}^{n}$ such that $b_{ij} = a_{ij}$ if $1 \leq i \leq j \leq n$. First we treat the case when the U-specified matrix A is not decomposed.

Theorem 5.1. *Let A be a not decomposed U-specified $n \times n$ matrix, and let $\alpha_1, \dots, \alpha_n$ be given complex numbers (not necessarily all different). Then there is a completion of A which has eigenvalues $\alpha_1, \dots, \alpha_n$ if and only if $\sum_{i=1}^n \alpha_i =$ trace A_V.*

PROOF. First assume that we have a completion with eigenvalues $\alpha_1, \dots, \alpha_n$. Since the trace of a matrix is the sum of its eigenvalues and all completions share the same main diagonal, it is clear that $\sum_{i=1}^n \alpha_i = \operatorname{trace} A$.

Conversely, assume that $\sum_{i=1}^n \alpha_i = \operatorname{trace} A$. It follows from Theorem 4.1 that there exists a U-specified matrix $\tilde{A}$ which is U-similar to A and has diagonal entries $\alpha_1, \dots, \alpha_n$. Choose L to be lower triangular and invertible and such that A and $L\tilde{A}L^{-1}$ have the same upper triangular part. Let $\tilde{B} = \left(\tilde{b}_{ij}\right)_{i,j=1}^n$ be a completion of $\tilde{A}$ with $\tilde{b}_{ij} = 0$ if $i > j$. Since $\tilde{B}$ is an upper triangular matrix, $\tilde{B}$ has eigenvalues $\alpha_1, \dots, \alpha_n$. Put $B = L\tilde{B}L^{-1}$. The matrix B has eigenvalues $\alpha_1, \dots, \alpha_n$. On the other hand the entries in the upper triangular part of B depend on the entries in the upper triangular part of $\tilde{B}$ and on L only. Therefore the upper triangular part of B is A. This proves that B is a completion of A. □

For a general U-specified matrix (not necessarily not decomposed) we have the following result.

Theorem 5.2. *Let A be a U-specified $n \times n$ matrix. Assume that the matrix A admits a block partitioning $A = \left(A_{pq}\right)_{p,q=1}^r$ with $r > 1$, where A_{pp} is a not decomposed U-specified $n_p \times n_p$ matrix and $A_{pq} = 0$ for $1 \le p < q \le r$. Let $\alpha_1, \dots, \alpha_n$ be given complex numbers (not necessarily all different). Then there exists a completion of A which has eigenvalues $\alpha_1, \dots, \alpha_n$ if and only if there is a permutation σ of $\{1, \dots, n\}$ such that*

$$\sum_{i=m_{p-1}+1}^{m_p} \alpha_{\sigma(i)} = \operatorname{trace} A_{pp}, \quad (p = 1, \dots, r), \tag{5.1}$$

where $m_0 = 0$ and $m_p = \sum_{q=1}^p n_q$ for $p = 1, \dots, r$.

PROOF. First we show the necessity of (5.1). Since $A_{pq} = 0$ if $p < q$, the characteristic polynomial of any completion of A is the product of the induced completions of A_{pp} for $p = 1, \dots, r$. Now use that the sum of the roots (counted with multiplicity) of the characteristic polynomial of A_{pp} is precisely the trace of A_{pp}.

Conversely, assume that (5.1) holds true. Apply Theorem 5.1 to each of the not decomposed U-specified matrices A_{pp} and the n_p numbers $\alpha_{\sigma(m_{p-1}+1)}, \dots, \alpha_{\sigma(m_p)}$. □

Note that in the generic case (i.e., when (4.5) holds) it is also possible to provide a proof of Theorem 5.1 based on the canonical form given in Theorem 3.1.

XV.6 Applications

In this section we give applications of Theorems 5.1 and 5.2. The first is the following result.

Theorem 6.1. *Let $d_1, \dots, d_n$ and $\alpha_1, \dots, \alpha_n$ be complex numbers. There exists a matrix with diagonal entries $d_1, \dots, d_n$ and with eigenvalues $\alpha_1, \dots, \alpha_n$ if and only if $\sum_{i=1}^n d_i = \sum_{i=1}^n \alpha_i$.*

PROOF. First we construct the not decomposed U-specified matrix A with entries a_{ij} such that $a_{ii} = d_i$ for $i = 1, \dots, n$ and $a_{ij} = 1$ if $1 \le i < j \le n$. Apply Theorem 5.1 to this matrix and find a completion B of A with eigenvalues $\alpha_1, \dots, \alpha_n$. Then A has the prescribed eigenvalues and diagonal entries. □

The second application concerns the problem to find the minimal possible value of the spectral radius of a completion of a U-specified matrix.

Theorem 6.2. *Let A be a U-specified $n \times n$ matrix. Assume that A admits a block partitioning $A = \left(A_{pq}\right)_{p,q=1}^r$ with A_{pp} a not decomposed U-specified $n_p \times n_p$ matrix and $A_{pq} = 0$ for $1 \le p < q \le r$. Then*

$$\min\{\rho(B) \mid B \text{ is a completion of } A\} = \max\left\{\frac{|\text{trace}\, A_{pp}|}{n_p} \mid p = 1, \dots, r\right\},$$

where $\rho(B)$ denotes the spectral radius of B.

PROOF. Put $m_p = \sum_{q=1}^p n_q$ for $p = 1, \dots, r$ and $m_0 = 0$. We define numbers $\alpha_1, \dots, \alpha_n$ by

$$\alpha_i = \frac{\text{trace}\, A_{pp}}{n_p}, \quad i = m_{p-1} + 1, \dots, m_p, \quad p = 1, \dots, r.$$

From Theorem 5.2 we conclude that there exists a completion B_0 of A such that the eigenvalues of B_0 are $\alpha_1, \dots, \alpha_n$. So we get

$$\rho(B_0) = \max\left\{\frac{|\text{trace}\, A_{pp}|}{n_p} \mid p = 1, \dots, r\right\}.$$

On the other hand let $B = \left(B_{pq}\right)_{p,q=1}^r$ be any completion of A, where B_{pp} has size $n_p \times n_p$ for each p. Then B is a block lower triangular matrix, and hence $\rho(B) \ge \rho(B_{pp})$ for $p = 1, \dots, r$. Use that A_{pp} and B_{pp} share the same diagonal to see that $\rho(B_{pp}) \ge \frac{1}{n_p} |\,\text{trace}\, A_{pp}|$. So we get for any completion B of A that

$$\rho(B) \ge \max\left\{\frac{|\,\text{trace}\, A_{pp}|}{n_p} \mid p = 1, \dots, r\right\}.$$ □

Notes

The first section is based mainly on Ball-Gohberg-Rodman-Shalom [1]. Sections 2 and 3 are taken from Gohberg-Rubinstein [1]. Sections 4, 5 and 6 again come from Ball-Gohberg-Rodman-Shalom [1]. The results of the first five sections provide partial answers for the two main problems (the description of U-similarity invariants and the U-canonical form). The full solution for the general case is not known. Theorem 6.1 is due to Mirsky [1]. Theorem 6.2 is motivated by an infinite dimensional result of Bercovici-Foias-Tannenbaum [1]. For related results in this direction see Gohberg-Rodman-Shalom-Woerdeman [1], Gurvits-Rodman-Shalom [1], [2], Rodman-Shalom [1], Friedland [1] and Krupnik-Rodman [1].

Appendix

This appendix provides prerequisites for the text on matrix polynomials and rational matrix functions in the Chapters X and XI. Sections 1–4 give a concise and self-contained exposition of the local spectral theory of regular analytic matrix functions based on the notion of root function. Section 5 develops the spectral theory of regular matrix functions that are analytic in a domain.

A.1 Root functions of regular analytic matrix functions

We consider a regular analytic matrix function L, that is, L is a square, say $m \times m$, analytic matrix function in one variable with a determinant that does not vanish identically. Throughout this appendix μ is a point in the domain of analyticity of L (for short L is analytic at μ). We assume that $\det L(\mu) = 0$. A $\mathbb{C}^m$-vector function ϕ, analytic in a neighbourhood of μ, is called a *(right) root function* of $L(\lambda)$ at μ if $\phi(\mu) \neq 0$ and $L(\mu)\phi(\mu) = 0$. The order k of μ as a zero of the analytic vector function $L\phi$ is called the *order* of the root function ϕ at μ. Thus an analytic vector function ϕ is a root function of order at least k if and only if $\phi(\mu) \neq 0$ and

$$L(\lambda)\phi(\lambda) = \sum_{j=k}^{\infty} (\lambda - \mu)^j y_j. \tag{1.1}$$

If the analytic vector function ϕ and the analytic matrix function L are given in a neighbourhood of μ by

$$\phi(\lambda) = \sum_{j=0}^{\infty} (\lambda - \mu)^j \phi_j, \quad L(\lambda) = \sum_{j=0}^{\infty} (\lambda - \mu)^j L_j, \tag{1.2}$$

then formula (1.1) implies

$$\sum_{j=0}^{r} L_j \phi_{r-j} = 0, \quad r = 0, \dots, k-1. \tag{1.3}$$

A sequence of $\mathbb{C}^m$-vectors $\phi_0, \phi_1, \dots, \phi_{k-1}$, with $\phi_0 \neq 0$, satisfies (1.3) if and only if the polynomial

$$\tilde{\phi}(\lambda) = \phi_0 + (\lambda - \mu)\phi_1 + \cdots + (\lambda - \mu)^{k-1}\phi_{k-1} \tag{1.4}$$

is a root function of L at μ of order at least k. Thus ϕ in (1.2) is a root function of order at least k, if and only if the same holds true for the polynomial $\tilde{\phi}$ in (1.4).

The order k of a root function ϕ at μ is at most the order of μ as a zero of $\det L(\mu)$. To see this, choose $f_2, \ldots, f_m$ such that $\phi(\mu), f_2, \ldots, f_m$ is a basis of $\mathbb{C}^m$, and consider the equality

$$\det L(\lambda)\ \det\begin{pmatrix}\phi(\lambda) & f_2 & \cdots & f_m\end{pmatrix} = \det\begin{pmatrix}L(\lambda)\phi(\lambda) & L(\lambda)f_2 & \cdots & L(\lambda)f_m\end{pmatrix}.$$

Since $\det\begin{pmatrix}\phi(\mu) & f_2 & \cdots & f_m\end{pmatrix} \neq 0$, the order of μ as a zero of the left hand side is the order r of μ as a zero of $\det L(\lambda)$. On the other hand in the right side of the equality $(\lambda-\mu)^k$ is a factor of the first column of the matrix and therefore also a factor of the determinant. So in the right hand side μ is a zero of at least order k. This proves that $k \leq r$.

We assume that $\det L(\mu) = 0$. Therefore there are infinitely many root functions at μ. We will select a special set of root functions as follows. Choose from all root functions at μ a root function ϕ_1 of the highest order κ_1. Since the orders of the root functions are bounded by the order of μ as a zero of $\det L(\lambda)$, such a function exists. Next choose from all root functions ϕ, with $\phi(\mu)$ not a multiple of $\phi_1(\mu)$, a root function ϕ_2 of the highest order, say κ_2. We proceed by induction. If the functions $\phi_1, \ldots, \phi_{k-1}$ are already chosen, we choose the next ϕ_k to be a root function of the highest order κ_k among all root functions ϕ such that $\phi(\mu)$ is independent of $\phi_1(\mu), \ldots, \phi_{k-1}(\mu)$. This process stops at the moment that the vectors $\phi_1(\mu), \ldots, \phi_s(\mu)$ span the finite dimensional space $\operatorname{Ker} L(\mu)$. Any set of root functions $\phi_1, \ldots, \phi_s$ obtained in this manner is called *a canonical system of (right) root functions* of L at μ. Such a canonical system of root functions is not unique. For instance we could replace $\phi_1(\lambda)$ by the function $\phi_1(\lambda)+(\lambda-\mu)^{(\kappa_1-\kappa_2)}\phi_2(\lambda)$. On the other hand the next lemma shows that the sequence of numbers $\kappa_1 \geq \kappa_2 \geq \cdots \geq \kappa_s$ is uniquely determined by L.

Lemma 1.1. *Let L be a regular matrix function which is analytic at μ, and let $\psi_1, \ldots, \psi_r$ be a set of root functions of L at μ with orders $\rho_1 \geq \cdots \geq \rho_r$ such that $\psi_1(\mu), \ldots, \psi_r(\mu)$ is an independent set in $\operatorname{Ker} L(\mu)$. If $\phi_1, \ldots, \phi_s$ is a canonical system of root functions of L at μ with orders $\kappa_1 \geq \cdots \geq \kappa_s$, then $s \geq r$ and $\kappa_i \geq \rho_i$ for $i = 1, \ldots, r$. Moreover, $\psi_1, \ldots, \psi_r$ is a canonical system of root functions of L at μ if and only if $r = s$ and $\kappa_i = \rho_i$ for $i = 1, \ldots, r$.*

Proof. Since $s = \dim \operatorname{Ker} L(\mu)$ and the set of vectors $\psi_1(\mu), \ldots, \psi_r(\mu)$ is independent in $\operatorname{Ker} L(\mu)$, it is immediate that $r \leq s$.

Suppose the set $\psi_1, \ldots, \psi_r$ is given. From the definition of the functions $\phi_1, \ldots, \phi_s$, we see that the root function with the largest order has at least order ρ_1. This proves that $\kappa_1 \geq \rho_1$. Let $1 < k \leq r$. We remark that $\psi_1(\mu), \ldots, \psi_k(\mu)$ is an independent set of vectors and therefore

$$\{\psi_1(\mu), \ldots, \psi_k(\mu)\} \not\subset \operatorname{span}\{\phi_1(\mu), \ldots, \phi_{k-1}(\mu)\}. \tag{1.5}$$

So at least one of the vectors $\psi_1(\mu), \ldots, \psi_k(\mu)$ is not in $\operatorname{span}\{\phi_1(\mu), \ldots, \phi_{k-1}(\mu)\}$. Since ϕ_k has maximal order among all root functions ϕ with $\phi(\mu)$ not in the span of $\{\phi_1(\mu), \ldots, \phi_{k-1}(\mu)\}$, the order of ϕ_k must be at least ρ_k. This proves that $\kappa_k \geq \rho_k$. We proved that $\kappa_k \geq \rho_k$, for $k = 1, \ldots, r$.

If $\psi_1, \ldots, \psi_r$ is a canonical system of root functions, then we may interchange the roles of the sets $\psi_1, \ldots, \psi_r$ and $\phi_1, \ldots, \phi_s$. This gives that also $s \leq r$ and $\rho_k \geq \kappa_k$ for $k = 1, \ldots, s$.

Assume that $\kappa_i = \rho_i$ for $i = 1, \ldots, r = s$ and $\psi_1, \ldots, \psi_s$ is not a canonical system of root functions. This means that for some i the function ψ_i is not a root function of maximal order among all the root functions ψ with $\psi(\mu) \notin \operatorname{span}\{\psi_1(\mu), \ldots, \psi_{i-1}(\mu)\}$. So we could choose an i-th root function with order larger than κ_i and thus construct a canonical system of root polynomials with orders $\nu_1, \ldots, \nu_r$ and $\nu_i > \kappa_i$. This contradicts the results obtained above. □

Let $\phi_1, \ldots, \phi_k$ be a canonical system of root functions of the analytic matrix function L at μ. We choose functions $\phi_{k+1}, \ldots, \phi_m$ such that the set of vectors

$$\phi_1(\mu), \ldots, \phi_k(\mu), \phi_{k+1}(\mu), \ldots, \phi_m(\mu) \tag{1.6}$$

is a basis for the space $\mathbb{C}^m$. We call the system $\phi_1, \ldots, \phi_k, \phi_{k+1}, \ldots, \phi_m$ an *extended canonical system of root functions.* The (possibly constant) functions $\phi_{k+1}, \ldots, \phi_m$ are not root functions in the strict sense of our definition. Nevertheless it is convenient to assign them an order as root function, and to put this order equal to 0. So in an extended canonical system of root functions each function has an order, and those of positive order are the genuine root functions. The orders of the root functions in an extended canonical system of L at μ are called the *partial multiplicities* of L at μ.

Theorem 1.2. *Let L be a regular $m \times m$ matrix function analytic at μ such that $\det L(\mu) = 0$, and let Φ be an $m \times m$ analytic matrix function such that its columns form an extended canonical system of root functions of L at μ. Then for λ in a neighbourhood of μ*

$$L(\lambda)\Phi(\lambda) = P(\lambda)D(\lambda), \tag{1.7}$$

where $D(\lambda)$ is a diagonal matrix with diagonal entries $(\lambda-\mu)^{\kappa_1}, \ldots, (\lambda-\mu)^{\kappa_m}$ and P is an $m \times m$ matrix function analytic at μ such that $\det P(\mu) \neq 0$. Furthermore, the exponents $\kappa_1, \ldots, \kappa_m$ in D are the partial multiplicities of L at μ.

Proof. Let $\phi_1, \ldots, \phi_m$ be an extended canonical system of root functions for L at μ with orders $\kappa_1 \geq \cdots \geq \kappa_m$. Put Φ the analytic matrix function with columns $\phi_1, \ldots, \phi_m$. Then we can write $L\Phi = PD$ with D as required and $\Phi(\mu)$ invertible. It remains to show that $P(\mu)$ is invertible. Assume that $P(\mu)x = 0$ for a vector $x \neq 0$. Without loss of generality we may assume that x has the form

$$x = \begin{pmatrix} 0 & \cdots & 0 & 1 & x_{j+1} & \cdots & x_m \end{pmatrix}^T.$$

Write

$$\tilde{\phi}_j(\lambda) = \phi_j(\lambda) + \sum_{i=j+1}^{m} (\lambda - \mu)^{\kappa_j - \kappa_i} x_i \phi_i(\lambda).$$

Then $\tilde{\phi}_j(\mu)$ is independent of the vectors $\phi_1(\mu), \ldots, \phi_{j-1}(\mu)$. Note that

$$L(\lambda)\tilde{\phi}_j(\lambda) = L(\lambda)\phi_j(\lambda) + \sum_{i=j+1}^{m} (\lambda - \mu)^{\kappa_j - \kappa_i} x_i L(\lambda)\phi_i(\lambda)$$
$$= (\lambda - \mu)^{\kappa_j}\big(p_j(\lambda) + \sum_{i=j+1}^{m} x_i p_i(\lambda)\big).$$

Here $p_i(\lambda)$ is the i-th column of $P(\lambda)$. Since $p_j(\mu) + \sum_{i=j+1}^{m} x_i p_i(\mu) = 0$, it follows that the order of $\tilde{\phi}_j$ as a root function of $L(\lambda)$ is at least $\kappa_j + 1$. If $\kappa_j > 0$, then this contradicts the choice of ϕ_j as a root function with maximal order among those that have a value at μ linearly independent of $\phi_1(\mu), \ldots, \phi_{j-1}(\mu)$. If $\kappa_j = 0$, then $j > s = \dim \operatorname{Ker} L(\mu)$ and thus $\tilde{\phi}_j(\mu)$ is independent of $\phi_1(\mu), \ldots, \phi_s(\mu)$, which implies that $\tilde{\phi}_j(\mu) \notin \operatorname{Ker} L(\mu)$. On the other hand $\tilde{\phi}_j$ is a genuine root function of order 1, and hence $\tilde{\phi}_j(\mu) \in \operatorname{Ker} L(\mu)$. We proved that $P(\mu)$ is invertible. □

The next result gives a characterization of an extended canonical system of root functions.

Theorem 1.3. *Let L, Φ, D and P be regular $m \times m$ matrix functions, analytic at μ, such that $L(\lambda)\Phi(\lambda) = P(\lambda)D(\lambda)$ for λ in a neighbourhood of μ. Assume that $\Phi(\mu)$ is invertible and that $D(\lambda)$ is a diagonal matrix with diagonal entries $(\lambda-\mu)^{\kappa_1}, \ldots, (\lambda-\mu)^{\kappa_m}$, where $\kappa_1 \geq \cdots \geq \kappa_m$. Then the following three conditions are equivalent:*

(1) *the columns $\phi_1, \ldots, \phi_m$ of the analytic matrix function Φ form an extended canonical system of root functions of $L(\lambda)$ at μ with orders $\kappa_1 \geq \cdots \geq \kappa_m$;*

(2) *$\det P(\mu) \neq 0$;*

(3) *$\sum_{i=1}^{m} \kappa_i$ is equal to the order of μ as a zero of $\det L(\lambda)$.*

PROOF. Assume that $\phi_1, \ldots, \phi_m$ is an extended canonical system of root functions for L with orders $\kappa_1 \geq \cdots \geq \kappa_m$. Then we conclude from Theorem 1.2 that (2) is satisfied.

We show that (2) and (3) are equivalent. Consider the equality

$$\det L(\lambda) \det \Phi(\lambda) = \det P(\lambda) \det D(\lambda).$$

If $\det P(\mu) \neq 0$, then $\det \Phi(\mu) \neq 0$ implies that the order of μ as a zero of $\det L(\lambda)$ is equal to the order of μ as a zero of $\det D(\lambda)$. This proves that $\sum_{i=1}^{m} \kappa_i$ is equal to the order of μ as a zero of $\det L(\lambda)$.

If $\sum_{i=1}^{m} \kappa_i$ is equal to the order of μ as a zero of $\det L(\lambda)$, then the orders of μ as a zero of $\det L(\lambda)$ and $\det D(\lambda)$ are equal. Since $\det \Phi(\mu) \neq 0$, this implies that $\det P(\mu) \neq 0$.

Finally we prove that (2) implies (1). Let the orders of an extended canonical system of root functions be $\rho_1 \geq \cdots \geq \rho_m$. We have proved that this implies that $\sum_{i=1}^m \rho_i$ is equal to the order of μ as a zero of $\det L(\lambda)$. Since $P(\mu)$ and $\Phi(\mu)$ are invertible, the orders of μ as a zero of $\det L(\lambda)$ and $\det D(\lambda)$ are equal. So $\sum_{i=1}^m \kappa_i = \sum_{i=1}^m \rho_i$. Let the j-th column of Φ be ϕ_j, and let the j-th column of P be p_j. Then $\phi_j(\mu) \neq 0$, $p_j(\mu) \neq 0$ and $L(\lambda)\phi_j(\lambda) = (\lambda-\mu)^{\kappa_j} p_j(\lambda)$, and thus ϕ_j is a root function of order κ_j. Apply Lemma 1.1 to see that this proves that $\kappa_j \leq \rho_j$ for $j = 1, \ldots, m$. Since $\sum_{i=1}^m \kappa_i = \sum_{i=1}^m \rho_i$, it follows that $\kappa_j = \rho_j$ for $j = 1, \ldots, m$, and thus $\phi_1, \ldots, \phi_m$ is an extended canonical system of root functions for L. □

From Theorems 1.2 and 1.3 it follows that the diagonal factor D in (1.7) is uniquely determined by L, provided that $\Phi(\mu)$ and $P(\mu)$ are invertible. One refers to D as the *local Smith* form of L.

Of course there exists a relation between the Smith canonical form of a matrix polynomial and this local Smith form of a regular analytic matrix function. To see this assume that L is a regular $m \times m$ matrix polynomial with $\det L(\mu) = 0$. Let $\operatorname{diag}\big(p_i(\lambda)\big)_{i=1}^m$ be the Smith canonical form of $L(\lambda)$, i.e., $p_1, \ldots, p_m$ are scalar polynomials, $p_i | p_{i-1}$ for $i = 2, 3, \ldots, m$, and L admits the factorization

$$L(\lambda) = E(\lambda) \operatorname{diag}\big(p_i(\lambda)\big)_{i=1}^m F(\lambda)$$

with E and F unimodular. Write $p_i(\lambda) = (\lambda-\mu)^{\kappa_i} q_i(\lambda)$ with $q_i(\mu) \neq 0$. So

$$L(\lambda) = E(\lambda) \operatorname{diag}\left((\lambda-\mu)^{\kappa_i}\right)_{i=1}^m \operatorname{diag}\left(q_i(\lambda)\right)_{i=1}^m F(\lambda).$$

The matrices $E(\mu)$ and $\operatorname{diag}\left(q_i(\mu)\right)_{i=1}^m F(\mu)$ are invertible, and hence the matrix functions $P(\lambda) := E(\lambda)$ and $\Phi(\lambda) := \left(\operatorname{diag}\big(q_i(\lambda)\big)_{i=1}^m F(\lambda)\right)^{-1}$ satisfy the conditions of Theorem 1.3. This shows that $\operatorname{diag}\left((\lambda-\mu)^{\kappa_i}\right)_{i=1}^m$ is the local Smith form of $L(\lambda)$ at μ.

A.2 Right Jordan pairs of regular analytic matrix functions

Let $\phi(\lambda) = \phi_0 + (\lambda-\mu)\phi_1 + \cdots + (\lambda-\mu)^{\ell-1}\phi_{\ell-1}$ be an $\mathbb{C}^m$-vector polynomial. With ϕ we associate the matrix

$$X_{\mu,\phi} = \begin{pmatrix} \phi_0 & \cdots & \phi_{\ell-1} \end{pmatrix}.$$

By $J_{\mu,\kappa}$ we denote the single $\kappa \times \kappa$ Jordan block with eigenvalue μ. So

$$J_{\mu,\kappa} = \begin{pmatrix} \mu & 1 & & & \\ 0 & \mu & \ddots & & \\ & \ddots & \ddots & \ddots & \\ & & \ddots & \mu & 1 \\ & & & 0 & \mu \end{pmatrix}.$$

The polynomial ϕ can be expressed in terms of the pair $(X_{\mu,\phi}, J_{\mu,\ell})$ by the following formula

$$\phi(\lambda) = X_{\mu,\phi}(\lambda I - J_{\mu,\ell})^{-1} e_{\ell\ell}(\lambda - \mu)^{\ell},$$

where $e_{\ell\ell}$ is the ℓ-th vector in the standard basis of $\mathbb{C}^{\ell}$. So $e_{\ell\ell} = (0 \ \ \dots \ \ 0 \ \ 1)^T$.

We now present a characterization of root functions in terms of the corresponding pair of matrices.

Lemma 2.1. *Let L be a regular $m \times m$ matrix function analytic at μ, and let*

$$\phi(\lambda) = \sum_{i=0}^{\infty} (\lambda - \mu)^i \phi_i$$

be an analytic $\mathbb{C}^m$-vector function with $\phi_0 \neq 0$. Write $X_{\mu,\phi} = \begin{pmatrix} \phi_0 & \cdots & \phi_{\kappa-1} \end{pmatrix}$. Then ϕ is a root function of L at μ of order at least κ if and only if the $m \times \kappa$ matrix function $L(\lambda)X_{\mu,\phi}(\lambda I - J_{\mu,\kappa})^{-1}$ is analytic at μ.

PROOF. For $j = 1, \dots, \kappa$ we write $\psi_j(\lambda) = \sum_{i=0}^{j-1} (\lambda - \mu)^i \phi_i$. One computes that

$$\psi_j(\lambda) = X_{\mu,\phi}(\lambda I - J_{\mu,\kappa})^{-1} e_{\kappa j}(\lambda - \mu)^j,$$

where $e_{\kappa j}$ is the j-th vector in the standard basis of $\mathbb{C}^{\kappa}$. So

$$\begin{pmatrix} \psi_1(\lambda) & \cdots & \psi_\kappa(\lambda) \end{pmatrix} = \\ X_{\mu,\phi}(\lambda I - J_{\mu,\kappa})^{-1} \begin{pmatrix} e_{\kappa 1} & \cdots & e_{\kappa\kappa} \end{pmatrix} \operatorname{diag}\big((\lambda - \mu), \dots, (\lambda - \mu)^{\kappa}\big)$$

and hence

$$L(\lambda) \begin{pmatrix} \psi_1(\lambda) & \cdots & \psi_\kappa(\lambda) \end{pmatrix} \operatorname{diag}\big((\lambda - \mu)^{-1} \dots, (\lambda - \mu)^{-\kappa}\big) =$$

$$L(\lambda) X_{\mu,\phi}(\lambda I - J_{\mu,\kappa})^{-1}.$$

Now assume that ϕ is a root function of order at least κ. Then $\psi_j(\lambda) = \sum_{i=0}^{j-1} (\lambda - \mu)^i \phi_i$ is a root function of at least order j for $j = 1, \dots, \kappa$. Hence $L(\lambda)\psi_j(\lambda)(\lambda - \mu)^{-j}$ is analytic at μ. This proves that each column in the matrix $L(\lambda)X_{\mu,\phi}(\lambda I - J_{\mu,\kappa})^{-1}$ is analytic at μ, and hence $L(\lambda)X_{\mu,\phi}(\lambda I - J_{\mu,\kappa})^{-1}$ is analytic at μ.

Conversely, assume that $L(\lambda)X_{\mu,\phi}(\lambda I - J_{\mu,\kappa})^{-1}$ is analytic at μ. In particular its last column is analytic at μ. This last column is $L(\lambda)\psi_\kappa(\lambda)(\lambda - \mu)^{-\kappa}$. It follows that ψ_κ is a root function of order at least κ. However we already noted that this implies that also ϕ is a root function of order $\geq \kappa$. □

Let L be a regular analytic matrix function, and let $\phi_1, \dots, \phi_s$ be a canonical system of root functions of L at μ of orders $\kappa_1 \geq \cdots \geq \kappa_s$. Write

$$\phi_i(\lambda) = \phi_{i0} + (\lambda - \mu)\phi_{i1} + \cdots + (\lambda - \mu)^{\kappa_i - 1}\phi_{i\kappa_i - 1} + \cdots \quad .$$

We define the matrix X_μ by

$$X_\mu = \begin{pmatrix} \phi_{10} & \cdots & \phi_{1\kappa_1 - 1} & \cdots & \phi_{s0} & \cdots & \phi_{s\kappa_s - 1} \end{pmatrix}. \tag{2.1}$$

With X_μ we associate the Jordan matrix

$$J_\mu = \operatorname{diag}(J_{\mu,\kappa_1}, \dots, J_{\mu,\kappa_s}). \tag{2.2}$$

The pair (X_μ, J_μ) is called a *right Jordan pair of $L(\lambda)$ at the point μ.*

Assume that (X_μ, J_μ) is a right Jordan pair of L at the point μ, with X_μ given by (2.1) and J_μ given by (2.2). For $i = 1, \dots, s$, put

$$\tilde{\phi}_i(\lambda) = \phi_{i0} + (\lambda - \mu)\phi_{i1} + \cdots + \phi_{i\kappa_i - 1}(\lambda - \mu)^{\kappa_i - 1}.$$

Then $\tilde{\phi}_i$ is the truncation to degree $\kappa_i - 1$ of the root function ϕ_i. So $\tilde{\phi}_i$ is itself a root function of L at μ of order at least κ_i. From Lemma 1.1 it follows that $\tilde{\phi}_1, \dots, \tilde{\phi}_s$ is a canonical system of root functions of L at μ. Extend this system to an extended canonical system of root functions. A system constructed in this way is called an *extended canonical system of root functions corresponding to* (X_μ, J_μ).

We have already seen that the matrix J_μ is fully determined by the analytic matrix function L and μ. Indeed the eigenvalue of J_μ is μ and the sizes of the Jordan blocks are given by the partial multiplicities of the analytic matrix function L at μ. A natural question is to what extent the matrix X_μ is determined by the function L. We return to this question in the next section. First we derive some other properties of a right Jordan pair.

Lemma 2.2. *Let (X_μ, J_μ) be a right Jordan pair of L at μ. Then*

$$\begin{pmatrix} \lambda I - J_\mu \\ X_\mu \end{pmatrix}$$

has full rank for each λ in $\mathbb{C}$.

PROOF. Put $R(\lambda) = \begin{pmatrix} \lambda I - J_\mu \\ X_\mu \end{pmatrix}$. Assume that $R(\nu)x = 0$, for some $x \neq 0$ and some ν. Write X_μ as in (2.1) and J_μ as in (2.2), and partition x as

$$x = \begin{pmatrix} x_{10} & \cdots & x_{1\kappa_1 - 1} & \cdots & x_{s0} & \cdots & x_{s\kappa_s - 1} \end{pmatrix}^T.$$

If $\nu \neq \mu$, then $x = 0$. So we have that $\nu = \mu$. Then $(\mu I - J_\mu)x = 0$ gives that $x_{i\alpha} = 0$ unless $\alpha = 0$. Thus $X_\mu x = 0$ implies that $x_{i0}\phi_{10} + \cdots + x_{s0}\phi_{s0} = 0$. The vectors $\phi_{10}, \dots, \phi_{s0}$ are the values at μ of a canonical set of root functions of $L(\lambda)$ at μ. Therefore $\phi_{10}, \dots, \phi_{s0}$ is an independent set of vectors, and thus also $x_{i\alpha} = 0$ if $\alpha = 0$. This contradicts our assumption on x. We proved that $R(\lambda)x = 0$ implies that $x = 0$. □

Let A be a $n \times n$ matrix and C be an $m \times n$ matrix. The pair (C, A) is called a zero kernel pair (cf., Section X.1) if $\bigcap_{i=0}^{n-1} \operatorname{Ker}(CA^i) = \{0\}$. From Theorem III.3.5 we know that this condition is equivalent to the requirement that for each $\lambda \in \mathbb{C}$

$$\operatorname{Ker}\begin{pmatrix} \lambda I - A \\ C \end{pmatrix} x = \{0\}.$$

This notion also coincides with the notion of an observable pair from mathematical systems theory. The next result gives a first characterization of a right Jordan pair.

Theorem 2.3. *Let L be a regular $m \times m$ analytic matrix function, and let J be a right Jordan matrix with μ as its only eigenvalue and Jordan blocks of sizes $\kappa_1 \geq \cdots \geq \kappa_s$. Then (X, J) is a right Jordan pair at $\mu \in \mathbb{C}$ if and only if the following three conditions hold true:*

(1) *the order of J is equal to the order of μ as a zero of* $\det L(\lambda)$*;*

(2) *the pair (X, J) is a zero kernel pair;*

(3) *$L(\lambda)X(\lambda I - J)^{-1}$ is analytic at μ.*

PROOF. Assume that (X, J) is a right Jordan pair at μ. Then statement (1) follows from Theorem 1.3, statement (2) is proved in Lemma 2.2, and (3) follows from Lemma 2.1.

Assume that the pair (X, J) fulfills the conditions (1), (2) and (3). We shall show that (X, J) is a right Jordan pair. Write $J = \operatorname{diag}(J_{\mu,\kappa_1}, \ldots, J_{\mu,\kappa_s})$ and decompose X correspondingly as $X = (X_1 \quad \cdots \quad X_s)$. Put

$$\phi_i(\lambda) = X_i(\lambda I - J_{\mu,\kappa_i})^{-1} e_{\kappa_i \kappa_i} (\lambda - \mu)^{\kappa_i}.$$

From condition (3) and Lemma 2.1 it follows that the polynomial ϕ_i is a root function of order at least κ_i. Condition (2) gives that the matrix

$$\begin{pmatrix} \mu I - J \\ X \end{pmatrix}$$

has independent columns. In particular the columns with numbers

$$1, \kappa_1 + 1, \ldots, \kappa_1 + \cdots + \kappa_{s-1} + 1$$

are independent. Column number $\kappa_1 + \cdots + \kappa_{j-1} + 1$ has the form

$$\begin{pmatrix} 0 \\ \phi_j(\mu) \end{pmatrix}.$$

So we see that $\phi_1(\mu), \ldots, \phi_s(\mu)$ are linearly independent. Next let $\nu_1, \ldots, \nu_k$ be the orders of a canonical system of root polynomials of L at μ. It follows from Lemma 1.1 that $\kappa_i \leq \nu_i$ and $s \leq k$. From Theorem 1.3 we obtain that

$$\kappa_1 + \cdots + \kappa_s \leq \nu_1 + \cdots + \nu_k = r, \tag{2.3}$$

where r is the order of μ as a zero of $\det L(\lambda)$. Now (1) gives that, in fact, in (2.3) one has equalities and $k = s$. Again apply Lemma 1.1 to conclude that (X, J) is a right Jordan pair. □

We single out one partial result from the previous proof.

Corollary 2.4. *Let L be a regular $m \times m$ analytic matrix function, and let J be a Jordan matrix with μ as its only eigenvalue and blocks of sizes $\kappa_1 \geq \cdots \geq \kappa_s$. If conditions (2) and (3) in Theorem 2.3 are fulfilled, then the order of J is at most the order of μ as a zero of $\det L(\lambda)$, and (X, J) is a right Jordan pair at μ if these orders are equal.*

Note that for a regular $m \times m$ matrix polynomial

$$L(\lambda) = A_0 + \lambda A_1 + \cdots + \lambda^\ell A_\ell$$

condition (3) in Theorem 2.3 is equivalent to

$$A_0 X + A_1 X J + \cdots + A_\ell X J^\ell = 0. \tag{2.4}$$

Let us check this. First note that

$$(\lambda I - J)^{-1} = \lambda^{-1}(I - \lambda^{-1} J)^{-1} = \lambda^{-1} + \lambda^{-2} J + \cdots + \lambda^{-k} J^{k-1} + \cdots$$

for $|\lambda| > |\mu|$. Therefore

$$L(\lambda) X (\lambda I - J)^{-1} = R(\lambda) + \sum_{\nu=1}^{\infty} \lambda^{-\nu} (A_0 X + A_1 X J + \cdots + A_\ell X J^\ell) J^{\nu-1}, \quad |\lambda| > |\mu|, \tag{2.5}$$

where $R(\lambda)$ is a polynomial in λ. Recall that $L(\lambda)X(\lambda I - J)^{-1}$ has no singularities except maybe at μ. It follows that $L(\lambda)X(\lambda I - J)^{-1}$ is analytic at μ if and only if (2.4) holds.

Assume that we have a zero kernel pair (X, J), where J is a right Jordan matrix with a single eigenvalue μ and Jordan blocks of sizes $\kappa_1 \geq \cdots \geq \kappa_s > 0$. A natural question is whether or not there exists a regular matrix function $L(\lambda)$ analytic at μ, such that the pair (X, J) is a right Jordan pair of $L(\lambda)$ at μ. The answer is positive; in fact, as the next theorem shows one can always find a matrix polynomial with this property.

Theorem 2.5. *Let X be a $m \times n$ matrix, and let J be an $n \times n$ matrix with a single eigenvalue μ and Jordan blocks of sizes $\kappa_1 \geq \cdots \geq \kappa_s > 0$. If the pair (X, J) is a zero kernel pair, then there exists a matrix polynomial L such that the pair (X, J) is a right Jordan pair of L at μ.*

PROOF. Let $J = \operatorname{diag}(J_{\mu,\kappa_1}, \ldots, J_{\mu,\kappa_s})$, and let the corresponding decomposition of X be given by $X = \begin{pmatrix} X_1 & \cdots & X_s \end{pmatrix}$. Write

$$X_i = \begin{pmatrix} \phi_{i0} & \cdots & \phi_{i\kappa_i - 1} \end{pmatrix}$$

and

$$\phi_i(\lambda) = \phi_{i0} + (\lambda - \mu)\phi_{i1} + \cdots + (\lambda - \mu)^{\kappa_i - 1}\phi_{i\kappa_i - 1}.$$

Since (X, J) is a zero kernel pair, the values $\phi_1(\mu), \ldots, \phi_s(\mu)$ are linearly independent (cf., the proof of Theorem 2.3). Choose vectors $\phi_{s+1}, \ldots, \phi_m$ in $\mathbb{C}^m$ such that

$$\phi_1(\mu), \ldots, \phi_s(\mu), \phi_{s+1}, \ldots, \phi_m$$

is a basis for $\mathbb{C}^m$. Write

$$\Phi(\lambda) = \begin{pmatrix} \phi_1(\lambda) & \ldots & \phi_s(\lambda) & \phi_{s+1} & \ldots & \phi_m \end{pmatrix}.$$

Then $\Phi(\mu)$ is invertible. Consider the $m \times m$ matrix

$$D(\lambda) = \operatorname{diag}\bigl((\lambda - \mu)^{\kappa_1}, \ldots, (\lambda - \mu)^{\kappa_s}, 1, \ldots, 1\bigr).$$

Let $p(\lambda)$ be the denominator of the scalar rational function $\det\bigl(D(\lambda)\Phi(\lambda)^{-1}\bigr)$. We may choose $p(\mu) = 1$, because $\Phi(\mu)$ is invertible. With this choice of $p(\lambda)$ the function $L(\lambda) = p(\lambda)D(\lambda)\Phi(\lambda)^{-1}$ is a matrix polynomial. Since $L(\lambda)\Phi(\lambda) = (p(\lambda)I)D(\lambda)$, it follows from Theorem 1.3 that the columns of $\Phi(\lambda)$ form an extended canonical system of root functions of L at μ. This proves that (X, J) is a right Jordan pair of $L(\lambda)$ at μ. □

The following theorem describes to what extent a right Jordan pair at a point determines the analytic matrix function.

Theorem 2.6. *Two regular $m \times m$ matrix functions L_1 and L_2, analytic at μ, have the same right Jordan pair at $\mu \in \mathbb{C}$ if and only if $L_2(\lambda)L_1(\lambda)^{-1}$ is analytic and invertible at μ.*

PROOF. Assume that $L_2(\lambda)L_1(\lambda)^{-1}$ is analytic and invertible at μ. Let (X, J) be a right Jordan pair of $L_1(\lambda)$ at μ. We shall prove that (X, J) is also a right Jordan pair of $L_2(\lambda)$ at μ. From our hypothesis it follows that the order of μ as a zero of $\det L_2(\lambda)$ is equal to the order of μ as a zero of $\det L_1(\lambda)$, and hence is equal to the order of J. Since $L_1(\lambda)X(\lambda I - J)^{-1}$ is analytic at μ, also $L_2(\lambda)X(\lambda I - J)^{-1}$ is analytic at μ. So it follows from Theorem 2.3 that (X, J) is a right Jordan pair of $L_2(\lambda)$ at μ.

Conversely, assume that (X, J) is a right Jordan pair of $L_1(\lambda)$ and $L_2(\lambda)$ at μ. Let $\Phi(\lambda)$ be the matrix of which the columns form an extended canonical system of root functions corresponding to (X, J). Thus (by Theorem 1.2)

$$L_1(\lambda)\Phi(\lambda) = P_1(\lambda)D(\lambda), \quad L_2(\lambda)\Phi(\lambda) = P_2(\lambda)D(\lambda)$$

with $P_1(\lambda)$ and $P_2(\lambda)$ analytic matrix functions such that $P_1(\mu)$ and $P_2(\mu)$ are invertible. It follows that $L_2(\lambda)L_1(\lambda)^{-1} = P_2(\lambda)P_1(\lambda)^{-1}$. The right hand side of this equality clearly is analytic and invertible at μ. □

A.3 Left Jordan pairs

In this section we present the analogies of the Sections 1 and 2 for row vector functions instead of column vector functions. The row vector function ψ is called a *left root function* of the regular analytic $m \times m$ matrix function L at μ if the transposed column vector ψ^T is a right root function of the transposed matrix function L^T. The row vector functions $\psi_1, \ldots, \psi_s$ are called a *canonical system of left root functions* of L at μ if $\psi_1^T, \ldots, \psi_s^T$ is a canonical system of right root functions of L^T. Analogously one defines an *extended canonical system of left root functions.* The order of the left root function ψ is just the order of the right root function ψ^T. By using these transposition relations between right root functions and left root functions, we derive from Theorem 1.2 the following result.

Theorem 3.1. *Let L be a regular $m \times m$ matrix function analytic at μ such that $\det L(\mu) = 0$, and let Φ be an $m \times m$ analytic matrix function such that its rows form an extended canonical system of left root functions of L at μ. Then for λ in a neighbourhood of μ*

$$\Psi(\lambda)L(\lambda) = D(\lambda)Q(\lambda),$$

where $D(\lambda)$ is a diagonal matrix with on the diagonal $(\lambda-\mu)^{\kappa_1}, \ldots, (\lambda-\mu)^{\kappa_m}$ and Q is an $m \times m$ matrix function, analytic at μ, such that $\det Q(\mu) \neq 0$. Furthermore, the exponents $\kappa_1, \ldots, \kappa_m$ in $D(\lambda)$ are the orders of an extended canonical system of left root functions of L at μ.

Also by transposing we obtain the following analogue to Theorem 1.3.

Theorem 3.2. *Let L, Ψ, D and Q be regular $m \times m$ matrix functions, analytic at μ, such that $\Psi(\lambda)L(\lambda) = D(\lambda)Q(\lambda)$ for λ in a neighbourhood of μ. Assume that $\Psi(\mu)$ is invertible and that $D(\lambda)$ is a diagonal matrix with diagonal entries $(\lambda-\mu)^{\kappa_1}, \ldots, (\lambda-\mu)^{\kappa_m}$, where $\kappa_1 \geq \cdots \geq \kappa_m$. Then the following three conditions are equivalent:*

(1) *the rows $\psi_1, \ldots, \psi_m$ of the analytic matrix function Ψ form an extended canonical system of left root functions of L at μ with orders $\kappa_1 \geq \cdots \geq \kappa_m$;*

(2) *$\det Q(\mu) \neq 0$;*

(3) *$\sum_{i=1}^m \kappa_i$ is equal to the order of μ as a zero of $\det L(\lambda)$.*

Let $\phi_1, \ldots, \phi_m$ be an extended canonical system of right root functions for L at μ. With

$$\Phi(\lambda) = \begin{pmatrix} \phi_1(\lambda) & \cdots & \phi_m(\lambda) \end{pmatrix}$$

one has, according to Theorem 1.3, that $L(\lambda)\Phi(\lambda) = P(\lambda)D(\lambda)$ with $\det \Phi(\mu) \neq 0$ and $\det P(\mu) \neq 0$. Thus the matrix functions $\Psi(\lambda) = P(\lambda)^{-1}$ and $Q(\lambda) = \Phi(\lambda)^{-1}$ are analytic at μ and $\Psi(\lambda)L(\lambda) = D(\lambda)Q(\lambda)$. So Theorem 3.2 gives that the rows of Ψ form an extended system of left root functions of L at μ. We conclude that each extended canonical system of right root functions gives rise to an extended system of left root functions and vice versa.

The next step is to introduce left Jordan pairs. Here simply transposing is not good enough because we want to stick to our choice for upper triangular Jordan matrices in the left Jordan pairs. Let

$$\psi(\lambda) = \psi_0 + (\lambda - \mu)\psi_1 + \cdots + (\lambda - \mu)^{\ell-1}\psi_{\ell-1}$$

be a row vector function with m components. Define the $\ell \times m$ matrix

$$Y_{\mu,\psi} = \begin{pmatrix} \psi_{\ell-1}^T & \cdots \psi_1^T & \psi_0^T \end{pmatrix}^T = \begin{pmatrix} \psi_{\ell-1} \\ \vdots \\ \psi_1 \\ \psi_0 \end{pmatrix}.$$

Let E_ℓ be the $\ell \times \ell$ permutation matrix given by

$$E_\ell = \begin{pmatrix} & & 1 \\ & \cdot^{\cdot^{\cdot}} & \\ 1 & & \end{pmatrix}.$$

Write $X_{\mu,\psi^T} = (Y_{\mu,\psi})^T E_\ell$. Then ψ^T is a right root function of L^T if and only if $L(\lambda)^T X_{\mu,\psi^T}(\lambda I - J_{\mu,\ell})^{-1}$ is an analytic $m \times \ell$ matrix function at μ. Now use that $(J_{\mu,\ell})^T = E_\ell J_{\mu,\ell} E_\ell$, and conclude that ψ is a left root function of L at μ if and only if $(\lambda I - J_{\mu,\ell})^{-1} Y_{\mu,\psi} L(\lambda)$ is an analytic $\ell \times m$ matrix matrix function at μ.

Let $\psi_1, \ldots, \psi_s$ be a canonical system of left root functions of L at μ of orders $\kappa_1 \geq \cdots \geq \kappa_s$. Write

$$\psi_i(\lambda) = \psi_{i0} + (\lambda - \mu)\psi_{i1} + \cdots + (\lambda - \mu)^{\kappa_i - 1}\psi_{i\kappa_i-1} + \cdots \quad .$$

Define the matrix Y_μ by

$$Y_\mu = \begin{pmatrix} \phi_{1\kappa_1-1}^T & \cdots & \phi_{10}^T & \cdots & \phi_{s\kappa_s-1}^T & \cdots & \phi_{s0}^T \end{pmatrix}^T,$$

and J_μ by (2.2). The pair (J_μ, Y_μ) is called a *left Jordan pair of* $L(\lambda)$ *at the point* μ. Put

$$E = \begin{pmatrix} E_{\kappa_1} & & \\ & \ddots & \\ & & E_{\kappa_s} \end{pmatrix}.$$

Then (J_μ, Y_μ) is a left Jordan pair of L at μ if and only if $(Y_\mu^T E, J_\mu)$ is a right Jordan pair of L^T at μ. With this observation at hand the results that follow can easily be derived from their counterparts in Section 2.

Lemma 3.3. *Let* (J_μ, Y_μ) *be a left Jordan pair of* $L(\lambda)$ *at* μ. *Then* $\begin{pmatrix} \lambda I - J_\mu & Y_\mu \end{pmatrix}$ *has full rank for each* λ *in* $\mathbb{C}$.

Theorem 3.4. *Let L be a regular $m \times m$ analytic matrix function, and let J be a Jordan matrix with μ as its only eigenvalue and Jordan blocks of sizes $\kappa_1 \geq \cdots \geq \kappa_s$. Then (J, Y) is a left Jordan pair at $\mu \in \mathbb{C}$ if and only if the following three conditions hold true:*

(1) *the order of J is equal to the order of μ as a zero of $\det L(\lambda)$;*

(2) *the pair (J, Y) is a full range pair;*

(3) *$(\lambda I - J)^{-1} Y L(\lambda)$ is analytic at μ.*

Theorem 3.5. *Two regular $m \times m$ matrix functions L_1 and L_2, analytic at μ, have a common left Jordan pair at $\mu \in \mathbb{C}$ if and only if $L_1(\lambda)^{-1} L_2(\lambda)$ is analytic and invertible at μ.*

From Section X.1 we know that the condition that $\begin{pmatrix} \lambda I - J_\mu & Y_\mu \end{pmatrix}$ has full rank for each λ in $\mathbb{C}$, which appears in Lemma 3.3, is equivalent to the requirement that the the pair (J_μ, Y_μ) is a full range pair. Theorem 3.4 is the left analogue of Theorem 2.3. If $L(\lambda) = A_0 + \lambda A_1 + \cdots + \lambda^\ell A_\ell$, then condition (3) in Theorem 3.4 is equivalent to

$$Y A_0 + J Y A_1 + \cdots + J^\ell X A_\ell = 0. \tag{3.1}$$

Theorem 3.5 is the left analogue of Theorem 2.6.

A.4 Jordan pairs and Laurent principal parts

In Section 1 we constructed the local Smith form from right root functions. In the Sections 2 and 3 we used right root functions to construct a right Jordan pair and left root functions to construct a left Jordan pair. The next step is to use the local Smith form to simultaneously construct a left Jordan pair and a right Jordan pair.

Theorem 4.1. *Let L be a regular $m \times m$ matrix function, analytic at μ, and such that $\det L(\mu) = 0$. Let*

$$\begin{pmatrix} \psi_1(\lambda) \\ \vdots \\ \psi_m(\lambda) \end{pmatrix} L(\lambda) \begin{pmatrix} \phi_1(\lambda) & \cdots & \phi_m(\lambda) \end{pmatrix} = \operatorname{diag}\left((\lambda - \mu)^{\kappa_i} \right)_{i=1}^m \tag{4.1}$$

with $\phi_1, \ldots, \phi_m, \psi_1, \ldots, \psi_m$ analytic at μ and such that $\phi_1(\mu), \ldots, \phi_m(\mu)$ and $\psi_i^T(\mu), \ldots, \psi_m^T(\mu)$ are bases of $\mathbb{C}^m$, and $\kappa_1 \geq \cdots \geq \kappa_s > 0 = \kappa_{s+1} = \cdots = \kappa_m$. Write

$$\phi_i(\lambda) = \phi_{i0} + (\lambda - \mu)\phi_{i1} + \cdots + (\lambda - \mu)^{\kappa_i - 1} \phi_{i\kappa_i - 1},$$

$$\psi_i(\lambda) = \psi_{i0} + (\lambda - \mu)\psi_{i1} + \cdots + (\lambda - \mu)^{\kappa_i - 1} \psi_{i\kappa_i - 1},$$

$$X_\mu = \begin{pmatrix} \phi_{10} & \cdots & \phi_{1\kappa_1 - 1} & \cdots & \phi_{s0} & \cdots & \phi_{s\kappa_s - 1} \end{pmatrix}, \tag{4.2}$$

$$Y_\mu = \begin{pmatrix} \psi_{1\kappa_1 - 1}^T & \cdots & \psi_{10}^T & \cdots & \psi_{s\kappa_s - 1}^T & \cdots & \psi_{s0}^T \end{pmatrix}^T, \tag{4.3}$$

and

$$J_\mu = \bigoplus_{i=1}^{s} J_{\mu,\kappa_i}.$$

Then (X_μ, J_μ) is a right Jordan pair of $L(\lambda)$ at μ, (J_μ, Y_μ) is a left Jordan pair of $L(\lambda)$ at μ, and $X_\mu(\lambda I - J_\mu)^{-1}Y_\mu$ is the Laurent principal part of $L(\lambda)^{-1}$ at μ.

PROOF. Write

$$\Phi = \begin{pmatrix} \phi_1 & \cdots & \phi_m \end{pmatrix}, \quad \Psi = \begin{pmatrix} \psi_1 \\ \vdots \\ \psi_m \end{pmatrix}$$

and

$$D(\lambda) = \operatorname{diag}\left((\lambda - \mu)^{\kappa_i}\right)_{i=1}^{m}.$$

Then $L(\lambda)\Phi(\lambda) = \Psi(\lambda)^{-1}D(\lambda)$, and we obtain from Theorem 1.3 that $\phi_1, \ldots, \phi_m$ is an extended canonical system of right root functions. Hence (X_μ, J_μ) is a right Jordan pair by definition. Also $\Psi(\lambda)L(\lambda) = D(\lambda)\Phi(\lambda)^{-1}$. From Theorem 3.2 we get that $\psi_1, \ldots, \psi_m$ is an extended canonical system of left root functions, and hence (J_μ, Y_μ) is a left Jordan pair. Next note that $L(\lambda)^{-1} = \Phi(\lambda)D(\lambda)^{-1}\Psi(\lambda)$. Thus we get

$$L(\lambda)^{-1} = \sum_{i=1}^{m} \phi_i(\lambda)(\lambda - \mu)^{-\kappa_i}\psi_i(\lambda).$$

Now the Laurent principal part $r_i(\lambda)$ of $\phi_i(\lambda)(\lambda - \mu)^{-\kappa_i}\psi_i(\lambda)$ is given by

$$r_i(\lambda) = \begin{pmatrix} \phi_{i0} & \cdots & \phi_{i\kappa_i-1} \end{pmatrix}(\lambda I - J_{\mu,\kappa_i})^{-1}\begin{pmatrix} \psi_{i\kappa_i-1}^T & \cdots & \psi_{i0}^T \end{pmatrix}^T$$

for $\kappa_i > 0$, and is $r_i(\lambda) = 0$ if $\kappa_i = 0$. To finish the proof remark that the Laurent principal part of $L(\lambda)^{-1}$ is the sum of the Laurent principal parts of $\phi_i(\lambda)(\lambda - \mu)^{-\kappa_i}\psi_i(\lambda)$ for $i = 1, \ldots, s$, and $X_\mu(\lambda I - J_\mu)^{-1}Y_\mu$ is the sum of the functions $r_i(\lambda)$ for $i = 1, \ldots, s$. □

From the previous theorem we see that some left and right Jordan pairs together represent the Laurent principal part of L at μ. The next theorem characterizes right Jordan pairs and explains how to find a left Jordan pair (J_μ, Y_μ) from a given right Jordan pair (X_μ, J_μ) such that $X_\mu(\lambda I - J_\mu)^{-1}Y_\mu$ is the principal part of the Laurent expansion of $L(\lambda)^{-1}$ at μ.

Theorem 4.2. *Let L be an $m \times m$ regular matrix function, analytic at μ, such that $\det L(\mu) = 0$, and let (X, J) be a zero kernel pair, where the matrix $J = \operatorname{diag}(J_{\mu,\kappa_1}, \ldots, J_{\mu,\kappa_s})$, with $\kappa_1 \geq \cdots \geq \kappa_s$, is an $n \times n$ matrix and X is an $m \times n$ matrix. The pair (X, J) is a right Jordan pair of $L(\lambda)$ at μ if and only if there exists a matrix Y such that $X(\lambda I - J)^{-1}Y$ is the principal part of the Laurent* *expansion of $L(\lambda)^{-1}$ at μ and the pair (J, Y) is a full range pair. In that case the matrix Y is uniquely determined and is given by*

$$Y = \begin{pmatrix} Y_1 \\ \vdots \\ Y_s \end{pmatrix}, \quad Y_i = \begin{pmatrix} q_{i1} \\ \vdots \\ q_{i\kappa_i} \end{pmatrix}, \tag{4.4}$$

where for $i = 1, \ldots, s$ the row vector function

$$q_i(\lambda) = (\lambda - \mu)^{-1} q_{i1} + \cdots + (\lambda - \mu)^{-\kappa_i} q_{i\kappa_i}, \tag{4.5}$$

is the i-th row of the Laurent principal part of $\big(L(\lambda)\Phi(\lambda)\big)^{-1}$ at μ. Here Φ is any matrix whose columns form an extended canonical system of right root functions of L at μ corresponding to (X, J). Moreover, the pair (J, Y) is a left Jordan pair.

PROOF. Assume that (X, J) is a right Jordan pair of L at μ. We shall use Theorem 4.1 to show that with the matrix Y given by (4.4) the function $X(\lambda I - J)^{-1}Y$ is the principal part of the Laurent expansion of L at μ and the pair (J, Y) is a left Jordan pair. Let

$$\phi_1, \ldots, \phi_s, \phi_{s+1}, \ldots, \phi_m$$

be an extended canonical system of right root functions corresponding to (X, J). The orders of the system therefore are $\kappa_1 \geq \cdots \geq \kappa_s > \kappa_{s+1} = \cdots = \kappa_m = 0$. Write $\Phi(\lambda)$ for the matrix which has the vectors $\phi_1(\lambda), \ldots, \phi_m(\lambda)$ as columns. Put

$$D(\lambda) = \operatorname{diag}\big((\lambda - \mu)^{\kappa_1}, \ldots, (\lambda - \mu)^{\kappa_m}\big).$$

Then $L(\lambda)\Phi(\lambda) = P(\lambda)D(\lambda)$ and $P(\mu)$ is invertible. Now apply Theorem 4.1 with $\psi_1, \ldots, \psi_m$ given by

$$P(\lambda)^{-1} = \begin{pmatrix} \psi_1(\lambda) \\ \vdots \\ \psi_m(\lambda) \end{pmatrix}.$$

Since $\big(L(\lambda)\Phi(\lambda)\big)^{-1} = D(\lambda)^{-1}P(\lambda)^{-1}$,we see that the Laurent principal part of $(\lambda - \mu)^{-\kappa_i}\psi_i(\lambda)$ is $q_i(\lambda)$. Thus $q_{ij} = \psi_{i,\kappa_i - j}$ for $j = 1, \ldots, \kappa_i$, and hence for the case considered here the matrix Y_μ in (4.3) is equal to the matrix Y given by (4.4). The construction of Φ gives that $X = X_\mu$, where X_μ is given by (4.2), and that $J = J_\mu$. We conclude that $X(\lambda I - J)^{-1}Y$ is the principal part of the Laurent expansion of L at μ and that the pair (J, Y) is a left Jordan pair. In particular, (J, Y) is a full range pair.

Conversely, assume that Y is such that $X(\lambda I - J)^{-1}Y$ is the principal part of the Laurent expansion of $L(\lambda)^{-1}$ at μ and (J, Y) is a full range pair. We shall prove that (X, J) is a right Jordan pair of L at μ. Let (X_0, J_0) be a right Jordan pair at μ of L. Then, by the first part of the proof, there exists a matrix Y_0 such that $X_0(\lambda I - J_0)^{-1}Y_0$ is the principal part of the Laurent expansion of $L(\lambda)^{-1}$ at μ and (J_0, Y_0) is a full range pair. So, the order of J_0 is equal to the order of μ as a zero of $\det L(\lambda)$. Moreover, $X_0(\lambda I - J_0)^{-1}Y_0 = X(\lambda I - J)^{-1}Y$. Apply the state space isomorphism theorem from systems theory, Theorem IX.4.2, to see that there exists an invertible transformation S such that $X = X_0S$ and $J_0S = SJ$. In fact the ordering of the blocks with descending sizes in both J and J_0 gives that $J = J_0$. It remains to prove that this implies the pair (X, J) is a right Jordan pair of the function $L(\lambda)$ at the point μ. Let $J = \mathrm{diag}(J_{\mu,\kappa_1}, \ldots, J_{\mu,\kappa_s})$, and let the corresponding decompositions of X_0 and X be given by

$$X = \begin{pmatrix} X_{\kappa_1} & \cdots & X_{\kappa_s} \end{pmatrix}, \quad X_0 = \begin{pmatrix} X_{0\kappa_1} & \cdots & X_{0\kappa_s} \end{pmatrix}.$$

Apply Lemma 2.1 to see that

$$L(\lambda)X_{\kappa_i}(\lambda I - J_{\mu,\kappa_i})^{-1} = L(\lambda)X_{0\kappa_i}(\lambda I - J_{\mu,\kappa_i})^{-1}S$$

is analytic at μ. Again apply Lemma 2.1 to conclude that

$$\phi_i(\lambda) = X_{\kappa_i}(\lambda I - J_{\mu,\kappa_i})^{-1}e_{\kappa_i\kappa_i}(\lambda - \mu)^{\kappa_i}$$

is a right root function of order at least κ_i. Since

$$\begin{pmatrix} \lambda I - J \\ X \end{pmatrix} = \begin{pmatrix} \lambda I - J \\ X_0 \end{pmatrix} S,$$

it follows from Lemma 2.2 and the fact that the pair (X_0, J) is a right Jordan pair that the pair (X, J) is a zero kernel pair. So we see that the vectors $\phi_1(\mu), \ldots, \phi_s(\mu)$ are linearly independent. Now apply Lemma 1.1 to obtain that the system $\phi_1, \ldots, \phi_s$ is a canonical system of right root functions and therefore the pair (X, J) is a right Jordan pair at μ.

To prove that X, J and L determine Y uniquely assume that

$$X(\lambda I - J)^{-1}Y_1 = X(\lambda I - J)^{-1}Y_2.$$

Theorem IX.4.2 gives that there exists an invertible S such that $XS = X$, $SJ = JS$ and $Y_1 = SY_2$. It follows from Proposition IX.4.6 that if $XS_1 = X$ and $J = S_1^{-1}JS_1$, then $S = S_1$. So from $XI = X$ and $JI = IJ$ we conclude that $S = I$ and therefore $Y_1 = Y_2$. □

In the following example we compute for a right Jordan pair (X, J) the corresponding matrix Y using the formula in Theorem 4.2. Let

$$L(\lambda) = \begin{pmatrix} -(\lambda - 1)^3 & \lambda \\ 0 & \lambda + 1 \end{pmatrix}.$$

Then $\phi_1(\lambda) = \begin{pmatrix} 1 \\ 0 \end{pmatrix}$ is a right root function of order 3 at 1. Each other right root function ϕ at 1 has the property that $\phi(1)$ is a multiple of $\phi_1(1)$ and has order at most 3. So the pair

$$X = \begin{pmatrix} 1 & 0 & 0 \\ 0 & 0 & 0 \end{pmatrix}, \quad J = \begin{pmatrix} 1 & 1 & 0 \\ 0 & 1 & 1 \\ 0 & 0 & 1 \end{pmatrix}$$

is a right Jordan pair of L at 1. Now construct a matrix Y such that

$$L(\lambda) - X(\lambda - J)^{-1}Y$$

is analytic at $\lambda = 1$. Choose $\phi_2 = \begin{pmatrix} 0 \\ 1 \end{pmatrix}$. Then ϕ_1, ϕ_2 is an extended canonical system of right root functions of L at 1. Put $\Phi = (\,\phi_1 \quad \phi_2\,)$. So

$$\left(L(\lambda)\Phi(\lambda)\right)^{-1} = \begin{pmatrix} -(\lambda-1)^{-3} & (\lambda-1)^{-3}\lambda(\lambda+1)^{-1} \\ 0 & (\lambda+1)^{-1} \end{pmatrix},$$

and the Laurent principal part $Q(\lambda)$ of $\left(L(\lambda)\Phi(\lambda)\right)^{-1}$ at 1 is

$$Q(\lambda) = \begin{pmatrix} -(\lambda-1)^{-3} & -\frac{1}{8}(\lambda-1)^{-1} + \frac{1}{4}(\lambda-1)^{-2} + \frac{1}{2}(\lambda-1)^{-3} \\ 0 & 0 \end{pmatrix} = \begin{pmatrix} q_1(\lambda) \\ 0 \end{pmatrix}.$$

Thus we put

$$Y = \begin{pmatrix} 0 & -\frac{1}{8} \\ 0 & \frac{1}{4} \\ -1 & \frac{1}{2} \end{pmatrix}.$$

It follows that the Laurent principal part of $L(\lambda)^{-1}$ at 1 is $X(\lambda - J)^{-1}Y$. Remark that different choices of the vector ϕ_2 are possible, but in the end the matrix Y does not to depend on the choice of ϕ_2.

Corollary 4.3. *Let L be a regular $m \times m$ matrix function, analytic at μ, and such that $\det L(\mu) = 0$. Let (X, J) be a zero kernel pair, where the matrix $J = \operatorname{diag}(J_{\mu,\kappa_1}, \dots, J_{\mu,\kappa_s})$, with $\kappa_1 \geq \dots \geq \kappa_s$, is an $n \times n$ matrix and X is an $m \times n$ matrix. The pair (X, J) is a right Jordan pair of L at μ if and only if*

(1) *the order n of J is equal to the order of μ as a zero of $\det L(\lambda)$;*

(2) *there exists a matrix Y such that $L(\lambda)^{-1} - X(\lambda I - J)^{-1}Y$ is analytic at μ.*

PROOF. If (X, J) is a right Jordan pair, then (1) and (2) are satisfied according to Theorem 2.3 and Theorem 4.2. Conversely, assume that (1) and (2) are fulfilled. It is sufficient to show that (J, Y) is a full range pair. Consider the rational matrix function $R(\lambda) = X(\lambda I - J)^{-1}Y$. By Theorem IX.4.3 the realization

$(J, Y, X, 0)$ of $R(\lambda)$ is a dilation of a minimal realization $(A, B, C, 0)$ of $R(\lambda)$. Let $A = SJ_0S^{-1}$, $B = SX_0$ and $C = Y_0S^{-1}$, where J_0 is a Jordan matrix. Then

$$R(\lambda) = X_0(\lambda I - J_0)^{-1}Y_0.$$

Thus (X_0, J_0) fulfills the conditions of Theorem 4.2. Hence (X_0, J_0) is a right Jordan pair. Now Theorem 2.3 gives that the order of J_0 is equal the order of μ as a zero of $\det L(\lambda)$. Therefore the orders of A, J_0 and J are equal. Thus $A = J$ and $B = X$, and since $(A, B, C, 0)$ is a minimal realization, we have that (J, Y) is a full range pair. □

The next result shows to what extent a right Jordan pair is unique.

Theorem 4.4. *Let (X_1, J) be a right Jordan pair at μ of the $m \times m$ analytic matrix function L. Then the pair (X_2, J) is a right Jordan pair of L at μ if and only if there exists an invertible matrix S such that $X_1 = X_2S$ and $SJ = JS$.*

PROOF. Assume that (X_2, J) is a right Jordan pair of L at μ. There exist matrices Y_1 and Y_2 such that the Laurent principal part of L^{-1} at μ is equal to $X_1(\lambda I - J)^{-1}Y_1$ and to $X_2(\lambda I - J)^{-1}Y_2$ and the pairs (J, Y_1) and (J, Y_2) are full range. From Theorem IX.4.2 it follows that there exists an S as desired.

Conversely, assume that $X_1 = X_2S$ and $SJ = JS$. The pair (X_2, J) is a zero kernel pair because

$$\begin{pmatrix} \lambda I - J \\ X_2 \end{pmatrix} = \begin{pmatrix} S & 0 \\ 0 & I \end{pmatrix} \begin{pmatrix} \lambda I - J \\ X_1 \end{pmatrix} S^{-1}$$

and (X_1, J) is a zero kernel pair. Furthermore, there exists a matrix Y_1 such that the Laurent principal part of $L(\lambda)^{-1}$ is $X_1(\lambda I - J)^{-1}Y_1$ and the pair (J, Y_1) is a full range pair. Put $Y_2 = S^{-1}Y_1$. Then the Laurent principal part of $L(\lambda)^{-1}$ is $X_2(\lambda I - J)^{-1}Y_2$. Moreover the pair (J, Y_2) is a full range pair as is easily seen from the equality

$$\begin{pmatrix} \lambda I - J & Y_2 \end{pmatrix} = S^{-1}\begin{pmatrix} \lambda I - J & Y_1 \end{pmatrix} \begin{pmatrix} S & 0 \\ 0 & I \end{pmatrix}$$

and the fact that (J, Y_1) is a full range pair. □

Note that Theorem 4.2, Corollary 4.3 and Theorem 4.4 are presented for a right Jordan pair. Similar results for left Jordan pairs also hold. We omit the details.

A.5 Global spectral data for regular analytic matrix functions

We extend the local theory developed so far to a global theory for domains, i.e., for open and connected subsets in the complex plane. The notion of a right Jordan pair of a regular matrix function analytic at a point is extended to the notion of (right) null pair of a regular matrix function with respect a subset of its domain of analyticity.

Let L be a regular $m \times m$ matrix function, which is analytic in the domain Ω of the complex plane. Assume that $\det L(\lambda)$ has finitely many zeros in Ω, say $\lambda_1, \dots, \lambda_k$. For $i = 1, \dots, k$ let $(X_{\lambda_i}, J_{\lambda_i})$ be a right Jordan pair of L at the point λ_i. Put

$$X_\Omega = \begin{pmatrix} X_{\lambda_1} & \cdots & X_{\lambda_k} \end{pmatrix}, \tag{5.1}$$

$$J_\Omega = \operatorname{diag}(J_{\lambda_1}, \dots, J_{\lambda_k}). \tag{5.2}$$

The pair (X_Ω, J_Ω) is called a *right Jordan pair of L with respect to the set Ω*. The restriction that $\det L(\lambda)$ has only finitely many zeros in Ω is automatically fulfilled in the important special case when L is a rational matrix function. A pair (C, A) is called a *right null pair of L with respect to Ω* if there exists a right Jordan pair (X_Ω, J_Ω) of L with respect to Ω and an invertible linear transformation S such that $C = X_\Omega S$ and $SA = J_\Omega S$. For $i = 1, \dots, k$ let $(J_{\lambda_i}, Y_{\lambda_i})$ be a left Jordan pair of L at μ. Put

$$Y_\Omega = \begin{pmatrix} Y_{\lambda_1} \\ \vdots \\ Y_{\lambda_k} \end{pmatrix}.$$

Then (J_Ω, Y_Ω) is called a *left Jordan pair of L with respect to the set Ω*. A pair of matrices (A, B) is called a *left null pair of L with respect to Ω* if there exists a left Jordan pair (J_Ω, Y_Ω) of L with respect to Ω and an invertible linear transformation S such that $SA = J_\Omega S$ and $SB = Y_\Omega$. The next results characterize right null pairs of a regular analytic matrix function with respect to a set Ω.

Theorem 5.1. *Let L be a regular $m \times m$ matrix function analytic on the domain Ω, and assume that $\det L(\lambda)$ has only finitely many zeros in Ω. Let C be an $m \times n$ matrix and A be an $n \times n$ matrix. Then (C, A) is a right null pair of L with respect to Ω if and only if the following three conditions are satisfied:*

(1) *A has all its eigenvalues in Ω and the order n of A is equal to the sum of the orders of the zeros of $\det L(\lambda)$ in Ω;*

(2) *the pair (C, A) is a zero kernel pair;*

(3) *$L(\lambda)C(\lambda I - A)^{-1}$ is analytic in Ω.*

Before we prove this theorem we formulate and prove a simple lemma.

Lemma 5.2. *Let the pair (X, J) be given by*

$$X = \begin{pmatrix} X_1 & \cdots & X_k \end{pmatrix}, \tag{5.3}$$

$$J = \operatorname{diag}(J_1, \dots, J_k), \tag{5.4}$$

where X_i is an $m_i \times n_i$ matrix, and J_i is an $n_i \times n_i$ matrix with one eigenvalue, λ_i, for $i = 1, \dots, k$. Assume $\lambda_i \neq \lambda_j$ if $i \neq j$. Then (X, J) is a zero kernel pair if and only if (X_i, J_i) is a zero kernel pair for $i = 1, \dots, k$.

PROOF. Put

$$R(\lambda) = \begin{pmatrix} \lambda I - J \\ X \end{pmatrix}.$$

Assume that the pair (X, J) is not a zero kernel pair. So $R(\mu)x = 0$, for some $x \neq 0$ and some μ. Decompose x as $x = \mathrm{col}(x_1, \dots, x_k)$ corresponding to the decompositions (5.3) and (5.4). The first block row of $R(\mu)x = 0$ gives that μ is one of the eigenvalues of J. Assume that $\mu = \lambda_j$. Then we see that $x_i = 0$ if $i \neq j$. Also it follows that

$$\begin{pmatrix} \lambda I - J_j \\ X_j \end{pmatrix} x_j = 0, \tag{5.5}$$

with $x_j \neq 0$. So (X_j, J_j) is not a zero kernel pair.

Conversely, assume that (X_j, J_j) is not a zero kernel pair. Then there exists a vector $x_j \neq 0$ such that (5.5) holds true. Put $x_i = 0$ if $i \neq j$. Then $R(\lambda_j)x = 0$, where $x = \mathrm{col}(x_1, \dots, x_k) \neq 0$. Thus (X, J) is not a zero kernel pair. □

From this lemma it follows that a right Jordan pair is a zero kernel pair. Similarly one can prove that a left Jordan pair is a full range pair.

PROOF OF THEOREM 5.1. Let (C, A) be a right null pair with respect to Ω. So there exists a right Jordan pair (X_Ω, J_Ω) and an invertible S such that $C = X_\Omega S$ and $J_\Omega S = SA$. In particular, the order of A is equal to the order of J_Ω. Recall that X_Ω is represented by (5.1) and J_Ω by (5.2), with $\lambda_1, \dots, \lambda_k$ the zeros of $\det L(\lambda)$ in Ω. Statement (1) now follows from Theorem 2.3. Since for $i = 1, \dots, k$ the pair $(X_{\lambda_i}, J_{\lambda_i})$ is a zero kernel pair, Lemma 5.2 shows that (X_Ω, J_Ω) is a zero kernel pair. Use

$$\begin{pmatrix} \lambda I - J_\Omega \\ X_\Omega \end{pmatrix} S = \begin{pmatrix} S & 0 \\ 0 & I \end{pmatrix} \begin{pmatrix} \lambda I - A \\ C \end{pmatrix}$$

to see that this implies that (C, A) is a zero kernel pair. To prove (3) it is sufficient to prove that $L(\lambda)X_\Omega(\lambda I - J_\Omega)^{-1}$ is analytic at the points $\lambda_1, \dots, \lambda_k$. However, the latter follows from Theorem 2.3 if one uses the representations (5.1) and (5.2).

Assume that the pair (C, A) fulfills the conditions (1), (2) and (3). Let J be the Jordan canonical form of A. So there exists an invertible S such that $JS = SA$. Define X by $C = XS$. Then (X, J) satisfies (1), (2) and (3) with C replaced by X and A by J. It is sufficient to prove that (X, J) is a right Jordan pair of L with respect to Ω. Write $J = \mathrm{diag}(J_{\lambda_1}, \dots, J_{\lambda_k})$ with $\lambda_1, \dots, \lambda_k$ the different eigenvalues of J. Decompose X correspondingly as $X = (\, X_1 \ \ \dots \ \ X_k \,)$. Then it follows from Lemma 5.2 that the pairs (X_i, J_{λ_i}) are zero kernel pairs for $i = 1, \dots, k$. Also $L(\lambda)X_i(\lambda I - J_{\lambda_i})^{-1}$ is analytic on Ω. So from Corollary 2.4 we conclude that the order α_i of J_{λ_i} is at most the order ℓ_i of λ_i as a zero of $\det L(\lambda)$. From (1) we know that $\alpha_1 + \cdots + \alpha_k$, the order of J, is equal to the sum of the orders of the zeros $\lambda_1, \dots, \lambda_k, \dots, \lambda_s$ of $\det L(\lambda)$ in Ω. This means that $s = k$ and $\alpha_i = \ell_i$, for $i = 1, \dots, k$. Applying again Corollary 2.4 we get that $\lambda_1, \dots, \lambda_k$ are the zeros of $\det L(\lambda)$ and (X_i, J_{λ_i}) is a right Jordan pair at λ_i for L. This proves that (X, J) is a right Jordan pair of L with respect to Ω. □

Theorem 5.3. *Let L be a regular $m \times m$ matrix function, analytic on the domain Ω, and assume that $\det L(\lambda)$ has only finitely many zeros in Ω. Let (C, A) be a zero kernel pair, where C is an $m \times n$ matrix, and A is an $n \times n$ matrix. If (C, A) is a right null pair of L with respect to the set $\Omega \subset \mathbb{C}$, then there exists an $n \times m$ matrix B such that the sum of the Laurent principal parts of $L(\lambda)^{-1}$ at the points of Ω is equal to $C(\lambda I - A)^{-1}B$ and the pair (A, B) is a left null pair. Conversely, if there exists an $n \times m$ matrix B such that the sum of the Laurent principal parts of $L(\lambda)^{-1}$ at the points of Ω is equal to $C(\lambda I - A)^{-1}B$ and the pair (A, B) is a full range pair, then (C, A) is a right null pair of L with respect to the set Ω.*

PROOF. Assume that (C, A) is a right null pair of L with respect to Ω. Then there exists an invertible matrix S such that $(CS, S^{-1}AS)$ is a right Jordan pair of L with respect to Ω. Let $\lambda_1, \dots, \lambda_k$ be the zeros of the determinant of L in Ω. Put $J = S^{-1}AS$ and $X = CS$. Then $J = \operatorname{diag}(J_{\lambda_1}, \dots, J_{\lambda_k})$ and $X = \begin{pmatrix} X_1 & \cdots & X_k \end{pmatrix}$. With the right Jordan pair (X_j, J_{λ_j}) at λ_j we may associate (use Theorem 4.1) a matrix Y_j such that $X_j(\lambda I - J_{\lambda_j})^{-1}Y_j$ is the principal part of the Laurent expansion of $L(\lambda)^{-1}$ at λ_j. Write

$$Y = \begin{pmatrix} Y_1 \\ \vdots \\ Y_s \end{pmatrix}.$$

So we get that $X(\lambda I - J)^{-1}Y$ is the sum of the Laurent principal parts of $L(\lambda)^{-1}$ at the points of Ω. Put $B = S^{-1}Y$. Then $X(\lambda I - J)^{-1}Y = C(\lambda I - A)^{-1}B$. It remains to show that (A, B) is a left null pair. Note that (J_{λ_j}, Y_j) is a left Jordan pair of L at λ_j (by Theorem 4.2), and hence (J, Y) is a left Jordan pair with respect to Ω. Remark that the matrix $\begin{pmatrix} B & \lambda I - A \end{pmatrix}$ has the same rank as $\begin{pmatrix} Y & \lambda I - J \end{pmatrix}$ for each λ in $\mathbb{C}$. So it is sufficient to prove that the pair (J, Y) is a full range pair. According to Lemma 5.2 this follows from the fact that (J_{λ_i}, Y_i) is full range for $1 = 1, \dots, k$.

Conversely, assume that there exists a $n \times m$ matrix B such that the sum of the principal parts of the Laurent expansions of $L(\lambda)^{-1}$ at the points of Ω is $C(\lambda I - A)^{-1}B$ and (A, B) is a full range pair. We have to prove that (C, A) is a right null pair of L with respect to Ω. Let (X_0, J_0) be a right Jordan pair of L with respect to Ω. Then there exists a matrix Y_0 such that $X_0(\lambda I - J_0)^{-1}Y_0$ is the sum of the principal parts of the Laurent expansion of $L(\lambda)^{-1}$ at the zeros of $\det L(\lambda)$ in Ω. So $X_0(\lambda I - J_0)^{-1}Y_0 = C(\lambda I - A)^{-1}B$. Furthermore the pairs (C, A) and (X_0, J_0) are zero kernel pairs and the pairs (A, B) and (J_0, Y_0) are full range pairs. Apply Theorem IX.4.2. The conclusion is that there exists an invertible matrix S such that $X_0 S = C$ and $J_0 S = AS$. This proves that the pair (C, A) is a right null pair of L. □

Theorem 5.4. *Let L be a regular $m \times m$ matrix function, analytic on the domain Ω, and assume that $\det L(\lambda)$ has only finitely many zeros in Ω. Let (C_1, A_1) be a right null pair of L on the set Ω. Then the pair (C_2, A_2) is a right null pair of L*

on the set Ω if and only if there exists an invertible matrix S such that $C_1 = C_2S$ and $SA_1 = A_2S$.

PROOF. Assume that (C_2, A_2) is a right null pair. Then there exists a matrix B_2 such that $C_2(\lambda I - A_2)^{-1}B_2$ is the sum of the principal parts of $L(\lambda)^{-1}$ in the zeros of $\det L(\lambda)$ in Ω and (A_2, B_2) is a full range pair. Since there also exists a matrix B_1 such that $C_1(\lambda I - A_1)^{-1}B_1$ is the sum of the principal parts of $L(\lambda)^{-1}$ in the zeros of $\det L(\lambda)$ in Ω, we see that $C_1(\lambda I - A_1)^{-1}B_1 = C_2(\lambda I - A_2)^{-1}B_2$. Apply Theorem IX.4.2 to conclude that there exists an invertible S such that $C_1 = C_2S$ and $SA_1 = A_2S$.

Conversely, assume that there exists an invertible matrix S such that $C_1 = C_2S$ and $SA_1 = A_2S$. It is immediate from the definition of a right null pair of L for the set Ω, that also (C_2, A_2) is a right null pair of L for the set Ω. □

The next theorem describes to what extent a right null pair determines the function.

Theorem 5.5. *Let L_1 and L_2 be regular $m \times m$ matrix functions, analytic on the domain Ω, such that $\det L_1(\lambda)$ and $\det L_2(\lambda)$ have finitely many zeros in Ω. Then L_1 and L_2 have a common right null pair for $\Omega \subset \mathbb{C}$ if and only if $L_2(\lambda)L_1(\lambda)^{-1}$ and $L_1(\lambda)L_2(\lambda)^{-1}$ are analytic in each point of Ω.*

PROOF. Assume that $L_2(\lambda)L_1(\lambda)^{-1}$ and $L_1(\lambda)L_2(\lambda)^{-1}$ are analytic in each point of Ω. Let (C, A) be a right null pair of L_1. To prove that (C, A) is a right null pair of L_2, we reason as in the proof of Theorem 2.6, with Theorem 5.1 replacing Theorem 2.3.

Conversely, assume that (C, A) is a right null pair of L_1 and L_2 for Ω. Without loss of generality we may assume that

$$C = \begin{pmatrix} X_1 & \cdots & X_k \end{pmatrix}, \quad A = \operatorname{diag}(J_{\mu_1}, \ldots, J_{\mu_k}),$$

with $\mu_1, \ldots, \mu_k$ the eigenvalues of A in Ω, and (X_i, J_{μ_i}) a right Jordan pair of L_1 and L_2 at μ_i for $i = 1, \ldots, k$. From Theorem 2.6 we get that $L_1(\lambda)L_2(\lambda)^{-1}$ is analytic and invertible in the points $\mu_1, \ldots, \mu_k$. Since in the other points of Ω both $\det L_1(\lambda)$ and $\det L_2(\lambda)$ are nonzero, this proves that $L_2(\lambda)L_1(\lambda)^{-1}$ and $L_1(\lambda)L_2(\lambda)^{-1}$ are analytic in each point of Ω. □

The following result provides a useful sufficient condition for a pair (C, A) to be a right null pair and a pair (A, B) to be a left null pair.

Proposition 5.6. *Let L be a regular $m \times m$ matrix function, analytic on the domain Ω, and assume that $\det L(\lambda)$ has only finitely many zeros in Ω. Let C, A and B be matrices such that $L(\lambda)^{-1} - C(\lambda I - A)^{-1}B$ is analytic on $\mathbb{C}$. If the order of A is less than or equal to the number of zeros of $\det L(\lambda)$ (multiplicities taken into account) in Ω, then (C, A) is a right null pair of L and (A, B) is a left null pair.*

PROOF. We apply Proposition IX.4.5 to obtain submatrices C_0 of C, A_0 of A and B_0 of B, such that (C_0, A_0) is a zero kernel pair, (A_0, B_0) is a full range pair, and $C_0(\lambda I - A_0)^{-1}B_0 = C(\lambda I - A)^{-1}B$. So $L(\lambda)^{-1} - C_0(\lambda I - A_0)^{-1}B_0$ is analytic on the full complex plane. Therefore $C_0(\lambda I - A_0)^{-1}B_0$ is equal to the sum of the Laurent principal parts of $L(\lambda)^{-1}$. Theorem 5.3 gives that in this case the pair (C_0, A_0) is a right null pair and (A_0, B_0) is a left null pair of L. From Theorem 5.1 it follows that the order of A_0 is the sum of the orders of the zeros of $\det L(\lambda)$ in Ω. This proves that the order of A_0 is equal to the order of A. Thus $A_0 = A$, $C_0 = C$ and $B_0 = B$. □

The results of this section apply in particular to regular matrix polynomials with $\Omega = \mathbb{C}$, because such functions are analytic on $\mathbb{C}$ and their determinant have only finitely many zeros. The following theorem is a variant of Theorem 5.1 for matrix polynomials.

Theorem 5.7. *Let $L(\lambda) = L_0 + \lambda L_1 + \cdots + \lambda^\ell L_\ell$ be a regular $m \times m$ matrix polynomial. Let C be an $m \times n$ matrix and A be an $n \times n$ matrix. Then (C, A) is a right null pair with respect to Ω of L if and only if the following three conditions are satisfied:*

(1) *the order n of A is equal to the sum of the orders of the zeros of $\det L(\lambda)$ in Ω and A has all its eigenvalues in Ω;*

(2) *the pair (C, A) is a zero kernel pair;*

(3') $L_0C + L_1CA + \cdots + L_\ell CA^\ell = 0$.

PROOF. We only have to prove that (3') is equivalent to condition (3) in Theorem 5.1. This can be done in the same way as it is done in the proof of the analogous statement in Section 2 (see the paragraph after Corollary 2.4). □

Notes

The material in this Appendix is mainly taken from Gohberg-Kaashoek-Van Schagen [6] and has its origin in Gohberg-Lancaster-Rodman [1]. Theorem 4.2 develops further Theorem 7.1 in Gohberg-Sigal [1].

Bibliography

J.A. Ball, I. Gohberg

[1] Shift invariant subspaces, factorization, and interpolation for matrices. I. The canonical case, *Linear Algebra Appl.* **74** (1986), 87–150.

J.A. Ball, I. Gohberg, L. Rodman

[1] *Interpolation of Rational Matrix Functions*, Operator Theory: Advances and Applications Vol. 45, Birkhäuser, Basel, 1990.

J.A. Ball, I. Gohberg, L. Rodman, T. Shalom

[1] On the eigenvalues of matrices with given upper triangular part, *Integral Equations Operator Theory* **13** (1990), 488–497.

J.A. Ball, G. Groenewald, M.A. Kaashoek, J. Kim

[1] Column reduced rational matrix functions with given null-pole data in the complex plane, *Linear Algebra Appl.* **203/204** (1994), 67–110.

H. Bart, I. Gohberg, M.A. Kaashoek

[1] *Minimal Factorization of Matrix and Operator Functions*, Operator Theory: Advances and Applications Vol. 1, Birkhäuser Verlag, Basel, 1979.

H. Bercovici, C. Foias, A. Tannenbaum

[1] A Spectral Commutant Lifting Theorem, *Trans. Amer. Math. Soc.* **325** (1991), 741–763.

P. Brunovsky

[1] Classification of linear controllable systems, *Kybernetika* **3** (6) (1970), 173–188.

I. Cabral, F.C. Silva

[1] Similarity invariants of completions of Submatrices, *Linear Algebra Appl.* **169** (1992), 151–161.

C. Davis, W.M. Kahan, H.F. Weinberger

[1] Norm preserving dilations and their applications to optimal error bounds, *SIAM J. Numer. Anal.* **19** (1982), 444–469.

H. Dym

[1] *Contractive Matrix Functions, Reproducing Kernel Hilbert Spaces and Interpolation*, CBMS Regional Conference Series 71, American Math. Soc., Providence RI, 1989.

H. Dym, I. Gohberg

[1] Extensions of band matrices with band inverses, *Linear Algebra Appl.* **36** (1981), 1–24.

G. Eckstein

[1] Exact controllability and spectrum assignment, in *Topics in Modern Operator Theory*, Operator Theory: Advances and Applications Vol. 2, Birkhäuser Verlag, Basel, 1981, pp. 81–94.

R.L. Ellis, D.C. Lay

[1] Rank-preserving extensions of band matrices, *Linear Multilinear Algebra* **26** (1990), 147–179.

J. Ferrer, F. Puerta

[1] Similarity of non-everywhere defined linear maps, *Linear Algebra Appl.* **168** (1992), 27–55.

C. Foias, A.E. Frazho

[1] *The Commutant Lifting Approach to Interpolation Problems*, Operator Theory: Advances and Applications Vol. 44, Birkhäuser Verlag, Basel, 1990.

S. Friedland

[1] Inverse eigenvalue problems, *Linear Algebra Appl.* **17** (1977), 15–51.

F.R. Gantmacher

[1] *The Theory of Matrices*, Nauka, Moscow, 1967, English translation of the 1st ed., Chelsea, New York, vol. 1, 2, 1959, 1960.

I. Gohberg, S. Goldberg, M.A. Kaashoek

[1] *Classes of Linear Operators, Vol. 1*, Operator Theory: Advances and Applications Vol. 49, Birkhäuser Verlag, Basel, 1990.

[2] *Classes of Linear Operators, Vol. 2*, Operator Theory: Advances and Applications Vol. 63, Birkhäuser Verlag, Basel, 1993.

I. Gohberg, M.A. Kaashoek

[1] An inverse spectral problem for rational matrix functions and minimal divisibility, *Integral Equations Operator Theory* **10** (1987), 437–465.

I. Gohberg, M.A. Kaashoek, A.C.M. Ran

[1] Interpolation problems for rational matrix functions with incomplete data and Wiener-Hopf factorization, in *Topics in Interpolation Theory of Rational Matrix-valued Functions*, Operator Theory: Advances and Applications, Vol. 33, Birkhäuser Verlag, Basel, 1988, pp. 73–108.

[2] Matrix polynomials with prescribed zero structure in the finite complex plane, in *Topics in Matrix and Operator Theory*, Operator Theory: Advances and Applications, Vol. 50, Birkhäuser Verlag, Basel, 1991, pp. 241–266.

I. Gohberg, M.A. Kaashoek, F. van Schagen

[1] Similarity of operator blocks and canonical forms. I. General results, feedback equivalence and Kronecker indices, *Integral Equations Operator Theory*, **3** (1980), 350–396.

[2] Similarity of operator blocks and canonical forms. II. Infinite dimensional case and Wiener-Hopf factorization, in *Topics in Modern Operator Theory*, Operator Theory: Advances and Applications Vol. 2, Birkhäuser Verlag, Basel, 1981, pp. 121–170.

[3] Rational matrix and operator functions with prescribed singularities, *Integral Equations Operator Theory* **5** (1982), 673–717.

[4] Eigenvalues of completions of submatrices, *Linear Multilinear Algebra* **25** (1989), 55–70.

[5] The eigenvalue completion problem for blocks and related control problems, *Linear Algebra Appl.* **170** (1992), 201–206.

[6] On the local theory of regular analytic matrix functions, *Linear Algebra Appl.* **182** (1993), 9–26.

I. Gohberg, P. Lancaster, L. Rodman

[1] *Matrix Polynomials*, Academic Press, New York, 1982.

[2] *Invariant Subspaces of Matrices with Applications*, John Wiley and Sons, New York, 1986.

I. Gohberg, L. Lerer, L. Rodman

[1] On canonical factorization of operator polynomials, spectral divisors and Toeplitz matrices, *Integral Equations Operator Theory* **1** (1978), 176–214.

I. Gohberg, L. Rodman, T. Shalom, H. Woerdeman

[1] Bounds for eigenvalues and singular values of completions of partially specified matrices. *Linear and Multilinear Algebra*, **33** (1993), 233–250.

I. Gohberg, S. Rubinstein

[1] A classification of upper equivalent matrices: the generic case, *Integral Equations Operator Theory* **14** (1991), 533–543.

R. Grone, C.R. Johnson, E. Marques de Sà, H. Wolkowicz

[1] Positive definite completions of partial Hermitian matrices, *Linear Algebra Appl.* **58** (1984), 109–124.

L. Gurvits, L. Rodman, T. Shalom

[1] Controllability by completion of partial upper triangular matrices, *Mathematics of Control, Signals and Systems*, **6** (1993), 30–40.

[2] Controllability by completion of partial upper triangular matrices over rings, *Linear Algebra Appl.* **172** (1992), 135–149.

L. Gurvits, L. Rodman, I. Spitkovsky

[1] Spectrum assignment for Hilbert space operators *Houston J. Math.*, **17** (1991), 501–523.

J.W. Helton

[1] *Operator Theory, Analytic Functions, Matrices and Electrical Engineering*, CBMS Regional Conference Series 68, American Math. Soc., Providence RI, 1987.

M.A. Kaashoek, C.V.M. van der Mee, L. Rodman

[1] Analytic operator functions with compact spectrum. I. Spectral nodes linearization and equivalence, *Integral Equations Operator Theory* **4** (1981), 504–547.

[2] Analytic operator functions with compact spectrum. II. Spectral pairs and factorization, *Integral Equations Operator Theory* **5** (1982), 791–827.

[3] Analytic operator functions with compact spectrum. III. Hilbert space case: inverse problem and applications, *Journal of Operator Theory* **10** (1983), 229–336.

M.A. Kaashoek, H.J. Woerdeman

[1] Unique minimal rank extensions of triangular operators, *J. Math. Anal. Appl.* **131** (1988), 501–516.

T. Kailath

[1] *Linear Systems*, Prentice-Hall, Englewood Cliffs, 1980.

R.E. Kalman, P.L. Falb, M.A. Arbib

[1] *Topics in mathematical system theory*, McGraw-Hill, New York, 1969.

L. Kronecker

[1] Algebraische Reduction der Schaaren bilinearen Formen, *Sitz. Ber. Akad. Wiss. Phys.-Math. Klasse Berlin* (1890), 763–776.

M. Krupnik, L. Rodman

[1] Completions of partial Jordan and Hessenberg matrices, *Linear Algebra and Appl.* **212/213** (1994), 267–287.

P. Lancaster, M. Tismenetsky

[1] *The Theory of Matrices*, Academic Press, Orlando, 1985.

Ya.B. Lopatinskii

[1] Factorization of a polynomial matrix, *Nauch. Zap. L'vov. Politech. Inst., Ser. Fiz.-Mat.* **38** (1956), 3–9 (in Russian).

A.S. Markus, V.R. Olshevsky

[1] Complete controllability and spectral assignment in infinite dimensional spaces, *Integral Equations Operator Theory* **17** (1993), 107–122.

A. Marshall and I. Olkin

[1] *Inequalities : Theory of Majorization and Its Applications*, Academic Press, New York, 1979.

L. Mirsky

[1] Matrices with prescribed characteristic roots and diagonal elements, *J. London Math Soc.* **33** (1958), 14–21.

G.N. de Oliveira

[1] Matrices with prescribed characteristic polynomial and a prescribed submatrix III, *Monatsh. Math.* **75** (1971), 441–446.

[2] Matrices with prescribed characteristic polynomial and several prescribed submatrices, *Linear Multilinear Algebra* **2** (1975), 357–364.

[3] Matrices with prescribed characteristic polynomial and principal blocks II, *Linear Algebra Appl.* **47** (1982), 35–40.

S. Parrott

[1] On a quotient norm and the Sz.-Nagy-Foias lifting theorem, *J. Functional Analysis* **30** (1978), 311–328.

L. Rodman

[1] *An Introduction to operator polynomials*, Operator Theory: Advances and Applications Vol. 38, Birkhäuser Verlag, Basel, 1989.

[2] On exact controllability of operators, *Rocky Mountain Journal of Math.* **20** (1990), 549–560.

L. Rodman, T. Shalom

[1] Jordan form of completions of partial upper triangular matrices, *Linear Algebra Appl.* **168** (1992), 221–249.

W.H. Rosenbrock

[1] *State space and multivariable theory*, Nelson, London (1970).

B. Rowley

[1] Wiener-Hopf factorization of operator polynomials, *Integral Equations Operator Theory* **3** (1980), 427–462.

E. Marques de Sà

[1] Imbedding conditions for λ-matrices, *Linear Algebra Appl.* **24** (1979), 33–50.

F.C. Silva

[1] Matrices with prescribed characteristic polynomial and submatrices, *Portugaliae Math.* **44** (1987), 261–264.

[2] Matrices with prescribed eigenvalues and principal submatrices, *Linear Algebra Appl.*, **92** (1987), 241–250.

[3] Matrices with prescribed similarity class and a prescribed nonprincipal submatrix. *Portugaliae Math.* **47** (1990), 103–113.

K. Takahashi

[1] Exact controllability and spectrum assignment, *J. of Math. Anal. and Appl.* **104** (1984), 537–545.

R.C. Thompson

[1] Interlacing inequalities for invariant factors, *Linear Algebra Appl.* **24** (1979), 1–32.

H.K. Wimmer

[1] Existenzsätze in der Theorie der Matrizen und lineare Kontrolltheorie, *Monatshefte für Mathematik* **78** (1974), 256–263.

H.J. Woerdeman

[1] Minimal rank completions for block matrices, *Linear Algebra Appl.* **121** (1989), 105–122.

[2] *Matrix and Operator Extensions*, Ph.D. Thesis, Vrije Universiteit Amsterdam 1989 = CWI Tract 68, Centre for Mathematics and Computer Science, Amsterdam 1989.

W.M. Wonham

[1] *Linear Multivariable Control: a Geometric Approach*, Applications of Mathematics 10, Springer Verlag, New York, 1979.

I. Zaballa

[1] Matrices with prescribed rows and invariant factors, *Linear Algebra Appl.* **87** (1987), 113–146.

[2] Interlacing inequalities and control theory, *Linear Algebra Appl.* **101** (1988), 9–31.

[3] Interlacing and majorization in invariant factor assignment problems, *Linear Algebra Appl.* **121** (1989), 409–421.

[4] Invariant factor assignment on higher order systems using state feedback, *SIAM J. Matrix Anal. Appl.* **10** (1989), 147–154.

[5] Matrices with prescribed invariant factors and off-diagonal submatrices, *Linear Multilinear Algebra* **25** (1989), 29–54.

List of Notations

A^*	dual of the operator A
A^+	generalized inverse of the operator A, 218
$A\|_{\mathcal{W}}$	restriction of the operator A to the subspace $\mathcal{W}$
$A[\mathcal{W}]$	image of the subset $\mathcal{W}$ under the operator A
$A(\mathcal{X} \to \mathcal{Y})$	operator A defined on a subspace of $\mathcal{X}$
$\sigma(A)$	spectrum of the operator A
$\operatorname{Im} A$	image of the operator A
$\operatorname{Ker} A$	kernel of the operator A
$\operatorname{col}(A_j)_{j=0}^k$	block column with entries $A_0, A_1, \ldots, A_k$
$\operatorname{row}(A_j)_{j=0}^k$	block row with entries $A_0, A_1, \ldots, A_k$
$\operatorname{Im}(A\|B)$	image of the pair (A, B), 146
$\operatorname{Ker}(C\|A)$	kernel of the pair (C, A), 153
$I_{\mathcal{X}}$	identity operator on the space $\mathcal{X}$
I_n	identity operator on the space $\mathbb{C}^n$ or the $n \times n$ unit matrix
$J_{\mu,k}$	$k \times k$ Jordan block with eigenvalue μ, 301
$\mathcal{V}/\mathcal{W}$	quotient space of the space $\mathcal{V}$ with respect to the subspace $\mathcal{W}$
$(B; P, Q)$	operator block, 11
$\oplus$	direct sum
$\#W$	number of elements of the set W
$\mathcal{M}^\perp$	annihilator of the set $\mathcal{M}$
$\mathbb{C}_\infty$	extended complex plane $\mathbb{C} \cup \{\infty\}$
$p\|q$	the polynomial p divides the polynomial q
g.c.d.	greatest common divisor
l.c.m.	least common multiple
$p^\#(\lambda)$	reversed polynomial $\lambda^n p(\lambda^{-1})/p(0)$, 79
$\mu^\#$	dual sequence, 54

Index

— O —

— P —

— R —

Titles previously published in the series

OPERATOR THEORY: ADVANCES AND APPLICATIONS

BIRKHÄUSER VERLAG

20. **D. Amir:** Isometric characterization of Inner Product Spaces, 1986, (3-7643-1774-4)
21. **I. Gohberg, M.A. Kaashoek** (Eds.): Constructive Methods of Wiener-Hopf Factorization, 1986, (3-7643-1826-0)
22. **V.A. Marchenko:** Sturm-Liouville Operators and Applications, 1986, (3-7643-1794-9)
23. **W. Greenberg, C. van der Mee, V. Protopopescu:** Boundary Value Problems in Abstract Kinetic Theory, 1987, (3-7643-1765-5)
24. **H. Helson, B. Sz.-Nagy, F.-H. Vasilescu, D. Voiculescu, Gr. Arsene** (Eds.): Operators in Indefinite Metric Spaces, Scattering Theory and Other Topics, 1987, (3-7643-1843-0)
25. **G.S. Litvinchuk, I.M. Spitkovskii:** Factorization of Measurable Matrix Functions, 1987, (3-7643-1883-X)
26. **N.Y. Krupnik:** Banach Algebras with Symbol and Singular Integral Operators, 1987, (3-7643-1836-8)
27. **A. Bultheel:** Laurent Series and their Pade Approximation, 1987, (3-7643-1940-2)
28. **H. Helson, C.M. Pearcy, F.-H. Vasilescu, D. Voiculescu, Gr. Arsene** (Eds.): Special Classes of Linear Operators and Other Topics, 1988, (3-7643-1970-4)
29. **I. Gohberg** (Ed.): Topics in Operator Theory and Interpolation, 1988, (3-7634-1960-7)
30. **Yu.I. Lyubich:** Introduction to the Theory of Banach Representations of Groups, 1988, (3-7643-2207-1)
31. **E.M. Polishchuk:** Continual Means and Boundary Value Problems in Function Spaces, 1988, (3-7643-2217-9)
32. **I. Gohberg** (Ed.): Topics in Operator Theory. Constantin Apostol Memorial Issue, 1988, (3-7643-2232-2)
33. **I. Gohberg** (Ed.): Topics in Interplation Theory of Rational Matrix-Valued Functions, 1988, (3-7643-2233-0)
34. **I. Gohberg** (Ed.): Orthogonal Matrix-Valued Polynomials and Applications, 1988, (3-7643-2242-X)
35. **I. Gohberg, J.W. Helton, L. Rodman** (Eds.): Contributions to Operator Theory and its Applications, 1988, (3-7643-2221-7)
36. **G.R. Belitskii, Yu.I. Lyubich:** Matrix Norms and their Applications, 1988, (3-7643-2220-9)
37. **K. Schmüdgen:** Unbounded Operator Algebras and Representation Theory, 1990, (3-7643-2321-3)
38. **L. Rodman:** An Introduction to Operator Polynomials, 1989, (3-7643-2324-8)
39. **M. Martin, M. Putinar:** Lectures on Hyponormal Operators, 1989, (3-7643-2329-9)

40. **H. Dym, S. Goldberg, P. Lancaster, M.A. Kaashoek** (Eds.): The Gohberg Anniversary Collection, Volume I, 1989, (3-7643-2307-8)
41. **H. Dym, S. Goldberg, P. Lancaster, M.A. Kaashoek** (Eds.): The Gohberg Anniversary Collection, Volume II, 1989, (3-7643-2308-6)
42. **N.K. Nikolskii** (Ed.): Toeplitz Operators and Spectral Function Theory, 1989, (3-7643-2344-2)
43. **H. Helson, B. Sz.-Nagy, F.-H. Vasilescu, Gr. Arsene** (Eds.): Linear Operators in Function Spaces, 1990, (3-7643-2343-4)
44. **C. Foias, A. Frazho:** The Commutant Lifting Approach to Interpolation Problems, 1990, (3-7643-2461-9)
45. **J.A. Ball, I. Gohberg, L. Rodman:** Interpolation of Rational Matrix Functions, 1990, (3-7643-2476-7)
46. **P. Exner, H. Neidhardt** (Eds.): Order, Disorder and Chaos in Quantum Systems, 1990, (3-7643-2492-9)
47. **I. Gohberg** (Ed.): Extension and Interpolation of Linear Operators and Matrix Functions, 1990, (3-7643-2530-5)
48. **L. de Branges, I. Gohberg, J. Rovnyak** (Eds.): Topics in Operator Theory. Ernst D. Hellinger Memorial Volume, 1990, (3-7643-2532-1)
49. **I. Gohberg, S. Goldberg, M.A. Kaashoek:** Classes of Linear Operators, Volume I, 1990, (3-7643-2531-3)
50. **H. Bart, I. Gohberg, M.A. Kaashoek** (Eds.): Topics in Matrix and Operator Theory, 1991, (3-7643-2570-4)
51. **W. Greenberg, J. Polewczak** (Eds.): Modern Mathematical Methods in Transport Theory, 1991, (3-7643-2571-2)
52. **S. Prössdorf, B. Silbermann:** Numerical Analysis for Integral and Related Operator Equations, 1991, (3-7643-2620-4)
53. **I. Gohberg, N. Krupnik:** One-Dimensional Linear Singular Integral Equations, Volume I, Introduction, 1992, (3-7643-2584-4)
54. **I. Gohberg, N. Krupnik:** One-Dimensional Linear Singular Integral Equations, Volume II, General Theory and Applications, 1992, (3-7643-2796-0)
55. **R.R. Akhmerov, M.I. Kamenskii, A.S. Potapov, A.E. Rodkina, B.N. Sadovskii:** Measures of Noncompactness and Condensing Operators, 1992, (3-7643-2716-2)
56. **I. Gohberg** (Ed.): Time-Variant Systems and Interpolation, 1992, (3-7643-2738-3)
57. **M. Demuth, B. Gramsch, B.W. Schulze** (Eds.): Operator Calculus and Spectral Theory, 1992, (3-7643-2792-8)
58. **I. Gohberg** (Ed.): Continuous and Discrete Fourier Transforms, Extension Problems and Wiener-Hopf Equations, 1992, (3-7643-2809-6)
59. **T. Ando, I. Gohberg** (Eds.): Operator Theory and Complex Analysis, 1992, (3-7643-2824-X)
60. **P.A. Kuchment:** Floquet Theory for Partial Differential Equations, 1993, (3-7643-2901-7)

61. **A. Gheondea, D. Timotin, F.-H. Vasilescu** (Eds.): Operator Extensions, Interpolation of Functions and Related Topics, 1993, (3-7643-2902-5)
62. **T. Furuta, I. Gohberg, T. Nakazi** (Eds.): Contributions to Operator Theory and its Applications. The Tsuyoshi Ando Anniversary Volume, 1993, (3-7643-2928-9)
63. **I. Gohberg, S. Goldberg, M.A. Kaashoek:** Classes of Linear Operators, Volume 2, 1993, (3-7643-2944-0)
64. **I. Gohberg** (Ed.): New Aspects in Interpolation and Completion Theories, 1993, (3-7643-2948-3)
65. **M.M. Djrbashian:** Harmonic Analysis and Boundary Value Problems in the Complex Domain, 1993, (3-7643-2855-X)
66. **V. Khatskevich, D. Shoiykhet:** Differentiable Operators and Nonlinear Equations, 1993, (3-7643-2929-7)
67. **N.V. Govorov †:** Riemann's Boundary Problem with Infinite Index, 1994, (3-7643-2999-8)
68. **A. Halanay, V. Ionescu:** Time-Varying Discrete Linear Systems Input-Output Operators. Riccati Equations. Disturbance Attenuation, 1994, (3-7643-5012-1)
69. **A. Ashyralyev, P.E. Sobolevskii:** Well-Posedness of Parabolic Difference Equations, 1994, (3-7643-5024-5)
70. **M. Demuth, P. Exner, G. Neidhardt, V. Zagrebnov** (Eds): Mathematical Results in Quantum Mechanics. International Conference in Blossin (Germany), May 17-21, 1993, 1994, (3-7643-5025-3)
71. **E.L. Basor, I. Gohberg** (Eds): Toeplitz Operators and Related Topics. The Harold Widom Anniversary Volume. Workshop on Toeplitz and Wiener-Hopf Operators, Santa Cruz, California, September 20–22, 1992, 1994 (3-7643-5068-7)
72. **I. Gohberg, L.A. Sakhnovich** (Eds): Matrix and Operator Valued Functions. The Vladimir Petrovich Potapov Memorial Volume, 1994, (3-7643-5091-1)
73. **A. Feintuch, I. Gohberg** (Eds): Nonselfadjoint Operators and Related Topics. Workshop on Operator Theory and Its Applications, Beersheva, February 24–28, 1994, (3-7643-5097-0)
74. **R. Hagen, S. Roch, B. Silbermann:** Spectral Theory of Approximation Methods for Convolution Equations, 1994, (3-7643-5112-8)
75. **C.B. Huijsmans, M.A. Kaashoek, B. de Pagter**: Operator Theory in Function Spaces and Banach Lattices. The A.C. Zaanen Anniversary Volume, 1994, (ISBN 3-7643-5146-2)
76. **A.M. Krasnosel'skii**: Asymptotics of Nonlinearities and Operator Equations, 1995 (ISBN 3-7643-5175-6)
77. **J. Lindenstrauss, V.D. Milman** (Eds): Geometric Aspects of Functional Analysis Israel Seminar GAFA 1992-94, 1995, (ISBN 3-7643-5207-8)
78. **M. Demuth, B.-W. Schulze** (Eds): Partial Differential Operators and Mathematical Physics: International Conference in Holzhau (Germany), July 3-9, 1994, 1995, (ISBN 3-7643-5208-6)